C5325H

D0082325

DIFFUSION

Mass transfer in fluid systems

DIFFUSION
Mass transfer in fluid systems

E. L. CUSSLER
University of Minnesota

CAMBRIDGE
UNIVERSITY PRESS

Published by the Press Syndicate of the University of Cambridge
The Pitt Building, Trumpington Street, Cambridge CB2 1RP
40 West 20th Street, New York, NY 10011-4211, USA
10 Stamford Road, Oakleigh, Victoria 3166, Australia

© Cambridge University Press 1984

First published 1984
First paperback edition 1985
Reprinted 1986, 1987, 1988, 1989, 1991, 1992

Printed in the United States of America

Library of Congress Cataloging in Publication Data

Cussler, E. L.
Diffusion, mass transfer in fluid systems.
I. Diffusion.   2. Mass transfer.   3. Fluids.
1. Title.
TP156.D47C878   1984   660.2′8423   83-1905

ISBN 0 521 29846 6 paperback

For Jason, Liz, Sarah, and Varick
who wonder what I do all day

CONTENTS

PART II. DIFFUSION COEFFICIENTS

PART III. MASS TRANSFER

PREFACE

The purpose of this book is to provide a clear description of diffusion useful to engineers, chemists, and life scientists. Diffusion is a fascinating subject, as central to our daily lives as to the chemical industry. Diffusion equations describe the transport in living cells, the efficiency of distillation, and the dispersal of pollutants. Diffusion is responsible for gas absorption, for the fog formed by rain on snow, and for the dyeing of wool. Problems like these are easy to identify and fun to study.

Diffusion has the reputation of being a difficult subject, much harder than, say, fluid mechanics or solution thermodynamics. In fact, it is relatively simple. To prove this to yourself, try to explain a diffusion flux, a shear stress, and a chemical potential to some friends who have little scientific training. I can easily explain a diffusion flux: It is how much diffuses per area per time. I have more trouble with a shear stress. Whether I say it is a momentum flux or the force in one direction caused by motion in a second direction, my friends look blank. I suspect that I have never clearly explained chemical potentials to anyone.

However, past books on diffusion have enhanced its reputation as a difficult subject. These books fall into two distinct groups that are hard to read for different reasons. The first group is the traditional engineering text. Such texts are characterized by elaborate algebra, very complex examples, and turgid writing. Students cheerfully hate these books; moreover, they remember what they have learned as scattered topics, not as an organized subject.

The second group of books consists of texts on transport processes. These books present diffusion by analogy with fluid flow and heat transfer. They are organized around a tight mathematical framework. They avoid excessive detail, even though teaching by analogy does produce some redundancy. For those in engineering with substantial mathematical aptitude, these books can work well. They are a sharp improvement over traditional texts.

The trouble with teaching by analogy is that cases in which no analogy exists tend to be omitted or deemphasized. Such cases occur frequently for diffusion, especially in situations of simultaneous diffusion and chemical reaction. Moreover, books that teach by analogy usually present diffusion last, so that fluid mechanics and heat transfer must be at least superficially understood before diffusion can be learned. This approach effectively excludes those outside of engineering who have little interest in these other phenomena. Those in engineering find difficult problems emphasized, because the simple ones have already been covered for heat transfer. Whether they are engineers or not, all conclude that diffusion must be difficult.

In this book, I have tried to present a description of diffusion that retains the clarity of the books on transport processes but that includes important topics from the traditional texts. I have tried to maximize physical insight and to minimize mathematical tedium. I have discussed basic concepts in great detail, without assuming prior knowledge of other phenomena. I have included a large number of simple examples to emphasize these basics. I have written the book in an informal style because I find that this works best for most students.

This book can be effectively used as a text. For a graduate course in chemical engineering, I teach the material in Chapters 1–4, 11–15, and 17. For a similar chemistry course, I have taught Chapters 1–3, 5–8, and 13–15. For an undergraduate chemical engineering course, I combine material on unit operations with that in Chapters 1–3, 5, and 9–11. I have also taught short courses to physicians based on Chapters 1, 2, 4, 9, and 15. For those in the biological sciences who seek only an introduction, I recommend Chapters 2, 15, and 9. I think that the last is especially important, because it explains engineering techniques valuable, but rarely used, for the study of living systems. The book is less suitable for students in materials science, for the treatment of diffusion in solids is not extensive. To facilitate the use of this book as a text, I have also included some numerical exercises; many of these include answers.

I am indebted to many who helped me write this book. My principal debt is to Professor D. Fennell Evans of the University of Minnesota, who has been a collaborator, a critic, and a confidant for almost 20 years. I am indebted to Dr. Peter J. Dunlop of the University of Adelaide, who taught me how to be careful, and to Professor H. L. Toor of Carnegie-Mellon University, who taught me when to be careless. Professor John Quinn of the University of Pennsylvania, Professor Howard Brenner of the Massachusetts Institute of Technology, and Dr. William J. Ward of the General Electric Company have been stimulating mentors, and my colleagues here at Minnesota have provided the synergism of academic research. Those who have critically read the manuscript and suggested corrections include H. T. Cullinan, Christie Geankopolis, Stevin Gehrke, Kevin Hodgeson, Anthony Klos, Kwanmin Jem, Enrico Martinez, Shantilal Mohnot, Richard Noble, R. J. Ragagoplatan, and J. A. Shaeiwitz. Errors that remain are my fault. Professor M. A. Dayananda checked and extended Table 8.4-3, and J. L. Anderson provided unpublished material for Chapter 12. Kathleen Jones did most of the typing, including the final draft and the near-endless corrections; Dolores Dlugokecki typed the earlier chapters. Finally, my wife Betsy has been a constant source of support, encouragement, and skepticism, each when it was needed.

1 MODELS FOR DIFFUSION

If a few crystals of a colored material like copper sulfate are placed at the bottom of a tall bottle filled with water, the color will slowly spread through the bottle. At first the color will be concentrated in the bottom of the bottle. After a day it will penetrate upward a few centimeters. After several years the solution will appear homogeneous.

The process responsible for the movement of the colored material is diffusion, the subject of this book. Diffusion is caused by random molecular motion that leads to complete mixing. It can be a slow process. In gases, diffusion progresses at a rate of about 10 cm in a minute; in liquids, its rate is about 0.05 cm/min; in solids, its rate may be only about 0.00001 cm/min. In general, it varies less with temperature than do many other phenomena.

This slow rate of diffusion is responsible for its importance. In many cases, diffusion occurs sequentially with other phenomena. When it is the slowest step in the sequence, it limits the overall rate of the process. For example, diffusion often limits the efficiency of commercial distillations and the rate of industrial reactions using porous catalysts. It limits the speed with which acid and base react and the speed with which the human intestine absorbs nutrients. It controls the growth of microorganisms producing penicillin, the rate of the corrosion of steel, and the release of flavor from food.

In gases and liquids, the rates of these diffusion processes can often be accelerated by agitation. For example, the copper sulfate in the tall bottle can be completely mixed in a few minutes if the solution is stirred. This accelerated mixing is not due to diffusion alone, but to the combination of diffusion and stirring. Diffusion still depends on random molecular motions that take place over small molecular distances. The agitation or stirring is not a molecular process, but a macroscopic process that moves portions of the fluid over much larger distances. After this macroscopic motion, diffusion mixes newly adjacent portions of the fluid. In other cases, such as the dispersal of pollutants, the agitation of wind or water produces effects qualitatively similar to diffusion; these effects, called dispersion, will be treated separately.

The description of diffusion involves a mathematical model based on a fundamental hypothesis or "law." Interestingly, there are two common choices for such a law. The more fundamental, Fick's law of diffusion, uses a diffusion coefficient. This is the law that is commonly cited in descriptions

1

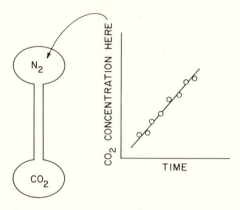

Fig. 1.1-1. A simple diffusion experiment. Two bulbs initially containing different gases are connected with a long thin capillary. The change of concentration in each bulb is a measure of diffusion and can be analyzed in two different ways.

of diffusion. The second, which has no formal name, involves a mass transfer coefficient, a type of reversible rate constant.

Choosing between these two models is the subject of this first chapter. Choosing Fick's law leads to descriptions common to physics, physical chemistry, and biology. These descriptions are explored and extended in Chapters 2–8. Choosing mass transfer coefficients produces correlations developed explicitly in chemical engineering and used implicitly in chemical kinetics and in medicine. These correlations are described in Chapters 9–12. Both approaches are used in Chapters 13–17.

We discuss the differences between the two models in Section 1.1 of this chapter. In Section 1.2 we show how the choice of the most appropriate model is determined. In Section 1.3 we conclude with additional examples to illustrate how the choice between the models is made.

Section 1.1. The two basic models

In this section we want to illustrate the two basic ways in which diffusion can be described. To do this, we first imagine two large bulbs connected by a long thin capillary (Fig. 1.1-1). The bulbs are at constant temperature and pressure and are of equal volumes. However, one bulb contains carbon dioxide, and the other is filled with nitrogen.

To find how fast these two gases will mix, we measure the concentration of carbon dioxide in the bulb that initially contains nitrogen. We make these measurements when only a trace of carbon dioxide has been transferred, and we find that the concentration of carbon dioxide varies linearly with time. From this, we know the amount transferred per unit time.

We want to analyze this amount transferred to determine physical properties that will be applicable not only to this experiment but also in other experiments. To do this, we first define the flux:

$$\text{(carbon dioxide flux)} = \left(\frac{\text{amount of gas removed}}{\text{time (area capillary)}}\right) \quad (1.1\text{-}1)$$

In other words, if we double the cross-sectional area, we expect the amount transported to double. Defining the flux in this way is a first step in removing the influences of our particular apparatus and making our results more general. We next assume that the flux is proportional to the gas concentration:

$$\text{(carbon dioxide flux)} = k \left(\begin{array}{c}\text{carbon dioxide}\\ \text{concentration}\\ \text{difference}\end{array}\right) \quad (1.1\text{-}2)$$

The proportionality constant k is called a mass transfer coefficient. Its introduction signals one of the two basic models of diffusion. Alternatively, we can recognize that increasing the capillary's length will decrease the flux, and we can then assume that

$$\text{(carbon dioxide flux)} = D \left(\frac{\text{carbon dioxide concentration difference}}{\text{capillary length}}\right)$$

$$(1.1\text{-}3)$$

The new proportionality constant D is the diffusion coefficient. Its introduction implies the other model for diffusion, the model often called Fick's law.

These assumptions may seem arbitrary, but they are similar to those made in many other branches of science. For example, they are similar to those used in developing Ohm's law, which states that

$$\left(\begin{array}{c}\text{current, or}\\ \text{area times flux}\\ \text{of electrons}\end{array}\right) = \left(\frac{1}{\text{resistance}}\right)\left(\begin{array}{c}\text{voltage, or}\\ \text{potential}\\ \text{difference}\end{array}\right) \quad (1.1\text{-}4)$$

Thus, the mass transfer coefficient k is analogous to the reciprocal of the resistance. An alternative form of Ohm's law is

$$\left(\begin{array}{c}\text{current density}\\ \text{or flux of}\\ \text{electrons}\end{array}\right) = \left(\frac{1}{\text{resistivity}}\right)\left(\begin{array}{c}\text{potential}\\ \text{difference}\\ \text{length}\end{array}\right) \quad (1.1\text{-}5)$$

The diffusion coefficient D is analogous to the reciprocal of the resistivity.

Neither the equation using the mass transfer coefficient k nor that using the diffusion coefficient D is always successful. This is because of the assumptions made in their development. For example, the flux may not be proportional to the concentration difference if the capillary is very thin or if the two gases react. In the same way, Ohm's law is not always valid at very high voltages. But these cases are exceptions; both diffusion equations work well in most practical situations, just as Ohm's law does.

The parallels with Ohm's law also provide a clue about how the choice between diffusion models is made. The mass transfer coefficient in Eq. 1.1-2 and the resistance in Eq. 1.1-4 are simpler, best used for practical situations and rough measurements. The diffusion coefficient in Eq. 1.1-3 and the resistivity in Eq. 1.1-5 are more fundamental, involving physical properties like those found in handbooks. How these differences guide the choice between the two models is the subject of the next section.

Section 1.2. Choosing between the two models

The choice between the two models outlined in the previous section represents a compromise between ambition and experimental resources. Obviously, we would like to express our results in the most general and fundamental ways possible. This suggests working with diffusion coefficients. However, in many cases our experimental measurements will dictate a more approximate and phenomenological approach. Such approximations often imply mass transfer coefficients, but they usually still permit us to reach our research goals.

This choice and the resulting approximations are best illustrated by two examples. In the first, we consider hydrogen diffusion in metals. This diffusion substantially reduces a metal's ductility, so much so that parts made from the embrittled metal frequently fracture. To study this embrittlement, we might expose the metal to hydrogen under a variety of conditions and measure the degree of embrittlement versus these conditions. Such empiricism would be a reasonable first approximation, but it would quickly flood us with uncorrelated information that would be difficult to use effectively.

As an improvement, we can undertake two sets of experiments. In the first, we can saturate metal samples with hydrogen and determine their degrees of embrittlement. Thus, we know metal properties versus hydrogen

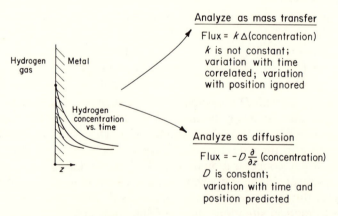

Fig. 1.2-1. Hydrogen diffusion into a metal. This process can be described with either a mass transfer coefficient k or a diffusion coefficient D. The description with a diffusion coefficient correctly predicts the variation of concentration with position and time, and so is superior.

concentration. Second, we can measure hydrogen uptake versus time, as suggested in Fig. 1.2-1, and correlate our measurements as mass transfer coefficients. Thus, we know average hydrogen concentration versus time.

To our dismay, the mass transfer coefficients in this case will be difficult to interpret. They are anything but constant. At zero time, they approach infinity; at large time, they approach zero. At all times, they vary with the hydrogen concentration in the gas surrounding the metal. They are an inconvenient way to summarize our results. Moreover, the mass transfer coefficients give only the *average* hydrogen concentration in the metal. They ignore the fact that the hydrogen concentration very near the metal's surface will reach saturation, but the concentration deep within the bar will remain zero. As a result, the metal near the surface may be very brittle, but that within may be essentially unchanged.

We can include these details in the diffusion model described in the previous section. This model assumed that

$$\begin{pmatrix} \text{hydrogen} \\ \text{flux} \end{pmatrix} = D \frac{\begin{pmatrix} \text{hydrogen} \\ \text{concentration at } z = 0 \end{pmatrix} - \begin{pmatrix} \text{hydrogen} \\ \text{concentration at } z = l \end{pmatrix}}{(\text{thickness at } z = l) \quad - (\text{thickness at } z = 0)}$$

$$(1.2\text{-}1)$$

or, symbolically,

$$j_1 = D \frac{c_1|_{z=0} - c_1|_{z=l}}{l - 0} \tag{1.2-2}$$

where the subscript 1 symbolizes the diffusing species. In these equations, the distance l is that over which diffusion occurs. In the previous section, the length of the capillary was appropriately this distance; but in this case, it seems uncertain what the distance should be. If we assume that it is very small,

$$j_1 = D \lim_{l \to 0} \frac{c_1|_{z=z} - c_1|_{z=z+l}}{z|_{z+l} - z|_z} = -D \frac{dc_1}{dz} \tag{1.2-3}$$

We can use this relation and the techniques developed later in this book to correlate our experiments with only one parameter, the diffusion coefficient D. We then can correctly predict the hydrogen uptake versus time and the hydrogen concentration in the gas. As a dividend, we get the hydrogen concentration at all positions and times within the metal.

Thus, the model based on the diffusion coefficient gives results of more fundamental value than the model based on mass transfer coefficients. In mathematical terms, the diffusion model is said to have distributed parameters, for the dependent variable (the concentration) is allowed to vary with all independent variables (like position and time). In contrast, the mass transfer model is said to have lumped parameters (like the average hydrogen concentration in the metal).

These results would appear to imply that the diffusion model is superior to the mass transfer model and so should always be used. However, in many interesting cases, the models are equivalent. To illustrate this, imagine that

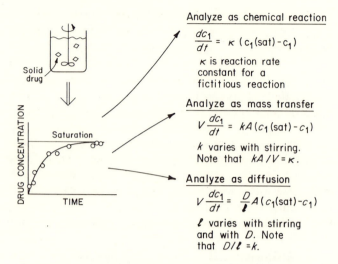

Fig. 1.2-2. Rates of drug dissolution. In this case, describing the system with a mass transfer coefficient k is best because it easily correlates the solution's concentration versus time. Describing the system with a diffusion coefficient D gives a similar correlation, but introduces an unnecessary parameter, the film thickness l. Describing the system with a reaction rate constant κ also works, but this rate constant is a function not of chemistry but of physics.

we are studying the dissolution of a solid drug suspended in water, as schematically suggested by Fig. 1.2-2. The dissolution of this drug is known to be controlled by the diffusion of the dissolved drug away from the solid surface of the undissolved material. We measure the drug concentration versus time as shown, and we want to correlate these results in terms of as few parameters as possible.

One way to correlate the dissolution results is to use a mass transfer coefficient. To do this, we write a mass balance on the solution:

$$\begin{pmatrix} \text{accumulation} \\ \text{of drug in} \\ \text{solution} \end{pmatrix} = \begin{pmatrix} \text{total rate of} \\ \text{dissolution} \end{pmatrix}$$

$$V\frac{dc_1}{dt} = Aj_1$$

$$= Ak[c_1(\text{sat}) - c_1] \tag{1.2-4}$$

where V is the volume of solution, A is the total area of the drug particles, $c_1(\text{sat})$ is the drug concentration at saturation and at the solid's surface, and c_1 is the concentration in the bulk solution. Integrating this equation allows quantitatively fitting our results with one parameter, the mass transfer coefficient k. This quantity is independent of drug solubility, drug area, and solution volume, but it does vary with physical properties like stirring rate and solution viscosity. Correlating the effects of these properties turns out to be straightforward.

The alternative to mass transfer is diffusion theory, for which the mass balance is

$$V \frac{dc_1}{dt} = A\left(\frac{D}{l}\right)[c_1(\text{sat}) - c_1] \qquad (1.2\text{-}5)$$

in which l is an unknown parameter, equal to the average distance across which diffusion occurs. This unknown, called a "film" or "unstirred layer" thickness, is a function not only of flow and viscosity but also of the diffusion coefficient itself.

Equations 1.2-4 and 1.2-5 are equivalent, and they share the same successes and shortcomings. In the former, we must determine the mass transfer coefficient experimentally; in the latter, we determine instead the thickness l. Those who like a scientific veneer prefer to measure l, for it genuflects toward Fick's law of diffusion. Those who are more pragmatic prefer explicitly recognizing the empirical nature of the mass transfer coefficient.

The choice between the mass transfer and diffusion models is thus often a question of taste rather than precision. The diffusion model is more fundamental and is appropriate when concentrations are measured or needed versus both position and time. The mass transfer model is simpler and more approximate and is especially useful when only average concentrations are involved. The additional examples in the next section should help us decide which model is appropriate for our purposes.

Before going on to the next section, we should mention a third way to correlate the results, other than the two diffusion models. This third way is to assume that dissolution is a first-order, reversible chemical reaction. Such a reaction might be described by

$$\frac{dc_1}{dt} = \kappa c_1(\text{sat}) - \kappa c_1 \qquad (1.2\text{-}6)$$

In this equation, the quantity $\kappa c_1(\text{sat})$ represents the rate of dissolution, κc_1 stands for the rate of precipitation, and κ is a rate constant for this process. This equation is mathematically identical with Eqs. 1.2-4 and 1.2-5, and so is equally successful. However, the idea of treating dissolution as a chemical reaction is flawed. Because the reaction is hypothetical, the rate constant is a composite of physical factors rather than chemical factors. We do better to consider the physical process in terms of a diffusion or mass transfer model.

Section 1.3. Examples

In this section we give examples that illustrate the choice between diffusion coefficients and mass transfer coefficients. This choice is often difficult, a juncture where many have trouble. I often do. I think my trouble comes from evolving research goals, from the fact that as I understand the problem better, the questions that I am trying to answer tend to change. I notice the same evolution in my peers, who routinely start work with one model and switch to the other model before the end of their research.

We shall not solve the following examples. Instead, we want only to discuss which diffusion model we would initially use for their solution. The examples given certainly do not cover all types of diffusion problems, but they are among those about which I have been asked in the last year.

Example 1.3-1: Ammonia scrubbing

Ammonia, the major material for fertilizer, is made by reacting nitrogen and hydrogen under pressure. The product gas is washed with water to dissolve the ammonia and separate it from other unreacted gases. How can you correlate the dissolution rate of ammonia during washing?

Solution. The easiest way is to use mass transfer coefficients. If you use diffusion coefficients, you must somehow specify the distance across which diffusion occurs. This distance is unknown unless the detailed flows of gases and the water are known; they rarely are (see Chapters 9 and 11).

Example 1.3-2: Reactions in porous catalysts

Many industrial reactions use catalysts containing small amounts of noble metals dispersed in a porous inert material like silica. The reactions on such a catalyst are sometimes slower in large pellets than in small ones. This is because the reagents take longer to diffuse into the pellet than they do to react. How should you model this effect?

Solution. You should use diffusion coefficients to describe the simultaneous diffusion and reaction in the pores in the catalyst. You should not use mass transfer coefficients because you cannot easily include the effect of reaction (see Sections 13.1 and 14.3).

Example 1.3-3: Corrosion of marble

Industrial pollutants in urban areas like Venice cause significant corrosion of marble statues. You want to study how these pollutants penetrate marble. Which diffusion model should you use?

Solution. The model using diffusion coefficients is the only one that will allow you to predict concentration versus position in the marble. The model using mass transfer coefficients will only correlate how much pollutant enters the statue, not what happens to the pollutant (see Sections 2.3 and 9.1).

Example 1.3-4: Protein size in solution

You are studying a variety of proteins that you hope to purify and use as food supplements. You want to characterize the size of the proteins in solution. How can you use diffusion to do this?

Solution. Your aim is determining the molecular size of the protein molecules. You are not interested in the protein mass transfer except as a route to these molecular properties. As a result, you should measure the protein's diffusion coefficient, not its mass transfer coefficient. The protein's diffusion coefficient will turn out to be proportional to its radius in solution (see Sections 5.2 and 6.1).

Example 1.3-5: Antibiotic production

Many drugs are made by fermentations in which microorganisms are grown in a huge stirred vat of a dilute nutrient solution or "beer." Many of these fermentations are aerobic, so the nutrient solution requires aeration. How should you model oxygen uptake in this type of solution?

Solution. Practical models use mass transfer coefficients. The complexities of the problem, including changes in air bubble size, flow effects of the non-Newtonian solution, and foam caused by biological surfactants, all inhibit more careful study (see Chapter 9).

Example 1.3-6: Facilitated transport across membranes

Some membranes contain a mobile carrier, a reactive species that reacts with diffusing solutes, facilitating their transport across the membrane. Such membranes are used to concentrate copper ions from industrial wastes and to remove carbon dioxide from coal gas. Similar membranes are believed to exist in the human intestine and liver. Diffusion across these membranes does not vary linearly with the concentration difference across them. The diffusion can be highly selective, but it is often easily poisoned. Should this diffusion be described with mass transfer coefficients or with diffusion coefficients?

Solution. This system includes not only diffusion but also chemical reaction. Diffusion and reaction couple in a nonlinear way to give the unusual behavior observed. Understanding such behavior will certainly require the more fundamental model of diffusion coefficients (see Section 15.3).

Example 1.3-7: Flavor retention

When food products are spray-dried, they lose a lot of flavor. However, they lose less than would be expected on the basis of the relative vapor pressures of water and the flavor compounds. The reason apparently is that the drying food often forms a tight gellike skin across which diffusion of the flavor compounds is inhibited. What diffusion model should you use to study this effect?

Solution. Because spray drying is a complex, industrial-scale process, it is usually modeled using mass transfer coefficients. However, in this case, you are interested in the inhibition of diffusion. Such inhibition will involve the sizes of pores in the food and of molecules of the flavor compounds. Thus, you should use the more basic diffusion model, which includes these molecular factors (see Section 7.4).

Example 1.3-8: The smell of marijuana

Recently, a large shipment of marijuana was seized in the Minneapolis–St. Paul airport. The police said their dog smelled it. The owners claimed that it was too well wrapped in plastic to smell and that the police had conducted an illegal search without a search warrant. How could you tell who was right?

Solution. In this case, you are concerned with the diffusion of odor across the thin plastic film. The diffusion rate is well described by either mass transfer or diffusion coefficients. However, the diffusion model explicitly isolates the effect of the solubility of the smell in the film, which dominates the transport. This solubility is the dominant variable (see Section 2.2). In this case, the search was illegal.

Example 1.3-9: Scale-up of wet scrubbers

You want to use a wet scrubber to remove sulfur oxides from the flue gas of a large power plant. A wet scrubber is essentially a large piece of pipe set on its end and filled with inert ceramic material. You pump the flue gas up from the bottom of the pipe and pour a lime slurry down from the top. In the scrubber, there are various reactions, such as

$$CaO + SO_2 \rightarrow CaSO_3$$

The lime reacts with the sulfur oxides to make an insoluble precipitate, which is discarded. You have been studying a small unit and want to use these results to predict the behavior of a larger unit. Such an increase in size is called a "scale-up." Should you make these predictions using a model based on diffusion or mass transfer coefficients?

Solution. This situation is complex because of the chemical reactions and the irregular flows within the scrubber. Your first try at correlating your data should be a simple model based on mass transfer coefficients. Should these correlations prove unreliable, you may be forced to use the more difficult diffusion model (see Chapters 10, 13, and 14).

Section 1.4. Conclusions

This chapter discusses the two common models used to describe diffusion and suggests how you can choose between these models. For fundamental studies where you want to know concentration versus position and time, use diffusion coefficients. For practical problems where you want to use one experiment to tell how a similar one will behave, use mass transfer coefficients. The former approach is the distributed-parameter model used in chemistry, and the latter is the lumped-parameter model used in engineering. Both approaches are used in medicine and biology, but not always explicitly.

The rest of this book is organized in terms of these two models. Chapters 2–4 present the basic model of diffusion coefficients, and Chapters 5–8 review the values of the diffusion coefficients themselves. Chapters 9–12 discuss the model of mass transfer coefficients, including their relation to diffusion coefficients. Chapters 13–15 explore the coupling of diffusion with heterogeneous and homogeneous chemical reactions, using both models. Chapters 16–18 explore the simpler coupling between diffusion and heat transfer.

In these following chapters, keep both models in mind. People involved in basic research tend to be overcommitted to diffusion coefficients, whereas those with broader objectives tend to emphasize mass transfer coefficients. Each group should recognize that the other has a complementary approach that may be more helpful for the case in hand.

PART I

Fundamentals of diffusion

2 DIFFUSION IN DILUTE SOLUTIONS

In this chapter we consider the basic law that underlies diffusion and its application to several simple examples. The examples that will be given are restricted to dilute solutions. Results for concentrated solutions are deferred until Chapter 3.

This focus on the special case of dilute solutions may seem strange. Surely, it would seem more sensible to treat the general case of all solutions and then see mathematically what the dilute-solution limit is like. Most books use this approach. Indeed, because concentrated solutions are complex, these books often describe heat transfer or fluid mechanics first and then teach diffusion by analogy. The complexity of concentrated diffusion then becomes a mathematical cancer grafted onto equations of energy and momentum.

I have rejected this approach for two reasons. First, the most common diffusion problems do take place in dilute solutions. For example, diffusion in living tissue almost always involves the transport of small amounts of solutes like salts, antibodies, enzymes, or steroids. Thus, many who are interested in diffusion need not worry about the complexities of concentrated solutions; they can work effectively and contentedly with the simpler concepts in this chapter.

Second and more important, diffusion in dilute solutions is easier to understand in physical terms. A diffusion flux is the rate per unit area at which mass moves. A concentration profile is simply the variation of the concentration versus time and position. These ideas are much more easily grasped than concepts like momentum flux, which is the momentum per area per time. This seems particularly true for those whose backgrounds are not in engineering, those who need to know about diffusion but not about other transport phenomena.

This emphasis on dilute solutions is found in the historical department of the basic laws involved, as described in Section 2.1. The second and third sections of this chapter focus on two simple cases of diffusion: steady-state diffusion across a thin film and unsteady-state diffusion into an infinite slab. This focus is a logical choice because these two cases are so common. For example, diffusion across thin films is basic to membrane transport, and diffusion in slabs is important in the strength of welds and in the decay of teeth. These two cases are the two extremes in nature, and they bracket the

15

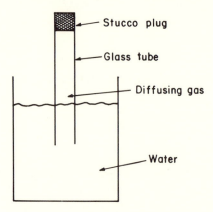

Fig. 2.1-1. Graham's diffusion tube for gases. This apparatus was used in the best early study of diffusion. As a gas like hydrogen diffuses out through the plug, the tube is lowered to ensure that there will be no pressure difference.

behavior observed experimentally. In Section 2.4 these ideas are extended to other examples that demonstrate mathematical ideas useful for other situations.

Section 2.1. Pioneers in diffusion

Thomas Graham

Our modern ideas on diffusion are largely due to two men, Thomas Graham and Adolf Fick. Graham was the elder. Born on December 20, 1805, Graham was the son of a successful manufacturer. At 13 years of age he entered the University of Glasgow with the intention of becoming a minister, and there his interest in science was stimulated by Thomas Thomson.

Graham's research on the diffusion of gases, largely conducted during the years 1828 to 1833, depended strongly on the apparatus shown in Fig. 2.1-1 (Graham, 1829, 1833). This apparatus, a "diffusion tube," consists of a straight glass tube, one end of which is closed with a dense stucco plug. The tube is filled with hydrogen, and the end is sealed with water, as shown. Hydrogen diffuses through the plug and out of the tube, while air diffuses back through the plug and into the tube.

Because the diffusion of hydrogen is faster than the diffusion of air, the water level in this tube will rise during the process. Graham saw that this change in water level would lead to a pressure gradient that in turn would alter the diffusion. To avoid this pressure gradient, he continually lowered the tube so that the water level stayed constant. His experimental results then consisted of a volume-change characteristic of each gas originally held in the tube. Because this volume change was characteristic of diffusion, "the diffusion or spontaneous intermixture of two gases in contact is effected by an interchange of position of infinitely minute volumes, being, in the case of

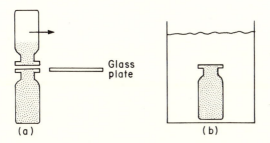

Fig. 2.1-2. Graham's diffusion apparatus for liquids. The equipment in (a) is the ancestor of free diffusion experiments; that in (b) is a forerunner of the capillary method.

each gas, inversely proportional to the square root of the density of the gas'' (Graham, 1833, p. 222). Graham's original experiment was unusual because the diffusion took place at constant pressure, not at constant volume (Mason, 1970).

Graham also performed important experiments on liquid diffusion using the equipment shown in Fig. 2.1-2 (Graham, 1850); in these experiments he worked with dilute solutions. In one series of experiments, he connected two bottles that contained solutions at different concentrations; he waited several days and then separated the bottles and analyzed their contents. In another series of experiments, he placed a small bottle containing a solution of known concentration in a larger jar containing only water. After waiting several days, he removed the bottle and analyzed its contents.

Graham's results were simple and definitive. He showed that diffusion in liquids was at least several thousand times slower than diffusion in gases. He recognized that the diffusion process got still slower as the experiment progressed, that "diffusion must necessarily follow a diminishing progression." Most important, he concluded from the results in Table 2.1-1 that "the quantities diffused appear to be closely in proportion . . . to the quantity of salt in the diffusion solution" (Graham, 1850, p. 6). In other words, the flux caused by diffusion is proportional to the concentration difference of the salt.

Adolf Fick

The next major advance in the theory of diffusion came from the work of Adolf Eugen Fick. Fick was born on September 3, 1829, the youngest of five children. His father, a civil engineer, was a superintendent of buildings. During his secondary schooling, Fick was delighted by mathematics, especially the work of Poisson. He intended to make mathematics his career. However, an older brother, a professor of anatomy at the University of Marlburg, persuaded him to switch to medicine.

In the spring of 1847, Fick went to Marlburg, where he was occasionally tutored by Carl Ludwig. Ludwig strongly believed that medicine, and indeed

Table 2.1-1. *Graham's results for liquid diffusion*

Weight percent of sodium chloride	Relative flux
1	1.00
2	1.99
3	3.01
4	4.00

Source: Data from Graham (1850).

life itself, must have a basis in mathematics, physics, and chemistry. This attitude must have been especially appealing to Fick, who saw the chance to combine his real love, mathematics, with his chosen profession, medicine.

In the fall of 1849, Fick's education continued in Berlin, where he did a considerable amount of clinical work. In 1851 he returned to Marlburg, where he received his degree. His thesis dealt with the visual errors caused by astigmatism, again illustrating his determination to combine science and medicine (Fick, 1852). In the fall of 1851, Carl Ludwig became professor of anatomy in Zurich, and in the spring of 1852 he brought Fick along as a prosector. Ludwig moved to Vienna in 1855, but Fick remained in Zurich until 1868.

Paradoxically, the majority of Fick's scientific accomplishments do not depend on diffusion studies at all, but on his more general investigations of physiology (Fick, 1903). He did outstanding work in mechanics (particularly as applied to the functioning of muscles), in hydrodynamics and hemorheology, and in the visual and thermal functioning of the human body. He was an intriguing man. However, in this discussion we are interested only in his development of the fundamental laws of diffusion.

In his first diffusion paper, Fick (1855a) codified Graham's experiments through an impressive combination of qualitative theories, casual analogies, and quantitative experiments. His paper, which is refreshingly straightforward, deserves reading today. Fick's introduction of his basic idea is almost casual: "[T]he diffusion of the dissolved material . . . is left completely to the influence of the molecular forces basic to the same law . . . for the spreading of warmth in a conductor and which has already been applied with such great success to the spreading of electricity" (Fick, 1855a, p. 65). In other words, diffusion can be described on the same mathematical basis as Fourier's law for heat conduction or Ohm's law for electrical conduction. This analogy remains a useful pedagogical tool.

Fick seemed initially nervous about his hypothesis. He buttressed it with a variety of arguments based on kinetic theory. Although these arguments are now dated, they show physical insights that would be exceptional in medicine today. For example, Fick recognized that diffusion is a dynamic molecular process. He understood the difference between a true equilibrium and a steady state, possibly as a result of his studies with muscles (Fick, 1856). Later, Fick became more confident as he realized his hypothesis was consistent with Graham's results (Fick, 1855*b*).

Using this basic hypothesis, Fick quickly developed the laws of diffusion by means of analogies with Fourier's work (Fourier, 1822). He defined a total one-dimensional flux J_1 as

$$J_1 = Aj_1 = -AD \frac{\partial c_1}{\partial z} \tag{2.1-1}$$

where A is the area across which diffusion occurs, j_1 is the flux per unit area, c_1 is concentration, and z is distance. This is the first suggestion of what is now known as Fick's law. The quantity D, which Fick called "the constant depending on the nature of the substances," is, of course, the diffusion coefficient. Fick also paralleled Fourier's development to determine the more general conservation equation

$$\frac{\partial c_1}{\partial t} = D \left(\frac{\partial^2 c_1}{\partial z^2} + \frac{1}{A} \frac{\partial A}{\partial z} \frac{\partial c_1}{\partial z} \right) \tag{2.1-2}$$

When the area A is a constant, this becomes the basic equation for one-dimensional unsteady-state diffusion, sometimes called Fick's second law.

Fick next had to prove his hypothesis that diffusion and thermal conduction can be described by the same equations. He was by no means immediately successful. First, he tried to integrate Eq. 2.1-2 for constant area, but he became discouraged by the numerical effort required. Second, he tried to measure the second derivative experimentally. Like many others, he found that second derivatives are difficult to measure: "the second difference increases exceptionally the effect of [experimental] errors."

His third effort was more successful. He used a glass cylinder containing crystalline sodium chloride in the bottom and a large volume of water in the top, shown as the upper apparatus in Fig. 2.1-3. By periodically changing the water in the top volume, he was able to establish a steady-state concentration gradient in the cylindrical cell. He found that this gradient was linear, as shown in Fig. 2.1-3. Because this result can be predicted either from Eq. 2.1-1 or from Eq. 2.1-2, this was a triumph.

But this success was by no means complete. After all, Graham's data for liquids anticipated Eq. 2.1-1, so that any analogy with thermal conduction was not exact. To try to strengthen the analogy, Fick used the lower apparatus shown in Fig. 2.1-3. In this apparatus, he established the steady-state concentration profile in the same manner as before. He measured this profile and then tried to predict these results using Eq. 2.1-2, in which the funnel area A available for diffusion varied with the distance z. When Fick compared his calculations with his experimental results, he found the good agreement shown in Fig. 2.1-3. These results were the initial verification of Fick's law.

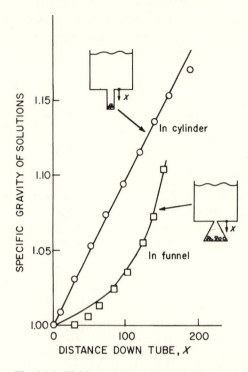

Fig. 2.1-3. Fick's experimental results. The crystals in the bottom of each apparatus saturate the adjacent solution, so that a fixed concentration gradient is established along the narrow, lower part of the apparatus. Fick's calculation of the curve for the funnel was his best proof of Fick's law.

Table 2.1-2. *Fick's law for diffusion without convection*[a]

For one-dimensional diffusion in Cartesian coordinates	$-j_1 = D \dfrac{dc_1}{dz}$
For radial diffusion in cylindrical coordinates	$-j_1 = D \dfrac{dc_1}{dr}$
For radial diffusion in spherical coordinates	$-j_1 = D \dfrac{dc_1}{dr}$

[a] More general equations are given in Table 3.2-1.

Forms of Fick's law

Useful forms of Fick's law in dilute solutions are shown in Table 2.1-2. Each equation closely parallels that suggested by Fick: Eq. 2.1-1. Each involves the same phenomenological diffusion coefficient. Each will be combined with mass balances to analyze the problems central to the rest of this chapter.

One must remember that these flux equations imply no convection in the same direction as the one-dimensional diffusion. They are thus special cases of the general equations given in Table 3.2-1. This lack of convection often indicates a dilute solution. In fact, the assumption of a dilute solution is more restrictive than necessary, for there are many concentrated solutions for which these simple equations can be used without inaccuracy. Nonetheless, for the novice, I suggest thinking of diffusion in a dilute solution.

Section 2.2. Steady diffusion across a thin film

In the previous section we detailed the development of Fick's law, the basic relation for diffusion. Armed with this law, we can now attack the simplest example: steady diffusion across a thin film. In this attack, we want to find both the diffusion flux and the concentration profile. In other words, we want to determine how much solute moves across the film and how the solute concentration changes within the film.

This problem is very important. It is one extreme of diffusion behavior, a counterpoint to diffusion in an infinite slab. Every reader, whether casual or diligent, should try to master this problem now. Many will fail because film diffusion is too simple mathematically. Please do not dismiss this important problem; it is mathematically straightforward but physically subtle. Think about it carefully.

The physical situation

Steady diffusion across a thin film is illustrated schematically in Fig. 2.2-1. On each side of the film is a well-mixed solution of one solute, species 1. Both these solutions are dilute. The solute diffuses from the fixed higher concentration, located at $z \leq 0$ on the left-hand side of the film, into the fixed, less concentrated solution, located at $z \geq l$ on the right-hand side.

We want to find the solute concentration profile and the flux across this film. To do this, we first write a mass balance on a thin layer Δz, located at some arbitrary position z within the thin film. The mass balance in this layer is

$$\begin{pmatrix} \text{solute} \\ \text{accumulation} \end{pmatrix} = \begin{pmatrix} \text{rate of diffusion} \\ \text{into the layer at } z \end{pmatrix} - \begin{pmatrix} \text{rate of diffusion} \\ \text{out of the layer} \\ \text{at } z + \Delta z \end{pmatrix}$$

Because the process is in steady state, the accumulation is zero. The diffusion rate is the diffusion flux times the film's area A. Thus,

$$0 = A(j_1|_z - j_1|_{z+\Delta z}) \tag{2.2-1}$$

Dividing this equation by the film's volume, $A\Delta z$, and rearranging,

$$0 = - \left(\frac{j_1|_{z+\Delta z} - j_1|_z}{(z + \Delta z) - z} \right) \tag{2.2-2}$$

When Δz becomes very small, this equation becomes the definition of the derivative

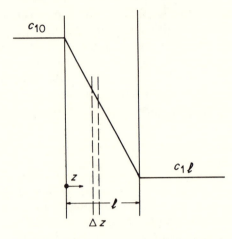

Fig. 2.2-1. Diffusion across a thin film. This is the simplest diffusion problem, basic to perhaps 80% of what follows. Note that the concentration profile is independent of the diffusion coefficient.

$$0 = -\frac{d}{dz} j_1 \tag{2.2-3}$$

Combining this equation with Fick's law,

$$-j_1 = D \frac{dc_1}{dz} \tag{2.2-4}$$

we find, for a constant diffusion coefficient D,

$$0 = D \frac{d^2}{dz^2} c_1 \tag{2.2-5}$$

This differential equation is subject to two boundary conditions:

$$z = 0, \quad c_1 = c_{10} \tag{2.2-6}$$
$$z = l, \quad c_1 = c_{1l} \tag{2.2-7}$$

Again, because this system is in steady state, the concentrations c_{10} and c_{1l} are independent of time. Physically, this means that the volumes of the adjacent solutions must be much greater than the volume of the film.

Mathematical results

The desired concentration profile and flux are now easily found. First, we integrate Eq. 2.2-5 twice to find

$$c_1 = a + bz \tag{2.2-8}$$

The constants a and b can be found from Eqs. 2.2-6 and 2.2-7; so the concentration profile is

$$c_1 = c_{10} + (c_{1l} - c_{10}) \frac{z}{l} \tag{2.2-9}$$

This linear variation was, of course, anticipated by the sketch in Fig. 2.2-1.

The flux is found by differentiating this profile:

$$j_1 = -D \frac{dc_1}{dz} = \frac{D}{l}(c_{10} - c_{1l}) \tag{2.2-10}$$

Because the system is in steady state, the flux is a constant.

As mentioned earlier, this case is easy mathematically. Although it is very important, it is often underemphasized because it seems trivial. Before you conclude this, try some of the examples that follow to make sure you understand what is happening.

Example 2.2-1: Membrane diffusion

Derive the concentration profile and the flux for a single solute diffusing across a thin membrane. As in the preceding case of a film, the membrane separates two well-stirred solutions. Unlike the film, the membrane is chemically different from these solutions.

Solution. As before, we first write a mass balance on a thin layer Δz:

$$0 = A(j_1|_z - j_1|_{z+\Delta z})$$

This leads to a differential equation identical with Eq. 2.2-5:

$$0 = D \frac{d^2 c_1}{dz^2}$$

However, this new mass balance is subject to somewhat different boundary conditions:

$$z = 0, \quad c_1 = HC_{10}$$
$$z = l, \quad c_1 = HC_{1l}$$

where H is a partition coefficient, the concentration in the membrane divided by that in the adjacent solution. This partition coefficient is an equilibrium property; so its use implies that equilibrium exists across the membrane surface.

The concentration profile that results from these relations is

$$c_1 = HC_{10} + H(C_{1l} - C_{10})\frac{z}{l}$$

which is analogous to Eq. 2.2-9. This result looks harmless enough. However, it suggests concentration profiles like those in Fig. 2.2-2, which contain sudden discontinuities at the interface. If the solute is more soluble in the membrane than in the surrounding solutions, then the concentration increases. If the solute is less soluble in the membrane, then its concentration drops. Either case produces enigmas. For example, at the left-hand side of the membrane in Fig. 2.2-2(a), solute diffuses from the solution at c_{10} into the membrane at *higher* concentration.

This apparent quandary is resolved when we think carefully about the solute's diffusion. Diffusion often can occur from a region of low concentration into a region of high concentration; indeed, this is the basis of many liquid–liquid extractions. Thus, the jumps in concentration in Fig. 2.2-2 are

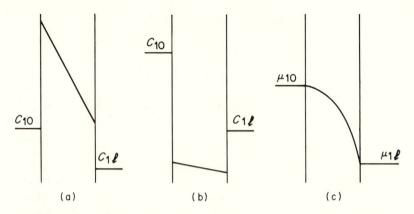

Fig. 2.2-2. Concentration profiles across thin membranes. In (a), the solute is more soluble in the membrane than in the adjacent solutions; in (b), it is less so. Both cases correspond to a chemical potential gradient like that in (c).

not as bizarre as they might appear; rather, they are graphic accidents that result from using the same scale to represent concentrations inside and outside the membrane.

This type of diffusion can also be described in terms of the solute's energy or, more exactly, in terms of its chemical potential. The solute's chemical potential does not change across the membrane's interface, because equilibrium exists there. Moreover, this potential, which drops smoothly with concentration, as shown in Fig. 2.2-2(c), is the driving force responsible for the diffusion. The exact role of this driving force is discussed more completely in Sections 7.3 and 8.2.

The flux across a thin membrane can be found by combining the foregoing concentration profile with Fick's law:

$$j_1 = \left[\frac{DH}{l}\right](C_{10} - C_{1l})$$

This is parallel to Eq. 2.2-10. The quantity in square brackets in this equation is called the permeability, and it is often reported experimentally. Sometimes this same term is called the permeability per unit length. The partition coefficient H is found to vary more widely than the diffusion coefficient D; so differences in diffusion tend to be much less important than the differences in solubility. This has important implications for membrane transport and is discussed in more detail in Chapter 15.

Example 2.2-2: Porous-membrane diffusion

Determine how the results of the previous example are changed if the homogeneous membrane is replaced by a microporous layer. The pores of this layer are filled with the same material that was used before.

Solution. The difference between this case and the previous one is that diffusion is no longer one-dimensional; it now wiggles along the tortuous pores that make up the membrane. Rather than try to treat this problem exactly, you can assume an effective diffusion coefficient that encompasses all ignorance of the pores' geometry. All the earlier answers are then adopted; for example, the flux is

$$j_1 = \left[\frac{D_{\text{eff}}H}{l}\right](C_{10} - C_{1l})$$

where D_{eff} is a new, "effective" diffusion coefficient. Such a quantity is a function not only of solute and solvent but also of the local geometry.

Example 2.2-3: Membrane diffusion with fast reaction

Imagine that while a solute is diffusing steadily across a thin membrane, it can rapidly and reversibly react with other immobile solutes fixed within the membrane. Find how this fast reaction affects the solute's flux.

Solution. The answer is surprising: The reaction has no effect. This is an excellent example because it requires careful thinking. Again, we begin by writing a mass balance on a layer Δz located within the membrane:

$$\begin{pmatrix} \text{solute} \\ \text{accumulation} \end{pmatrix} = \begin{pmatrix} \text{solute diffusion in} \\ \text{minus that out} \end{pmatrix} + \begin{pmatrix} \text{amount produced} \\ \text{by chemical reaction} \end{pmatrix}$$

Because the system is in steady state, this leads to

$$0 = A(j_1|_z - j_1|_{z+\Delta z}) - r_1 A\Delta z$$

or

$$0 = -\frac{d}{dz}j_1 - r_1$$

where r_1 is the rate of disappearance of the mobile species 1 in the membrane. A similar mass balance for the immobile product 2 gives

$$0 = -\frac{d}{dz}j_2 + r_1$$

But because the product is immobile, j_2 is zero, and hence r_1 is zero. As a result, the mass balance for species 1 is identical with Eq. 2.2-3, leaving the flux and concentration profile unchanged.

This result is easier to appreciate in physical terms. After the diffusion reaches a steady state, the local concentration is everywhere in equilibrium with the appropriate amount of the fast reaction's product. Because these local concentrations do not change with time, the amounts of the product do not change either. Diffusion continues unaltered.

This case in which a chemical reaction does not affect diffusion is unusual. For almost any other situation, the reaction can engender dramatically different mass transfer. If the reaction is irreversible, the flux can be increased many orders of magnitude, as shown in Section 14.1. If the reaction

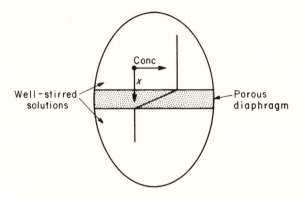

Fig. 2.2-3. A diaphragm cell for measuring diffusion coefficients. Because the diaphragm has a much smaller volume than the adjacent solutions, the concentration profile within the diaphragm has essentially the linear, steady-state value.

product diffuses, highly selective membrane transport occurs, as described in Section 15.3. If the diffusion is not steady, the apparent diffusion coefficient can be much less than expected; such effects occur in the dyeing of wool discussed in Section 6.3. However, in the case described in this example, the chemical reaction does not affect diffusion.

Example 2.2-4: Diaphragm-cell diffusion

One of the best ways to measure diffusion coefficients is the diaphragm cell, shown in Fig. 2.2-3. These cells consist of two well-stirred volumes separated by a thin porous barrier or diaphragm. In the more accurate experiments, the diaphragm is often a sintered glass frit; in many successful experiments, it is just a piece of filter paper (see Section 5.5). To measure a diffusion coefficient with this cell, we fill the lower compartment with a solution of known concentration and the upper compartment with solvent. After a known time, we sample both upper and lower compartments and measure their concentrations.

Find an equation that uses the known time and the measured concentrations to calculate the diffusion coefficient.

Solution. An exact solution to this problem is elaborate and unnecessary. Such a solution is known, but never used (Barnes, 1934). The useful approximate solution depends on the assumption that the flux across the diaphragm quickly reaches its steady-state value (Robinson & Stokes, 1960). This steady-state flux is approached even though the concentrations in the upper and lower compartments are changing with time. The approximations introduced by this assumption will be considered again later.

In this pseudosteady state, the flux across the diaphragm is that given for membrane diffusion:

$$j_1 = \left[\frac{DH}{l}\right] (C_{1,\text{lower}} - C_{1,\text{upper}})$$

Here, the quantity H includes the fraction of the diaphragm's area that is available for diffusion. We next write an overall mass balance on the adjacent compartments:

$$V_{\text{lower}} \frac{dC_{1,\text{lower}}}{dt} = -Aj_1$$

$$V_{\text{upper}} \frac{dC_{1,\text{upper}}}{dt} = +Aj_1$$

where A is the diaphragm's area. If these mass balances are divided by V_{lower} and V_{upper}, respectively, and the equations are subtracted, one can combine the result with the flux equation to obtain

$$\frac{d}{dt}(C_{1,\text{lower}} - C_{1,\text{upper}}) = D\beta(C_{1,\text{upper}} - C_{1,\text{lower}})$$

in which

$$\beta = \frac{AH}{l}\left(\frac{1}{V_{\text{lower}}} + \frac{1}{V_{\text{upper}}}\right)$$

which is a geometrical constant characteristic of the particular diaphragm cell being used. This differential equation is subject to the obvious initial condition

$$t = 0, \qquad C_{1,\text{lower}} - C_{1,\text{upper}} = C^0_{1,\text{lower}} - C^0_{1,\text{upper}}$$

If the upper compartment is initially filled with solvent, then its initial solute concentration will be zero.

Integrating the differential equation subject to this condition gives the desired result:

$$\frac{C_{1,\text{lower}} - C_{1,\text{upper}}}{C^0_{1,\text{lower}} - C^0_{1,\text{upper}}} = e^{-\beta Dt}$$

or

$$D = \frac{1}{\beta t} \ln\left(\frac{C^0_{1,\text{lower}} - C^0_{1,\text{upper}}}{C_{1,\text{lower}} - C_{1,\text{upper}}}\right)$$

We can measure the time t and the various concentrations directly. We can also determine the geometric factor β by calibration of the cell with a species whose diffusion coefficient is known. Then we can determine the diffusion coefficients of unknown solutes.

There are two major ways in which this analysis can be questioned. First, the diffusion coefficient used here is an effective value altered by the tortuosity in the diaphragm. Theoreticians occasionally assert that different solutes will have different tortuosities, so that the diffusion coefficients measured will apply only to that particular diaphragm cell and will not be generally usable. Experimentalists have cheerfully ignored these assertions by writing

$$D = \frac{1}{\beta' t} \ln\left(\frac{C^0_{1,\text{lower}} - C^0_{1,\text{upper}}}{C_{1,\text{lower}} - C_{1,\text{upper}}}\right)$$

where β' is a new calibration constant that includes any tortuosity effects. So far, the experimentalists have gotten away with this because diffusion coefficients measured with the diaphragm cell do agree with those measured by other methods.

The second major question about this analysis comes from the combination of the steady-state flux equation with an unsteady-state mass balance. You may find this combination to be one of those areas where superficial inspection is reassuring, but where careful reflection is disquieting. I have been tempted to skip over this point, but have decided that I had better not. Here goes:

The adjacent compartments are much larger than the diaphragm itself because they contain much more material. Their concentrations change slowly, ponderously, as a result of the transfer of a lot of solute. In contrast, the diaphragm itself contains relatively little material. Changes in its concentration profile occur quickly. Thus, even if this profile is initially very different from steady state, it will approach a steady state before the concentrations in the adjacent compartments can change much. As a result, the profile across the diaphragm will always be close to its steady value, even though the compartment concentrations are time-dependent.

These ideas can be placed on a more quantitative basis by comparing the relaxation time of the diaphragm, l^2/D, with that of the compartments, $1/(D\beta)$. The analysis used here will be accurate when (Mills et al., 1968)

$$1 \gg \frac{l^2 D_{\text{eff}} H}{1/(\beta D_{\text{eff}})} = V_{\substack{\text{diaphragm} \\ \text{voids}}} \left(\frac{1}{V_{\text{lower}}} + \frac{1}{V_{\text{upper}}} \right)$$

This type of "pseudosteady-state approximation" is very common and will be found to underlie most mass transfer coefficients.

Example 2.2-5: Concentration-dependent diffusion

In all the examples thus far, we have assumed that the diffusion coefficient is constant. However, in some cases this is not true; the diffusion coefficient can suddenly drop from a high value to a much lower one (Stannet et al., 1979). Such changes can occur for water diffusion across films and in detergent solutions.

Find the flux across a thin film in which diffusion varies sharply. To keep the problem simple, assume that below some critical concentration c_{1c}, diffusion is fast, but above this concentration it is suddenly much slower.

Solution. This problem is best idealized as two films that are stuck together (Fig. 2.2-4). The interface between these films occurs when the concentration equals c_{1c}. In either film, a steady-state mass balance leads to the same equation:

$$0 = -\frac{dj_1}{dz}$$

As a result, the flux j_1 is a constant everywhere in the film. However, in the left-hand film the high concentration produces a small diffusion coefficient:

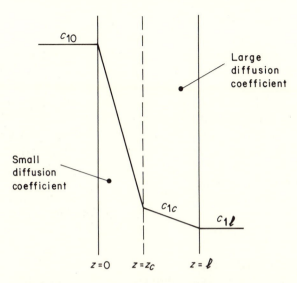

Fig. 2.2-4. Concentration-dependent diffusion across a thin film. Above the concentration c_{1c}, the diffusion coefficient is small; below this critical value, it is larger.

$$j_1 = -\mathbf{D}\,\frac{dc_1}{dz}$$

This result is easily integrated:

$$\int_0^{z_c} j_1\,dz = -\mathbf{D}\int_{c_{10}}^{c_{1c}} dc_1$$

giving the result

$$j_1 = \frac{\mathbf{D}}{z_c}(c_{10} - c_{1c})$$

In the right-hand film, the concentration is small, and the diffusion coefficient is large:

$$j_1 = -\mathbf{D}\,\frac{dc_1}{dz}$$

$$= \frac{\mathbf{D}}{l - z_c}(c_{1c} - c_{1l})$$

The unknown position z_c can be found by recognizing that the flux is the same across both films:

$$z_c = \frac{l}{\dfrac{\mathbf{D}(c_{1c} - c_{1l})}{\mathbf{D}(c_{10} - c_{1c})}}$$

The flux becomes

$$j_1 = \frac{\mathbf{D}(c_{10} - c_{1c}) + \mathbf{D}(c_{1c} - c_{1l})}{l}$$

If the critical concentration equals the average of c_{10} and c_{1l}, then the apparent diffusion coefficient will be the arithmetic average of the two diffusion coefficients.

In passing, we should recognize that the concentration profile shown in Fig. 2.2-4 implicitly gives the ratio of the diffusion coefficients. The flux across the film is constant and is proportional to the concentration gradient. Because the gradient is larger on the left, the diffusion coefficient is smaller. Because the gradient is smaller on the right, the diffusion coefficient is larger. To test your understanding of this point, you should consider what the concentration profile will look like if the diffusion coefficient suddenly *decreases* as the concentration drops. Such considerations will help you understand the next and final example in this section.

Example 2.2-6: Skin diffusion

The diffusion of inert gases through the skin can cause itching, burning rashes, which in turn can lead to vertigo and nausea. These symptoms are believed to occur because gas permeability and diffusion in skin are variable. Indeed, skin behaves as if it consists of two layers, each of which has a different permeability (Idicula et al., 1976). Explain how these two layers can lead to the rashes observed clinically.

Solution. This problem is similar to Examples 2.2-1 and 2.2-5, but the solution is very complex in terms of concentration. We can reduce this complexity by defining a new variable: *the gas pressure that would be in equilibrium with the local concentration*. The "concentration profiles" across skin are much simpler in terms of this pressure, even though it may not exist physically. To make these ideas more specific, we label the two layers of skin A and B. For layer A,

$$p_1 = p_{1,\text{gas}} + \frac{z}{l_A}(p_{1i} - p_{1,\text{gas}})$$

and for layer B,

$$p_1 = p_{1i} + \frac{z - l_A}{l_B - l_A}(p_{1,\text{tissue}} - p_{1i})$$

The interfacial pressure

$$p_{1i} = \frac{\left(\dfrac{D_A H_A}{l_A}\right) p_{1,\text{gas}} + \left(\dfrac{D_B H_B}{l_B}\right) p_{1,\text{tissue}}}{\dfrac{D_A H_A}{l_A} + \dfrac{D_B H_B}{l_B}}$$

can be found from the fact that the flux through layer A equals that through layer B.

These profiles, which are shown in Fig. 2.2-5, imply why rashes form in the skin. In particular, these graphs illustrate the transport of gas 1 from the

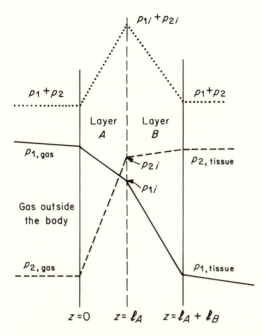

Fig. 2.2-5. Gas diffusion across skin. The gas pressures shown are those in equilibrium with the actual concentrations. In the specific case considered here, gas 2 is more permeable in layer *B,* and gas 1 is more permeable in layer *A.* The resulting total pressure can have major physiologic effects.

surroundings into the tissue and the simultaneous diffusion of gas 2 across the skin in the opposite direction. Gas 1 is more permeable in layer *A* than in layer *B;* as a result, its pressure and concentration gradients fall less sharply in layer *A* than in layer *B.* The reverse is true for gas 2; it is more permeable in layer *B* than in *A.*

These different permeabilities lead to a total pressure that will have a maximum at the interface between the two skin layers. This total pressure, shown by the dotted line in Fig. 2.2-5, may exceed the surrounding pressure outside the skin and within the body. If it does so, gas bubbles will form around the interface between the two skin layers. These bubbles produce the medically observed symptoms. Thus, this condition is a consequence of unequal diffusion (or, more exactly, unequal permeabilities) across different layers of skin.

The examples in this section show that diffusion across thin films can be difficult to understand. The difficulty does not derive from mathematical complexity; the calculation is easy and essentially unchanged. The simplicity of the mathematics is the reason why diffusion across thin films tends to be discussed superficially in mathematically oriented books. The difficulty in thin-film diffusion comes from adapting the same mathematics to widely varying situations with different chemical and physical effects. This is what is difficult to understand about film diffusion. It is an understanding that you must gain before you can do creative work on harder mass transfer problems.

Section 2.3. Unsteady diffusion in a semi-infinite slab

We now turn to a discussion of diffusion in a semi-infinite slab. We consider a volume of solution that starts at an interface and extends a very long way. Such a solution can be a gas, liquid, or solid. We want to find how the concentration varies in this solution as a result of a concentration change at its interface. In mathematical terms, we want to find the concentration and flux as functions of position and time.

This type of mass transfer is often called "free diffusion" (Gosting, 1956), simply because this is briefer than "unsteady diffusion in a semi-infinite slab." At first glance, this situation may seem rare, because no solution can extend an infinite distance. The previous thin-film example made more sense, because we can think of many more thin films than semi-infinite slabs. Thus, we might conclude that this semi-infinite case is not common. That conclusion would be a serious error.

The important case of an infinite slab is common because any diffusion problem will behave as if the slab is infinitely thick at short enough times. For example, imagine that one of the thin membranes discussed in the previous section separates two identical solutions, so that it initially contains a solute at constant concentration. Everything is quiescent, at equilibrium. Suddenly the concentration on the left-hand interface of the membrane is raised, as shown in Fig. 2.3-1. Just after this sudden increase, the concentration near this left interface rises rapidly on its way to a new steady state. In these first few seconds, the concentration at the right interface remains unaltered, ignorant of the turmoil on the left. The left might as well be infinitely far away; the membrane, for these first few seconds, might as well be infinitely thick. Of course, at larger times, the system will slither into the steady-state limit in Fig. 2.3-1(c). But in those first seconds, the membrane does behave like a semi-infinite slab.

This example points to an important corollary, which states that cases involving an infinite slab and a thin membrane will bracket the observed behavior. At short times, diffusion will proceed as if the slab is infinite; at long times, it will occur as if the slab is thin. By focusing on these limits, we can bracket the possible physical responses to different diffusion problems.

The physical situation

The diffusion in a semi-infinite slab is schematically sketched in Fig. 2.3-2. The slab initially contains a uniform concentration of solute $c_{1\infty}$. At some time, chosen as time zero, the concentration at the interface is suddenly and abruptly increased, although the solute is always present at high dilution. This increase produces the time-dependent concentration profile that develops as solute penetrates into the slab.

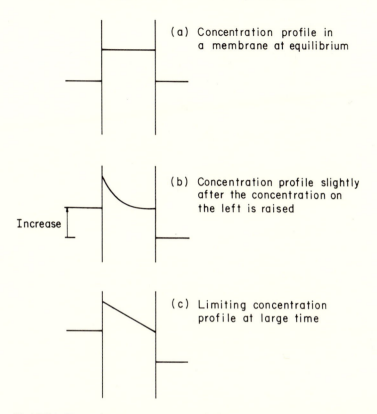

(a) Concentration profile in a membrane at equilibrium

(b) Concentration profile slightly after the concentration on the left is raised

Increase

(c) Limiting concentration profile at large time

Fig. 2.3-1. Unsteady- versus steady-state diffusion. At small times, diffusion will occur only near the left-hand side of the membrane. As a result, at these small times, the diffusion will be the same as if the membrane was infinitely thick. At large times, the results become those in the thin film.

We want to find the concentration profile and the flux in this situation, and so again we need a mass balance written on the thin layer of volume $A\Delta z$:

$$\begin{pmatrix} \text{solute accumulation} \\ \text{in volume } A\Delta z \end{pmatrix} = \begin{pmatrix} \text{rate of diffusion} \\ \text{into the layer} \\ \text{at } z \end{pmatrix} - \begin{pmatrix} \text{rate of diffusion} \\ \text{out of the layer} \\ \text{at } z + \Delta z \end{pmatrix} \quad (2.3\text{-}1)$$

In mathematical terms, this is

$$\frac{\partial}{\partial t}(A\Delta z c_1) = A(j_1|_z - j_1|_{z+\Delta z}) \quad (2.3\text{-}2)$$

We divide by $A\Delta z$ to find

$$\frac{\partial c_1}{\partial t} = -\left(\frac{j_1|_{z+\Delta z} - j_1|_z}{(z+\Delta z) - z}\right) \quad (2.3\text{-}3)$$

We then let Δz go to zero and use the definition of the derivative

$$\frac{\partial c_1}{\partial t} = -\frac{\partial j_1}{\partial z} \quad (2.3\text{-}4)$$

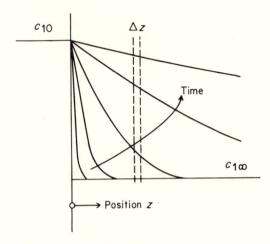

Fig. 2.3-2. Free diffusion. In this case, the concentration at the left is suddenly increased to a higher constant value. Diffusion occurs in the region to the right. This case and that in Fig. 2.2-1 are basic to most diffusion problems.

Combining this equation with Fick's law, and assuming that the diffusion coefficient is independent of concentration, we get

$$\frac{\partial c_1}{\partial t} = D \frac{\partial^2 c_1}{\partial z^2} \qquad (2.3\text{-}5)$$

This equation is sometimes called Fick's second law, and it is often referred to as one example of a "diffusion equation." In this case, it is subject to the following conditions:

$$t = 0, \quad \text{all } z, \qquad c_1 = c_{1\infty} \qquad (2.3\text{-}6)$$
$$t > 0, \quad z = 0, \qquad c_1 = c_{10} \qquad (2.3\text{-}7)$$
$$z = \infty, \qquad c_1 = c_{1\infty} \qquad (2.3\text{-}8)$$

Note that both $c_{1\infty}$ and c_{10} are taken as constants. The concentration $c_{1\infty}$ is constant because it is so far from the interface as to be unaffected by events there; the concentration c_{10} is kept constant by adding material at the interface.

Mathematical solution

The solution of this problem is easiest using the method of "combination of variables." This method is easy to follow, but it must have been difficult to invent. Fourier, Graham, and Fick failed in the attempt; it required Boltzmann's tortured imagination (Boltzmann, 1894).

The trick to solving this problem is to define a new variable

$$\zeta = \frac{z}{\sqrt{4Dt}} \qquad (2.3\text{-}9)$$

The differential equation can then be written as

$$\frac{dc_1}{d\zeta}\left(\frac{\partial \zeta}{\partial t}\right) = D\,\frac{d^2 c_1}{d\zeta^2}\left(\frac{\partial \zeta}{\partial x}\right)^2 \tag{2.3-10}$$

or

$$\frac{d^2 c_1}{d\zeta^2} + 2\zeta\,\frac{dc_1}{d\zeta} = 0 \tag{2.3-11}$$

In other words, the partial differential equation has been almost magically transformed into an ordinary differential equation. The magic also works for the boundary conditions; from Eq. 2.3-7,

$$\zeta = 0, \quad c_1 = c_{10} \tag{2.3-12}$$

and from Eqs. 2.3-6 and 2.3-8,

$$\zeta = \infty, \quad c_1 = c_{1\infty} \tag{2.3-13}$$

With the method of combination of variables, the transformation of the initial and boundary conditions is often more critical than the transformation of the differential equation.

The solution is now straightforward. One integration of Eq. 2.3-11 gives

$$\frac{dc_1}{d\zeta} = ae^{-\zeta^2} \tag{2.3-14}$$

where a is an integration constant. A second integration and use of the boundary conditions gives

$$\frac{c_1 - c_{10}}{c_{1\infty} - c_{10}} = \text{erf}\ \zeta \tag{2.3-15}$$

where

$$\text{erf}\ \zeta = \frac{2}{\sqrt{\pi}}\int_0^\zeta e^{-t^2}\,dt \tag{2.3-16}$$

which is the error function of ζ. This is the desired concentration profile giving the variation of concentration with position and time.

In many practical problems, the flux in the slab is of greater interest than the concentration profile itself. This flux can again be found by combining Fick's law with Eq. 2.3-15:

$$j_1 = -D\,\frac{\partial c_1}{\partial z} = \sqrt{D/\pi t}\ e^{-z^2/4Dt}(c_{10} - c_{1\infty}) \tag{2.3-17}$$

One particularly useful limit is the flux across the interface at $z = 0$:

$$j_1|_{z=0} = \sqrt{D/\pi t}\ (c_{10} - c_{1\infty}) \tag{2.3-18}$$

This result is of special value in the calculation of mass transfer coefficients in Chapter 11.

At this point, I have the same pedagogical problem I had in the previous section: I must convince you that the apparently simple results in Eqs. 2.3-15 and 2.3-18 are valuable. These results are exceeded in importance only by Eqs. 2.2-9 and 2.2-10. Fortunately, the mathematics may be difficult enough to spark thought and reflection; if not, the examples that follow should do so.

Example 2.3-1: Diffusion across an interface

The picture of the process in Fig. 2.3-2 implies that the concentration at $z = 0$ is continuous. This would be true, for example, if when $z \geq 0$ there was a swollen gel, and when $z < 0$ there was a highly dilute solution.

However, a much more common case occurs when there is a gas–liquid interface at $z = 0$. Ordinarily, the gas at $z < 0$ will be well mixed, but the liquid will not. How will this interface affect the results given earlier?

Solution. Basically, it will have no effect. The only change will be a new boundary condition, replacing Eq. 2.3-7:

$$z = 0, \qquad c_1 = cx_1 = c\,\frac{p_{10}}{H}$$

where c_1 is the concentration of solute in the liquid, x_1 is its mole fraction, p_{10} is its partial pressure in the gas phase, H is the solute's Henry's law constant, and c is the total molar concentration in the liquid.

The difficulties caused by a gas–liquid interface are another result of the plethora of units in which concentration can be expressed. These difficulties require concern about units, but they do not demand new mathematical weapons. The changes required for a liquid–liquid interface can be similarly subtle.

Example 2.3-2: Free diffusion into a porous slab

How would the foregoing results be changed if the semi-infinite slab was a porous solid? The diffusion in the gas-filled pores is much faster than in the solid.

Solution. This problem involves diffusion in all three directions as the solute moves through the tortuous pores. The common method of handling this is to define an effective diffusion coefficient D_{eff} and treat the problem as one-dimensional. The concentration profile is then

$$\frac{c_1 - c_{10}}{c_{1\infty} - c_{10}} = \text{erf}\,\frac{z}{\sqrt{4D_{\text{eff}}t}}$$

and the interfacial flux is

$$j_1|_{z=0} = \sqrt{D_{\text{eff}}/\pi t}\,(c_{10} - c_{1\infty})$$

This type of approximation often works well if the distances over which diffusion occurs are large compared with the size of the pores.

Example 2.3-3: Free diffusion with fast chemical reaction

In many problems, the diffusing solutes react rapidly and reversibly with surrounding material. The surrounding material is stationary and cannot diffuse. For example, in the dyeing of wool, the dye can react quickly with the wool as it diffuses into the fiber. How does such a rapid chemical reaction change the results obtained earlier?

Solution. In this case, the chemical reaction can radically change the process by reducing the apparent diffusion coefficient and increasing the interfacial flux of solute (Crank, 1975). These radical changes stand in stark contrast to the steady-state result, where the chemical reaction produces no effect.

To solve this example, we first recognize that the solute is effectively present in two forms: (1) free solute that can diffuse and (2) reacted solute fixed at the point of reaction. If this reaction is faster than diffusion,

$$c_2 = K c_1$$

where c_2 is the concentration of the solute that has already reacted, c_1 is the concentration of the unreacted solute that can diffuse, and K is the equilibrium constant of the reaction. If the reaction is minor, K will be small; as the reaction becomes irreversible, K will become very large.

With these definitions, we now write a mass balance for each solute form. These mass balances should have the form

$$\begin{pmatrix} \text{accumulation} \\ \text{in } A\Delta z \end{pmatrix} = \begin{pmatrix} \text{diffusion in} \\ \text{minus that out} \end{pmatrix} + \begin{pmatrix} \text{amount produced by} \\ \text{reaction in } A\Delta z \end{pmatrix}$$

For the diffusing solute, this is

$$\frac{\partial}{\partial t} [A\Delta z c_1] = A(j_1|_z - j_1|_{z+\Delta z}) + r_1 A\Delta z$$

where r_1 is the rate of production per volume of species 1, the diffusing solute. By arguments analogous to Eqs. 2.3-2 to 2.3-5, this becomes

$$\frac{\partial c_1}{\partial t} = D \frac{\partial^2 c_1}{\partial z^2} + r_1$$

The term on the left-hand side is the accumulation; the first term on the right is the diffusion in minus the diffusion out; the term r_1 is the effect of chemical reaction.

When we write a similar mass balance on the second species, we find

$$\frac{\partial}{\partial t} [A\Delta z c_2] = -r_1 A\Delta z$$

or

$$\frac{\partial c_2}{\partial t} = -r_1$$

We do not get a diffusion term because the reacted solute cannot diffuse. We get a reaction term that has a different sign but the same magnitude, because any solute that disappears as species 1 reappears as species 2.

To solve these questions, we first add them to eliminate the reaction term:

$$\frac{\partial}{\partial t} (c_1 + c_2) = D \frac{\partial^2 c_1}{\partial z^2}$$

We now use the fact that the chemical reaction is at equilibrium:

$$\frac{\partial}{\partial t}(c_1 + Kc_1) = D\frac{\partial^2 c_1}{\partial z^2}$$

$$\frac{\partial c_1}{\partial t} = \frac{D}{1 + K}\frac{\partial^2 c_1}{\partial z^2}$$

This result is subject to the same initial and boundary conditions as before in Eqs. 2.3-6, 2.3-7, and 2.3-8. As a result, the only difference between this example and the earlier problem is that $D/(1 + K)$ replaces D.

This is intriguing. The chemical reaction has left the mathematical form of the answer unchanged, but it has altered the diffusion coefficient. The concentration profile now is

$$\frac{c_1 - c_{10}}{c_{1\infty} - c_{10}} = \mathrm{erf}\,\frac{z}{\sqrt{4[D/(1 + K)]t}}$$

and the interfacial flux is

$$j_1|_{z=0} = \sqrt{D(1 + K)/\pi t}\,(c_{10} - c_{1\infty})$$

The flux has been increased by the chemical reaction.

These effects of chemical reaction can easily be several orders of magnitude. As will be detailed in Chapter 5, diffusion coefficients tend to fall in fairly narrow ranges. Those coefficients for gases are around 0.3 cm^2/sec; those in ordinary liquids cluster about 10^{-5} cm^2/sec. Deviations from these values of more than an order of magnitude are unusual. However, differences in the equilibrium constant K of 1 million or more occur frequently. Thus, a fast chemical reaction can tremendously influence the unsteady diffusion process.

Example 2.3-4: Determining diffusion coefficients from free diffusion experiments

The analysis of diffusion in a semi-infinite slab provides a partial basis for the most accurate measurements of diffuson coefficients. These measurements use interferometers. Two of the most effective interferometers are the Rayleigh and the Gouy. Both utilize a rectangular cell in which there is an initial step function in a refractive index. The decay of this step function with time is followed by shining collimated light through the cell to produce a pattern of interference fringes, dark lines on a photographic plate. Some further details are given in Section 5.5; only some of the mathematics are given here.

The difference between these two instruments is in the optics. In the Rayleigh interferometer, the interference fringes obtained give a record of the refractive index versus location in the cell. In the Gouy interferometer, the interference fringes are elaborate functions of the derivative of the refractive index. These elaborate functions, rife with wave optical corrections, are known; they give a direct measure of the refractive index gradient.

Find equations that allow information from these instruments to be used to calculate diffusion coefficients.

Solution. The concentration profiles established in the diffusion cell closely approach the profiles calculated earlier for a semi-infinite slab. The cell now effectively contains two semi-infinite slabs joined together at $z = 0$. The concentration profile is unaltered from Eq. 2.3-15:

$$\frac{c_1 - c_{10}}{c_{1\infty} - c_{10}} = \text{erf} \frac{z}{\sqrt{4Dt}}$$

where c_{10} [$= (c_{1\infty} - c_{1-\infty})/2$] is the average concentration between the two ends of the cell. How accurate this equation is depends on how exactly the initial change in concentration can be realized; in practice, this change can routinely be within 10 sec of a true step function (Kahn & Polson, 1947).

For the Rayleigh interferometer, we must convert the concentration and cell position into the experimentally measured refractive index and camera position. The refractive index n is linearly proportional to the concentration:

$$n = n_{\text{solvent}} + bc_1$$

where n_{solvent} is the refractive index of the solvent. Each position in the camera is proportional to a position in the diffusion cell:

$$Z = mz$$

where m is the magnification of the apparatus. It is experimentally convenient not to measure the position of one fringe, but rather to measure the intensity minima of many fringes (Longsworth, 1950). These minima occur when

$$\frac{n - n_0}{n_\infty - n_0} = \frac{j}{J/2}$$

where n_∞ and n_0 are the refractive indices at $z = \infty$ and $z = 0$, respectively; J is the total number of interference fringes, and j is an integer called the fringe number. This number is most conveniently defined as zero at $z = 0$, the center of the cell. Combining these equations,

$$\frac{j}{J/2} = \text{erf} \frac{Z_j}{m \sqrt{4Dt}}$$

where Z_j is the intensity minimum associated with the jth fringe. Because m and t are experimentally accessible, measurements of $Z_j(j,J)$ can be used to find the diffusion coefficient D. The development for the Gouy interferometer is similar but much more complex.

Section 2.4. Three other examples

The two previous sections describe diffusion across thin films and in semi-infinite slabs. In this section we turn to discussing mathematical variations of diffusion problems. This mathematical emphasis changes both the pace and

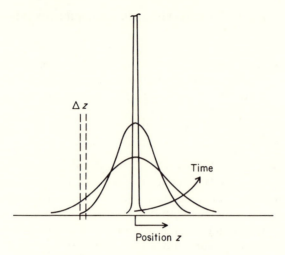

Fig. 2.4-1. Diffusion of a pulse. The concentrated solute originally located at $z = 0$ diffuses as the Gaussian profile shown. This is the third of the three most important cases, along with those in Figs. 2.2-1 and 2.3-2.

the tone of this book. Up to now, we have consistently stressed the physical origins of the problems, constantly harping on natural effects like changing liquid to gas or replacing a homogeneous fluid with a porous solid. Now we shift to the more common textbook composition, a sequence of equations sometimes as jarring as a twelve-tone concerto.

In these examples, we have three principal goals:

1   We want to show how the differential equations describing diffusion are derived.
2   We want to examine the effects of spherical and cylindrical geometries.
3   We want to supply a mathematical primer for solving these different diffusion equations.

In all three examples, we continue to assume dilute solutions. The three problems examined next are physically important and will be referred to again and again in this book. However, they are introduced largely to achieve these mathematical goals.

Decay of a pulse (Laplace transforms)

As a first example, we consider the diffusion away from a sharp pulse of solute like that shown in Fig. 2.4-1. The initially sharp concentration gradient relaxes by diffusion in the z direction into the smooth curves shown (Crank, 1975). We want to calculate the shape of these curves. This calculation illustrates the development of a differential equation and its solution using Laplace transforms.

As usual, our first step is to make a mass balance on the differential volume $A\Delta z$ as shown:

$$\begin{pmatrix} \text{solute} \\ \text{accumulation} \\ \text{in } A\Delta z \end{pmatrix} = \begin{pmatrix} \text{solute} \\ \text{diffusion into} \\ \text{this volume} \end{pmatrix} - \begin{pmatrix} \text{solute} \\ \text{diffusion out of} \\ \text{this volume} \end{pmatrix} \quad (2.4\text{-}1)$$

In mathematical terms, this is

$$\frac{\partial}{\partial t}[A\Delta z c_1] = Aj_1|_z - Aj_1|_{z+\Delta z} \quad (2.4\text{-}2)$$

Dividing by the volume and taking the limit as Δz goes to zero gives

$$\frac{\partial c_1}{\partial t} = -\frac{\partial j_1}{\partial z} \quad (2.4\text{-}3)$$

Combining this relation with Fick's law of diffusion,

$$\frac{\partial c_1}{\partial t} = D\frac{\partial^2 c_1}{\partial z^2} \quad (2.4\text{-}4)$$

This is the same differential equation basic to the free diffusion considered in the previous section. The boundary conditions on this equation are as follows. First, far from the pulse, the solute concentration is zero:

$$t > 0, \quad z = \infty, \quad c_1 = 0 \quad (2.4\text{-}5)$$

Second, because diffusion occurs at the same speed in both directions, the pulse is symmetric:

$$t > 0, \quad z = 0, \quad \frac{\partial c_1}{\partial z} = 0 \quad (2.4\text{-}6)$$

This is equivalent to saying that at $z = 0$, the flux has the same magnitude in the positive and negative directions.

The initial condition for the pulse is more interesting in that all the solute is initially located at $z = 0$:

$$t = 0, \quad c_1 = \frac{M}{A}\delta(z) \quad (2.4\text{-}7)$$

where A is still the cross-sectional area over which diffusion is occurring, M is the total amount of solute in the system, and $\delta(z)$ is the Dirac function. This can be shown to be a reasonable condition by a mass balance:

$$\int_{-\infty}^{\infty} c_1 A \, dz = \int_{-\infty}^{\infty} \frac{M}{A}\delta(z)A \, dz = M \quad (2.4\text{-}8)$$

In this integration, we should remember that $\delta(z)$ has dimensions of $(\text{length})^{-1}$.

To solve this problem, we first take the Laplace transform of Eq. 2.4-4 with respect to time:

$$s\bar{c}_1 - c_1(t = 0)^{\nearrow 0} = D\frac{d^2\bar{c}_1}{dz^2} \quad (2.4\text{-}9)$$

where $\bar{c}_1$ is the transformed concentration. The boundary conditions are

$$z = 0, \quad \frac{d\bar{c}_1}{dz} = -\frac{M/A}{2D} \tag{2.4-10}$$

$$z = \infty, \quad \bar{c}_1 = 0 \tag{2.4-11}$$

The first of these reflects the properties of the Dirac function, but the second is routine. Equation 2.4-9 can then easily be integrated to give

$$\bar{c}_1 = a e^{\sqrt{s/D}\, z} + b e^{-\sqrt{s/D}\, z} \tag{2.4-12}$$

where a and b are integration constants. Clearly, a is zero by Eq. 2.4-11. Using Eq. 2.4-10, we find b and hence $\bar{c}_1$:

$$\bar{c}_1 = \frac{M/A}{2D} \sqrt{D/s}\, e^{-\sqrt{s/D}\, z} \tag{2.4-13}$$

The inverse Laplace transform of this function gives

$$c_1 = \frac{M/A}{\sqrt{4\pi Dt}}\, e^{-z^2/4Dt} \tag{2.4-14}$$

which is a Gaussian curve. You may wish to integrate the concentration over the entire system to check that the total solute present is M.

This solution can be used to solve many unsteady diffusion problems that have unusual initial conditions (Crank, 1975). More important, it is often used to correlate the dispersion of pollutants, especially in the air, as discussed in Chapter 4.

Steady dissolution of a sphere (spherical coordinates)

Our second example, which is easier mathematically, is the steady dissolution of a spherical particle, as shown in Fig. 2.4-2. The sphere is of a sparingly soluble material, so that the sphere's size does not change much. However, this material quickly dissolves in the surrounding solvent, so that the solute's concentration at the sphere's surface is saturated. Because the sphere is immersed in a very large fluid volume, the concentration far from the sphere is zero.

The goal is to find both the dissolution rate and the concentration profile around the sphere. Again, the first step is a mass balance. In contrast with the previous examples, this mass balance is most conveniently made in spherical coordinates originating from the center of the sphere (Sherwood et al., 1975). Then we can make a mass balance on a spherical shell of thickness Δr located at some arbitrary distance r from the sphere. This spherical shell is like the rubber of a balloon of surface area $4\pi r^2$ and thickness Δr.

A mass balance on this shell has the same general form as those used earlier:

$$\begin{pmatrix} \text{solute accumulation} \\ \text{within the shell} \end{pmatrix} = \begin{pmatrix} \text{diffusion} \\ \text{into the shell} \end{pmatrix} - \begin{pmatrix} \text{diffusion} \\ \text{out of the shell} \end{pmatrix} \tag{2.4-15}$$

In mathematical terms, this is

$$\frac{\partial}{\partial t} (4\pi r^2 \Delta r c_1) = 0 = (4\pi r^2 j_1)_r - (4\pi r^2 j_1)_{r+\Delta r} \tag{2.4-16}$$

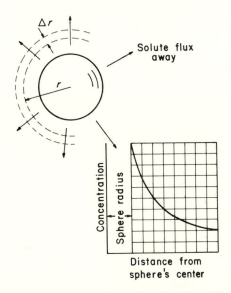

Fig. 2.4-2. Steady dissolution of a sphere. This problem represents an extension of diffusion theory to a spherically symmetric situation. In actual physical situations, this dissolution can be complicated by free convection caused by diffusion (see Chapter 12).

The accumulation on the left-hand side of this mass balance is zero, because diffusion is steady, not varying with time. Novices frequently make a serious error at this point by canceling the r^2 out of both terms on the right-hand side. This is wrong. The term $r^2 j_1$ is evaluated at r in the first term; that is, it is $r^2(j_1|_r)$. This term is evaluated at $(r + \Delta r)$ in the second term; so it equals $(r + \Delta r)^2(j_1|_{r+\Delta r})$.

If we divide both sides of this equation by the spherical shell's volume and take the limit as $\Delta r \rightarrow 0$, we find

$$0 = -\frac{1}{r^2}\frac{d}{dr}(r^2 j_1) \tag{2.4-17}$$

Combining this with Fick's law and assuming that the diffusion coefficient is constant,

$$0 = \frac{D}{r^2}\frac{d}{dr}r^2\frac{dc_1}{dr} \tag{2.4-18}$$

This basic differential equation is subject to two boundary conditions:

$$r = R_0, \quad c_1 = c_1(\text{sat}) \tag{2.4-19}$$
$$r = \infty, \quad c_1 = 0 \tag{2.4-20}$$

If the sphere were dissolving in a partially saturated solution, this second condition would be changed, but the basic mathematical structure would remain unaltered. One integration of Eq. 2.4-18 yields

$$\frac{dc_1}{dr} = \frac{a}{r^2} \tag{2.4-21}$$

where a is an integration constant. A second integration gives

$$c_1 = b - \frac{a}{r} \tag{2.4-22}$$

Use of the two boundary conditions gives the concentration profile

$$c_1 = c_1(\text{sat}) \frac{R_0}{r} \tag{2.4-23}$$

The dissolution flux can then be found from Fick's law:

$$j_1 = -D \frac{dc_1}{dr} = \frac{DR_0}{r^2} c_1(\text{sat}) \tag{2.4-24}$$

which, at the sphere's surface, is

$$j_1 = \frac{D}{R_0} c_1(\text{sat}) \tag{2.4-25}$$

If the sphere is twice as large, the dissolution rate per unit area is only half as large, though the total dissolution rate over the entire surface is doubled.

This example forms the basis for such varied phenomena as the growth of fog droplets and the dissolution of drugs. It is included here to illustrate the derivation and solution of differential equations describing diffusion in spherical coordinate systems. Different coordinate systems are also basic to the final example in this section.

Unsteady diffusion into cylinders (cylindrical coordinates and separation of variables)

The final example, probably the hardest of the three, concerns the diffusion of a solute into the cylinder shown in Fig. 2.4-3. The cylinder initially contains no solute. At time zero, it is suddenly immersed in a well-stirred solution that is of such enormous volume that its solute concentration is constant. The solute diffuses into the cylinder symmetrically. Problems like this are important in the chemical treatment of wood and in the dyeing of wool.

We want to find the solute's concentration in this cylinder as a function of time and location (Bird et al., 1960). As in the previous examples, the first step is a mass balance; in contrast, this mass balance is made on a cylindrical shell located at r, of area $2\pi Lr$, and of volume $2\pi Lr\Delta r$. The basic balance

$$\begin{pmatrix} \text{solute accumulation} \\ \text{in this cylindrical shell} \end{pmatrix} = \begin{pmatrix} \text{solute diffusion} \\ \text{into the shell} \end{pmatrix} - \begin{pmatrix} \text{solute diffusion} \\ \text{out of the shell} \end{pmatrix}$$

$$\tag{2.4-26}$$

becomes in mathematical terms

$$\frac{\partial}{\partial t} (2\pi rL\Delta rc_1) = (2\pi rLj_1)_r - (2\pi rLj_1)_{r+\Delta r} \tag{2.4-27}$$

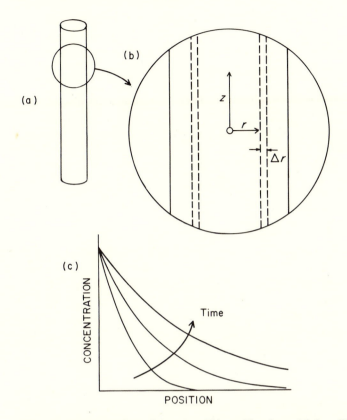

Fig. 2.4-3. Waterproofing a fence post. This problem is modeled as diffusion in an infinite cylinder, and so represents an extension to a cylindrically symmetric situation. In reality, the ends of the post must be considered, especially because diffusion with the grain is faster than across the grain.

We can now divide by the shell's volume and take the limit as Δr becomes small:

$$\frac{\partial}{\partial t} c_1 = - \frac{1}{r} \frac{\partial}{\partial r} r j_1 \qquad (2.4\text{-}28)$$

Combining this expression with Fick's law gives the required differential equation

$$\frac{\partial c_1}{\partial t} = \frac{D}{r} \frac{\partial}{\partial r} r \frac{\partial c_1}{\partial r} \qquad (2.4\text{-}29)$$

which is subject to the following conditions:

$$t \le 0, \quad \text{all } r, \qquad c_1 = 0 \qquad (2.4\text{-}30)$$

$$t > 0, \quad r = R_0, \quad c_1 = c_1(\text{surface}) \qquad (2.4\text{-}31)$$

$$r = 0, \quad \frac{\partial c_1}{\partial r} = 0 \qquad (2.4\text{-}32)$$

In these equations, c_1(surface) is the concentration at the cylinder's surface, and R_0 is the cylinder's radius. The first of the boundary conditions results from the large volume of surrounding solution, and the second reflects the symmetry of the concentration profiles.

Problems like this are often algebraically simplified if they are written in terms of dimensionless variables. This is standard practice in many advanced textbooks. I often find this procedure confusing, because for me it produces only a small gain in algebra at the expense of a large loss in physical insight. Nonetheless, we shall follow this procedure here to illustrate the simplification possible. We first define three new variables:

$$\text{dimensionless concentration:} \quad \theta = 1 - \frac{c_1}{c_1(\text{surface})} \tag{2.4-33}$$

$$\text{dimensionless position:} \quad \xi = \frac{r}{R_0} \tag{2.4-34}$$

$$\text{dimensionless time:} \quad \tau = \frac{Dt}{R_0^2} \tag{2.4-35}$$

The differential equation and boundary conditions now become

$$\frac{\partial \theta}{\partial \tau} = \frac{1}{\xi} \frac{\partial}{\partial \xi} \xi \frac{\partial \theta}{\partial \xi} \tag{2.4-36}$$

subject to

$$\tau = 0, \quad \theta = 1 \tag{2.4-37}$$

$$\xi = 1, \quad \theta = 0 \tag{2.4-38}$$

$$\xi = 0, \quad \frac{\partial \theta}{\partial \xi} = 0 \tag{2.4-39}$$

For the novice, this manipulation can be more troublesome than it looks.

To solve these equations, we first assume that the solution is the product of two functions, one of time and one of radius:

$$\theta(\tau, \xi) = g(\tau) f(\xi) \tag{2.4-40}$$

When Eqs. 2.4-36 and 2.4-40 are combined, the resulting tangle of terms can be separated by division with $g(\tau) f(\xi)$:

$$f(\xi) \frac{dg(\tau)}{d\tau} = \frac{g(\tau)}{\xi} \frac{d}{d\xi} \xi \frac{df(\xi)}{d\xi}$$

$$\frac{1}{g(\tau)} \frac{dg(\tau)}{d\tau} = \frac{1}{\xi f(\xi)} \frac{d}{d\xi} \xi \frac{df(\xi)}{d\xi} \tag{2.4-41}$$

Now, if one fixes ξ and changes τ, $f(\xi)$ remains constant, but $g(\tau)$ varies. As a result,

$$\frac{1}{g(\tau)} \frac{dg(\tau)}{d\tau} = -\alpha^2 \tag{2.4-42}$$

where α is a constant. Similarly, if we hold τ constant and let ξ change, we realize

$$\frac{1}{\xi f(\xi)} \frac{d}{d\xi} \xi \frac{df(\xi)}{d\xi} = -\alpha^2 \tag{2.4-43}$$

Thus, the partial differential equation 2.4-36 has been converted into two ordinary differential equations 2.4-42 and 2.4-43.

The solution of the time-dependent part of this result is easy:

$$g(\tau) = a'e^{-\alpha^2\tau} \tag{2.4-44}$$

where a' is an integration constant. The solution for $f(\xi)$ is more complicated, but straightforward:

$$f(\xi) = aJ_0(\alpha\xi) + bY_0(\alpha\xi) \tag{2.4-45}$$

where J_0 and Y_0 are Bessell functions and a and b are two more constants. From Eq. 2.4-39 we see that $b = 0$. From Eq. 2.4-38 we see that

$$0 = aJ_0(\alpha) \tag{2.4-46}$$

Because a cannot be zero, we recognize that there must be an entire family of solutions for which

$$J_0(\alpha_n) = 0 \tag{2.4-47}$$

The most general solution must be the sum of all solutions of this form found for different integral values of n:

$$\theta(\tau, \xi) = \sum_{n=1}^{\infty} (aa')_n J_0(\alpha_n\xi)e^{-\alpha_n^2\tau} \tag{2.4-48}$$

We now use the initial condition in Eq. 2.4-37 to find the remaining integration constant $(aa')_n$:

$$1 = \sum_{n=1}^{\infty} (aa')_n J_0(\alpha_n\xi) \tag{2.4-49}$$

We multiply both sides of this equation by $J_1(\alpha_n\xi)$ and integrate $\xi = 0$ to $\xi = 1$ to find (aa'). The total result is then

$$\theta = \sum_{n=1}^{\infty} \left[\frac{2}{\alpha_n J_1(\alpha_n)} \right] J_0(\alpha_n\xi)e^{-\alpha_n\tau} \tag{2.4-50}$$

or, in terms of our original variables,

$$\frac{c_1}{c_1(\text{surface})} = 1 - 2 \sum_{n=1}^{\infty} \frac{e^{-D\alpha_n^2 t/R^2} {}_0 J_0(\alpha_n r/R_0)}{\alpha_n J_1(\alpha_n r/R_0)} \tag{2.4-51}$$

The α_n must still be found from Eq. 2.4-47. This is the desired result.

This problem clearly involves a lot of work. The serious reader should certainly work one more problem of this type to get a feel for the idea of separation of variables and for the practice of evaluating integration constants. Even the serious reader probably will embrace the ways of avoiding this work described in the next chapter.

Section 2.5. Convection and dilute diffusion

In many practical problems, both diffusion and convective flow occur. In some cases, especially in fast mass transfer in concentrated solutions, the diffusion itself causes the convection. This type of mass transfer, the subject of Chapter 3, requires more complicated physical and mathematical analyses.

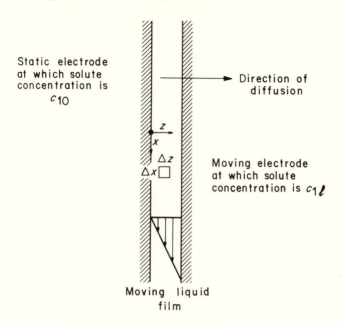

Fig. 2.5-1. Steady diffusion in a moving film. This case is mathematically the same as diffusion across a stagnant film, shown in Fig. 2.2-1. It is basic to the film theory of mass transfer described in Section 11.1.

There is another group of important problems in which diffusion and convection can be more easily handled. These problems arise when diffusion and convection occur normal to each other. In other words, diffusion occurs in one direction, and convective flow occurs in a perpendicular direction. Two of these problems are examined in this section. The first, diffusion across a thin flowing film, parallels Section 2.2; the second, diffusion into a liquid film, is a less obvious analogue to Section 2.3. These two examples tend to bracket the observed experimental behavior, and they are basic to theories relating diffusion and mass transfer coefficients (see Chapter 11).

Steady diffusion across a falling film

The first of the problems of concern here, sketched in Fig. 2.5-1, involves diffusion across a thin, moving liquid film. The concentrations on both sides of this film are fixed by electrochemical reactions, but the film itself is moving steadily. I have chosen this example not because it occurs often but because it is simple. I hope that readers oriented toward the practical will wait for later examples for results of greater applicability.

To solve this problem, we make three key assumptions:

1   The liquid solution is dilute. This assumption is the axiom for this entire chapter.

2   The liquid is the only resistance to mass transfer. This implies that the electrode reactions are fast.

3   Mass transport is by diffusion in the z direction and by convection in the x direction. Transport by other mechanisms is negligible.

It is the last of these assumptions that is most critical. It implies that convection is negligible in the z direction. In fact, diffusion in the z direction automatically generates convection in this direction, but this convection is small in a dilute solution. The last assumption also suggests that there is no diffusion in the x direction. There may be such diffusion, but it is assumed much slower and hence much less important in the x direction than convection.

This problem can be solved by writing a mass balance on the differential volume $W\Delta x\Delta z$, where W is the width of the liquid film, normal to the plane of the paper:

$$\begin{pmatrix} \text{solute accumulation} \\ \text{in } W\Delta x\Delta z \end{pmatrix} = \begin{pmatrix} \text{solute diffusing in at } z \text{ minus} \\ \text{solute diffusing out at } z + \Delta z \end{pmatrix}$$

$$+ \begin{pmatrix} \text{solute flowing in at } x \text{ minus} \\ \text{solute flowing out at } x + \Delta x \end{pmatrix} \quad (2.5\text{-}1)$$

or, in mathematical terms,

$$\frac{\partial}{\partial t}(c_1 W\Delta x\Delta z) = [(j_1 W\Delta x)_z - (j_1 W\Delta x)_{z+\Delta z}]$$

$$+ [(c_1 v_x W\Delta z)_x - (c_1 v_x W\Delta z)_{x+\Delta x}] \quad (2.5\text{-}2)$$

The term on the left-hand side is zero because of the steady state. The second term in square brackets on the right-hand side is also zero, because neither c_1 nor v_x changes with x. The concentration c_1 does not change with x, because the film is long, and there is nothing that will cause the concentration to change in the x direction. The velocity v_x certainly varies with how far we are across the film (i.e., with z), but it does not vary with how far we are along the film (i.e., with x).

After dividing by $W\Delta x\Delta z$ and taking the limit as this volume goes to zero, the mass balance in Eq. 2.5-2 becomes

$$0 = -\frac{dj_1}{dz} \quad (2.5\text{-}3)$$

This can be combined with Fick's law to give

$$0 = D\frac{d^2 c_1}{dz^2} \quad (2.5\text{-}4)$$

This equation is subject to the boundary conditions

$$z = 0, \quad c_1 = c_{10} \quad (2.5\text{-}5)$$
$$z = l, \quad c_1 = c_{1l} \quad (2.5\text{-}6)$$

When these results are combined with Fick's law, we have exactly the same problem as that in Section 2.2. The answers are

$$c_1 = c_{10} + (c_{1l} - c_{10})\frac{z}{l} \quad (2.5\text{-}7)$$

$$j_1 = \frac{D}{l}(c_{10} - c_{1l}) \quad (2.5\text{-}8)$$

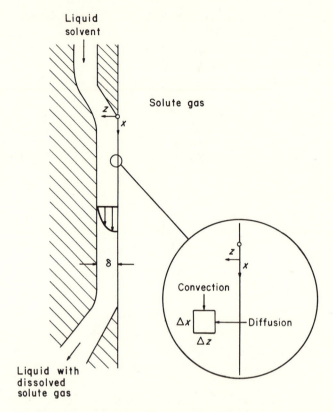

Fig. 2.5-2. Unsteady-state diffusion into a falling film. This analysis turns out to be mathematically equivalent to free diffusion (see Fig. 2.3-2). It is basic to the penetration theory of mass transfer described in Section 11.2.

The flow has no effect. Indeed, the answer is the same as if the fluid was not flowing.

This answer is typical of many problems involving diffusion and flow. When the solutions are dilute, the diffusion and convection often are perpendicular to each other, and the solution is straightforward. You may almost feel gypped; you girded yourself for a difficult problem and found an easy one. Rest assured that more difficult problems follow.

Diffusion into a falling film

The second problem of interest is illustrated schematically in Fig. 2.5-2 (Bird et al., 1960). A thin liquid film flows slowly and without ripples down a flat surface. One side of this film wets the surface; the other side is in contact with a gas, which is sparingly soluble in the liquid. We want to find out how much gas has dissolved in the liquid.

To solve this problem, we again go through the increasingly familiar litany; we write a mass balance as a differential equation, combine this with Fick's law, and then integrate this to find the desired result. We do this subject to four key assumptions:

1 The solutions are always dilute.
2 Mass transport is by z diffusion and x convection.
3 The gas is pure.
4 The contact between gas and liquid is short.

The first two assumptions are identical with those given in the earlier example. The third means that there is no resistance to diffusion in the gas phase, only in the liquid. The final assumption simplifies the analysis.

We now make a mass balance on the differential volume W in width, shown in the inset in Fig. 2.5-2:

$$\begin{pmatrix} \text{mass accumulation} \\ \text{within } W\Delta x\Delta z \end{pmatrix} = \begin{pmatrix} \text{mass diffusing in at } z \text{ minus} \\ \text{mass diffusing out at } z + \Delta z \end{pmatrix}$$

$$+ \begin{pmatrix} \text{mass flowing in at } z \text{ minus} \\ \text{mass flowing out at } z + \Delta z \end{pmatrix} \quad (2.5\text{-}9)$$

This result is parallel to those found in earlier sections:

$$\left[\frac{\partial}{\partial t} (c_1 \Delta x \Delta z W) \right] = [(W\Delta x j_1)_z - (W\Delta x j_1)_{z+\Delta z}]$$

$$+ [(W\Delta z c_1 v_x)_x - (W\Delta z c_1 v_x)_{x+\Delta x}] \quad (2.5\text{-}10)$$

When the system is at steady state, the accumulation is zero. Therefore, the left-hand side of the equation is zero. No other terms are zero, because j_1 and c_1 vary with both z and x. If we divide by the volume $W\Delta x\Delta z$ and take the limit as this volume goes to zero, we find

$$0 = -\frac{\partial j_1}{\partial z} - \frac{\partial}{\partial x} c_1 v_x \quad (2.5\text{-}11)$$

We now make two further manipulations; we combine this with Fick's law and set v_x equal to its maximum value, a constant. This second change reflects the assumption of short contact times. At such times, the solute barely has a chance to cross the interface, and it diffuses only slightly into the fluid. In this interfacial region, the fluid velocity reaches the maximum suggested in Fig. 2.5-2; so the use of a constant value is probably not a serious assumption. Thus, the mass balance is

$$\frac{\partial c_1}{\partial (x/v_{\max})} = D \frac{\partial^2 c_1}{\partial z^2} \quad (2.5\text{-}12)$$

The left-hand side of this equation represents the solute flow out minus that in; the right-hand side is the diffusion in minus the diffusion out.

This mass balance is subject to the following conditions:

$$x = 0, \quad \text{all } z, \quad c_1 = 0 \quad (2.5\text{-}13)$$

$$x > 0, \quad z = 0, \quad c_1 = c_1(\text{sat}) \quad (2.5\text{-}14)$$

$$z = l, \quad c_1 = 0 \quad (2.5\text{-}15)$$

where $c_1(\text{sat})$ is the concentration of dissolved gas in equilibrium with the gas itself, and l is the thickness of the falling film in Fig. 2.5-2. The last of these three boundary conditions is replaced with

$$x > 0, \quad z = \infty, \quad c_1 = 0 \qquad (2.5\text{-}16)$$

This again reflects the assumption that the film is exposed only a very short time. As a result, the solute can diffuse only a short way into the film. Its diffusion is then unaffected by the exact location of the other wall, which, from the standpoint of diffusion, might as well be infinitely far away.

This problem is described by the same differential equation and boundary conditions as diffusion in a semi-infinite slab. The sole difference is that the quantity x/v_{max} replaces the time t. Because the mathematics is the same, the solution is the same. The concentration profile is

$$\frac{c_1}{c_1(\text{sat})} = 1 - \text{erf} \frac{z}{\sqrt{4Dx/v_{max}}} \qquad (2.5\text{-}17)$$

and the flux at the interface is

$$j_1|_{z=0} = \sqrt{Dv_{max}/\pi x} \; c_1(\text{sat}) \qquad (2.5\text{-}18)$$

These are the answers to this problem.

These answers appear abruptly because we can adopt the mathematical results of Section 2.2. Those studying this material for the first time often find this abruptness jarring. Stop and think about this problem. It is an important problem, basic to the penetration theory of mass transfer discussed in Section 11.2. To supply a forum for further discussion, we shall now consider this problem from another viewpoint.

The alternative viewpoint involves changing the differential volume on which we make the mass balance. In the foregoing problem, we chose a volume fixed in space, a volume through which liquid was flowing. This volume accumulated no mass; so its use led to a steady-state differential equation. Alternatively, we can choose a differential volume floating along with the fluid at a speed v_{max}. The use of this volume leads to an unsteady-state differential equation like Eq. 2.3-5. Which viewpoint is correct?

The answer is that both are correct; both eventually lead to the same answer. The fixed-coordinate method used earlier is often dignified as "Eulerian," and the moving-coordinate picture is described as "Lagrangian." The difference between them can be illustrated by the situation of watching fish swimming upstream in a fast-flowing river. If we watch the fish from a bridge, we may see only slow movement, but if we watch the fish from a freely floating canoe, we realize that the fish are moving rapidly.

Section 2.6. A final perspective

This chapter is very important, a keystone of this book. It introduces Fick's law for dilute solutions and shows how this law can be combined with mass balances to calculate concentrations and fluxes. The mass balances are made on thin shells. When these shells are very thin, the mass balances become the differential equations necessary to solve the various problems. Thus, the bricks from which this chapter is built are largely mathematical: shell balances, differential equations, and integrations in different coordinate systems.

However, we must also see a different and broader blueprint based on physics, not mathematics. This blueprint includes the two limiting cases of diffusion across a thin film and diffusion in a semi-infinite slab. Most diffusion problems fall between these two limits. The first, the thin film, is a steady-state problem, mathematically easy and sometimes physically subtle. The second, the unsteady-state problem of the thick slab, is a little harder to calculate mathematically, and it is the limit at short times.

In many cases we can use a simple criterion to decide which of the two central limits is more closely approached. This criterion hinges on the magnitude of the variable

$$\frac{(\text{length})^2}{\left(\dfrac{\text{diffusion}}{\text{coefficient}}\right)(\text{time})}$$

This variable is the argument of the error function of the semi-infinite slab; it determines the standard deviation of the decaying pulse; and it is central to the time dependence of diffusion into the cylinder. In other words, it is a key to all the foregoing unsteady-state problems. Indeed, it can be easily isolated by dimensional analysis.

This variable can be used to estimate the appropriate two limiting cases. If it is much larger than unity, we can assume an infinite slab. If it is much less than unity, we should expect a steady state or an equilibrium. If it is approximately unity, we may be forced to make a fancier analysis. For example, imagine that we are testing a membrane for an industrial separation. The membrane is 0.01 cm thick, and the diffusion coefficient in it is 10^{-7} cm^2/sec. If our experiments take only 10 sec, we have an unsteady-state problem like the infinite slab; if they take three hours we approach a steady-state situation.

In unsteady-state problems, this variable may also be used to estimate how far or how long mass transfer has occurred. Basically, the process is significantly advanced when this variable equals unity. For example, imagine that we want to guess how far gasoline has evaporated into the stagnant air in a glass-fiber filter. The evaporation has been going on about 10 min, and the diffusion coefficient is about 0.1 cm^2/sec. Thus,

$$\frac{(\text{length})^2}{(0.1 \text{ cm}^2/\text{sec})(600 \text{ sec})} = 1$$

$$\text{length} = 8 \text{ cm}$$

Alternatively, suppose we find that hydrogen has penetrated about 0.1 cm into nickel. Because the diffusion coefficient in this case is about 10^{-8} cm^2/sec, we can estimate how long this process has been going on:

$$\frac{(10^{-1} \text{ cm})^2}{(10^{-8} \text{ cm}^2/\text{sec})(\text{time})} = 1$$

$$\text{time} = 10 \text{ days}$$

This sort of heuristic argument is often successful.

A second important perspective between these two limiting cases results from comparing their interfacial fluxes given in Eqs. 2.2-10 and 2.3-18:

$$j_1 = \frac{D}{l} \Delta c_1 \quad \text{(thin film)}$$

$$j_1 = \sqrt{D/\pi t} \, \Delta c_1 \quad \text{(thick slab)}$$

Although the quantities D/l and $(D/\pi t)^{1/2}$ vary differently with diffusion coefficients, they both have dimensions of velocity; in fact, in the life sciences, they sometimes are called "the velocity of diffusion." In later chapters we shall discover that these quantities are equivalent to the mass transfer coefficients used at the beginning of this book.

References

Barnes, C. (1934). *Physics*, **5**, 4.
Bird, R. B., Stewart, W. E., & Lightfoot, E. N. (1960). *Transport Phenomena*. New York: Wiley.
Boltzmann, L. (1894). *Annalen der Physik und Chemie*, **53**, 959.
Crank, J. (1975). *The Mathematics of Diffusion*, 2nd ed. Oxford: Clarendon Press.
Fick, A. E. (1852). *Zeitschrift für Rationelle Medicin*, **2**, 83.
Fick, A. E. (1855a). *Poggendorff's Annelen der Physik*, **94**, 59.
Fick, A. E. (1855b). *Philosophical Magazine*, **10**, 30.
Fick, A. E. (1856). *Medizinische Physik*. Brunswick.
Fick, A. E. (1903). *Gesammelte Abhandlungen*. Würzburg.
Fourier, J. B. J. (1822). *Théorie analytique de la chaleur*. Paris.
Gosting, L. J. (1956). *Advances in Protein Chemistry*, **11**, 429.
Graham, T. (1829). *Quarterly Journal of Science, Literature and Art*, **27**, 74.
Graham, T. (1833). *Philosophical Magazine*, **2**, 175, 222, 351.
Graham, T. (1850). *Philosophical Transactions of the Royal Society of London*, **140**, 1.
Idicula, J., Graves, D. J., Quinn, J. A., & Lambertsen, C. J. (1976). In: *Underwater Physiology, Vol. 5*, ed. C. J. Lambertsen, p. 335. New York: Academic.
Kahn, D. S., & Polson, H. (1947). *Journal of Physical and Colloid Chemistry*, **51**, 816.
Longsworth, L. G. (1950). *Review of Scientific Instruments*, **21**, 524.
Mason, E. A. (1970). *Philosophical Journal*, **7**, 99.
Mills, R., Woolf, L. A., & Watts, R. O. (1968). *American Institute of Chemical Engineers Journal*, **14**, 671.
Robinson, R. A., & Stokes, R. H. (1960). *Electrolyte Solutions*. London: Butterworth.
Sherwood, T. K., Pigford, R. L., & Wilke, C. R. (1975). *Mass Transfer*. New York: McGraw-Hill.
Stannet, V. T., Koros, W. J., Paul, D. R., Lonsdale, H. K., & Baker, R. W. (1979). *Advances in Polymer Science*, **32**, 69.

3 DIFFUSION IN CONCENTRATED SOLUTIONS

Diffusion causes convection. To be sure, convective flow can have many causes. For example, it can occur because of pressure gradients or temperature differences. However, even in isothermal and isobaric systems, diffusion will always produce convection. This was clearly stated by Maxwell more than 100 years ago: "Mass transfer is due partly to the motion of translation and partly to that of agitation" (Maxwell, 1860). In more modern terms, we would say that any mass flux may include both convection and diffusion.

This combination of convection and diffusion can complicate our analysis. The easier analyses occur in dilute solutions, in which the convection caused by diffusion is vanishingly small. This dilute limit provides the framework within which most people analyze diffusion. This is the framework presented in Chapter 2.

In some cases, however, our dilute-solution analyses do not successfully correlate our experimental observations. Consequently, we must use more elaborate equations. This elaboration is best initiated with the physically based examples given in Section 3.1. This is followed by a catalogue of general flux equations in Section 3.2. These flux equations form the basis for the simple analyses of diffusion and convection in Section 3.3 that parallel those in the previous chapter.

After simple analyses, we move to Section 3.4 and more general mass balances, sometimes called the general continuity equations. These equations involve the various coordinate systems introduced in Chapter 2. They allow solutions for the more difficult problems that arise from the more complicated physical situation. Fortunately, the complexities inherent in these examples can often be dodged by effectively exploiting selected readings. A guide to these readings is given in Section 3.5.

The material in this chapter is more complicated than that in Chapter 2 and is unnecessary for many who are not trying to pass exams in advanced courses. Nonetheless, this material has fascinating aspects, as well as some tedious ones. Those studying these aspects often tend to substitute mathematical manipulation for thought. Make sure that the intellectual framework in Chapter 2 is secure before starting this more advanced material.

Section 3.1. Diffusion with convection

The statement by Maxwell quoted earlier suggests that diffusion and convection always occur together, that one cannot occur without the other. This fact sets diffusion apart from many other phenomena. For example, thermal conduction can certainly occur without convection. In contrast, diffusion generates its own convection, so that understanding the process can be much more complicated, especially in concentrated solutions.

A qualitative example

To illustrate how diffusion and convection are interrelated, we consider the example shown in Fig. 3.1-1. The physical system consists of a large reservoir of benzene connected to a large volume of air by means of a capillary tube. Benzene evaporates and moves through the capillary into the surrounding air.

At room temperature, not much benzene evaporates because its vapor pressure is low. Benzene vapor moves slowly up the tube because of Brownian motion, that is, because of thermally induced agitation of the molecules. This is the process basic to diffusion studied in the previous chapter.

At the boiling point, the situation is completely different. The liquid benzene boils into vapor, and the vapor rushes up the capillary. This rush is clearly a pressure-driven flow, a convection caused by the sharply increased volume of the vapor as compared with the liquid. It has little to do with diffusion.

At intermediate temperatures, both diffusion and convection will be important, because the processes take place simultaneously. To understand such intermediate cases, we must look at how mass transport works.

Separating convection from diffusion

The complete description of mass transfer requires separating the contributions of diffusion and convection. The usual way of effecting this separation is to assume that these two effects are additive:

$$\begin{pmatrix} \text{total mass} \\ \text{transported} \end{pmatrix} = \begin{pmatrix} \text{mass transported} \\ \text{by diffusion} \end{pmatrix} + \begin{pmatrix} \text{mass transported} \\ \text{by convection} \end{pmatrix} \quad (3.1\text{-}1)$$

In more exact terms, we define the total mass flux $\mathbf{n}_1$ as the mass transported per area per time relative to fixed coordinates. This flux, in turn, is used to define an average solute velocity $\mathbf{v}_1$:

$$\mathbf{n}_1 = c_1 \mathbf{v}_1 \quad (3.1\text{-}2)$$

where c_1 is the local concentration. We then divide $\mathbf{v}_1$ into two parts:

$$\mathbf{n}_1 = c_1(\mathbf{v}_1 - \mathbf{v}^a) + c_1 \mathbf{v}^a = \mathbf{j}_1^a + c_1 \mathbf{v}^a \quad (3.1\text{-}3)$$

where $\mathbf{v}^a$ is some convective "reference" velocity. The first term $\mathbf{j}_1^a$ on the right-hand side of this equation represents the diffusion flux, and the second term $c_1 \mathbf{v}^a$ describes the convection.

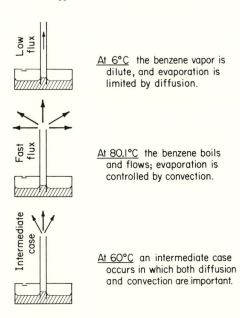

Fig. 3.1-1. Evaporation of benzene. This process is dominated by diffusion in dilute solutions, but it includes both diffusion and convection in concentrated solutions.

Interestingly, there is no clear choice for what this convective reference velocity should be. It might be the mass average velocity that is basic to the equations of motion, which in turn are a generalization of Newton's second law. It might be the velocity of the solvent, because that species is usually present in excess. We cannot automatically tell. We only know that we should choose v^a so that v^a is zero as frequently as possible. By doing so, we eliminate convection essentially by definition, and we are left with a substantially easier problem.

To see which reference velocity is easiest to use, we consider the diffusion apparatus shown in Fig. 3.1-2. This apparatus consists of two bulbs, each of which contains a gas or liquid solution of different composition. The two bulbs are connected by a long thin capillary containing a stopcock. At time zero, the stopcock is opened; after an experimentally desired time, the stopcock is closed. The solutions in the two bulbs are then analyzed, and the concentrations are used to calculate the diffusion coefficient. The equations used in these calculations are identical with those used for the diaphragm cell.

Here, we examine this apparatus to elucidate the interaction of diffusion and convection, not to measure the diffusion coefficient. The examination is easiest for the special cases of gases and liquids. For gases, we imagine that one bulb is filled with nitrogen and the other with hydrogen. During the experiment, the number of moles in the left bulb always equals the number of

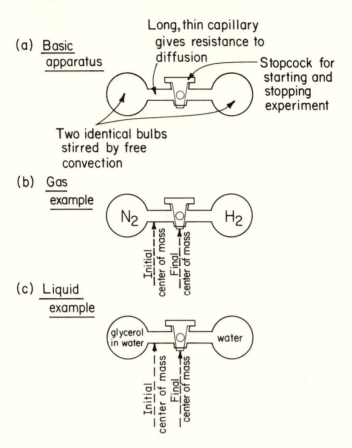

Fig. 3.1-2. An example of reference velocities. Descriptions of diffusion imply reference to a velocity relative to the system's mass or volume. Whereas the mass usually has a nonzero velocity, the volume often shows no velocity. Hence, diffusion is best referred to the volume's average velocity.

moles in the identical right bulb, because isothermal and isobaric ideal gases have a constant number of moles per volume. The volume of the left bulb equals the volume of the right bulb, because the bulbs are rigid. Thus, the average velocity of the moles $\mathbf{v}^*$ and the average velocity of the volume $\mathbf{v}^0$ are both zero.

In contrast, the average velocity of the mass $\mathbf{v}$ in this system is not zero. To see why this is so, imagine balancing the apparatus on a knife edge. This edge will initially be located left of center, as in Fig. 3.1-2(b), because the nitrogen on the left is heavier than the hydrogen on the right. As the experiment proceeds, the knife edge must be shifted toward the center, because the densities in the two bulbs will become more nearly equal.

Thus, in gases, the molar and volume average velocities are zero, but the mass average velocity is not. Therefore, the molar and volume average velocities allow a simpler description of diffusion in gases than the mass average velocity.

We now turn to the special case of liquids, shown in Fig. 3.1-2(c). The volume of the solution is very nearly constant during diffusion, so that the volume average velocity is very nearly zero. This approximation holds whenever there is no significant volume change after mixing. In my experience, this is true except for some alcohol–water systems, and even in those systems it is not a bad approximation.

The other two velocities are more difficult to estimate. If the densities of both solutions are the same, then the center of mass does not move, and the mass average velocity is zero. This is often a reasonable approximation. If the molar concentration is constant, as in an ideal gas, then the center of moles is immobile, and the molar average velocity is zero.

To estimate the various reference velocities, imagine allowing 50 weight percent glycerol to diffuse into water. The volume changes less than 0.1% during this mixing, so that the volume average velocity is very nearly zero. The glycerol solution has a density of about 1.1 g/cm^3, as compared with water at 1 g/cm^3, so that the mass density changes about 10%. In contrast, the glycerol solution has a molar density of about 33 moles/liter, as compared with water at 55 moles/liter; so the molar concentration changes about 50%. Thus, the mass average velocity will be nearer to zero than the molar average velocity.

In summary, the molar and volume average velocities are zero for ideal gases, and the volume and mass average velocities are closer to zero for liquids. The mass average velocity is often inappropriate for gases, and the molar average velocity is rarely used for liquids. The volume average velocity is appropriate most frequently, and so it will be emphasized in this book.

Section 3.2. Different forms of the diffusion equation

Before we begin the analysis of diffusion and simultaneous convection, we should summarize the various diffusion equations currently used in the literature. These equations illustrate the different ways in which diffusion and convection can be separated. Their presentation is abstract, so that those who are mathematically proficient will be more at home than those who think in physical terms.

The different forms of the flux equations are given in Table 3.2-1 (Tyrrell, 1961; deGroot & Mazur, 1962; Katchalsky & Curran, 1967). All these equations are equivalent; each can be rearranged in the mathematical form of the others. Unlike the flux equations in Table 2.1-2, each of these new equations includes convection. Each is vectorial and therefore describes diffusion in three dimensions. However, not all define the same diffusion coefficients, so that translation among these forms is often complex.

The first three flux equations in Table 3.2-1 are defined relative to mass average velocity, molar average velocity, and volume average velocity.

Table 3.2-1. *Different forms of the diffusion equation*

Choice	Total flux (diffusion + convection)	Diffusion equation	Reference velocity	Where best used
Mass	$\mathbf{n}_1 = \mathbf{j}_1^m + c_1\mathbf{v}$	$\mathbf{j}_1^m = \rho_1(\mathbf{v}_1 - \mathbf{v})$ $= -D\rho\nabla\omega_1$	$\mathbf{v} = \omega_1\mathbf{v}_1 + \omega_2\mathbf{v}_2$ $\rho\mathbf{v} = \mathbf{n}_1 + \mathbf{n}_2$	Constant-density liquids; coupled mass and momentum transport
Molar	$\mathbf{n}_1 = \mathbf{j}_1^* + c_1\mathbf{v}^*$	$\mathbf{j}_1^* = c_1(\mathbf{v}_1 - \mathbf{v}^*)$ $= -Dc\nabla y_1$	$\mathbf{v}^* = y_1\mathbf{v}_1 + y_2\mathbf{v}_2$ $c\mathbf{v}^* = \mathbf{n}_1 + \mathbf{n}_2$	Ideal gases where the molar concentration c is constant
Volume	$\mathbf{n}_1 = \mathbf{j}_1 + c_1\mathbf{v}^0$	$\mathbf{j}_1 = c_1(\mathbf{v}_1 - \mathbf{v}^0)$ $= -D\nabla c_1$	$\mathbf{v}^0 = c_1\bar{V}_1\mathbf{v}_1 + c_2\bar{V}_2\mathbf{v}_2$ $= \bar{V}_1\mathbf{n}_1 + \bar{V}_2\mathbf{n}_2$	Best overall; good for constant-density liquids and for ideal gases; may use either mass or mole concentration
Solvent	$\mathbf{n}_1 = \mathbf{j}_1^{(2)} + c_1\mathbf{v}_2$	$\mathbf{j}_1^{(2)} = c_1(\mathbf{v}_1 - \mathbf{v}_2)$ $= -D_1\nabla c_1$	$\mathbf{v}_2$	Rare except for some membranes; note that $D_1 \neq D_2 \neq D$
Stefan–Maxwell		$\nabla y_1 = \dfrac{y_1 y_2}{D}(\mathbf{v}_2 - \mathbf{v}_1)$	None	Frequent theoretical result; difficult to use in practice

They will be discussed momentarily. The remaining two flux equations will be mentioned only briefly. The flux equation relative to the solvent velocity is used only in the description of reverse osmosis, where its use leads to concepts like filtration and reflection coefficients, discussed in Section 15.1.

The last equation in Table 3.2-1, sometimes called the Stefan–Maxwell equation, is the only form that separates diffusion from convection in a simple way (Maxwell, 1952; Lightfoot, 1974). It does so by replacing the usual flux equation with a difference in species velocities. This is a real advantage. However, the Stefan–Maxwell equation is almost never used because it is difficult to solve mathematically, even in the simplest cases.

We stated earlier that the flux equation relative to the volume average velocity often gives the simplest solution to practical problems. This relation is identical with the flux equation relative to the mass average velocity in systems of constant density. To prove this contention, we note that

$$c_1\bar{V}_1 = c_1\tilde{M}_1\left(\frac{\partial V}{\partial(\mathfrak{N}_1\tilde{M}_1)}\right)_{\mathfrak{N}_2} = \rho_1\left(\frac{\partial V}{\partial m_1}\right)_{m_2} \tag{3.2-1}$$

This derivative is the change in volume with a change in mass of species 1. If the system has constant density, this change is merely the reciprocal of the density:

$$c_1 \bar{V}_1 = \rho_1/\rho = \omega_1 \tag{3.2-2}$$

Then, for constant ρ,

$$\mathbf{v}^0 = \sum_{i=1}^{2} c_i \bar{V}_i \mathbf{v}_i = \sum_{i=1}^{2} \omega_i \mathbf{v}_i = \mathbf{v} \tag{3.2-3}$$

The volume and mass average velocities are the same for a system of constant density.

The volume average velocity is equal to the molar average velocity for ideal gases. Here,

$$c_1 \bar{V}_1 = c_1 \left[\frac{\partial}{\partial \mathfrak{N}_1} \left(\frac{RT(\mathfrak{N}_1 + \mathfrak{N}_2)}{p} \right) \right]_{\mathfrak{N}_2} \tag{3.2-4}$$

$$= c_1 \left(\frac{RT}{p} \right) = \frac{c_1}{c} = y_1$$

For constant c,

$$\mathbf{v}^0 = \sum_{i=1}^{2} c_1 \bar{V}_1 \mathbf{v}_1 = \sum_{i=1}^{2} y_1 \mathbf{v}_1 = \mathbf{v}^* \tag{3.2-5}$$

The volume and molar average velocities are the same for systems that, like the ideal gas, have constant molar concentration.

Example 3.2-1: Diffusion-engendered flow

In the diffusion apparatus shown in Fig. 3.1-2(b), one bulb contains nitrogen and the other hydrogen. The temperature and pressure are such that the diffusion coefficient is 0.1 cm²/sec. Find v^0, v^*, and v at the average concentration in the system.

Solution. The volume in this system does not move; so v^0 is zero. If the gases are ideal, then the molar concentration is constant everywhere, and $v^* = 0$. Because of this, we can use the thin-film results from Section 2.2:

$$j_1 = c_1 v_1 = \frac{D}{l} (c_{10} - c_{1l})$$

If species 1 is nitrogen,

$$v_1 = \left[\frac{D}{l} \right] \left(\frac{c_{10} - c_{1l}}{c_1} \right) = \left[\frac{0.1 \text{ cm}^2/\text{sec}}{10 \text{ cm}} \right] \left(\frac{1 - 0}{0.5} \right) = 0.02 \text{ cm/sec}$$

By similar arguments, for hydrogen,

$$v_2 = -0.02 \text{ cm/sec}$$

Note that these velocities vary as the average concentration c_1 varies.

We next find the mass fractions of each species:

$$\omega_1 = \frac{c_1 \tilde{M}_1}{c_1 \tilde{M}_1 + c_2 \tilde{M}_2} = \frac{0.5(28)}{0.5(28) + 0.5(2)} = 0.933$$

Similarly,

$$\omega_2 = 0.067$$

Then the mass average velocity is

$$v = \omega_1 v_1 + \omega_2 v_2 = 0.933(0.020) + 0.067(-0.020) = 0.017 \text{ cm/sec}$$

The result is dominated by the nitrogen because of its higher molecular weight.

Example 3.2-2: The binary diffusion coefficient

Prove that if the partial molar volumes are constant, there is only one diffusion coefficient in a two-component mixture. In other words, if we define

$$-\mathbf{j}_1 = D_1 \nabla C_1$$

and

$$-\mathbf{j}_2 = D_2 \nabla C_2$$

prove that $D_1 = D_2$.

Solution. By definition, we know that

$$c_1 \bar{V}_1 \mathbf{v}_1 + c_2 \bar{V}_2 \mathbf{v}_2 = \mathbf{v}^0$$

Moreover, the volume fractions $c_i \bar{V}_i$ must sum to unity:

$$c_1 \bar{V}_1 + c_2 \bar{V}_2 = 1$$

Combining these results,

$$c_1 \bar{V}_1 (\mathbf{v}_1 - \mathbf{v}^0) + c_2 \bar{V}_2 (\mathbf{v}_2 - \mathbf{v}^0) = \mathbf{j}_1 \bar{V}_1 + \mathbf{j}_2 \bar{V}_2 = 0$$

From Eqs. 3.2-12 and 3.2-13,

$$\bar{V}_1 D_1 \nabla c_1 + \bar{V}_2 D_2 \nabla c_2 = 0$$

but from Eq. 3.2-15,

$$\bar{V}_1 \nabla c_1 + \bar{V}_2 \nabla c_2 = 0$$

Thus,

$$D_1 = D_2$$

There is only one diffusion coefficient if the partial molar volumes are constant. This result can be shown from the Gibbs–Duhem equation to be valid even when the $\bar{V}_i$ are not constant (deGroot & Mazur, 1962).

Section 3.3. Parallel diffusion and convection

We now want to combine the equations developed earlier with mass balances to calculate fluxes and concentration profiles. This is, of course, the same objective as in Chapter 2. The difference here is that both diffusion and convection are significant. The analysis of the more complicated problems of diffusion and convection is aided by the parallels in the cases of a thin film and an infinite slab around which Chapter 2 is organized. Such parallels lead to powerful pedagogy.

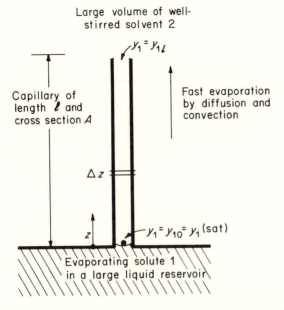

Fig. 3.3-1. Fast evaporation in a thin capillary. This problem is analogous to that shown in Fig. 2.2-1, but for a concentrated solution.

Fast diffusion through a stagnant film

The first problem that we consider involves the same rapid evaporation that was used as the key example in Section 3.1. We recall that at intermediate temperatures, the evaporation rate depended on both diffusion and convection up the tube.

We want to calculate the flux and the concentration profile where both diffusion and convection are important. To make this calculation, we must parallel our earlier scheme, but with a more exact physical understanding and a more complicated mathematical analysis. Just as before, the scheme starts with a mass balance, combines this balance with Fick's law, and then runs through the math to the desired result.

This conservation relation is written on the differential volume $A\Delta z$ shown in Fig. 3.3-1:

$$\begin{pmatrix} \text{solute accumulated} \\ \text{in volume } A\Delta z \end{pmatrix} = \begin{pmatrix} \text{solute transported} \\ \text{in at } z \end{pmatrix} - \begin{pmatrix} \text{solute transported} \\ \text{out at } z + \Delta z \end{pmatrix}$$

(3.3-1)

In mathematical terms, this is

$$\frac{\partial}{\partial t}(A\Delta z c_1) = An_1|_z - An_1|_{z+\Delta z}$$

(3.3-2)

If we divide by the volume $A\Delta z$ and take the limit as this volume goes to zero, we find

$$\frac{\partial c_1}{\partial t} = -\frac{\partial n_1}{\partial z} \tag{3.3-3}$$

At steady state, there is no accumulation; so

$$0 = \frac{\partial n_1}{\partial z} \tag{3.3-4}$$

This is easily integrated to show that n_1 is constant. This sensibly says that at steady state, the total flux up the tube is constant. Note that we have not shown that the diffusion flux is constant.

We now want to combine this result with Fick's law. However, because we are dealing with fast evaporation and a potentially concentrated solution, we must consider both diffusion and convection. For simplicity, we choose the volume average velocity; so, from Table 3.2-1,

$$n_1 = j_1 + c_1 v^0 = -D\frac{dc_1}{dz} + c_1(c_1 \bar{V}_1 v_1 + c_2 \bar{V}_2 v_2) \tag{3.3-5}$$

By definition, $c_1 v_1$ equals n_1, and $c_2 v_2$ equals n_2. If the solvent vapor is stagnant, its flux n_2 and its velocity v_2 must be zero. Thus,

$$n_1 = -D\frac{dc_1}{dz} + c_1 \bar{V}_1 n_1 \tag{3.3-6}$$

Moreover, if the vapors in the capillary are ideal, then the total molar concentration is a constant, and $\bar{V}_1$ equals $1/c$ (see Eq. 3.2-4). Thus, the differential equation we seek is

$$n_1(1 - y_1) = -Dc\frac{dy_1}{dz} \tag{3.3-7}$$

This is subject to the two boundary conditions

$$z = 0, \quad y_1 = y_{10} \tag{3.3-8}$$
$$z = l, \quad y_1 = y_{1l} \tag{3.3-9}$$

There are two boundary conditions for the first-order differential equation because n_1 is an unknown integration constant.

The flux and concentration profiles are now routinely found. The concentration profile is exponential:

$$\frac{1 - y_1}{1 - y_{10}} = \left(\frac{1 - y_{1l}}{1 - y_{10}}\right)^{z/l} \tag{3.3-10}$$

The total flux is constant and logarithmic:

$$n_1 = \frac{Dc}{l}\ln\left(\frac{1 - y_{1l}}{1 - y_{10}}\right) \tag{3.3-11}$$

but the diffusion flux varies with z:

$$j_1 = -Dc\frac{dy_1}{dz} = Dc\left(\frac{1 - y_{10}}{l}\right)\left(\frac{1 - y_{1l}}{1 - y_{10}}\right)^{z/l}\ln\left(\frac{1 - y_{1l}}{1 - y_{10}}\right) \tag{3.3-12}$$

The diffusion flux has different values within the capillary.

The foregoing results are in startling contrast to those found for a dilute solution. The contrast comes from the mathematics, which is a lot fancier. If the solutions are dilute, then y_1, y_{10}, and y_{1l} are always small, and we can simplify the results. To do this, we first remember that for small y_1,

$$(1 - y_1)^a \doteq 1 - ay_1 + \cdots \qquad (3.3\text{-}13)$$

$$\frac{1}{1 - y_1} \doteq 1 + y_1 + \cdots \qquad (3.3\text{-}14)$$

and

$$\ln(1 - y_1) \doteq -y_1 + \cdots \qquad (3.3\text{-}15)$$

The concentration profile in Eq. 3.3-10 thus becomes

$$1 - y_1 = (1 - y_{10})(1 - y_{1l} + y_{10} - \cdots)^{z/l} \qquad (3.3\text{-}16)$$

$$= 1 - y_{10} + \frac{z}{l}(y_{10} - y_{1l}) + \cdots$$

This can be rewritten in more familiar terms by multiplying both sides of the equation by the total concentration c and rearranging:

$$c_1 = c_{10} + (c_{1l} - c_{10})\frac{z}{l} \qquad (3.3\text{-}17)$$

In other words, the concentration profile becomes linear, not exponential, as the solution becomes dilute.

The total flux in dilute solution can be simplified in similar fashion:

$$n_1 = \frac{Dc}{l}[\ln(1 - y_{1l}) - \ln(1 - y_{10})]$$

$$\doteq \frac{Dc}{l}(y_{10} - y_{1l})$$

$$\doteq \frac{D}{l}(c_{10} - c_{1l}) \qquad (3.3\text{-}18)$$

which is, of course, the simple relation derived earlier in Eq. 2.2-10. The diffusion flux j_1 equals n_1 in this dilute limit. Thus, Eqs. 3.3-10 and 3.3-11 are equivalent to Eqs. 2.2-9 and 2.2-10 in dilute solution.

Fast diffusion into a semi-infinite slab

The second problem considered in this section is illustrated schematically in Fig. 3.3-2. In this problem, a volatile liquid solute evaporates into a long gas-filled capillary. The solvent gas in the capillary initially contains no solute. As solute evaporates, the interface between the vapor and the liquid solute drops. However, the gas is essentially insoluble in the liquid. We want to calculate the solute's evaporation rate, including the effects of diffusion-induced convection and the effects of the moving interface (Arnold, 1944; Bird et al., 1960).

In this problem, we first choose the origin of our coordinate system ($z = 0$) as the liquid–vapor interface. We then write a mass balance for the solute 1 on the differential volume $A\Delta z$, shown in Fig. 3.3-3:

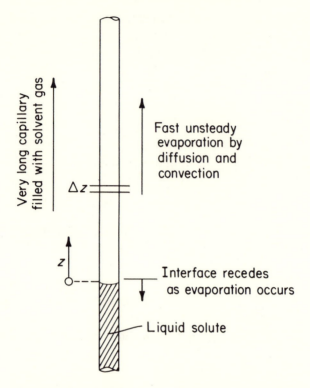

Fast unsteady
evaporation by
diffusion and
convection

Interface recedes
as evaporation occurs

Liquid solute

Fig. 3.3-2. Fast diffusion in a semiinfinite slab. This problem is analogous to that shown in Fig. 2.3-2, but for a concentrated solution. Because of this higher concentration, the liquid–vapor interface moves significantly, complicating the situation.

$$
\begin{pmatrix}
\text{solute} \\
\text{accumulation} \\
\text{in } A\Delta z
\end{pmatrix}
=
\begin{pmatrix}
\text{solute} \\
\text{transport} \\
\text{in}
\end{pmatrix}
-
\begin{pmatrix}
\text{solute} \\
\text{transport} \\
\text{out}
\end{pmatrix}
\tag{3.3-19}
$$

or, in symbolic terms,

$$
\frac{\partial}{\partial t}(c_1 A\Delta z) = (An_1)_z - (An_1)_{z+\Delta z}
\tag{3.3-20}
$$

Dividing by the differential volume and taking the limit as this volume goes to zero,

$$
\frac{\partial c_1}{\partial t} = -\frac{\partial n_1}{\partial z}
\tag{3.3-21}
$$

We then try to split the diffusion and convection:

$$
\frac{\partial c_1}{\partial t} = D\frac{\partial^2 c_1}{\partial z^2} - \frac{\partial}{\partial z}c_1 v^0
\tag{3.3-22}
$$

By definition, the volume average velocity is

$$
v^0 = c_1 \bar{V}_1 v_1 + c_2 \bar{V}_2 v_2 = \bar{V}_1 n_1 + \bar{V}_2 n_2
\tag{3.3-23}
$$

In the steady-state case treated earlier, we argued that the solvent was stagnant, so that n_2 was zero, and the problem was simple. Here, in an unsteady case, the solvent flux varies with position and time; therefore, no easy simplification is possible.

We must write a continuity equation for the solvent gas 2:

$$\frac{\partial c_2}{\partial t} = -\frac{\partial n_2}{\partial z} \tag{3.3-24}$$

If we multiply Eqs. 3.3-21 and 3.3-24 by the appropriate partial molar volumes and add them, we find

$$\frac{\partial}{\partial t}(\bar{V}_1 c_1 + \bar{V}_2 c_2) = -\frac{\partial}{\partial z}(\bar{V}_1 n_1 + \bar{V}_2 n_2) \tag{3.3-25}$$

But the quantity $\bar{V}_1 c_1 + \bar{V}_2 c_2$ always equals unity, making the left-hand side of this equation zero; thus, $\bar{V}_1 n_1 + \bar{V}_2 n_2$ must be independent of z. However, at the interface, n_2 is zero, because the solvent gas 2 is insoluble in the liquid. Thus,

$$\bar{V}_1 n_1 + \bar{V}_2 n_2 = \bar{V}_1 n_1|_{z=0} = \bar{V}_1\left(-D\frac{\partial c_1}{\partial z}\bigg|_{z=0} + c_1\bar{V}_1 n_1|_{z=0}\right) \tag{3.3-26}$$

When we combine this with Eq. 3.3-23, we find

$$\frac{\partial c_1}{\partial t} = D\frac{\partial^2 c_1}{\partial z^2} + \left(\frac{D\bar{V}_1(\partial c_1/\partial z)}{1 - c_1\bar{V}_1}\right)_{z=0}\frac{\partial c_1}{\partial z} \tag{3.3-27}$$

subject to the conditions

$$t = 0, \quad \text{all } z > 0, \quad c_1 = 0 \tag{3.3-28}$$

$$t > 0, \qquad z = 0, \quad c_1 = c_1(\text{sat}) \tag{3.3-29}$$

$$z = \infty, \quad c_1 = 0 \tag{3.3-30}$$

The solute concentration $c_1(\text{sat})$ is that in the vapor in equilibrium with the liquid.

Like the problem of diffusion in a semi-infinite slab, this problem is solved by defining the combined variable

$$\zeta = z/\sqrt{4Dt} \tag{3.3-31}$$

The differential equation now becomes

$$\frac{d^2 c_1}{d\zeta^2} + 2(\zeta - \Phi)\frac{dc_1}{d\zeta} = 0 \tag{3.3-32}$$

subject to the conditions

$$\zeta = 0, \quad c_1 = c_1(\text{sat}) \tag{3.3-33}$$

$$\zeta = \infty, \quad c_1 = 0 \tag{3.3-34}$$

and in which

$$\Phi = -\frac{1}{2}\left(\frac{\bar{V}_1(dc_1/d\zeta)}{1 - c_1\bar{V}_1}\right)_{\zeta=0} \tag{3.3-35}$$

The constant Φ, a dimensionless velocity, characterizes both the convection by diffusion and the movement of the interface. If Φ is zero, convection effects are zero.

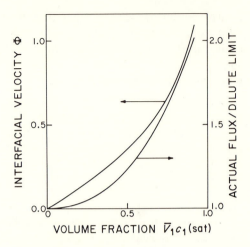

Fig. 3.3-3. Flux and interfacial movement. As the solution becomes dilute, the interfacial concentration $c_1(\text{sat})$ becomes small, the actual flux approaches the dilute-solution limit (see Eq. 2.3-18), and the interfacial velocity Φ becomes zero.

Equation 3.3-32 can be integrated once to give

$$\frac{dc_1}{d\zeta} = (\text{constant})e^{-(\zeta - \Phi)^2} \tag{3.3-36}$$

A second integration and evaluation of the conditions give

$$\frac{c_1}{c_1(\text{sat})} = \frac{1 - \text{erf}(\zeta - \Phi)}{1 + \text{erf }\Phi} \tag{3.3-37}$$

We can calculate Φ from this result and Eq. 3.3-35:

$$\bar{V}_1 c_1(\text{sat}) = \left(1 + \frac{1}{\sqrt{\pi}\,(1 + \text{erf }\Phi)\Phi e^{\Phi^2}}\right)^{-1} \tag{3.3-38}$$

A plot of Φ versus concentration is shown in Fig. 3.3-4. Note that when $c_1(\text{sat})$ is small, Φ goes to zero. In other words, when the solution is dilute, convection is unimportant.

We also want to calculate the interfacial flux. To find this, we must again split diffusion and convection, using Fick's law:

$$N_1 = n_1|_{z=0} = -\left(\frac{D(\partial c_1/\partial z)}{1 - c_1\bar{V}_1}\right)_{z=0} = \sqrt{D/\pi t}\,\frac{1}{1 - \bar{V}_1 c_1(\text{sat})}\,\frac{e^{-\Phi^2}}{1 + \text{erf }\Phi}\,c_1(\text{sat}) \tag{3.3-39}$$

where Φ is still found from Eq. 3.3-38 or Fig. 3.3-3. The increase of this flux beyond that in a dilute solution are also given in this figure.

Example 3.3-1: Errors caused by neglecting convection

Consider the experiments shown in Fig. 3.3-4. How much error is caused by calculating the rate of benzene evaporation if only diffusion is considered?

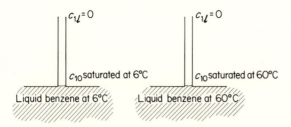

Fig. 3.3-4. Examples of benzene diffusion and convection. In the dilute solution at the left, the exact results are close to the approximate ones in Eq. 2.2-10. In the concentrated case at the right, they are not.

Solution. The sizes of the errors depend on the concentrations and thus on the temperature. At 6°C, the vapor pressure of benzene is about 37 mm Hg. If the total pressure is one atmosphere,

$$y_1 = \frac{c_1}{c} = \frac{p_1(\text{sat})}{p} = \frac{37}{760} = 0.049$$

The total flux is, from Eq. 3.3-11,

$$n_1 = \frac{Dc}{l} \ln\left(\frac{1-0}{1-0.049}\right)$$
$$= \frac{0.050Dc}{l}$$

The flux, assuming a dilute solution, calculated from Eq. 2.2-10, is

$$n_1 = j_1 = \frac{Dc}{l}(0.049 - 0)$$

This 2% difference is well within the needs of most practical calculations. Thus, the dilute-solution equations are more than adequate here.

At 60°C, the choice is less obvious, because the vapor pressure is about 395 mm Hg. When we calculate the mole fraction in the same way, we find

$$n_1 = \frac{Dc}{l} \ln\left(\frac{1-0}{1-(395/760)}\right) = 0.73 \frac{Dc}{l}$$

The dilute-solution estimate is

$$n_1 = j_1 = \frac{395}{760}\frac{Dc}{l} = 0.52 \frac{Dc}{l}$$

The dilute-solution equations underestimate the flux by a significant error of about 40%.

Section 3.4. "Generalized" mass balances

As the problems that we discuss in this chapter become more and more complex, the development of the differential equations becomes more and more tedious. Such tedium can be avoided by using the generalized mass

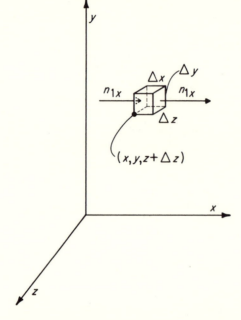

Fig. 3.4-1. The arbitrary volume for deriving the generalized mass balances. The fluxes in the z direction are shown in this figure; fluxes in other directions are also included in the derivation. The results are shown in Tables 3.4-1 and 3.4-2.

balances developed in this section. These mass balances automatically include both steady- and unsteady-state situations. They imply the usual variety of coordinate systems, and they reflect the vectorial nature of mass fluxes. They are excellent weapons (Bird et al., 1960; Schlichting, 1979).

However, like most weapons, the generalized mass balances can injure those trying to use them. Effective use requires uncommon skill in connecting the mathematical ideal and the physical reality. Some seem born with this skill; more seem to develop it over time. If you have trouble applying these equations, return to the shell balance method. It may take longer, but it is safer. You can check your equations by later comparing them with those found from the generalized results.

To find the generalized mass balances, we consider the small differential volume located at (x, y, z) shown in Fig. 3.4-1. We want to write a mass balance on this volume:

$$\begin{pmatrix} \text{mass of species 1} \\ \text{accumulating in} \\ \Delta x \Delta y \Delta z \end{pmatrix} = \begin{pmatrix} \text{mass flux of} \\ \text{species 1} \\ \text{in minus that out} \end{pmatrix} + \begin{pmatrix} \text{mass produced by} \\ \text{homogeneous} \\ \text{chemical reaction} \end{pmatrix}$$

The mass fluxes are relative to fixed coordinates and include transport in all three directions. For example, the mass flux out of the volume in the x

direction, shown in Fig. 3.4-1, is $n_{1x}\Delta y\Delta z$, where $\Delta y\Delta z$ is the area across which this flux occurs. In mathematical terms, the mass balance is then

$$\frac{\partial}{\partial t}(c_1\Delta x\Delta y\Delta z) = (n_{1x}\Delta y\Delta z)_x - (n_{1x}\Delta y\Delta z)_{x+\Delta x}$$
$$+ (n_{1y}\Delta x\Delta z)_y - (n_{1y}\Delta x\Delta z)_{y+\Delta y}$$
$$+ (n_{1z}\Delta x\Delta y)_z - (n_{1z}\Delta x\Delta y)_{z+\Delta z}$$
$$+ r_1\Delta x\Delta y\Delta z \tag{3.4-2}$$

where r_1 is the rate per unit volume of a homogeneous chemical reaction producing solute 1. Dividing by the differential volume $\Delta x\Delta y\Delta z$ and taking the limit as this volume goes to zero gives

$$\frac{\partial}{\partial t}c_1 = -\frac{\partial}{\partial x}n_{1x} - \frac{\partial}{\partial y}n_{1y} - \frac{\partial}{\partial z}n_{1z} + r_1 \tag{3.4-3}$$

or, in vectorial notation,

$$\frac{\partial}{\partial t}c_1 = -\nabla\cdot\mathbf{n}_1 + r_1 \tag{3.4-4}$$

We can also write the flux in terms of diffusion and convection:

$$\mathbf{n}_1 = -D\nabla c_1 + c_1\mathbf{v}^0 \tag{3.4-5}$$

where $\mathbf{v}^0$ is the volume average velocity. Combining,

$$\frac{\partial c_1}{\partial t} = D\nabla^2 c_1 - \nabla\cdot c_1\mathbf{v}^0 + r_1 \tag{3.4-6}$$

This equation is the general form of all the shell balances derived to date.

The species mass balance represented by Eq. 3.4-6 is often effectively complemented by the overall mass balance:

$$\begin{pmatrix} \text{total mass} \\ \text{accumulation} \\ \text{in } \Delta x\Delta y\Delta z \end{pmatrix} = \begin{pmatrix} \text{total mass} \\ \text{flux in minus} \\ \text{that out} \end{pmatrix} \tag{3.4-7}$$

This can be written in terms similar to those used earlier:

$$\frac{\partial}{\partial t}(\rho v_x\Delta x\Delta y\Delta z) = (\rho v_x\Delta y\Delta z)_x - (\rho v_x\Delta y\Delta z)_{x+\Delta x}$$
$$+ (\rho v_y\Delta x\Delta z)_y - (\rho v_y\Delta x\Delta z)_{y+\Delta y}$$
$$+ (\rho v_z\Delta x\Delta y)_z - (\rho v_z\Delta x\Delta y)_{z+\Delta z} \tag{3.4-8}$$

in which v_x, v_y, and v_z are components of the mass average velocity. Dividing by the volume $\Delta x\Delta y\Delta z$ and taking the limit as each difference becomes small, we find

$$\frac{\partial\rho}{\partial t} = -\frac{\partial}{\partial x}\rho v_x - \frac{\partial}{\partial y}\rho v_y - \frac{\partial}{\partial z}\rho v_z \tag{3.4-9}$$

In vectorial notation, this is

$$\frac{\partial\rho}{\partial t} = -\nabla\cdot\rho\mathbf{v} \tag{3.4-10}$$

Table 3.4-1. *Mass balance of species 1 in various coordinate systems[a]*

Rectangular coordinates

$$\frac{\partial c_1}{\partial t} = -\frac{\partial n_{1x}}{\partial x} - \frac{\partial n_{1y}}{\partial y} - \frac{\partial n_{1z}}{\partial z} + r_1 \tag{A}$$

Cylindrical coordinates

$$\frac{\partial c_1}{\partial t} = -\frac{1}{r}\frac{\partial}{\partial r}(rn_{1r}) - \frac{1}{r}\frac{\partial n_{1\theta}}{\partial \theta} - \frac{\partial n_{1z}}{\partial z} + r_1 \tag{B}$$

Spherical coordinates

$$\frac{\partial c_1}{\partial t} = \frac{1}{r^2}\frac{\partial}{\partial r}(r^2 n_{1r}) - \frac{1}{r\sin\theta}\frac{\partial}{\partial \theta}(n_{1\theta}\sin\theta) - \frac{1}{r\sin\theta}\frac{\partial n_{1\phi}}{\partial \phi} + r_1 \tag{C}$$

[a] The rate r_1 is for the production of species 1 per volume.

This result, called the continuity equation, has no reaction term, because no total mass is generated or destroyed by nonnuclear chemical reactions.

We would like to use the continuity equation to simplify the species mass balance. We cannot do so directly because the continuity equation contains the mass average velocity, and the species mass balance involves the volume average velocity. Although some investigators fuss about this difference, we should recognize that we can solve many problems where these velocities are the same. They are the same at constant density, as shown by Eq. 3.2-3.

If we assume constant density, the overall continuity equation becomes

$$0 = -\nabla \cdot \mathbf{v} = -\nabla \cdot \mathbf{v}^0 \tag{3.4-11}$$

We then multiply this equation by c_1 and subtract the result from Eq. 3.4-7:

$$\frac{\partial c_1}{\partial t} + \mathbf{v}^0 \cdot \nabla c_1 = D\nabla^2 c_1 + r_1 \tag{3.4-12}$$

This result is frequently useful for problems of diffusion and convection.

These generalized equations are shown in different coordinate systems in Tables 3.4-1 and 3.4-2. The overall mass balance is given in Table 3.4-3. These equations include the effects of chemical reaction, convection, and concentration-driven diffusion. However, they are not quite as general as their title suggests. For example, they do not include the effects of electric or magnetic forces. Nonetheless, they often provide a useful route to the differential equations for diffusion, as shown by the following examples.

Example 3.4.1: Fast diffusion through a stagnant film and into a semi-infinite slab

Find differential equations describing these two situations from the general equations in Tables 3.4-1 to 3.4-3. Compare your results with the shell-balance results in the previous section.

Table 3.4-2. *Mass balance for species 1 combined with Fick's law*[a]

Rectangular coordinates

$$\frac{\partial c_1}{\partial t} + v_x^0 \frac{\partial c_1}{\partial x} + v_y^0 \frac{\partial c_1}{\partial y} + v_z^0 \frac{\partial c_1}{\partial z} = D \left(\frac{\partial^2 c_1}{\partial x^2} + \frac{\partial^2 c_1}{\partial y^2} + \frac{\partial^2 c_1}{\partial z^2} \right) + r_1 \tag{A}$$

Cylindrical coordinates

$$\frac{\partial c_1}{\partial t} + v_r^0 \frac{\partial c_1}{\partial r} + \frac{v_\theta^0}{r} \frac{\partial c_1}{\partial \theta} + v_z^0 \frac{\partial c_1}{\partial z}$$

$$= D \left[\frac{1}{r} \frac{\partial}{\partial r} \left(r \frac{\partial c_1}{\partial r} \right) + \frac{1}{r^2} \frac{\partial^2 c_1}{\partial \theta^2} + \frac{\partial^2 c_1}{\partial z^2} \right] + r_1 \tag{B}$$

Spherical coordinates

$$\frac{\partial c_1}{\partial t} + v_r^0 \frac{\partial c_1}{\partial r} + v_\theta^0 \frac{\partial c_1}{\partial \theta} + \frac{v_\theta^0}{r \sin \theta} \frac{\partial c_1}{\partial \phi}$$

$$= D \left[\frac{1}{r^2} \frac{\partial}{\partial r} \left(r^2 \frac{\partial c_1}{\partial r} \right) + \frac{1}{r^2 \sin \theta} \frac{\partial}{\partial \theta} \left(\sin \theta \frac{\partial c_1}{\partial \theta} \right) + \frac{1}{r^2 \sin^2 \theta} \frac{\partial^2 c_1}{\partial \phi^2} \right] + r_1 \tag{C}$$

[a] The diffusion coefficient D and the density ρ are assumed constant. In this case, the mass average and volume average velocities are equal. Again, r_1 is the rate of production of species 1 per volume.

Table 3.4-3. *Total mass balance in several coordinate systems*[a]

Rectangular coordinates

$$\frac{\partial \rho}{\partial t} = -\frac{\partial}{\partial x} (\rho v_x) - \frac{\partial}{\partial y} (\rho v_y) - \frac{\partial}{\partial z} (\rho v_z) \tag{A}$$

Cylindrical coordinates

$$\frac{\partial \rho}{\partial t} = -\frac{1}{r} \frac{\partial}{\partial r} (\rho r v_r) - \frac{1}{r} \frac{\partial}{\partial \theta} (\rho v_\theta) - \frac{\partial}{\partial z} (\rho v_z) \tag{B}$$

Spherical coordinates

$$\frac{\partial \rho}{\partial t} = -\frac{1}{r^2} \frac{\partial}{\partial r} (\rho r^2 v_r) - \frac{1}{r \sin \theta} \frac{\partial}{\partial \theta} (\rho v_\theta \sin \theta) - \frac{1}{r \sin \theta} \frac{\partial}{\partial \phi} (\rho v_\phi) \tag{C}$$

[a] The velocity here is the mass average, and not the volume average commonly used with Fick's law.

Solution. The first of these cases, sketched in Fig. 3.1-1 or Fig. 3.3-1, concerns the fast evaporation of a liquid solute through a stagnant vapor. This evaporation is in steady state, has no chemical reaction, and occurs only in the z direction. Thus, Eq. B in Table 3.4-1 becomes

$$\frac{\partial n_{1z}}{\partial z} = 0$$

Alternatively, for constant density, Eq. B in Table 3.4-2 becomes

$$v_z^0 \frac{\partial}{\partial z} c = D \frac{\partial^2 c_1}{\partial z^2}$$

Either of these equations leads to a solution of the problem like that in Section 3.3.

The second example, shown schematically in Fig. 3.3-2, depends on the unsteady evaporation of a liquid solute into a solvent gas. Again, the process is one-dimensional, without chemical reaction. From Eq. B in Table 3.4-1, we find

$$\frac{\partial c_1}{\partial t} + \frac{\partial n_{1z}}{\partial z} = 0$$

Alternatively, for constant density, Eq. A in Table 3.4-2 becomes

$$\frac{\partial c_1}{\partial t} + v_z^0 \frac{\partial}{\partial z} c_1 = D \frac{\partial^2 c_1}{\partial z^2}$$

The first term on the left-hand side of this result represents accumulation, and the second is convection. The right-hand side represents diffusion. Again, the solution to these equations parallels those in the previous section.

The reader whose primary interest is in diffusion may question why these generalized equations are necessary and why the shell balances used before are not sufficient. I share this skepticism, and I also prefer the physical insight supplied by the shell-balance technique.

At the same time, students often plead to be taught the material in this section, even though later they may question its utility. The students' plea originates not from considerations of mass transfer but from their studies of fluid mechanics. In fluid mechanics, the generalized equations are extremely helpful, especially in cases of curved streamlines. Analogues of curved streamlines do not occur frequently in diffusion. Thus, the mathematics of diffusion are easier than those of fluid mechanics, but the physical chemistry is more difficult.

Example 3.4-2: The flux across a reverse-osmosis membrane

Imagine that sea water is separated from fresh water by a thin membrane made from cellulose acetate. The sea water is placed under a higher pressure than the osmotic pressure, so that flow occurs from the sea water into the fresh water (Scott, 1981). The membrane acts as a sort of molecular filter, stopping the salt from going through. As a result, the salt concentration near the membrane rises, increasing the local osmotic pressure and reducing the flow of pure water. This effect is shown for a rectangular channel in Fig. 3.4-2. Find differential equations for calculating the drop in flux caused by this concentration polarization (i.e., by the salt accumulation near the membrane surface).

Solution. We could find the desired equations by making a shell balance on a differential volume $\Delta x \Delta y$ located at (x, y). Instead, we use Eq.

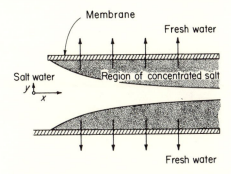

Fig. 3.4-2. Salt concentration in reverse osmosis. Water is forced through the semipermeable membrane shown. The result is a region of higher salt concentration near the membrane; such a region is described as "concentration polarization."

A of Table 3.4-2 to find

$$\frac{\partial}{\partial x}\, c_1 v_x + \frac{\partial}{\partial y}\, c_1 v_y = D \left(\frac{\partial^2 c_1}{\partial x^2} + \frac{\partial^2 c_1}{\partial y^2}\right)$$

The left-hand side represents convection in both the x and y directions; both these flows will be significant. The right-hand side again represents diffusion, but this time in two directions. Diffusion in the y direction is responsible for the concentration polarization that is of central importance in this problem. Diffusion in the x direction is dominated by x convection and is often neglected.

For me, this problem illustrates both the advantages and disadvantages of the generalized approach. The differential equation is found very quickly, and that is good. On the other hand, simplifying this equation requires physical insight, and that is difficult. In this case, if we look at Fig. 3.4-2, we see that the salt concentration c_1 is certainly a function of the distance x; $\partial^2 c_1/\partial x^2$ is certainly not zero. We can neglect this term only after invoking a physical argument: that diffusion in the x direction is negligible compared with the convection. Moreover, attaining the differential equation quickly is not very helpful if we cannot solve it. Indeed, in this case the solution must be numerical and must include the overall mass balance and the momentum equation (Soltanich & Gill, 1981).

Example 3.4-3: The flux near a spinning disc

The final example in this section is the spinning disc shown in Fig. 3.4-3. The disc is made of a sparingly soluble solute that slowly dissolves in the flowing solvent. This dissolution rate is diffusion-controlled. Calculate the rate at which the disc dissolves.

Solution. This problem requires both mathematical skill and physical intuition. The dissolution will reach a steady state only when the disc is

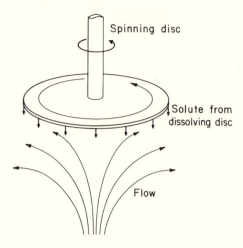

Fig. 3.4-3. Diffusion near a spinning disc. The amount dissolving per unit area is found to be the same everywhere on the disc's surface. Such simplicity makes this disc a powerful experimental tool.

rotating; if the disc is not rotating, the problem will be equivalent to the semi-infinite slab discussed in Section 3.4-2. To solve this problem, we choose cylindrical coordinates centered on the disc. The steady-state mass balance is found from Eq. B in Table 3.4-2:

$$v_r^0 \frac{\partial c_1}{\partial r} + \frac{\overset{0}{v_\theta^0}}{r} \frac{\partial c_1}{\partial \theta} + v_z^0 \frac{\partial c_1}{\partial z} = D \left[\frac{1}{r} \frac{\partial}{\partial r} \left(r \frac{\partial c_1}{\partial r} \right) + \frac{1}{r^2} \frac{\partial^2 c_1}{\partial \theta^2} + \frac{\partial^2 c_1}{\partial z^2} \right]$$

We recognize that the problem is angularly symmetric; so c_1 does not vary with θ. We also assume that the disc is infinitely wide, so that the concentration is a function only of z. I find this assumption mind-boggling, but it is justified by the success of the following calculations.

With these simplifications, the mass balance becomes

$$v_z \frac{dc_1}{dz} = D \frac{d^2 c_1}{dz^2}$$

subject to the conditions

$$z = 0, \quad c_1 = c_1(\text{sat})$$
$$z = \infty, \quad c_1 = 0$$

The first of these conditions implies equilibrium across the solid–fluid interface. Integration of the preceding equation gives

$$c_1 = a \int_0^z e^{-(1/D)[\int_0^r v_z(s)\, ds]}\, dr + b$$

where a and b are integration constants. From the foregoing conditions, b equals $c_1(\text{sat})$, so that

$$\frac{c_1}{c_1(\text{sat})} = 1 - \frac{\displaystyle\int_0^z e^{-(1/D)[\int_0^r v_z(s)\, ds]}\, dr}{\displaystyle\int_0^\infty e^{-(1/D)[\int_0^r v_z(s)\, ds]}\, dr}$$

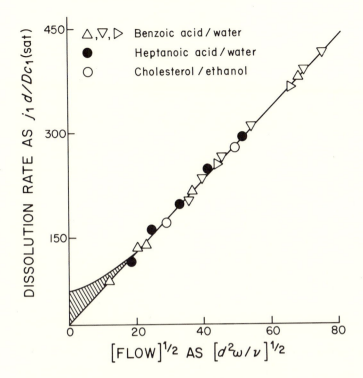

Fig. 3.4-4. Dissolution rate versus flow for a spinning disc. The dissolution rate and flow are described as Sherwood and Reynolds numbers, respectively. The data fit the form predicted by Eq. 3.4-13, which should be valid over the unshaded region. [Data from Tao (1974) and Huang (1980).]

If we know the velocity $v_z(z)$, we can find the concentration profile. We then use Fick's law to find the reaction rate.

The calculation of $v_z(z)$ is a problem in fluid mechanics beyond the scope of this book, but given in detail in the literature (Levich, 1962). When the values found for $v_z(z)$ are inserted into the previous equations, the result is

$$\frac{c_1}{c_1(\text{sat})} = \frac{\int_0^\Omega e^{-u^3}\, du}{\int_0^\infty e^{-u^3}\, du}$$

in which

$$\Omega = z \left(\frac{1.82 D^{1/3} \nu^{1/6}}{\omega^{1/2}} \right)^{-1}$$

and ω is the angular velocity of the disc. The diffusion flux is then

$$j_1|_{z=0} = -D\, \frac{\partial c_1}{\partial z}\bigg|_{z=0} = 0.62 \left(\frac{D^{2/3} \omega^{1/2}}{\nu^{1/6}} \right) c_1(\text{sat})$$

This result is often written in terms of dimensionless groups:

$$-j_1 = \left[0.62\, \frac{D}{d} \left(\frac{d^2 \omega \rho}{\mu} \right)^{1/2} \left(\frac{\mu}{\rho D} \right)^{1/3} \right] c_1(\text{sat})$$

where d is the disc diameter. The first term in parentheses is the Reynolds number, and the second is the Schmidt number.

To my delight, this analysis is verified by experiment. The dissolution varies with the square root of the Reynolds number, as shown in Fig. 3.4-4. As a result, the assumption that the flux is a function only of z is justified. Because the flux is independent of disc diameter, it has the same value near the disc's center and near its edge. Such a constant flux is uncommon, and it makes the interpretation of experimental results unusually straightforward. It is this feature that makes the rotating disc a popular experimental tool.

Section 3.5. A guide to previous work

In many cases, detailed solutions to diffusion problems can be adapted from calculations that have already been published, thus avoiding the mathematical detail presented in the earlier sections in this chapter. These calculations involve the same differential equations, with relatively minor changes of boundary conditions. They include elaborate but straightforward manipulations, like the integration of concentration profiles to find average concentrations.

Unfortunately, the published results are limited because they are often based on mathematical analogies with thermal conduction. Such analogies have merit; indeed, they provided the original stimulus for Fick's law of diffusion. However, in thermal conduction there is no analogy for diffusion-induced convection and rarely an analogy for an effect like chemical reaction. On the other hand, in diffusion there is no effect parallel to thermal radiation. These differences are commonly ignored by teachers because they want the pedagogical benefits of analogy. In fact, convection, chemical reaction, and radiation are frequently central in the problems studied.

Even with these limitations, the published solutions can be used to save considerable effort. Besides individual papers, there are two important books that have collected and compared this literature. The first, Crank's *The Mathematics of Diffusion* (1975), discusses aspects of chemical reactions. The second, Carslaw and Jaeger's *The Conduction of Heat in Solids* (1959), must be used by analogy, but it includes a more complete selection of boundary conditions. The notation used in these books is compared with that used here in Table 3.5-1.

In the remainder of this section we give examples illustrating how this literature can be used effectively.

Example 3.5-1: Diffusion through a polymer film

Imagine that we are studying a polymer film that is permeable to olefins like ethylene, but much less permeable to aliphatic hydrocarbons. Such a film could be used for selectively separating the ethylene produced by dehydrogenation reactions. As part of our study, we use the diaphragm cell shown in Fig. 3.5-1. This diaphragm cell consists of two compartments

Table 3.5-1. *Comparisons of notation between this book and two major references*

Variable	Our notation	Crank's notation	Carslaw and Jaeger analogue
Time	t	t	t
Position	x, y, z, r	x, y, z, r	x, y, z, r
Concentration	c_1	C	Temperature v
Concentration at boundary	$c_{10}, c_{1l}, C_{10}, \dots$	C_1, C_0	Temperature at boundary ϕ
Binary diffusion coefficient	D	D	"Thermometric conductivity" κ
Flux relative to reference velocity	$\mathbf{j}_1$	F	Heat flux f
Flux relative to fixed coordinates	$\mathbf{n}_1$	F	Heat flux f
Flux at boundary	N_1 or $n_1\|_{z=l}$	—	Heat flux at boundary F_0
Total amount diffusing from time 0 to t	M_t	M_t	—

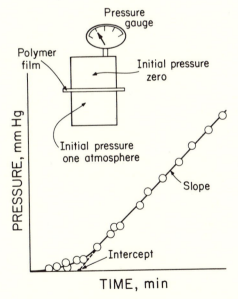

Fig. 3.5-1. Diffusion across a polymer film. When the pressure in the top compartment is varied as a function of time, the slope and intercept are measures of diffusion and solubility of gas in the polymer.

separated by the polymer film of interest. The top compartment is initially evacuated, but the lower one is filled with ethylene. We measure the ethylene concentration in the upper compartment as a function of time.

The data obtained for ethylene transport are exemplified by those shown in the figure. Initially, the pressure in the upper compartment varies in a complex way, but it will eventually approach that in the lower compartment. At the moderate times of most of our experiment, the pressure in the upper compartment is proportional to time, with a known slope and a definite intercept. How are this slope and intercept related to diffusion in the polymer film?

Solution. The basic differential equation for this problem is that for a slab:

$$\frac{\partial c_1}{\partial t} = D \frac{\partial^2 c_1}{\partial z^2}$$

subject to the conditions

$$t = 0, \quad \text{all } z, \quad c_1 = 0$$
$$t > 0, \quad z = 0, \quad c_1 = Hp_0$$
$$z = l, \quad c_1 = Hp_l \doteq 0$$

in which l is the film's thickness and H is a Henry's law coefficient relating ethylene pressure in the gas to ethylene concentration in the film. The solution to this equation and the boundary conditions are given by Crank (1975, p. 50, Eq. 4.22):

$$\frac{c_1}{Hp_0} = 1 - \frac{z}{l} - \frac{2}{\pi} \sum_{n=1}^{\infty} \left(\frac{\sin(n\pi z/l)}{n} \right) e^{-Dn^2\pi^2 t/l^2}$$

Thus, almost before we have started, we have the concentration profile that we need.

We now must cast the problem in terms of the actual experimental variables we are using. First, from a mass balance on the top compartment,

$$\frac{d\mathfrak{N}_1}{dt} = \frac{V}{RT} \frac{dp}{dt} = -AD \frac{\partial c_1}{\partial z}\bigg|_{z=l}$$

in which V and p are the volume and pressure of the upper compartment and A is the film's area. Combining this with the concentration profile, we integrate subject to the condition that the upper compartment's pressure is initially zero:

$$p = \frac{ARTp_0}{Vl} \left[HDt + \frac{2Hl^2}{\pi^2} \sum_{n=1}^{\infty} \frac{\cos n\pi}{n^2} (1 - e^{-Dn^2\pi^2 t/l^2}) \right]$$

At large time, the exponential terms become small, and this result becomes

$$p = \left[\frac{ARTp_0}{Vl} \right] \left(HDt - \frac{l^2 H}{6} \right)$$

The quantity in brackets is known experimentally. Thus, the intercept of the data in Fig. 3.5-1 is related to the Henry's law coefficient H; that is, it is related to the solubility of the ethylene in the membrane. The slope of these data is related to the permeability HD. Alternatively, because H is known from the intercept, the slope gives values of the ethylene diffusion coefficient in the polymer film. I am always delighted that an experiment like this gives both an equilibrium and a transport property.

This example has value well beyond the specific case studied. It shows how the mathematical complexities inherent in the problem can be circumvented by carefully using the literature. This circumventure focuses attention on the real difficulty of the problem, which is connecting the specific physical situation with the more general mathematical abstraction. This is the connection where most of you will have trouble. You can learn how to use the mathematics involved; you must think harder about connecting them with the actual situation.

Example 3.5-2: A dissolving pill

In pharmacology, we frequently wish to know how fast a particular medicine can permeate the body. This rate can be strongly influenced by physical and chemical factors. For example, preparations of some antibiotics include detergents whose function is to increase the amount of the drug dissolved in the gut. In other words, the detergent solubilizes the drug.

Imagine taking such a pill. After a short time, the dissolution of this pill will reach a steady rate, providing a steady flux of drug. We want to estimate the time required to produce this steady drug supply.

Solution. One way to make this estimate is to assume that the drug's dissolution is controlled by diffusion into the stagnant contents of the gut. Such an assumption has two parts: that dissolution is diffusion-controlled and that the surroundings are stagnant. The first part is easy to check. If drug dissolution varies significantly with gentle stirring, then dissolution must be affected by diffusion. The second part is more difficult to check, because it depends on how much flow there is in the gut in vivo. Because any stirring will increase the dissolution rate, our estimate of the time based on stagnant surroundings is a conservative one.

We begin this analysis either with a mass balance on a spherical shell around the pill or by reducing the generalized mass balances given earlier. We find that

$$\frac{\partial c_1}{\partial t} = \frac{D}{r^2} \frac{\partial}{\partial r} \left(r^2 \frac{\partial c_1}{\partial r} \right)$$

This result is subject to the three conditions

$$t = 0, \quad \text{all } r, \quad c_1 = 0$$
$$t > 0, \quad r = R_0, \quad c_1 = c_1(\text{sat})$$
$$r = \infty, \quad c_1 = 0$$

The third condition is the most interesting because it implies that far from the pill the drug is quickly removed by, for example, transport across the intestinal wall.

The solution to this problem, taken either from Crank (1975) or from Carslaw and Jaeger (1959), depends on the variable substitution

$$u = \frac{c_1}{r}$$

Use of this substitution converts the problem into the ubiquitous semi-infinite slab. The integration gives

$$\frac{c_1}{c_1(\text{sat})} = \frac{R_0}{r} \left(1 - \text{erf}\, \frac{r - R_0}{\sqrt{4Dt}} \right)$$

Similarly, because the drug is only sparingly soluble, the solution is dilute, and

$$n_1 \doteq j_1 = -D \left. \frac{\partial c_1}{\partial r} \right|_{r=R_0}$$

$$= \frac{Dc_1(\text{sat})}{R_0} \left(1 + \frac{R_0}{\sqrt{\pi Dt}} \right)$$

The term $R_0/\sqrt{\pi Dt}$ represents the effect of the unsteady state.

We now want to know when we reach the steady state. In other words, we want to know when the term $R_0/\sqrt{\pi Dt}$ is much less than unity, equal to perhaps 0.10. A 3-mm pill and a diffusion coefficient of 10^{-15} cm^2/sec give a time of about eight hours, much longer than the experimental result of about 10 minutes.

To find why our answer is too long, we return to the approximations made in the analysis. We have assumed that the pill is spherical, that the dissolution is diffusion-controlled, and that the surrounding fluid is stagnant. Replacing the approximation of spherical symmetry with a more realistic shape for the pill will change the numerical answer, but not dramatically. Relaxing the assumption of diffusion-controlled dissolution makes sense, because the dissolution of some drugs, including aspirin, is not diffusion-controlled. However, changing this assumption will make the time to reach equilibrium longer, not shorter.

The most likely reason that our answer is too long stems from our assumption of no flow. There is certainly peristaltic flow in the gut that sharply increases mass transfer. Even without this forced flow, there will be free convection driven by the density differences caused by the dissolution itself.

Thus, although the foregoing analysis is mathematically correct, the answer is wrong. I find myself reluctant to abandon such wrong answers when I have done a lot of mathematical work. I see the same reticence in my colleagues. If we all use the literature effectively, we should be emotionally able to discard such nonsense quickly and move on to more successful models.

Example 3.5-3: Effective diffusion coefficients in a porous catalyst pellet

Imagine that we have a porous catalyst pellet containing a dilute gaseous solution. We want to measure the effective diffusion of solute by dropping this pellet into a small, well-stirred bath of a solvent gas and measuring how fast the solute appears in this bath. How can we plot these measurements to find the effective diffusion coefficient?

Solution. Again, we begin with a mass balance, combine this with Fick's law and the appropriate boundary conditions, and then adapt the available mathematical hoopla to find the result. The only feature different from before is that we must do so for both the pellet and the bath.

Within the pellet, a mass balance on a spherical shell or one taken from Table 3.4-2 yields

$$\frac{\partial c_1}{\partial t} = \frac{D_{\text{eff}}}{r^2} \frac{\partial}{\partial r} r^2 \frac{\partial c_1}{\partial r}$$

This implicitly lumps any tortuous multidimensional diffusion into an "effective" one-dimensional diffusion coefficient D_{eff}. This equation is subject to

$$t = 0, \quad \text{all } r, \quad c_1 = c_{10}$$

$$t > 0, \quad r = 0, \quad \frac{\partial c_1}{\partial r} = 0$$

$$r = R_0, \quad c_1 = C_1(t)$$

where R_0 is the pellet radius and $C_1(t)$ is the bath concentration, a function of time. It is this coupling of the sphere and bath concentrations that makes this problem interesting.

We now make a mass balance on the solute in the bath of volume V_B:

$$V_B \frac{dC_1}{dt} = \pi R^2 n_1 |_{r=R_0} = -\pi R^2 D_{\text{eff}} \left(\frac{\partial c_1}{\partial r} \right) \Big|_{r=R_0}$$

subject to

$$t = 0, \quad C_1 = 0$$

This mass balance contains no diffusion term, because the bath is well mixed.

Problems that are mathematically analogous to this one are discussed by Carslaw and Jaeger (1959) and Crank (1975). The most useful result given is that for the concentration in the bath:

$$C_1 = \frac{c_{10}}{1 + B} - 6BV_B \sum_{n=1}^{\infty} \frac{e^{-D_{\text{eff}}\alpha_n^2 t}}{B^2 R_0^2 \alpha_n^2 + 9(B + 1)}$$

in which

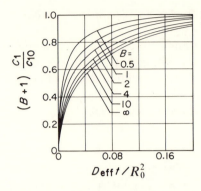

Fig. 3.5-2. Bath concentration versus time. A porous catalyst pellet containing a solute gas is dropped into a stirred bath of solvent gas. The solute concentration in the bath measured versus time provides a value for diffusion in the pellet. A similar graph for heat conduction is given by Carslaw and Jaeger (1959, p. 241).

$$\tan(R_0\alpha_n) = \frac{3R_0\alpha_n}{3 + BR_0^2\alpha_n^2}$$

and

$$B = \frac{V_B}{(4/3)\pi R_0^3 \varepsilon}$$

where ε is the void fraction in the sphere.

The answers are shown in Fig. 3.5-2. To find the diffusion coefficient, we first calculate B and $C_1(1 + B)/c_{10}$. We then read $D_{\text{eff}}t/R_0^2$ from the chart and calculate D_{eff}.

Section 3.6. Conclusions

Diffusion in concentrated solutions is complicated by the convection caused by the diffusion process. This convection must be handled with a more complete form of Fick's law, including a reference velocity. The best reference velocity is the volume average, for it is most frequently zero. The results in this chapter are valid for both concentrated and dilute solutions; so they are more complete than the limits of dilute solutions given in Chapter 2.

Nonetheless, those who study diffusion routinely think and work in terms of the dilute-solution limit. You should also. The dilute limit is easier to understand and easier to use for quick, qualitative calculations. It is the basis for finding how diffusion is related to chemical reaction, dispersion, or mass transfer coefficients. You should be aware of the problems that arise in nondilute cases; you should be able to work through them if necessary, but you need not recall their details. Think dilute.

References

Arnold, J. H. (1944). *Transactions of the American Institute of Chemical Engineers,* **40,** 361.

Bird, R. B., Stewart, W. E., & Lightfoot, E. N. (1960). *Transport Phenomena.* New York: Wiley.

Carslaw, H. S., & Jaeger, J. C. (1959). *The Conduction of Heat in Solids,* 2nd ed. Oxford: Clarendon Press.

Crank, J. (1975). *The Mathematics of Diffusion,* 2nd ed. Oxford: Clarendon Press.

deGroot, S. R., & Mazur, P. (1962). *Non-Equilibrium Thermodynamics.* Amsterdam: North-Holland.

Huang, C. (1980). Ph.D. thesis. Carnegie-Mellon University, Pittsburgh.

Katchalsky, A., & Curran, P. F. (1967). *Non Equilibrium Thermodynamics in Biophysics.* Cambridge: Harvard University Press.

Levich, V. (1962). *Physicochemical Hydrodynamics.* New York: Prentice-Hall.

Lightfoot, E. N. (1974). *Transport Phenomena in Living Systems.* New York: Wiley.

Maxwell, J. C. (1860). *Philosophical Magazine,* **19,** 19; **20,** 21.

Maxwell, J. C. (1952). *Scientific Papers Vol. 2,* ed. W. D. Niven, p. 629. New York: Dover.

Schlichting, H. (1979). *Boundary Layer Theory,* 7th ed. New York: McGraw-Hill.

Scott, J. (1981). *Desalination of Seawater by Reverse Osmosis.* Cleveland: Noyes Data Corporation.

Soltanich, M., & Gill, W. N. (1981). *Chemical Engineering Communications,* **12,** 279.

Tao, J. C.-C. (1974). Ph.D. thesis. Carnegie-Mellon University, Pittsburgh.

Tyrrell, H. J. V. (1961). *Diffusion and Heat Flow in Liquids.* London: Butterworth.

4 DISPERSION

Over the past 20 years the public has been justifiably concerned with the presence of chemicals in the environment. In some cases, chemicals like pesticides and perfumes are deliberately released; in other cases, chemicals like hydrogen sulfide and carbon dioxide are discharged as the result of manufacturing; in still others, chemicals like styrene and dioxin can be accidentally spilled. In all cases, the public worries about the long-term effects of such chemical challenges.

Public concern has led to legislation at federal, state, and local levels. This legislation usually is phrased in terms of regulation of chemical concentrations. These regulations take different forms. The maximum allowable concentration may be averaged over a day or over a year. The acid concentration (as pH) can be held within a particular range, or the number and size of particles going up a stack can be restricted. Those working with chemicals must be able to anticipate whether or not these chemicals can be adequately dispersed. They must consider the problems involved in locating a chemical plant on the shore of a lake or at the mouth of a river.

The theory for dispersion of these chemicals is introduced in this short chapter. As might be expected, dispersion is related to diffusion. The relation exists on two very different levels. First, dispersion is a form of mixing, and so on a microscopic level it involves diffusion of molecules. This microscopic dispersion is not understood in detail, but it takes place so rapidly that it is rarely the most important feature of the process. Second, dispersion and diffusion are described with very similar mathematics. This means that analyses developed for diffusion can often correlate results for dispersion.

In Section 4.1 we give a simple example of dispersion to illustrate the similarities to and differences from diffusion. In Section 4.2 we discuss how diffusion and flow interact to produce dispersion in laminar flow. In Section 4.3 we make similar calculations for turbulent flow. We then discuss dispersion coefficients for environmental and industrial situations in Section 4.4. Overall, the material is presented at an elementary level, partly because it is unevenly understood at any other level and partly because more detail seems outside the scope of this book.

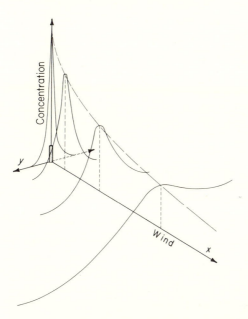

Fig. 4.1-1. Dispersion of smoke. Smoke discharged continuously from a stack has an average concentration that is approximately Gaussian. This shape can be predicted from a diffusion theory. However, the smoke is dispersed much more rapidly than would be expected from diffusion coefficients.

Section 4.1. Dispersion from a stack

Everyone has seen smoke pouring from a smokestack. On a cold, clear day, the plume will climb high into the sky, spreading and fading. In a high wind, the plume will be quickly dispersed, almost as if it never existed.

We want to explain these differences in dispersion so that we can anticipate the effects of wind, weather, and different amounts of smoke. To do so, we need to model the dispersion. Such a model should recognize the characteristics of the smoke as it moves downwind. For example, we might find characteristics like those in Fig. 4.1-1 for a plume in a 15-km/hr wind. The smoke concentration has a roughly Gaussian shape and has a width of about 1 km when it is 10 km downwind.

Before we begin to model this plume, we should consider what we mean by "smoke concentration." Such a concentration is clearly some arbitrary average over all components, be they present as molecules or as small particles. Such a concentration may affect people in different ways. For example, if the smoke has an odor, doubling the smoke concentration will make the odor less than twice as strong. If the smoke contains poisons, doubling its concentration may more than double its toxicity. We should remember to consider the effects of smoke concentration carefully, not just its magnitude.

The obvious model for a plume like that in Fig. 4.1-1 is that developed earlier for the one-dimensional decay from a pulse (Crank, 1975). In this model, we assume that x is the wind direction, z is the vertical height, and y is the horizontal direction normal to both the wind and the ground. On this basis, we can adapt the solution given in Section 2.4 to find

$$\bar{c}_1 = \frac{M/A}{\sqrt{4\pi Dt}} e^{-y^2/4Dt} \tag{4.1-1}$$

where $\bar{c}_1$ is the concentration averaged over time and height z. The quantity M/A was previously the amount of solute per area in the pulse; it now must be closely related to how much comes out of the stack. The distance y must be the amount that the pulse or plume has spread. The time t must be replaced by (x/v^0), the distance the smoke has traveled divided by the wind's velocity. So far, adapting this diffusion model looks promising.

However, this model is a disaster at predicting how much the plume spreads. From observation, we know that it actually has spread about 1 km. From the arguments in Section 2.6, we know that the width of this peak l should be about

$$l = \sqrt{4Dt} \tag{4.1-2}$$

In gases, diffusion coefficients are about 0.1 cm²/sec, and the time is about 10 km/(15 km/hr), or 40 min. On this basis, l should be about 30 cm, 3,000 times less than the observed width of 1 km. A factor of 3,000 is a big error, even for engineers.

The explanation for this major discrepancy is the wind. In previous chapters, mixing occurred by diffusion caused by Brownian motion. Here, mixing occurs as the wind blows the plume over woods, around hills, and across lakes. This mixing is more rapid than diffusion because of the flow.

We now are in something of a quandary. We have a good diffusion model in Eq. 4.1-1 that explains most of the qualitative features of the plume, but this model grossly underpredicts the effects. To resolve this, we assume that the plume can be described by

$$\bar{c}_1 \propto \frac{1}{\sqrt{E_y(x/v^0)}} e^{-y^2/4E_y t} \tag{4.1-3}$$

In this, E_y is a new "dispersion" coefficient, which must be measured experimentally. Like the diffusion coefficient, the dispersion coefficient has dimensions of (L^2/t). Unlike the diffusion coefficient, the dispersion coefficient is largely independent of chemistry. It will not be a strong function of molecular weight or structure, but will have close to the same values for carbon monoxide, styrene, and smoke. Unlike the diffusion coefficient, the dispersion coefficient will be a strong function of position. It will have different values in different directions. Thus, dispersion may look like diffusion, and it may be described by the same kinds of equations, but it is a different effect.

Using this idea of a dispersion coefficient, we can derive a variety of concentration profiles for different boundary conditions. Some of those that are most useful are shown in Table 4.1-1 (Csanady, 1973; Seinfeld, 1975). We shall use these results to solve some simple problems later in this chapter.

Table 4.1-1. *Concentration profiles for free dispersion*

Situation	Basic equation	
Instantaneous point source at $t = 0$; uniform flow v^0 in x direction; source strength $S[=]M$	$c_1 = \dfrac{S}{8(\pi t)^{3/2}(E_x E_y E_z)^{1/2}} \cdot \exp\left(-\dfrac{(x - v^0 t)^2}{4E_x t} - \dfrac{y^2}{4E_y t} - \dfrac{z^2}{4E_z t}\right)$	(A)
Continuous point source at $x = 0$; uniform flow v^0 in x direction; source strength $S[=]M/t$	$c_1 = \dfrac{S}{4\pi v^0 t(E_y E_z)^{1/2}} \cdot \exp\left(-\dfrac{y^2}{4E_y t} - \dfrac{z^2}{4E_z t}\right)$	(B)
Continuous point source at $x = 0$; uniform flow v^0 in x direction; impermeable boundary at $z = z_0$; source strength $S[=]M/t$	$c_1 = \dfrac{S}{4\pi v^0 t(E_y E_z)^{1/2}} \exp\left(-\dfrac{y^2}{4E_y t}\right) \cdot \left[\exp\left(-\dfrac{(z - z_0)^2}{4E_z t}\right)\right.$ $\left. + \exp\left(-\dfrac{(z + z_0)^2}{4E_z t}\right)\right]$	(C)
Continuous point source at $x = 0$; uniform flow v^0 in x direction; absorbing boundary at $z = z_0$; source strength $S[=]M/t$	$c_1 = \dfrac{S}{4\pi v^0 t(E_y E_z)^{1/2}} \exp\left(-\dfrac{y^2}{4E_y t}\right) \cdot \left[\exp\left(-\dfrac{(z - z_0)^2}{4E_z t}\right)\right.$ $\left. - \exp\left(-\dfrac{(z + z_0)^2}{4E_z t}\right)\right]$	(D)

Source: Csanady (1973); Seinfeld (1975).

The foregoing arguments may strike you as silly, a casual invention with a veneer of equations. After all, diffusion is based on a "law." To try to describe dispersion with a diffusion equation seems like cheating.

Nonetheless, this is how dispersion is described. In the rest of this chapter, we explore the details of this discription more carefully. These details often lead to less accurate predictions than those possible for diffusion. However, dispersion can be very important, so that even an approximate solution can have considerable practical value.

Section 4.2. Dispersion in laminar flow: Taylor dispersion

In the first section of this chapter, we argued that flow and diffusion can couple to produce dispersion. The result was an effect that seemed much like diffusion mathematically, but that had a very different physical origin.

In this section, we want to explore one example of the coupling between flow and diffusion. The example is very specialized, for the flow is laminar, characterized by smooth streamlines. It leads to an accurate prediction of the dispersion coefficient. In most natural situations, the flow is not laminar but turbulent, and predictions of dispersion coefficients are not possible. Nonetheless, this particular example is so instructive that it is worth including.

The specific example concerns the fate of a sharp pulse of solute injected into a long, thin tube filled with solvent flowing in laminar flow (Fig. 4.2-1)

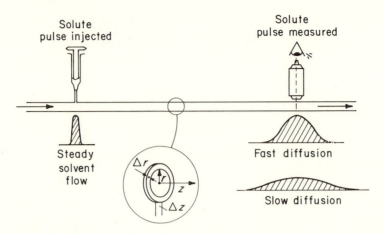

Fig. 4.2-1. Taylor dispersion. In this case, solvent is passing in steady laminar flow through a long, thin tube. A pulse of solute is injected near the tube's entrance. This pulse is dispersed by the solvent flow, as shown.

(Taylor, 1953, 1954). As the solute pulse moves through the tube, it is dispersed. We want to calculate the concentration profile resulting from this dispersion.

Because the complete analysis of this problem is complicated, we first give the results and then the derivation. The concentration of the pulse averaged across the tube's cross section will be shown to be (Taylor, 1953; Aris, 1956)

$$\bar{c}_1 = \frac{M/\pi R_0^2}{\sqrt{4\pi E_z t}} \, e^{-(z-v^0 t)^2/4E_z t} \tag{4.2-1}$$

in which M is the total solute in the pulse, R_0 is the tube's radius, z is the distance along the tube, v^0 is the fluid's velocity, and t is the time. This equation is a close parallel to Eq. 2.4-14, except that the diffusion coefficient D is replaced by the dispersion coefficient E_z, which can be shown explicitly to be

$$E_z = \frac{(R_0 v^0)^2}{48D} \tag{4.2-2}$$

Note that E_z depends *inversely* on the diffusion coefficient.

This fascinating result indicates that rapid diffusion leads to small dispersion and that slow diffusion produces large dispersion (Fig. 4.2-1). The reasons why this occurs are sketched in Fig. 4.2-2. The initial pulse is sharp, like that shown in (a), but the laminar flow quickly distorts the pulse, as in (b). If there is no diffusion, the distortion continues unabated, and the pulse is widely dispersed. If, instead, there is rapid diffusion, material in the center of the tube tends to diffuse outward, into a region of solvent that is moving more slowly. Simultaneously, material that is left behind near the tube walls tends to diffuse toward the center, into a region of faster flow. This radial diffusion thus inhibits the dispersion induced by axial convection.

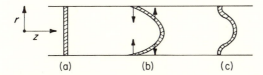

Fig. 4.2-2. Causes of Taylor dispersion. In Taylor dispersion, fast diffusion unexpectedly produces little dispersion, and vice versa. The reasons for this are shown here. The initial solute pulse (a) is deformed by flow (b). In the fast-flowing regions, diffusion occurs outward, and in the slow flow near the wall, diffusion occurs inward. Thus, diffusion in the radial direction inhibits dispersion caused by axial flow (c).

To apply these ideas quantitatively, we again write a balance, add Fick's law, and manipulate mathematically. In this instance, I am reminded of a cartoon by Thomas Nast, showing a virtuous soul laden with debt and responsibility, staggering along a tortuous path (Nast, 1974). To the left of the path, the ground drops away into ignorance; to the right, the ground disappears into chaos. In going through this next analysis, you may feel like that poor soul, treading a very narrow path.

We begin this analysis with three assumptions:

1   The solutions are dilute. This is assumed true even for the initial pulse.
2   The laminar flow is unchanged by the pulse. This means that the velocity varies only with radius.
3   Mass transport is by radial diffusion and axial convection. Other transport mechanisms are negligible.

The most important assumption is the last one, for it separates diffusion and convection. It is accurate if

$$7.2 \left(\frac{LD}{R_0^2 v^0} \right) \gg 1 \tag{4.2-3}$$

where L is the tube length (Aris, 1956). This condition is valid for long, thin tubes.

We now make a mass balance on the washer-shaped element shown in the inset in Fig. 4.2-1 to find

$$\frac{\partial c_1}{\partial t} = - \frac{1}{r} \frac{\partial}{\partial r} (r j_1) - \frac{\partial}{\partial z} (c_1 v_z) \tag{4.2-4}$$

The velocity v_z is the laminar result, and so is independent of z:

$$v_z = 2v^0 \left[1 - \left(\frac{r}{R_0} \right)^2 \right] \tag{4.2-5}$$

When Eqs. 4.2-4 and 4.2-5 are combined with Fick's law,

$$\frac{\partial c_1}{\partial t} = \frac{D}{r} \frac{\partial}{\partial r} r \frac{\partial c_1}{\partial r} - 2v^0 \left[1 - \left(\frac{r}{R_0} \right)^2 \right] \frac{\partial c_1}{\partial z} \tag{4.2-6}$$

This is subject to the conditions

$$t = 0, \quad z = 0, \quad c_1 = \left(\frac{M}{\pi R_0^2}\right) \delta(z) \tag{4.2-7}$$

$$t > 0, \quad r = R_0, \quad \partial c_1/\partial r = 0 \tag{4.2-8}$$

$$r = 0, \quad \partial c_1/\partial r = 0 \tag{4.2-9}$$

This initial condition is like that for the decay of a pulse.

We next define the new coordinates

$$\eta = \frac{r}{R_0} \tag{4.2-10}$$

$$\zeta = (z - v^0 t)/R_0 \tag{4.2-11}$$

In terms of these quantities, Eq. 4.2-6 becomes

$$\frac{D}{\eta} \frac{\partial}{\partial \eta} \left(\eta \frac{\partial c_1}{\partial \eta}\right) = 2 v^0 R_0 \left(\frac{1}{2} - \eta^2\right) \frac{\partial c_1}{\partial \zeta} \tag{4.2-12}$$

One solution to Eq. 4.2-12 that satisfies Eq. 4.2-8 is

$$c_1 = c_1 \Big|_{\eta=0} + \left[\frac{v^0 R_0}{D} \left(\frac{\partial c_1}{\partial \zeta}\right)\Big|_{\eta=0}\right]\left(\eta^2 - \frac{1}{2}\eta^4\right) \tag{4.2-13}$$

However, we want not the local concentration but the average across the tube:

$$\bar{c}_1(z) = \frac{1}{\pi R_0^2} \int_0^{R_0} 2\pi r c_1(r, z) \, dr$$

$$= 2 \int_0^1 \eta c_1 \, d\eta \tag{4.2-14}$$

Because of the pulse, the radial variations of concentration are small relative to the axial ones; so

$$\frac{\partial c_1}{\partial \zeta} \doteq \frac{\partial \bar{c}_1}{\partial \zeta} \tag{4.2-15}$$

We now can write a new overall mass balance in terms of this average concentration:

$$\frac{\partial \bar{c}_1}{\partial t} = -\frac{\partial J_1}{\partial(\zeta R_0)} \tag{4.2-16}$$

in which J_1 is the averaged flux

$$J_1 = \frac{1}{\pi R_0^2} \int_0^{R_0} 2\pi r(v_z - v^0)(c_1 - c_1^0) \, dr \tag{4.2-17}$$

This can be rewritten as

$$\frac{\partial \bar{c}_1}{\partial(t v^0/R_0)} = \frac{\partial \bar{c}_1}{\partial \tau} = -\frac{\partial(J_1/v^0)}{\partial \zeta}$$

$$= -\frac{\partial}{\partial \zeta}\left[4\int_0^1 \eta\left(\frac{1}{2} - \eta^2\right) c_1 \, d\eta\right] \tag{4.2-18}$$

Combining this result with Eqs. 4.2-13 and 4.2-14, we find, after some work, that

$$\frac{\partial \bar{c}_1}{\partial \tau} = \left(\frac{v^0 R_0}{48D}\right) \frac{\partial^2 \bar{c}_1}{\partial \zeta^2} \tag{4.2-19}$$

The quantity in parentheses is a Péclet number, giving the relative importance of axial convection and radial diffusion. The conditions are now

$$\tau = 0, \quad \text{all } \zeta, \qquad \bar{c}_1 = \frac{M}{\pi R_0^2} \delta(\zeta) \tag{4.2-20}$$

$$\tau > 0, \quad \zeta = \pm\infty, \quad \bar{c}_1 = 0 \tag{4.2-21}$$

$$\zeta = 0, \qquad \frac{\partial \bar{c}_1}{\partial \tau} = 0 \tag{4.2-22}$$

Equations 4.2-19 to 4.2-22 for Taylor dispersion have exactly the same mathematical form as those for the decay of a pulse in Section 2.4. As a result, they must have the same solution. This solution is that given in Eq. 4.2-1.

This example is truly an exception, because we can calculate the dispersion coefficient exactly. In some ways, it is like the friction factor for laminar flow in a pipe, which also can be calculated explicitly. In general, we should not expect such an exact result, just as we do not expect to calculate a priori the friction factors for laminar flow in packed beds or for turbulent flow in a pipe. In the next section we abandon efforts at exact calculation and return to more approximate models of dispersion.

Section 4.3. Dispersion in turbulent flow

In this section we want to extend the mathematical formalism used in earlier chapters to the description of dispersion in turbulent flow. This extension is a rationalization, invented to explain the more rapid mixing in cases like the smoke plume. In other words, we want to justify, in terms of our fundamental equations, why the smoke plume spreads much faster than expected.

Dispersion depends on flow. In most cases, this flow is turbulent, changing quickly by gusts and eddies. The mass balance in such a flowing system is given in Table 3.4-2:

$$\frac{\partial c_1}{\partial t} = D \left(\frac{\partial^2 c_1}{\partial x^2} + \frac{\partial^2 c_1}{\partial y^2} + \frac{\partial^2 c_1}{\partial z^2}\right) - \frac{\partial}{\partial x} c_1 v_x - \frac{\partial}{\partial y} c_1 v_y - \frac{\partial}{\partial z} c_1 v_z - \kappa c_1 c_2$$

$$\tag{4.3-1}$$

The left-hand side of this equation is the accumulation within a differential volume. The first three terms on the right-hand side describe the amount that enters by diffusion minus the amount that leaves by diffusion. The next three describe the same thing for convection. The last term on the right-hand side is the amount of solute consumed by a second-order chemical reaction, included for reasons that will become evident later. The quantity κ is the rate constant of this reaction.

In turbulent flow, we expect both velocity and concentration to fluctuate (Seinfeld, 1975; Schlichting, 1979). For the smoke plume, the velocity fluctuations are the wind gusts, and the concentration fluctuations can be reflected as sudden changes in odor. To rewrite this equation to include these fluctuations, we define

$$c_1 = \bar{c}_1 + c_1' \tag{4.3-2}$$

where c_1' is the fluctuation and $\bar{c}_1$ is the average value:

$$\bar{c}_1 = \frac{1}{\tau} \int_0^\tau c_1 \, dt \tag{4.3-3}$$

Note that the time average of c_1' is zero. By similar definitions,

$$v_x = \bar{v}_x + v_x' \tag{4.3-4}$$

where v_x' is the fluctuation, and

$$\bar{v}_x = \frac{1}{\tau} \int_0^\tau v_x \, dt \tag{4.3-5}$$

Again, the average of the fluctuations is zero. Definitions for v_y' and v_z' are similar.

We now insert these definitions into Eq. 4.3-1 and average this equation over the short time interval τ. In some cases, such a substitution is dull:

$$\frac{1}{\tau} \int_0^\tau \left(D \frac{\partial^2 \bar{c}_1}{\partial x^2} \right) dt = \frac{D}{\tau} \frac{\partial^2}{\partial x^2} \int_0^\tau c_1 \, dt = D \frac{\partial^2 \bar{c}_1}{\partial x^2}$$

In other cases, it is intriguing (Brodkey, 1975):

$$\frac{1}{\tau} \int_0^\tau k_R c_1 c_2 \, dt = \frac{\kappa}{\tau} \int_0^\tau (\bar{c}_1 + c_1')(\bar{c}_2 + c_2') \, dt$$

$$= \frac{\kappa}{\tau} \int_0^\tau (\bar{c}_1 \bar{c}_2 + \bar{c}_1 c_2' + c_1' \bar{c}_2 + c_1' c_2') \, dt$$

$$= \frac{\kappa}{\tau} [\bar{c}_1 \bar{c}_2 \tau + 0 + 0 + \int_0^\tau (c_1' c_2') \, dt] = \kappa(\bar{c}_1 \bar{c}_2 + \overline{c_1' c_2'}) \tag{4.3-6}$$

where the new term $\overline{c_1' c_2'}$ represents the time average of the product of the fluctuations. In practice, this new term may be as large as the term $\bar{c}_1 c_2$ but of opposite sign. These terms supply the theoretical underpinning for the stratified charge used to reduce pollution in automobile engines. In a similar fashion,

$$\frac{1}{\tau} \int_0^\tau \frac{\partial}{\partial x} v_x c_1 \, dt = \frac{\partial}{\partial x} \bar{v}_x \bar{c}_1 + \frac{1}{\tau} \frac{\partial}{\partial x} \int_0^\tau v_x' c_1' \, dt$$

$$= \frac{\partial}{\partial x} \bar{v}_x \bar{c}_1 + \frac{\partial}{\partial x} \overline{v_x' c_1'} \tag{4.3-7}$$

Again, we have the prospect of coupled fluctuations, analogous to the Reynolds stresses that are basic to theories of turbulent flow (Batchelor, 1967).

When we combine these averaged terms, we get the following mass balance:

$$\frac{\partial \bar{c}_1}{\partial t} = D \left(\frac{\partial^2 \bar{c}_1}{\partial x^2} + \frac{\partial^2 \bar{c}_1}{\partial y^2} + \frac{\partial^2 \bar{c}_1}{\partial z^2} \right) - \left(\frac{\partial}{\partial x} \bar{v}_x \bar{c}_1 + \frac{\partial}{\partial y} \bar{v}_y \bar{c}_1 + \frac{\partial}{\partial z} \bar{v}_z \bar{c}_1 \right)$$

$$- \left(\frac{\partial}{\partial x} \overline{v_x' c_1'} + \frac{\partial}{\partial y} \overline{v_y' c_1'} + \frac{\partial}{\partial z} \overline{v_z' c_1'} \right) - \kappa \bar{c}_1 \bar{c}_2 - \kappa \overline{c_1' c_2'} \tag{4.3-8}$$

Most of the terms are like those in Eq. 4.3-1, and they have the same physical significance. The underlined terms are new. The last one deals with changes in reaction rate effected by the fluctuations; it is introduced here to advertise its existence and will be discussed in Section 14.5. The other three describe the mixing caused by the turbulent flow, that is, by the dispersion. They are the focus of this chapter.

We next remember the origin of the diffusion terms, that

$$D \frac{\partial^2 \bar{c}_1}{\partial x^2} = - \frac{\partial}{\partial x} \bar{j}_1 \qquad (4.3\text{-}9)$$

or, more basically,

$$\bar{j}_1 = -D \frac{\partial \bar{c}_1}{\partial x} \qquad (4.3\text{-}10)$$

By analogy, because the flux $\overline{v'_x c'_1}$ has a physical meaning similar to the flux $\bar{j}_1$, we define

$$\overline{v'_x c'_1} = -E_x \frac{\partial \bar{c}_1}{\partial x} \qquad (4.3\text{-}11)$$

Definitions for other directions are made in similar ways. This always seems intellectually arrogant to me, because I know that we define E_x, E_y, and E_z so that we will get results that are mathematically parallel to diffusion. It seems a rationalization, jerry-built on top of that diffusion theory. It is these things. It also is the best first approximation of dispersion, the basis from which other theories proceed.

In the following section we shall discuss values of the dispersion coefficients and solve some simple problems. At this time it is worth anticipating two points. First, in turbulent flow, these coefficients are largely physical in origin; they are only weak functions of the different chemical properties of various solutes. Second, estimates of the dispersion coefficients are often inaccurate, and reliable values must be found from experiments.

Section 4.4. Values of dispersion coefficients

In this section we want to summarize experimental values of dispersion coefficients and to show how these values can be used to make estimates of dispersion in real situations. On the basis of our own experience, we might expect the dispersion coefficient to increase as the flow gets faster. On dimensional grounds, we could expect that

$$\frac{L v^0}{E} = \text{constant} \qquad (4.4\text{-}1)$$

where L is a characteristic distance. This equation says that the Péclet number for turbulent dispersion is a constant. This is shown by experiment to be roughly true. Going beyond this limit leads to the jungle of approximation and ambiguity described next.

Table 4.4-1. *Definition of six stability classes for use with the Pasquill–Gifford curves in Figs. 4.4-1 and 4.4-2*

Surface wind speed (m/sec)	Daytime insolation			Nighttime conditions	
	Strong	Moderate	Slight	Thin overcast or $\geq$ 4/8 cloudiness[a]	$\leq$ 3/8 cloudiness
<2	A	A–B	B		
2	A–B	B	C	E	F
4	B	B–C	C	D	E
6	C	C–D	D	D	D
>6	C	D	D	D	D

[a] The degree of cloudiness is defined as that fraction of the sky above the local apparent horizon that is covered by clouds.
[b] A, extremely unstable conditions; B, moderately unstable conditions; C, slightly unstable conditions; D, neutral conditions (applicable to heavy overcast, day or night); E, slightly stable conditions; F, moderately stable conditions.
Source: Data from Seinfeld (1975).

Dispersion of homogeneous fluids

The dispersion of airborne pollutants usually follows a Gaussian profile like those given in Table 4.1-1. In this table, the wind is taken as blowing in the x direction, the stack is erected in the vertical z direction, and horizontal position normal to these two is the y direction. The dispersion coefficients in the y and z directions are most commonly given as standard deviations σ of the Gaussian profile:

$$\sigma = \sqrt{2Ex/v^0} \qquad (4.4\text{-}2)$$

Values of σ_y and σ_z are given in Figs. 4.4-1 and 4.4-2 (Gifford, 1968; Pasquill & Smith, 1971). The different plume stabilities shown can be anticipated from the criteria in Table 4.4-1. Notice that, in many cases, the dispersion varies linearly with x. For example, for a slightly unstable plume, Fig. 4.4-1 predicts that the dispersion parallel to the ground is

$$\frac{xv^0}{E_y} = 200 \qquad (4.4\text{-}3)$$

The vertical dispersion is slightly slower.

Dispersion in lakes and rivers is sometimes modeled with equations like those in Table 4.1-1 to yield correlations for dispersion in water like those for air in Figs. 4.4-1 and 4.4-2 (Csanady, 1973). Such correlations are not always reliable, probably because of additional physical factors like the shape of the river bed and the effect of winds across the lake. Instead, dispersion in these cases is often modeled as a series of stirred tanks. Such models are accurate in the specific case, but difficult to generalize. They lead to mathematics different from those that describe diffusion; so they will not be discussed further here.

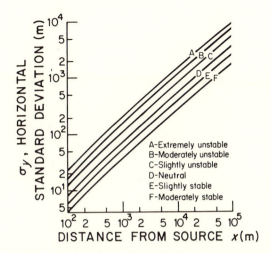

Fig. 4.4-1. Horizontal dispersion normal to the wind. This graph shows dispersion for various types of plumes defined in Table 4.4-1. The dispersion coefficient can be found from values of σ_y using Eq. 4.4-2. [From Seinfeld (1975), with permission.]

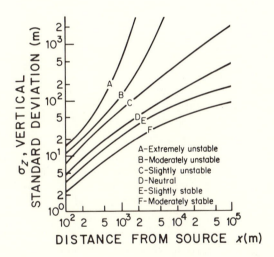

Fig. 4.4-2. Vertical dispersion normal to the wind. These results are comparable to those in Fig. 4.4-1, but for dispersion in the vertical direction. Again, the standard deviation σ_z is easily related to E_z by use of Eq. 4.4-2. [From Seinfeld (1975), with permission.]

Dispersion coefficients in pipelines containing air, water, and other fluids have been frequently measured. Some of the data obtained are summarized in Fig. 4.4-3 (Levenspiel, 1958). Obviously, whereas a particular investigation may give apparently precise results, different investigations routinely

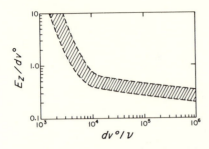

Fig. 4.4-3. Axial dispersion in pipes. At high Reynolds numbers, the Péclet number is about constant; so the dispersion coefficient is proportional to the flow. At lower Reynolds numbers, the behavior is complex. [Data from Sherwood et al. (1975).]

differ by a factor of 3 or more. At Reynolds numbers above 10,000, the axial dispersion coefficient E_z is approximately

$$\frac{dv^0}{E_z} = 2 \qquad (4.4\text{-}4)$$

On the other hand, the radial dispersion coefficient E_r is about

$$\frac{dv^0}{E_r} = 600 \qquad (4.4\text{-}5)$$

Thus, radial dispersion is slower.

Dispersion in porous media

The second major area where dispersion is important is in flow through porous materials. Such materials include filter cakes, chromatographic columns, oil-bearing sandstones, and reactors filled with solid catalysts. Dispersion in these systems reduces the separation in chromatography and alters reactor selectivity.

Radial and axial dispersion coefficients in packed beds are summarized in Fig. 4.4-4 (Sherwood et al., 1975). Again, the large number of experimental results available fall over a considerable range, so that the curves shown represent major approximations. At low flow, the dispersion becomes a function of the Schmidt number, because both flow and molecular diffusion are important in this region. At particle Reynolds numbers above about 10, the dispersion coefficients again vary linearly with flow. For example, for axial dispersion,

$$\frac{dv^0}{E_z} = 3 \qquad (4.4\text{-}6)$$

where d is now the diameter of the particle. Because under these conditions the particle diameter is about equal to the diameter of the flow channel, this result is very much like Eq. 4.4-4.

As a final point, we should mention that we have not discussed dispersion

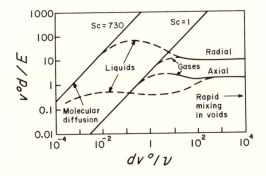

Fig. 4.4-4. Axial dispersion in packed beds. Again, at high flow, the Péclet number is about constant. At lower flow, dispersion becomes strongly affected by the Schmidt number (i.e., by diffusion). [Data from Sherwood et al. (1975).]

in terms of results like Prandtl's mixing-length model (Prandtl, 1925; Schlichting, 1979) or Deissler's correlation (Deissler, 1955). These results, developed for fluid flow and heat transfer, deal primarily with dispersion near solid walls. This is rarely an important case for dispersion of mass, and so it is not discussed here.

Example 4.4-1: Stability of a stack

In Section 4.1 we discussed dispersion from a stack. The resulting plume had spread about 1 km when it was 10 km from the source. The wind was 15 km/hr. How stable is a plume like this?

Solution. To solve this problem, we begin with Eq. C in Table 4.1-1. As a first approximation, we assume that the stack is small relative to the distance traveled, so that z_0 is zero and the concentration profile is given by

$$c_1 = \frac{S}{2xE_y}\, e^{-y^2/4E_y t}$$

This profile is Gaussian. Its approximate width occurs when

$$\frac{y^2}{4E_y t} = 1$$

$$= \frac{(0.5 \text{ km})^2}{4E_y \dfrac{10 \text{ km}}{15 \text{ km/hr}}}$$

Thus, E_y is about 0.1 km^2/hr. From Eq. 4.4-2, we see that

$$\sigma_y = \sqrt{2(0.1 \text{ km}^2/\text{hr})(10 \text{ km}/15 \text{ km/hr})}\ 1000 \text{ m/km}$$

$$\doteq 400 \text{ m}$$

From Fig. 4.4-1 we see that this value of σ_y and the value of x of 10 km means that the concentration is consistent with the slightly stable plume represented by curve E.

Example 4.4-2: Improving separations for liquid chromatography

We are trying to separate two optical isomers to be used for pharmaceuticals. The isomers are currently separated by chromatography with water through a 400-cm packed bed of 0.1-cm gel beads at a superficial velocity of 0.01 cm/sec. Although the isomer adsorption on the beads is selective, the separation is poor because of dispersion. How much will the use of beads 100 times smaller improve the separation?

Solution. At present, the Reynolds number in the bed is

$$\frac{dv^0}{\nu} = \frac{0.1 \text{ cm}(0.01 \text{ cm/sec})}{0.01 \text{ cm}^2/\text{sec}} = 0.1$$

Thus, from Fig. 4.4-4, for axial dispersion,

$$\frac{dv^0}{E_z} = 0.5$$

$$E_z = \frac{(0.1 \text{ cm})(0.01 \text{ cm/sec})}{0.5} = 0.002 \text{ cm}^2/\text{sec}$$

The approximate width of the peak is thus

$$z = \sqrt{4E_z t}$$
$$= \sqrt{4(0.002 \text{ cm}^2/\text{sec})[(400 \text{ cm})/(0.01 \text{ cm/sec})]}$$
$$= 18 \text{ cm}$$

If the particle z drops 100 times, the Reynolds number will as well. However, from Fig. 4.4-4, the Péclet number is unchanged. Thus, for the smaller particles,

$$\frac{dv^0}{E_z} = 0.5$$

$$E_z = \frac{(0.001 \text{ cm})(0.01 \text{ cm/sec})}{0.5}$$

$$= 0.00002 \text{ cm/sec}$$

The width of the peak is now about 1.8 cm. This is an improvement, but it may not be enough to justify the tremendous increase in pressure drop required to sustain this flow.

Example 4.4-3: Dispersion in a pipeline

We have a 10-cm pipeline 3 km long for moving reagent gases at 500 cm/sec from our wharf to our plant. We want to use this pipeline for different gases, one after the other. How much will the gases mix?

Solution. Imagine that we initially have the pipe filled with one gas, and then we suddenly start to pump in a second gas. Because the pipe has a much greater length than diameter, we can expect its contents to be well mixed radially. However, we do expect that there will be significant concentration changes in the axial direction. To describe these, we choose a coordinate system originally located at the initial interface between the gases, but moving with the average gas velocity. We then write a mass balance around this moving point:

$$\frac{\partial \bar{c}_1}{\partial t} = E_z \frac{\partial^2 \bar{c}_1}{\partial z^2}$$

This mass balance is subject to the conditions

$$t = 0, \quad z > 0, \quad \bar{c}_1 = \bar{c}_{1\infty}$$
$$t > 0, \quad z = 0, \quad \bar{c}_1 = \bar{c}_{10}$$
$$z = \infty, \quad \bar{c}_1 = \bar{c}_{1\infty}$$

in which $\bar{c}_{10}$ is the average concentration between the gases. The derivation of these relations is a complete parallel to that in Section 2.3. Indeed, the entire problem is mathematically identical with this earlier one, although the diffusion coefficient D used before is now replaced with the dispersion coefficient E_z. The results are, by analogy,

$$\frac{\bar{c}_1 - \bar{c}_{10}}{\bar{c}_{1\infty} - \bar{c}_{10}} = \text{erf} \frac{z}{\sqrt{4E_z t}}$$

The appropriate value for E_z is found from Fig. 4.4-3. In this case, assuming that the gases have the properties of air,

$$\frac{dv}{\nu} = \frac{(10 \text{ cm})(500 \text{ cm/sec})}{0.15 \text{ cm}^2/\text{sec}} = 30,000$$

The dispersion coefficient can then be found from Fig. 4.4-3:

$$\frac{E_z}{dv^0} = 0.4$$

$$E_z = 0.4(10 \text{ cm})(500 \text{ cm/sec}) = 2,000 \text{ cm}^2/\text{sec}$$

The concentration change is significant when

$$z = \sqrt{4E_z t}$$
$$= \sqrt{4(2,000 \text{ cm}^2/\text{sec})[(3 \text{ km})/(500 \text{ cm/sec})](1,000 \text{ m/km})(1 \text{ m}/100 \text{ cm})}$$
$$= 20 \text{ m}$$

Thus, about 1% of the pipeline will contain mixed gases.

Section 4.5. Conclusions

This chapter discusses dispersion, an important effect caused by the coupling of concentration differences and fluid flow. Dispersion frequently can be described by the same mathematics used so effectively for diffusion; in this sense, this chapter represents special cases of diffusion theory.

If you use the materials in this chapter, you should always remember that diffusion and dispersion have very different physical origins and proceed at very different speeds. Remembering this difference is especially important, because some refer to both processes as "diffusion." Physicians speak of "diffusion" of drugs in the bloodstream, and environmental engineers discuss "diffusion" of pollutants. Some of these processes may include the narrower definition of molecular diffusion used in this book, but the process dynamics cannot be predicted from diffusion theory alone. Be careful.

References

Aris, R. (1956). *Proceedings of the Royal Society of London, Series A,* **235,** 67.

Batchelor, G. K. (1967). *An Introduction to Fluid Dynamics.* Cambridge University Press.

Brodkey, R. S. (1975). *Turbulence in Mixing Operations.* New York: Academic.

Crank, J. (1975). *The Mathematics of Diffusion.* Oxford: Clarendon Press.

Csanady, G. T. (1973). *Turbulent Diffusion in the Environment.* Boston: Kluwer.

Deissler, R. G. (1955). NACA report 1210.

Gifford, F. A. (1968). In: *Meteorology and Atomic Energy,* ed. D. H. Slade. Oak Ridge: U. S. Atomic Energy Commission.

Levenspiel, O. (1958). *Industrial and Engineering Chemistry,* **50,** 343.

Nast, T. (1974). *Cartoons and Illustrations,* plate 110. New York: Dover.

Pasquill, F., & Smith, F. B. (1971). In: *Proceedings of the Second International Clean Air Congress.* New York: Academic.

Prandtl, L. (1925). *Zeitschrift für Angewandte Mathematik und Mechanik,* **5,** 136.

Schlichting, H. (1979). *Boundary Layer Theory,* 7th ed. New York: McGraw-Hill.

Seinfeld, J. H. (1975). *Air Pollution: Physical and Chemical Fundamentals.* New York: McGraw-Hill.

Sherwood, T. K., Pigford, R. L., & Wilke, C. R. (1975). *Mass Transfer.* New York: McGraw-Hill.

Taylor, G. I. (1953). *Proceedings of the Royal Society of London, Series A,* **219,** 186; (1954), **223,** 446.

PART II

Diffusion coefficients

5 VALUES OF DIFFUSION COEFFICIENTS

Until now, we have treated the diffusion coefficient as a proportionality constant, the unknown parameter appearing in Fick's law. We have found mass fluxes and concentration profiles in a broad spectrum of situations using this law. Our answers have always contained the diffusion coefficient as an adjustable parameter.

Now we want to calculate particular values of the flux and the concentration profile. For this, we need to know the diffusion coefficients in these particular situations. We must depend largely on experimental measurements of these coefficients, because no universal theory permits their accurate a priori calculation. Unfortunately, the experimental measurements are unusually difficult to make, and the quality of the results is variable. Accordingly, we must be able to evaluate how good our measurements are.

Before we begin, we should list the guidelines that tend to stick in everyone's mind. Diffusion coefficients in gases, which can be estimated theoretically, are about 0.1 cm^2/sec. Diffusion coefficients in liquids, which cannot be as reliably estimated, cluster around 10^{-5} cm^2/sec. Diffusion coefficients in solids are slower still, 10^{-10} cm^2/sec, and they vary strongly with temperature. Diffusion coefficients in polymers and glasses lie between liquid and solid values, say about 10^{-8} cm^2/sec, and these values can be strong functions of solute concentration.

The accuracy and origins of these guidelines are explored in this chapter. Gases, liquids, solids, and polymers are discussed in Sections 5.1 through 5.4, respectively. In these sections we give a selection of typical values as well as the most common method of estimating these values. After we sketch the sources of these estimations, we explore other common concerns, like the pressure dependence of diffusion in gases and the concentration variations of diffusion in liquids. Section 5.5 discusses the common experimental methods of measuring diffusion coefficients.

Section 5.1. Diffusion coefficients in gases

Diffusion coefficients in gases are illustrated by the values in Table 5.1-1 (Reid et al., 1977). At one atmosphere and near room temperature, these values lie between 0.1 and 1 cm^2/sec. To a first approximation, the coefficients are inversely proportional to pressure; so doubling the pressure cuts

105

Table 5.1-1. *Experimental values of diffusion coefficients in gases at one atmosphere*

Gas pair	Temperature (°K)	Diffusion coefficient (cm^2 sec^{-1})
Air–CH_4	273.0	0.196
Air–C_2H_5OH	273.0	0.102
Air–CO_2	276.2	0.142
	317.2	0.177
Air–H_2	273.0	0.611
Air–D_2	296.8	0.565
Air–H_2O	289.1	0.282
	298.2	0.260
	312.6	0.277
	333.2	0.3050
Air–He	276.2	0.6242
Air–O_2	273.0	0.1775
Air–*n*-hexane	294	0.080
Air–*n*-heptane	294	0.071
Air–benzene	298.2	0.096
Air–toluene	299.1	0.0860
Air–chlorobenzene	299.1	0.074
Air–aniline	299.1	0.074
Air–nitrobenzene	298.2	0.0855
Air–2-propanol	299.1	0.099
Air–butanol	299.1	0.087
Air–2-butanol	299.1	0.089
Air–2-pentanol	299.1	0.071
Air–ethylacetate	299.1	0.087
CH_4–Ar	298	0.202
CH_4–He	298	0.675
CH_4–H_2	298.0	0.726
CH_4–H_2O	307.7	0.292
CO–N_2	295.8	0.212
^{12}CO–^{14}CO	373	0.323
CO–H_2	295.6	0.7430
CO–D_2	295.7	0.5490
CO–He	295.6	0.7020
CO–Ar	295.7	0.1880
CO_2–H_2	298.0	0.6460
CO_2–N_2	298.2	0.165
CO_2–O_2	293.2	0.160
CO_2–He	298	0.612
CO_2–Ar	276.2	0.1326
CO_2–CO	296.1	0.1520
CO_2–H_2O	307.5	0.202
CO_2–N_2O	298.0	0.117
CO_2–SO_2	263	0.064
$^{12}CO_2$–$^{14}CO_2$	312.8	0.125
CO_2–propane	298.0	0.0863
CO_2–ethyleneoxide	298.0	0.0914
H_2–N_2	297.2	0.779
H_2–O_2	273.2	0.697
H_2–D_2	288.2	1.24

Table 5.1-1. (*cont.*)

Gas pair	Temperature (°K)	Diffusion coefficient (cm^2 sec^{-1})
H$_2$–He	298.2	1.132
H$_2$–Ar	287.9	0.828
H$_2$–Xe	341.2	0.751
H$_2$–SO$_2$	285.5	0.525
H$_2$–H$_2$O	307.1	0.915
H$_2$–NH$_3$	298	0.783
H$_2$–acetone	296	0.424
H$_2$–ethane	298.0	0.537
H$_2$–*n*-butane	287.9	0.361
H$_2$–*n*-hexane	288.7	0.290
H$_2$–cyclohexane	288.6	0.319
H$_2$–benzene	311.3	0.404
H$_2$–SF$_6$	286.2	0.396
H$_2$–*n*-heptane	303.2	0.283
H$_2$–*n*-decane	364.1	0.306
N$_2$–O$_2$	273.2	0.181
	293.2	0.22
N$_2$–He	298	0.687
N$_2$–Ar	293	0.194
N$_2$–NH$_3$	298	0.230
N$_2$–H$_2$O	307.5	0.256
N$_2$–SO$_2$	263	0.104
N$_2$–ethylene	298.0	0.163
N$_2$–ethane	298	0.148
N$_2$–*n*-butane	298	0.096
N$_2$–isobutane	298	0.0905
N$_2$–*n*-hexane	288.6	0.076
N$_2$–*n*-octane	303.1	0.073
N$_2$–2,2,4-trimethylpentane	303.3	0.071
N$_2$–*n*-decane	363.6	0.084
N$_2$–benzene	311.3	0.102
O$_2$–He (He trace)	298.2	0.737
(O$_2$ trace)	298.2	0.718
O$_2$–He	298	0.729
O$_2$–H$_2$O	308.1	0.282
O$_2$–CCl$_4$	296	0.075
O$_2$–benzene	311.3	0.101
O$_2$–cyclohexane	288.6	0.075
O$_2$–*n*-hexane	288.6	0.075
O$_2$–*n*-octane	303.1	0.071
O$_2$–2,2,4-trimethylpentane	303.0	0.071
He–D$_2$	295.1	1.250
He–Ar	298	0.742
He–H$_2$O	298.2	0.908
He–NH$_3$	297.1	0.842
He–*n*-hexane	417.0	0.1574
He–benzene	298.2	0.384
He–Ne	341.2	1.405
He–methanol	423.2	1.032
He–ethanol	298.2	0.494

Table 5.1-1. (*cont.*)

Gas pair	Temperature (°K)	Diffusion coefficient (cm^2 sec^{-1})
He–propanol	423.2	0.676
He–hexanol	423.2	0.469
Ar–Ne	303	0.327
Ar–Kr	303	0.140
Ar–Xe	329.9	0.137
Ar–NH$_3$	295.1	0.232
Ar–SO$_2$	263	0.077
Ar–*n*-hexane	288.6	0.066
Ne–Kr	273.0	0.223
Ethylene–H$_2$O	307.8	0.204
Ethane–*n*-hexane	294	0.0375
N$_2$O–propane	298	0.0860
N$_2$O–ethyleneoxide	298	0.0914
NH$_3$–SF$_6$	296.6	0.1090
Freon-12–H$_2$O	298.2	0.1050
Freon-12–benzene	298.2	0.0385
Freon-12–ethanol	298.2	0.0475

Source: Data from Hirschfelder et al. (1954) and Reid et al. (1977).

the diffusion in half. They vary with the 1.5 to 1.8 power of the temperature; so an increase in 300°K triples the coefficients. They vary in a more complicated fashion with factors like molecular weight.

The physical significance of diffusion coefficients of this size is best illustrated by remembering unsteady-state diffusion problems like the semi-infinite slab discussed in Chapter 2. In these problems, the key experimental variable is $z^2/4Dt$. When this variable equals unity, the diffusion process has proceeded significantly. In other words, where z^2 equals $4Dt$, the diffusion has penetrated a distance z in the time t.

In gases, this penetration distance is much larger than in other phases. For example, the diffusion coefficient of water vapor diffusing in air is about 0.3 cm^2/sec. In 1 sec, the diffusion will penetrate 0.5 cm; in 1 min, 4 cm; and in 1 hr, 30 cm.

Gaseous diffusion coefficients from the Chapman–Enskog theory

The most common method for theoretical estimation of gaseous diffusion is that developed independently by Chapman and by Enskog (Chapman & Cowling, 1970). This theory, accurate to an average of about 8%, leads to the equation

$$D = \frac{1.86 \cdot 10^{-3} T^{3/2} (1/\tilde{M}_1 + 1/\tilde{M}_2)^{1/2}}{p\sigma_{12}^2 \Omega} \tag{5.1-1}$$

in which D is the diffusion coefficient measured in centimeters squared per second, T is the absolute temperature in degrees Kelvin, p is the pressure in atmospheres, and the $\bar{M}_i$ are the molecular weights.

The quantities σ_{12} and Ω are molecular properties characteristic of the detailed theory. The collision diameter σ_{12}, given in angstroms, is the arithmetic average of the two species present:

$$\sigma_{12} = \frac{1}{2}(\sigma_1 + \sigma_2) \qquad (5.1\text{-}2)$$

Values of σ_1 and σ_2 are listed in Table 5.1-2. The dimensionless quantity Ω is more complex, but usually of order 1. Its detailed calculation depends on an integration of the interaction between the two species. This interaction is most frequently described by the Lennard–Jones 12-6 potential. The resulting integral varies with the temperature and the energy of interaction. This energy ε_{12} is a geometric average of contributions from the two species:

$$\varepsilon_{12} = \sqrt{\varepsilon_1 \varepsilon_2} \qquad (5.1\text{-}3)$$

Values of the ε_i are also given in Table 5.1-2. Once ε_{12} is known, Ω can be found as a function of ε_{12}/kT using the values in Table 5.1-3. The calculation of the diffusion coefficients now becomes straightforward if the σ_i and the ε_i are known.

The nature of kinetic theories

The results of the Chapman–Enskog theory are based on detailed analyses of molecular motion in dilute gases. These analyses depend on the assumption that molecular interactions involve collisions between only two molecules at a time (Fig. 5.1-1). Such interactions are much simpler than the lattice interactions in solids or the less regular and still more complex interactions in liquids.

The nature of theories of this type is best illustrated for a gas of rigid spheres of very small molecular dimensions (Jeans, 1921; Cunningham & Williams, 1980). For such a theory, the diffusion flux has the following form:

$$n_1 = -\frac{1}{3}\bar{v}l\frac{dc_1}{dz} + c_1 v^0 \qquad (5.1\text{-}4)$$

The second term on the right represents convection, and the first indicates diffusion. The diffusion term has three parts: $\bar{v}$, the average molecular velocity; l, the mean free path of the molecules; and dc_1/dz, the concentration gradient. This term makes physical sense; the flux will certainly increase if either the velocity of the molecules or the average distance they travel increases.

If we compare Eq. 5.1-4 with Fick's law, we find

$$D = \frac{1}{3}\bar{v}l \qquad (5.1\text{-}5)$$

Both the average velocity $\bar{v}$ and the mean free path l of the rigid spheres can be calculated. The average velocity is

Table 5.1-2. *Lennard–Jones potential parameters found from viscosities*

Substance		σ (Å)	ε/k (°K)
Ar	Argon	3.542	93.3
He	Helium	2.551	10.22
Kr	Krypton	3.655	178.9
Ne	Neon	2.820	32.8
Xe	Xenon	4.047	231.0
Air	Air	3.711	78.6
Br_2	Bromine	4.296	507.9
CCl_4	Carbon tetrachloride	5.947	322.7
CF_4	Carbon tetrafluoride	4.662	134.0
$CHCl_3$	Chloroform	5.389	340.2
CH_2Cl_2	Methylene chloride	4.898	356.3
CH_3Br	Methyl bromide	4.118	449.2
CH_3Cl	Methyl chloride	4.182	350
CH_3OH	Methanol	3.626	481.8
CH_4	Methane	3.758	148.6
CO	Carbon monoxide	3.690	91.7
CO_2	Carbon dioxide	3.941	195.2
CS_2	Carbon disulfide	4.483	467
C_2H_2	Acetylene	4.033	231.8
C_2H_4	Ethylene	4.163	224.7
C_2H_6	Ethane	4.443	215.7
C_2H_5Cl	Ethyl chloride	4.898	300
C_2H_5OH	Ethanol	4.530	362.6
CH_3OCH_3	Methyl ether	4.307	395.0
CH_2CHCH_3	Propylene	4.678	298.9
CH_3CCH	Methylacetylene	4.761	251.8
C_3H_6	Cyclopropane	4.807	248.9
C_3H_8	Propane	5.118	237.1
$n\text{-}C_3H_7OH$	*n*-Propyl alcohol	4.549	576.7
CH_3COCH_3	Acetone	4.600	560.2
CH_3COOCH_3	Methyl acetate	4.936	469.8
$n\text{-}C_4H_{10}$	*n*-Butane	4.687	531.4
iso-C_4H_{10}	Isobutane	5.278	330.1
$C_2H_5OC_2H_5$	Ethyl ether	5.678	313.8
$CH_3COOC_2H_5$	Ethyl acetate	5.205	521.3
$n\text{-}C_5H_{12}$	*n*-Pentane	5.784	341.1
$C(CH_3)_4$	2,2-Dimethylpropane	6.464	193.4
C_6H_6	Benzene	5.349	412.3
C_6H_{12}	Cyclohexane	6.182	297.1
$n\text{-}C_6H_{14}$	*n*-Hexane	5.949	399.3
Cl_2	Chlorine	4.217	316.0
F_2	Fluorine	3.357	112.6
HBr	Hydrogen bromide	3.353	449
HCN	Hydrogen cyanide	3.630	569.1
HCl	Hydrogen chloride	3.339	344.7
HF	Hydrogen fluoride	3.148	330
HI	Hydrogen iodide	4.211	288.7
H_2	Hydrogen	2.827	59.7
H_2O	Water	2.641	809.1
H_2O_2	Hydrogen peroxide	4.196	289.3
H_2S	Hydrogen sulfide	3.623	301.1

Table 5.1-2. (*cont.*)

Substance		σ (Å)	ε/k (°K)
Hg	Mercury	2.969	750
I_2	Iodine	5.160	474.2
NH_3	Ammonia	2.900	558.3
NO	Nitric oxide	3.492	116.7
N_2	Nitrogen	3.798	71.4
N_2O	Nitrous oxide	3.828	232.4
O_2	Oxygen	3.467	106.7
PH_3	Phosphine	3.981	251.5
SO_2	Sulfur dioxide	4.112	335.4
UF_6	Uranium hexafluoride	5.967	236.8

Source: Data from Hirschfelder et al. (1954).

Table 5.1-3. *The collision integral* Ω

kT/ε	Ω	kT/ε	Ω	kT/ε	Ω
0.30	2.662	1.65	1.153	4.0	0.8836
0.35	2.476	1.70	1.140	4.1	0.8788
0.40	2.318	1.75	1.128	4.2	0.8740
0.45	2.184	1.80	1.116	4.3	0.8694
0.50	2.066	1.85	1.105	4.4	0.8652
0.55	1.966	1.90	1.094	4.5	0.8610
0.60	1.877	1.95	1.084	4.6	0.8568
0.65	1.798	2.00	1.075	4.7	0.8530
0.70	1.729	2.1	1.057	4.8	0.8492
0.75	1.667	2.2	1.041	4.9	0.8456
0.80	1.612	2.3	1.026	5.0	0.8422
0.85	1.562	2.4	1.012	6	0.8124
0.90	1.517	2.5	0.9996	7	0.7896
0.95	1.476	2.6	0.9878	8	0.7712
1.00	1.439	2.7	0.9770	9	0.7556
1.05	1.406	2.8	0.9672	10	0.7424
1.10	1.375	2.9	0.9576	20	0.6640
1.15	1.346	3.0	0.9490	30	0.6232
1.20	1.320	3.1	0.9406	40	0.5960
1.25	1.296	3.2	0.9328	50	0.5756
1.30	1.273	3.3	0.9256	60	0.5596
1.35	1.253	3.4	0.9186	70	0.5464
1.40	1.233	3.5	0.9120	80	0.5352
1.45	1.215	3.6	0.9058	90	0.5256
1.50	1.198	3.7	0.8998	100	0.5130
1.55	1.182	3.8	0.8942	200	0.4644
1.60	1.167	3.9	0.8888	300	0.4360

Source: Data from Hirschfelder et al. (1954).

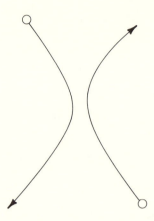

Fig. 5.1-1. Molecular motion in a dilute gas. In a gas, molecular collisions occur at low density, and so may be treated as bimolecular. This simplicity facilitates development of good kinetic theories for diffusion.

$$\bar{v} = \sqrt{8 k_B T / \pi (\tilde{M}_1 / \tilde{N})} \tag{5.1-6}$$

in which $(\tilde{M}_1 / \tilde{N})$ is the molecular mass. The mean free path is

$$l = \frac{k_B T / p}{\sqrt{2} \pi \sigma_{11}^2} \tag{5.1-7}$$

in which σ_{11} is the diameter of the spheres, and $\tilde{N} p / RT$ is the concentration in molecules per volume. Combining, we find

$$D = \left[\frac{2}{3} \left(\frac{k_B}{\pi} \right)^{3/2} \tilde{N}^{1/2} \right] \frac{T^{3/2} (1/\tilde{M}_1)^{1/2}}{p \sigma_{11}^2} \tag{5.1-8}$$

When we compare this result with Eq. 5.1-1, we see that the rigid-sphere theory predicts essentially the same dependence on temperature, pressure, molecular weight, and molecular size. The Chapman–Enskog theory is an improvement over the simple theory because the details of the collisions are explicitly included.

Gaseous diffusion coefficients from empirical correlations

Predictions from the Chapman–Enskog kinetic theory tend to be limited in two ways. First, the theory requires estimates of σ_{12} and ε_{12}; such estimates are not available for all gases. Second, the theory assumes nonpolar gases, and this excludes compounds like water and ammonia. These nonpolar interactions depend on replacing the Lennard–Jones potential used to characterize the collision with more exact potentials. Such replacement is often inhibited by numerical complexity.

Instead, many authors have developed empirical relations. One effective example (Fuller et al., 1966) is

$$D = 10^{-3} \frac{T^{1.75}(1/\tilde{M}_1 + 1/\tilde{M}_2)^{1/2}}{p \left[\left(\sum_i V_{i1} \right)^{1/3} + \left(\sum_i V_{i2} \right)^{1/3} \right]^2} \tag{5.1-9}$$

Table 5.1-4. *Atomic diffusion volumes for use in Eq. 5.1-12*

Atomic and structural diffusion-volume increments V_{ij}		Diffusion volumes for simple molecules $\Sigma\, V_{ij}$	
C	16.5	H_2	7.07
H	1.98	He	2.88
O	5.48	N_2	17.9
$(N)^a$	5.69	O_2	16.6
(Cl)	19.5	Air	20.1
(S)	17.0	Ar	16.1
Aromatic ring	−20.2	Kr	22.8
Heterocyclic ring	−20.2	CO	18.9
		CO_2	26.9
		N_2O	35.9
		NH_3	14.9
		H_2O	12.7
		(Cl_2)	37.7
		(SO_2)	41.1

[a] Parentheses indicate that the value is uncertain.
Source: Adapted from Fuller et al. (1966).

in which T is in degrees Kelvin, p is in atmospheres, and the V_{ij} are the volumes of parts of the molecule j, tabulated in Table 5.1-4. This correlation is slightly more successful than Eq. 5.1-1. To me, the impressive feature is the similarity between the two equations; the pressure and molecular-weight dependence is unchanged. The temperature dependence is not much different when we remember that Ω is a function of temperature. The term for diffusion volumes here parallels the term in σ^2. It is not surprising that the two equations have similar successes.

Gas diffusion at high pressure

The equations given earlier in this chapter allow prediction of diffusion coefficients in dilute gases to within an average of 8%. These predictions, which are about twice as accurate as those for liquids, are often hailed as a final answer. However, I have the nagging suspicion that their success is promulgated by those who have worked hard on these methods or who have become intimidated by the intellectual edifice erected by Maxwell, Enskog, and others. In fact, although these equations agree with experiment at low pressures, they are much less successful at high pressures. At higher pressures, few binary data are available; for self-diffusion, one sensible empirical suggestion (Reid et al., 1977) is

$$\rho D = \rho_0 D_0 \tag{5.1-10}$$

in which the subscript 0 indicates low pressure but the same temperature. The important feature is the inverse relation suggested between diffusion and density. This is consistent with Eq. 5.1-1 and is a good guideline. However, it should be used with caution, especially near the critical point.

Some other aspects of gaseous diffusion remain unexplored. For example, diffusion of molecules of very different sizes, like hydrogen and high molecular weight n-alkanes, has not been sufficiently studied. Concentration-dependent diffusion in gases, although a common phenomenon, has been largely ignored. These questions deserve careful inspection.

Example 5.1-1: Estimating diffusion with the Chapman–Enskog theory

Calculate the diffusion coefficient of argon in hydrogen at 1 atm and 175°C. The experimental value is 1.76 cm²/sec.

Solution. We first need to find σ_{12} and ε_{12}. From the values in Table 5.1-2,

$$\sigma_{12} = \frac{1}{2}(\sigma_1 + \sigma_2)$$

$$= \frac{1}{2}(3.42 + 2.92) = 3.17 \text{ Å}$$

and

$$\frac{\varepsilon_{12}}{k_B T} = \sqrt{(\varepsilon_1/k_B)(\varepsilon_2/k_B)}\Big/ T$$

$$= \frac{\sqrt{124(38.0)}}{448} = 0.153$$

From Table 5.1-3, we find that Ω is 0.80. Thus, from Eq. 5.1-1,

$$D = \frac{1.86 \cdot 10^{-3}T^{3/2}(1/\tilde{M}_1 + 1/\tilde{M}_2)^{1/2}}{p\sigma_{12}^2\Omega}$$

$$= \frac{1.86 \cdot 10^{-3}(448)^{3/2}(1/39.9 + 1/2.02)^{1/2}}{(1)(3.17)^2(0.80)}$$

$$= 1.58 \text{ cm}^2/\text{sec}$$

The theoretical prediction is about 10% below the experimental observation, and this is reasonable agreement.

Example 5.1-2: Comparing two estimates of gas diffusion

Use the Chapman–Enskog theory and the Fuller correlation to estimate the diffusion of hydrogen in nitrogen at 21°C and 2 atm. The experimental value is 0.38 cm²/sec.

Solution. For the Chapman–Enskog theory, the key parameters are
$$\sigma_{12} = \frac{1}{2}(\sigma_{H_2} + \sigma_{N_2}) = \frac{1}{2}(2.92 + 3.68) = 3.30 \text{ Å}$$

and

$$\frac{\varepsilon_{12}}{k_B T} = \frac{\sqrt{(\varepsilon_{H_2}/k_B)(\varepsilon_{N_2}/k_B)}}{T} = \frac{\sqrt{(38.0)(91.5)}}{294°K} = 0.201$$

This second value allows interpolation from Table 5.1-3:
$$\Omega = 0.842$$

Combining these results with Eq. 5.1-1 gives

$$D = \frac{1.86 \cdot 10^{-3}T^{3/2}(1/\tilde{M}_{H_2} + 1/\tilde{M}_{N_2})^{1/2}}{p\sigma^2\Omega}$$

$$= \frac{1.86 \cdot 10^{-3}(294)^{3/2}(1/2.01 + 1/28.0)^{1/2}}{2(3.30)^2(0.8422)}$$

$$= 0.37 \text{ cm}^2/\text{sec}$$

This value is about 3% low, a very solid estimate.

For the Fuller correlation, the appropriate volumes are found from Table 5.1-4. The results can then be combined with Eq. 5.1-12:

$$D = \frac{10^{-3}T^{1.75}(1/\tilde{M}_{H_2} + 1/\tilde{M}_{N_2})^{1/2}}{p[(V_{H_2})^{1/3} + (V_{N_2})^{1/3}]^2}$$

$$= \frac{10^{-3}(294)^{1.75}(1/2.01 + 1/28.0)^{1/2}}{2[(7.07)^{1/3} + (17.9)^{1/3}]^2}$$

$$= 0.37 \text{ cm}^2/\text{sec}$$

Again, the error is about 3%.

Example 5.1-3: Diffusion in supercritical carbon dioxide

Carbon dioxide, above its critical point, may become an important industrial solvent, because it is cheap, nontoxic, and nonexplosive. Estimate the diffusion of iodine in carbon dioxide at 0°C and 150 atm. The diffusion coefficient measured under these conditions is $8.0 \cdot 10^{-5}$ cm²/sec.

Solution. At the conditions given, the density of CO_2 is

$$c = 0.021 \text{ g-mol/cm}^3$$

The density and the binary diffusion coefficient at 0°C and 1 atm can be found from the ideal-gas law and Eq. 5.1-1, respectively:

$$c_0 = \text{g-mol}/(22.4 \cdot 10^3 \text{ cm}^3)$$
$$D_0 = 0.043 \text{ cm}^2/\text{sec}$$

Thus, from Eq. 5.1-10,

$$D = \frac{[1 \text{ g-mol}/(22.4 \cdot 10^3 \text{ cm}^3)](0.043 \text{ cm}^2/\text{sec})}{0.021 \text{ g-mol/cm}^3}$$

$$= 9 \cdot 10^{-5} \text{ cm}^2/\text{sec}$$

This is as accurate as we have any right to expect, especially because the critical point for carbon dioxide is close, at 30°C and 72 atm.

Section 5.2. Diffusion coefficients in liquids

Diffusion coefficients in liquids are exemplified by the values given in Tables 5.2-1 and 5.2-2. Most of these values fall close to 10^{-5} cm²/sec (Cussler, 1976; Reid et al., 1977). This is true for common organic solvents, mercury, and even molten iron. Exceptions occur for high-molecular-weight solutes

Table 5.2-1. *Diffusion coefficients at infinite dilution in water at 25°C*

Solute	D ($\cdot 10^{-5}$ cm^2/sec)
Argon	2.00
Air	2.00
Bromine	1.18
Carbon dioxide	1.92
Carbon monoxide	2.03
Chlorine	1.25
Ethane	1.20
Ethylene	1.87
Helium	6.28
Hydrogen	4.50
Methane	1.49
Nitric oxide	2.60
Nitrogen	1.88
Oxygen	2.10
Propane	0.97
Ammonia	1.64
Benzene	1.02
Hydrogen sulfide	1.41
Sulfuric acid	1.73
Nitric acid	2.60
Acetylene	0.88
Methanol	0.84
Ethanol	0.84
1-Propanol	0.87
2-Propanol	0.87
n-Butanol	0.77
Benzyl alcohol	0.821
Formic acid	1.50
Acetic acid	1.21
Propionic acid	1.06
Benzoic acid	1.00
Glycine	1.06
Valine	0.83
Acetone	1.16
Urea	$(1.380 - 0.0782c_1 + 0.00464c_1^2)^a$
Sucrose	$(0.5228 - 0.265c_1)^a$
Ovalbumin	0.078
Hemoglobin	0.069
Urease	0.035
Fibrinogen	0.020

[a] Known to very high accuracy, and so often used for calibration; c_1 is in moles per liter.
Source: Data from Cussler (1976) and Sherwood et al. (1975).

Table 5.2-2. *Diffusion coefficients at infinite dilation in nonaqueous liquids*

Solute[a]	Solvent	D ($\cdot 10^{-5}$ cm²/sec)
Acetone	Chloroform	2.35
Benzene		2.89
n-Butyl acetate		1.71
Ethyl alcohol (15°)		2.20
Ethyl ether		2.14
Ethyl acetate		2.02
Methyl ethyl ketone		2.13
Acetic acid	Benzene	2.09
Aniline		1.96
Benzoic acid		1.38
Cyclohexane		2.09
Ethyl alcohol (15°)		2.25
n-Heptane		2.10
Methyl ethyl ketone (30°)		2.09
Oxygen (29.6°)		2.89
Toluene		1.85
Acetic acid	Acetone	3.31
Benzoic acid		2.62
Nitrobenzene (20°)		2.94
Water		4.56
Carbon tetrachloride	n-Hexane	3.70
Dodecane		2.73
n-Hexane		4.21
Methyl ethyl ketone (30°)		3.74
Propane		4.87
Toluene		4.21
Benzene	Ethyl alcohol	1.81
Camphor (20°)		0.70
Iodine		1.32
Iodobenzene (20°)		1.00
Oxygen (29.6°)		2.64
Water		1.24
Carbon tetrachloride		1.50
Benzene	n-Butyl alcohol	0.988
Biphenyl		0.627
p-Dichlorobenzene		0.817
Propane		1.57
Water		0.56
Acetone (20°)	Ethyl acetate	3.18
Methyl ethyl ketone (30°)		2.93
Nitrobenzene (20°)		2.25
Water		3.20
Benzene	n-Heptane	3.40

[a] Temperature 25°C except as indicated.
Source: Data from Reid et al. (1977).

like albumin and polystyrene, where diffusion can be one hundred times slower.

These diffusion coefficients are about ten thousand times slower than those in dilute gases. To see what this means, we again calculate the penetration distance $\sqrt{4Dt}$, which was the distance we found central to unsteady diffusion. As an example, consider benzene diffusing into cyclohexane with a diffusion coefficient of about $2 \cdot 10^{-5}$ cm^2/sec. At time zero, we bring the benzene and cyclohexane into contact. After 1 sec, the diffusion has penetrated 0.004 cm, compared with 0.3 cm for gases; after 1 min, the penetration is 0.03 cm, compared with 4 cm; after 1 hr, it is 0.3 cm, compared with 30 cm.

The sloth characteristic of liquid diffusion means that diffusion often limits the overall rate of processes occurring in liquids. In chemistry, diffusion limits the rate of acid–base reactions; in physiology, diffusion limits the rate of digestion; in metallurgy, diffusion can control the rate of surface corrosion; in industry, diffusion is responsible for the rates of liquid–liquid extractions. Diffusion in liquids is important because it is slow.

Liquid diffusion coefficients from the Stokes–Einstein equation

The most common basis for estimating diffusion coefficients in liquids is the Stokes–Einstein equation. Coefficients calculated from this equation are accurate to only about 20% (Reid et al., 1977). Nonetheless, it remains the standard against which alternative correlations are judged.

The Stokes–Einstein equation is

$$D = \frac{k_B T}{f} = \frac{k_B T}{6\pi \mu R_0} \tag{5.2-1}$$

where f is the frictional coefficient of the solute, k_B is Boltzmann's constant, μ is the solvent viscosity, and R_0 is the solute radius. The temperature variation suggested by this equation is apparently correct, but it is much smaller than effects of solvent viscosity and solute radius. A discussion of these larger effects follows.

The diffusion coefficient varies inversely with viscosity when the ratio of solute to solvent radius exceeds 5, as shown in Fig. 5.2-1. This behavior is reassuring, because the Stokes–Einstein equation is derived by assuming a rigid solute sphere diffusing in a continuum of solvent. Thus, for a large solute in a small solvent, Eq. 5.2-1 seems correct.

When the solute size is less than five times that of the solvent, Eq. 5.2-1 breaks down (Chen et al., 1981). This failure becomes worse as solute size becomes smaller and smaller. Errors are especially large in high-viscosity solvents; the diffusion seems to vary with the $-\frac{2}{3}$ power of viscosity (Hiss & Cussler, 1973). In extremely high viscosity materials, diffusion becomes independent of viscosity; the diffusion of sugar in Jello is very nearly equal to the diffusion of sugar in water.

Two adaptations of the Stokes–Einstein equation deserve special mention. The first is for small solutes. For this case, the factor 6π in Eq. 5.2-1 is

(a) Actual situation

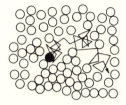

(b) Stokes – Einstein model

Fig. 5.2-1. Molecular motion in a liquid. In contrast with a gas, molecular motion in a liquid takes place at high density (a). Diffusion is complex, involving many interactions and vacancies. The available kinetic theories are good, but complex. To avoid this, many use the simple model of a solute sphere in a solvent continuum (b).

often replaced by a factor of 4π or of 2. The substitution of 4π can be rationalized on mechanical grounds as signifying solvent slipping past the surface of the solute molecule (Sutherland, 1905). The factor of 2 can be supported with the theory of absolute reaction rates (Glasstone et al., 1941). Neither substitution always works.

The second adaptation of the Stokes–Einstein equation is its frequent use to estimate the radius of macromolecules such as proteins that are present in dilute aqueous solution. Unfortunately, these estimates are compromised in two ways. First, if the solute is hydrated or solvated in some way, then the radius found will refer to the solute–solvent complex, not to the solute itself. Second, if the solute is not spherical, then the radius R_0 will represent some average over this shape. Specifically, if the solute is a prolate (football-shaped) ellipsoid, then (Perrin, 1936)

$$D\left(\begin{matrix}\text{prolate}\\\text{ellipsoid}\end{matrix}\right) = \frac{k_B T}{6\pi\mu\left[\dfrac{(a^2 - b^2)^{1/2}}{\ln\left(\dfrac{a + a^2 - b^2)^{1/2}}{b}\right)}\right]} \tag{5.2-2}$$

in which a and b are the major and minor axes of the ellipsoid. For an oblate (disc-shaped) ellipsoid,

$$D\left(\begin{matrix}\text{oblate}\\\text{ellipsoid}\end{matrix}\right) = \frac{k_B T}{6\pi\mu\left[\dfrac{(a^2 - b^2)^{1/2}}{\tan^{-1}\left[\left(\dfrac{a^2 - b^2}{b^2}\right)^{1/2}\right]}\right]} \tag{5.2-3}$$

These relations reduce to Eq. 5.2-1 for spheres, when a equals b.

Deriving the Stokes–Einstein equation

To predict diffusion in liquids, we do not account for molecular motion as in the kinetics theories used for gases. Instead, we idealize our system as a single rigid solute sphere moving slowly through a continuum of solvent (Fig. 5.2-1). We expect that the net velocity of this sphere will be proportional to the force acting on it:

$$\text{force} = (f)\mathbf{v}_1 \qquad (5.2\text{-}4)$$

where f is defined as the friction coefficient. Because the sphere moves slowly, this friction coefficient can be found from Stokes's law to be $6\pi\mu R_0$ (Stokes, 1850). The force was taken by Einstein to be the chemical potential gradient (Einstein, 1905). Thus, Eq. 5.2-4 can be rewritten:

$$-\nabla\mu_1 = (6\pi\mu R_0)\mathbf{v}_1 \qquad (5.2\text{-}5)$$

The chemical potential gradient, defined per molecule (not per mole), is often described as a "virtual force," a thermodynamic parallel to mechanical or electrostatic forces.

Because the solution is dilute, we can assume that it is ideal:

$$\mu_1 = \mu_1^0 + k_B T \ln x_1$$

$$= \mu_1^0 + k_B T \ln \frac{c_1}{c_1 + c_2} \doteq \mu_1^0 + k_B T \ln c_1 - k_B T \ln c_2 \qquad (5.2\text{-}6)$$

In this result, we recognize that solvent concentration c_2 far exceeds solute concentration c_1; so c_2 is approximately constant. The gradient is then

$$-\nabla\mu_1 = \frac{k_B T}{c_1} \nabla c_1 \qquad (5.2\text{-}7)$$

Combining this with Eq. 5.2-5, we find

$$\mathbf{j}_1 \doteq \mathbf{n}_1 = c_1\mathbf{v}_1 = -\frac{k_B T}{6\pi\mu R_0} \nabla c_1 \qquad (5.2\text{-}8)$$

Comparison with Fick's law produces the Stokes–Einstein equation, Eq. 5.2-1.

The interesting assumption in this analysis is the way in which the velocity or flux is assumed to vary with chemical potential gradient. This type of assumption is made frequently in studies of diffusion. It is central to the development of irreversible thermodynamics, and so it is at the core of the theories of multicomponent diffusion described in Chapter 8. Interestingly, it is known experimentally to be wrong in the highly nonideal solutions near critical points (see Section 7.2).

Liquid diffusion coefficients from empirical correlations

Because the Stokes–Einstein equation is limited to cases in which the solute is larger than the solvent, many investigators have developed correlations for cases in which solute and solvent are similar in size. Five of these are tabulated in Table 5.2-3. The impressive aspect of these efforts is their similarity to the Stokes–Einstein equation. All five show the same temperature and viscosity dependence. Future searches for superior correla-

Table 5.2-3. *Alternatives to Stokes–Einstein equation for diffusion in liquids*[a,b]

Authors	Origin	Basic equation	Viscosity variation	Solute size variation	Remarks
Sutherland (1905)	Parallel to Stokes–Einstein, but "no stick" at sphere's surface	$D = \dfrac{k_B T}{4\pi\mu R_0}$	μ^{-1}	R_0^{-1}	Always mentioned but rarely used
Glasstone et al. (1941)	Diffusion as a rate process	$D = \dfrac{k_B T}{2\mu R_0}$	μ^{-1}	R_0^{-1}	Smaller coefficient; closer to some experimental results
Scheibel (1954)[c,d]	Empirical	$D = \dfrac{AT}{\mu(\tilde{V}_1)^{1/3}}\left[1 + \left(\dfrac{3\tilde{V}_2}{\tilde{V}_1}\right)^{2/3}\right]$	μ^{-1}	Equivalent to R_0^{-1} for large solutes and R_0^{-3} for small ones	Variation with solute size is the interesting feature of this equation
Wilke & Chang (1955)[c,e]	Empirical	$D = \dfrac{7.4 \cdot 10^{-8}(\phi M_2)^{1/2}T}{\mu_2 \tilde{V}_1^{0.6}}$	μ^{-1}	Equivalent to $R_0^{1.8}$	Factor ϕ for solute–solvent interaction
King et al. (1965)[c]	Empirical	$D = 4.4 \cdot 10^{-8}\,\dfrac{T}{\mu_2}$ $\cdot \left(\dfrac{\tilde{V}_2}{\tilde{V}_1}\right)^{1/6}\left(\dfrac{\Delta H_{vap,2}}{\Delta H_{vap,1}}\right)$	μ^{-1}	R_0	Not suitable for viscous solvents or aqueous systems

[a] The subscripts 1 and 2 indicate the solute and solvent, respectively.

[b] These relations are accurate within about 10% for water and 20% for most organics, but they are often inaccurate for alcohols and other hydrogen-bonded solvents.

[c] Specific units implied are $D [=] cm^2/sec$; $T [=] °K$; $\mu [=] 10^{-2} g/cm\text{-}sec$; $V_i [=] cm^3/g\text{-}mol$.

[d] The $\tilde{V}$ are the molar volumes at the boiling points. The constant A equals $8.2 \cdot 10^{-8}$, except as follows: $25.2 \cdot 10^{-8}$ for water if $\tilde{V}_1 < \tilde{V}_2$; $18.9 \cdot 10^{-8}$ for benzene when $\tilde{V}_1 < 2\tilde{V}_2$; $17.5 \cdot 10^{-8}$ for others if $\tilde{V}_1 < 2.5\tilde{V}_2$.

[e] The factor ϕ has the following values: 2.26 for water, 1.9 for methanol, 1.5 for ethanol, and 1.0 for non-hydrogen-bonded solvents.

tions should question whether or not the Stokes–Einstein equation is a good starting point, especially in view of the experimentally observed viscosity dependence.

The common conclusion is to bemoan the accuracy of the predictions in liquids and to praise the accuracy of those in gases. In fact, the predictions in liquids are only twice as inaccurate as those in gases, even though the complexity of solute–solvent interactions in liquids is much greater. As a result, I do not share the frequent despair about these estimates, but feel that care and good judgment can lead to successful predictions.

Diffusion in concentrated solutions

The Stokes–Einstein equation and its empirical extensions are limited to infinitely dilute solutions. In fact, the diffusion coefficient in liquids varies with solute concentration, frequently by several hundred percent and sometimes with a maximum or minimum. We need a means of estimating these concentration-dependent diffusion coefficients.

One such means extends the Stokes–Einstein equation to include hydrodynamic interaction among different spheres. In this extension, the movement of one sphere is imagined to create a flow field that alters the movement of other spheres. The result of these extensions is, to my surprise, under dispute. The correction to the frictional coefficient f is known to be (Batchelor, 1972; Reed & Anderson, 1980)

$$f = 6\pi\mu R_0(1 + 6.5\phi_1 + \cdots) \tag{5.2-9}$$

in which ϕ_1 is the volume fraction of the solute. Note that this result indicates that two spheres will move more rapidly than a single sphere. This frictional coefficient and a chemical potential gradient must be inserted into Eq. 5.2-4, in a way parallel to Eqs. 5.2-5 to 5.2-8. The result is

$$D = \frac{k_B T}{6\pi\mu R_0}(1 + 1.5\phi_1 + \cdots) \tag{5.2-10}$$

This seems consistent with the available data for dilute suspensions of spheres (Batchelor, 1976; Russel, 1981). However, this predicted variation is much smaller than those observed experimentally for small solutes; so extending the Stokes–Einstein equation seems impractical.

Empirical explanations of the concentration dependence of diffusion usually involve two steps. First, we assume that Eq. 5.2-4 can be written

$$-\mathbf{v}_1 = \frac{1}{f}\nabla\mu_1 = \frac{D_0}{RT}\nabla\mu_1 \tag{5.2-11}$$

where D_0 is a new transport coefficient. For a nonideal solution,

$$\mu_1 = \mu_1^0 + k_B T \ln c_1\gamma_1 \tag{5.2-12}$$

where γ_1 is an activity coefficient. Combining these two equations, we find

$$\mathbf{n}_1 \doteq \mathbf{j}_1 \doteq c_1\mathbf{v}_1 = -\left[D_0\left(1 + \frac{\partial \ln \gamma_1}{\partial \ln c_1}\right)\right]\nabla c_1 \tag{5.2-13}$$

The quantity in brackets is the diffusion coefficient. This first step is a restatement of the idea that the velocity of diffusion varies with the gradient of chemical potential.

The second step consists of empirical estimations of the quantity D_0. These estimates are based on diffusion coefficients in dilute solutions. One of the most frequently cited estimates, used by Darken (1948), Hartley & Crank (1949), and others, is the arithmetic average:

$$D_0 = x_1 D_0(x_1 = 1) + x_2 D_0(x_2 = 1) \tag{5.2-14}$$

One of the most successful estimates, suggested by Vignes (1966), Kosanovich & Cullinan (1976), and others, is the geometric average:

$$D_0 = [D_0(x_1 = 1)]^{x_1}[D_0(x_2 = 1)]^{x_2} \tag{5.2-15}$$

Both forms are supported by proud defenders and scrutinized by incessant improvers. The improvers change various viscosity corrections, replace mole fractions with volume or mass fractions, or quarrel about reference velocities. Because the improvers can always justify their improvements by choosing a different data set than their competitors, we should not be excited by these changes.

Example 5.2-1: Oxygen diffusion in water

Estimate the diffusion at 25°C for oxygen dissolved in water using the Stokes–Einstein, Scheibel, and Wilke–Chang correlations. Compare your results with the experimental value of $1.80 \cdot 10^{-5}$ cm^2/sec.

Solution. For the Stokes–Einstein equation, the chief problem is to estimate the radius of the oxygen molecule. If we assume that this is half the collision diameter in the gas, then, from Table 5.1-2,

$$R_0 = \tfrac{1}{2}\sigma_1 = 1.72 \cdot 10^{-8} \text{ cm}$$

When we insert this into the Stokes–Einstein equation,

$$D = \frac{k_B T}{6\pi\mu R_0}$$
$$= \frac{(1.38 \cdot 10^{-16} \text{ g-cm}^2/\text{sec}^2\text{-}°\text{K})298°\text{K}}{6\pi(0.01 \text{ g/cm-sec})1.72 \cdot 10^{-8} \text{ cm}} = 1.27 \cdot 10^{-5} \text{ cm}^2/\text{sec}$$

This value is 30% low. Using the Sutherland correlation gives a more accurate result; using the Eyring approach gives much too high a value.

For the Scheibel correlation, we need the molar volume of oxygen at its boiling point, which is about 25 cm^3/g-mol. Thus,

$$D = \frac{8.2 \cdot 10^{-8} T}{\mu_{H_2O} \tilde{V}_{O_2}^{1/3}} \left[1 + \left(\frac{3\tilde{V}_{H_2O}}{\tilde{V}_{O_2}} \right)^{2/3} \right]$$
$$= \frac{8.2 \cdot 10^{-8}(298°\text{K})}{1 \text{ cp}(25 \text{ cm}^3/\text{g-mol})^{1/3}} \left[1 + \left(\frac{3(18 \text{ cm}^3/\text{g-mol})}{25 \text{ cm}^3/\text{g-mol}} \right)^{2/3} \right]$$
$$= 2.2 \cdot 10^{-5} \text{ cm}^2/\text{sec}$$

This is 20% too high. The Wilke–Chang correlation is somewhat better:

$$D = \frac{7.4 \cdot 10^{-8}(\phi \tilde{M}_{H_2O})^{1/2}T}{\mu_{H_2O}\, \tilde{V}_{O_2}^{0.6}}$$

$$= \frac{7.4 \cdot 10^{-8}[2.26(18 \text{ cm}^3/\text{g-mol})]^{1/2}298°\text{K}}{1 \text{ cp}(25 \text{ cm}^3/\text{g-mol})^{0.6}}$$

$$= 2.0 \cdot 10^{-5} \text{ cm}^2/\text{sec}$$

This is 10% high. Notice that neither empirical method is a dramatic improvement on the Stokes–Einstein estimate.

Example 5.2-2: Estimating molecular size from diffusion

Fibrinogen has a diffusion coefficient of about $2.0 \cdot 10^{-7}$ cm^2/sec at 37°C. It is believed to be rod-shaped, about 30 times longer than it is wide. How large is the molecule?

Solution. Because the molecule is rod-shaped, it can be approximated as a prolate ellipsoid. Thus, from Eq. 5.2-2,

$$D = \frac{k_B T}{6\pi\mu a \left[\dfrac{[1 - (b/a)^2]^{1/2}}{\ln\left\{\dfrac{a}{b}\left[1 + \left(\dfrac{a^2}{b^2} - 1\right)^{1/2}\right]\right\}} \right]}$$

$$2.0 \cdot 10^{-7} \text{ cm}^2/\text{sec} = \frac{(1.38 \cdot 10^{-16} \text{ g-cm}^2/\text{sec}^2\text{-}°\text{K})(310°\text{K})}{6\pi(0.00695 \text{ g/cm-sec})\, a \left[\dfrac{[1 - (1/30)^2]^{1/2}}{\ln\{30[1 + (30^2 - 1)^{1/2}]\}} \right]}$$

Solving, we find that a equals 1,100 Å and b equals 37 Å. If fibrinogen were a sphere, its radius would be about 160 Å.

Example 5.2-3: Diffusion in an acetone–water mixture

Estimate the diffusion coefficient in a 50-mole-percent mixture of acetone (1) and water (2). This solution is highly nonideal, so that $\partial \ln \gamma_1 / \partial \ln x_1$ equals -0.69. In pure acetone, the diffusion coefficient is $1.26 \cdot 10^{-5}$ cm^2/sec; in pure water, it is $4.68 \cdot 10^{-5}$ cm^2/sec. The experimental value in the mixture is $0.79 \cdot 10^{-5}$ cm^2/sec, less than both limits.

Solution. We first must estimate D_0. Because Eq. 5.2-15 is most often successful, we use it here:

$$D_0 = [D_0(x_1 = 1)]^{x_1}[D_0(x_2 = 1)]^{x_2}$$

$$= (1.26 \cdot 10^{-5} \text{ cm}^2/\text{sec})^{0.5}(4.68 \cdot 10^{-5} \text{ cm}^2/\text{sec})^{0.5}$$

$$= 2.43 \cdot 10^{-5} \text{ cm}^2/\text{sec}$$

From Eq. 5.2-13,

$$D = D_0\left(1 + \frac{\partial \ln \gamma_1}{\partial \ln c_1}\right)$$

$$= 2.43 \cdot 10^{-5} \text{ cm}^2/\text{sec}(1 - 0.69)$$
$$= 0.75 \cdot 10^{-5} \text{ cm}^2/\text{sec}$$

The agreement with the experimental value is unusually good.

Section 5.3. Diffusion in solids

Typical values for diffusion coefficients in solids are shown in Table 5.3-1 (Cussler, 1976; Smithells, 1976). One outstanding characteristic of these values is their small size, usually thousands of times less than those in a liquid, which are in turn 10,000 times less than those in a gas. A second important characteristic is the huge range of values reported. In gases, most values fall within a power of 10; in ordinary liquids, most values do the same. For solids, diffusion coefficients can differ by more than 10^{10}. For example, the diffusion coefficient of cadmium in copper is 10^{15} greater than that of aluminum in copper.

Two more characteristics of the values in Table 5.3-1 are the temperature dependence and the range of materials involved. First, the temperature dependence is sharp and nonlinear, as exemplified by the first three values in the table. Such dependence is characteristic of chemical reactions, and it suggests treating diffusion in solids as a chemical rate process. Second, the range of materials cited includes metals, ionic and molecular solids, and a few noncrystalline materials. Such a range naturally produces the wider spectrum of coefficients.

The penetration distance of hydrogen in iron provides a more definite example of these effects. After 1 sec, hydrogen penetrates about 1 μ, 70 times less than in liquids. After 1 min, it penetrates 6 μ, and after 1 hr, only 50 μ. Still, hydrogen diffuses much more rapidly than almost any other solute.

Diffusion coefficients in metals from lattice theory

The estimation of diffusion coefficients in solids is not accurate. In almost every case, one must use experimental results. Methods for rough estimates based on the theory for face-centered-cubic metals are the standard by which other theories are judged, just as the Stokes–Einstein equation is the standard for liquids. The diffusion coefficient in this theory is (Franklin, 1975; Stark, 1976)

$$D = a_0^2 N\omega \tag{5.3-1}$$

in which a_0 is the spacing between atoms, N is the fraction of sites vacant in the crystal, and ω is the jump frequency, the number of jumps per unit time from one position to the next. Values for a_0 are guessed from crystallographic data, and the fraction N is commonly estimated from the Gibbs free energy of mixing. The frequency ω is estimated by reaction-rate theories for the concentration of activated complexes, atoms midway between adjacent

Table 5.3-1. *Diffusion coefficients in the solid state*

System	T (°C)	Diffusivity (cm²/sec)
Hydrogen in iron	10	1.66×10^{-9}
	50	11.4×10^{-9}
	100	124×10^{-9}
Hydrogen in nickel	85	1.16×10^{-8}
	165	10.5×10^{-8}
Carbon monoxide in nickel	950	4×10^{-8}
	1,050	14×10^{-8}
Aluminum in copper	850	2.2×10^{-9}
Uranium in tungsten	1,727	1.3×10^{-11}
Cerium in tungsten	1,727	95×10^{-11}
Yttrium in tungsten	1,727	$1,820 \times 10^{-11}$
Tin in lead	285	1.6×10^{-10}
Gold in lead	285	4.6×10^{-6}
Gold in silver	760	3.6×10^{-10}
Antimony in silver	20	3.5×10^{-21}
Zinc in aluminum	500	2×10^{-9}
Silver in aluminum	50	1.2×10^{-9}
Bismuth in lead	20	1.1×10^{-16}
Aluminum in copper	20	1.3×10^{-30}
Cadmium in copper	20	2.7×10^{-15}
Carbon in iron	800	1.5×10^{-8}
	1,100	45×10^{-8}
Helium in SiO_2	20	4.0×10^{-10}
	500	7.8×10^{-8}
Hydrogen in SiO_2	200	6.5×10^{-10}
	500	1.3×10^{-8}
Helium in Pyrex	20	4.5×10^{-11}
	500	2×10^{-8}

Source: Data from Barrer (1941), American Society for Metals (1973).

sites. The results of these calculations are commonly expressed as

$$D = D_0 e^{-\Delta H/RT} \tag{5.3-2}$$

where D_0 and ΔH are estimated empirically. Qualitative arguments suggest that

$$D_0 = a_0^2 \nu e^{-\beta \Delta H_{melt}/RT_{melt}} \tag{5.3-3}$$

where ΔH_{melt} is the enthalpy of melting, T_{melt} is the melting temperature, and β is an empirical parameter approximately equal to 0.4. The frequency ν is estimated from the relation

$$\nu = \left(\frac{\Delta H}{2\tilde{M}a_0^2} \right)^{1/2} \tag{5.3-4}$$

where $\tilde{M}$ is the molecular weight and ΔH is given by the empirical rule

$$\Delta H = (36 \text{ cal/g-mol-°K}) T_{melt} \tag{5.3-5}$$

Quite obviously, these rules should be used only when direct experimental information is not available.

(a) Interstitial diffusion

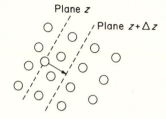

(b) Vacancy diffusion

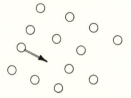

(c) Anisotropic diffusion

Fig. 5.3-1. Molecular motion in a crystalline solid. Motion takes place at high density, but now in a regular lattice. This lattice can have many forms, leading to a wide variety of diffusion mechanisms.

The sources of lattice theories

The foregoing approximate calculations originate in the detailed calculations of atomic motions within the crystal lattice. Because such calculations are difficult, we present only an outline to indicate the physical reasoning used.

We consider a face-centered-cubic crystal in which diffusion occurs by means of the interstitial mechanism shown in Fig. 5.3-1(a) (Stark, 1976). The net diffusion flux is the flux of atoms from z to $(z + \Delta z)$ minus the flux from $(z + \Delta z)$ to z:

$$\begin{pmatrix} \text{net} \\ \text{flux } j_1 \end{pmatrix} = 4N\omega \left[\begin{pmatrix} \text{number of atoms} \\ \text{per unit area at } z \end{pmatrix} - \begin{pmatrix} \text{number of atoms} \\ \text{per unit area at } z + \Delta z \end{pmatrix} \right]$$

$$(5.3\text{-}6)$$

where N is the average number of vacant sites and ω is the rate of jumps. The factor of 4 reflects the fact that the face-centered-cubic structure has four sites into which jumps can occur. Other crystal structures have different numbers of sites. In writing this equation, we have implied isotropic diffusion, so that the number of vacant sites and the jump rate do not vary with concentration or with position. The number of atoms per unit area is simply related to the concentration:

$$\left(\begin{array}{c}\text{number of atoms} \\ \text{per unit area at } z\end{array}\right) = \left(\frac{a_0}{2}\right) c_1|_z \tag{5.3-7}$$

The difference in these concentrations is close to the concentration gradient:

$$c_1|_z - c_1|_{z+\Delta z} = -\frac{a_0}{2}\frac{\partial c_1}{\partial z} \tag{5.3-8}$$

When we combine these relations, we find, in symbolic terms,

$$j_1 = -\left[4N\omega\left(\frac{a_0}{2}\right)^2\right]\frac{\partial c_1}{\partial z} \tag{5.3-9}$$

Comparing this with Fick's law gives an equation close to Eq. 5.3-1, the basic relation for the diffusion coefficient in a solid lattice. The assumptions in this analysis seem to me to be frequent and major; a specific lattice is assumed, and some (but not all) lattice characteristics are taken as independent of concentration. Nonetheless, the analysis for solids has power, because each explicit assumption can, at least in principle, be removed.

More complicated cases of solid diffusion

The simple face-centered-cubic lattice used earlier is only one of a tremendous variety of solid structures. Some of these structures can allow different routes for diffusion and so merit specific citation.

In some cases, diffusion does not involve movement within the interstitial spaces in the crystal, but rather depends on vacancies between the missing atoms or ions in the crystal (Farrington & Briant, 1979). The differences between these cases are illustrated in Figs. 5.3-1(a) and 5.3-1(b). Vacancy diffusion can be fast, as shown in Fig. 5.3-2. In some cases the diffusion astonishingly approaches the speed of transport in aqueous solutions, but in other cases it is very slow indeed. The different speeds of diffusion shown in Fig. 5.3-2 result from different types of vacancies in the crystals. When roughly equal numbers of positive and negative vacancies are formed, the disorder is said to be of the Schottky type. This is the case for NaCl and NaBr. When many more vacancies of one charge exist than those of the other charge, the disorder is called that of the Frenkel type. This is the case for AgBr.

Another complication of diffusion in solids is anisotropy; the diffusion coefficients are not the same in all directions, because of an anisotropic crystal lattice. For example, the diffusion of nickel in $(MgFe)_2SiO_4$ is 10 times faster along one crystalline axis than along the other two (Clark & Long, 1971). For noncrystalline solids like wood, diffusion with the grain is almost 20 times faster than diffusion across the grain.

Other complications are like those in liquids and gases. For example, diffusion in solids varies with composition, as it does in liquids. Mass transfer in solids can result from forces other than concentration gradient, just as mass transfer in gases can be caused by pressure-driven flows. In solids, the additional force may come from a stress that locally increases atomic energies. Such a stress can make important the definition of reference frames that separate diffusion and convection.

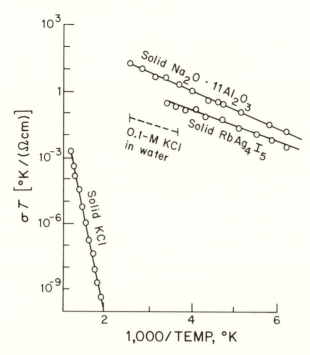

Fig. 5.3-2. Diffusion in several types of ionic solids. The electrical conductivity σ, a measure of ionic diffusion, varies widely in ionic solids. In some special cases, like $RbAg_4I_5$, mass transfer via vacancies can be faster than that in 0.1-M aqueous solutions.

Example 5.3-1: Diffusion of carbon in iron

Estimate the value of D_0 for carbon in iron (Shewmon, 1963). For iron, a_0 is approximately 2.9 Å, ΔH_{melt} is 20 kcal/g-mol, and T_{melt} is 1,800°K. From spectra measurements, ν is believed to be about $4 \cdot 10^{12}$/sec. The experimentally observed value of D_0 is $2 \cdot 10^{-4}$ cm²/sec.

Solution. From Eq. 5.3-3, we find

$$D_0 = a_0^2 \nu e^{-\beta \Delta H_{melt}/RT_{melt}}$$

$$= (2.9 \cdot 10^{-8} \text{ cm})^2 \left(\frac{4 \cdot 10^{12}}{\text{sec}}\right) \exp\left(\frac{0.4(20,000 \text{ cal/g-mol})}{(2 \text{ cal/g-mol-°K})1,800°K}\right)$$

$$= 4 \cdot 10^{-4} \text{ cm}^2/\text{sec}$$

This is high by a factor of 2, but it still represents unusually good agreement.

Section 5.4. Diffusion in polymers

Typical diffusion coefficients for synthetic high polymers are shown in Fig. 5.4-1. The values of these coefficients, which vary strongly with concentration, lie between the coefficients of liquids and those of solids. Naturally

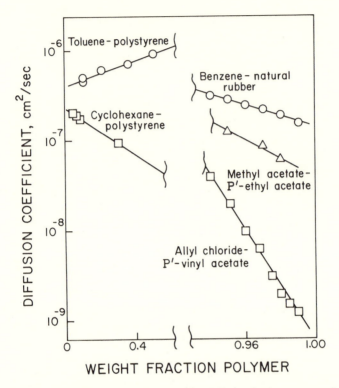

Fig. 5.4-1. Diffusion of high polymers. Diffusion in these systems has two interesting limits: at very low and at very high polymer concentrations. Interestingly, the diffusion coefficients in these two limits may not be very different, even though the viscosity change is tremendous.

occurring polymers like proteins are not included in Fig. 5.4-1, because these species are best handled with the dilute-solution arguments in Section 5.2.

The results in Fig. 5.4-1 imply that two very different limits exist. The first of these limits occurs in dilute solution, where a polymer molecule is easily imagined as a solute sphere moving through a continuum of solvent. The second limit is in highly concentrated solution, where small solvent molecules like benzene can be imagined to squeeze through a polymer matrix.

Polymer solutes in dilute solution

A polymer molecule dissolved in a solvent is best imagined to be like a necklace consisting of spherical beads connected by string that does not have any resistance to flow (Kirkwood, 1967; Vrentas & Duda, 1977, 1980). The necklace is floating in a neutrally buoyant solvent continuum. If the solution is very dilute, the polymer molecules are greatly separated, so that they do not interact with each other, but only with the solvent. In some cases the solvent will greatly expand the polymer necklace in the solution; such a solvent is referred to as "good." In other cases the solvent and polymer will not strongly interact, and the polymer necklace will shrink into a small, introspective blob; such a solvent is called "poor" (Ferry, 1980).

Between these two extremes, the polymer and solvent can interact just enough so that the segments of the polymer necklace will be randomly distributed. This limit of a "random coil" of polymer is conventionally chosen as the "ideal" polymer solution, and a solvent showing these characteristics is called a θ solvent. Under these conditions, the diffusion of the polymer can be calculated as a correction to the Stokes–Einstein equation:

$$D = \frac{k_B T}{6\pi\mu R_e} \tag{5.4-1}$$

where R_e is the equivalent radius of the polymer. This radius is calculated to be

$$R_e = 0.676\langle R^2\rangle^{1/2} \tag{5.4-2}$$

in which $\langle R^2\rangle^{1/2}$ is the root-mean-square radius of gyration, the common measure of the size of the polymer molecule in solution. This root-mean-square radius can be measured in a variety of ways; one common method is by light scattering.

Equations 5.4-1 and 5.4-2 are confirmed by experiment. The measured ratio of equivalent radius to root-mean-square radius is 0.68, which is very close to the 0.676 suggested theoretically. Because I am routinely skeptical of theories, I am especially impressed by this success. I should point out that the theory is successful only in a θ solvent; it is significantly in error when the polymer is not a random coil in solution (Cowie & Cussler, 1967).

Small solutes in a polymer solvent

The other limiting case of polymer diffusion, shown at the right in Fig. 5.4-1, occurs in very concentrated polymer solution. In this limit, small solute molecules are assumed to move through a solidlike polymer matrix.

No single theory for this situation is universally accepted. To me, the most appealing theory is that of "reptation," where a diffusing molecule is assumed to wiggle, snakelike, through the polymer melt (deGennes, 1979). Most older theories depend on the concept of "free volume" or "hole fraction" – that part of the polymer not occupied by the polymer molecules themselves (Vrentas & Duda, 1980; Ferry, 1980; Duda & Vrentas, 1982). This concept is roughly equivalent to the idea of vacancies used in solid-state diffusion. It leads to relations like the following:

$$D = \begin{pmatrix} \text{molecular} \\ \text{spacing} \end{pmatrix} \cdot \begin{pmatrix} \text{solute} \\ \text{velocity} \end{pmatrix} \cdot \exp\left[(\text{constant}) \frac{\text{volume of solute}}{\text{free volume of polymer}}\right] \tag{5.4-3}$$

The free volume is assumed to increase with temperature; so diffusion should follow an Arrhenius temperature dependence. This is observed experimentally. These arguments are now effective for correlating data, but they seem risky for reliable extrapolation.

One curious effect, called "non-Fickian diffusion" or "type II transport," sometimes occurs in the dissolution of high polymers by a good solvent. In these cases, diffusion may not follow Fick's law. For example, the speed with which the solvent penetrates into a thick polymer slab may not be proportional to the square root of time, which is the behavior expected from Fick's law (see Section 2.3).

This effect is believed to result from configurational changes in the polymer. As the solvent penetrates, the polymer molecules relax from their greatly hindered configuration as a partially crystalline solid into the more randomly coiled shape characteristic of a polymer dissolved in dilute solution. This relaxation process can be slower than the diffusion process; so the overall dissolution is controlled by the relaxation kinetics, not by Fick's law. Although the process does not involve any phase boundaries, it is similar to an interfacial chemical reaction followed by diffusion (Frisch, 1980).

At this point, you may decide that the various correlations available are not sufficiently accurate for your purposes. If so, you may decide to measure diffusion coefficients yourself. The best methods for these measurements are described in the next section.

Section 5.5. The measurement of diffusion coefficients

In this section we want to discuss the most convenient ways in which diffusion coefficients can be measured. This section is the counterpoint to the previous ones. Whereas the focus has been on using past experience to guide predictions, this section replaces the hope of prediction with the necessity of accurate measurements.

Measuring diffusion coefficients is reputed to be very difficult. For example, Tyrrell (1961) stated that "this is not an easy field of study in any sense. It took eighty years from the time when Thomas Graham worked on diffusion before precise data on diffusion coefficients began to be collected." This suggests that measurements of diffusion are a Holy Grail requiring noble knights who dedicate their lives to the quest.

In fact, although measurements are rarely routine, diffusion coefficients usually can be determined to within about 5% or 10% accuracy without excessive effort. Because such accuracy is sufficient for most situations, we should always consider measuring the coefficients we need. The pall over diffusion measurements stems from inherent masochists, like me, who make many of the experiments. We are never satisfied. When we attain coefficients accurate to 10%, we want 2%; when we achieve 2%, we want 0.5%.

If we have decided that measurements are essential, we must decide how to make them. There are many methods available, all described in glowing terms by their proponents (Dunlop et al., 1972). An exhaustive description of these methods could fill this book.

Instead of such a mastodonic list, we shall consider only those methods of measuring diffusion that are reasonably accurate, that are easy to use, or that have some special advantage. I have tried to state concisely the advantages and disadvantages of each method. I want to give the flavor of the laboratories themselves, and not just the polished publications that result.

The most useful methods of studying diffusion are shown in Table 5.5-1. The first four on this list are used most frequently. These four methods lead to solid research at accuracies sufficient for most practical purposes. They will be described in greater detail in the following paragraphs.

Diaphragm cell

The Stokes diaphragm cell is probably the best tool to start research on diffusion in gases or liquids or across membranes. It is inexpensive to build, rugged enough to use in an undergraduate lab, and yet capable of accuracies as high as 0.2%.

Diaphragm cells consist of two compartments separated either by a glass frit [Fig. 5.5-1(a)] or by a porous membrane [Fig. 5.5-1(b)] (Stokes, 1950; Mills et al., 1968; Choy et al., 1974). The two compartments are most commonly stirred at about 60 rpm with a magnet rotating around the cell. Initially the two compartments are filled with solutions of different concentrations. When the experiment is complete, the two compartments are emptied and the two solution concentrations are measured. The diffusion coefficient D is then calculated from the equation

$$D = \frac{1}{\beta t} \ln \left[\frac{(c_{1,\text{bottom}} - c_{1,\text{top}})_{\text{initial}}}{(c_{1,\text{bottom}} - c_{1,\text{top}})_{\text{at time } t}} \right] \qquad (5.5\text{-}1)$$

in which β (in cm^2) is a diaphragm-cell constant, t is the time, and c_1 is the solute concentration under the various conditions given. The detailed derivation of this equation is given in Example 2.2-4.

Four points about the diaphragm cell deserve emphasis. First, calculation of the diffusion coefficients requires accurate knowledge of the concentration differences, not the concentrations themselves. This means that very accurate chemical analyses may be required. For example, imagine we are measuring the diffusion of anthracene in hot decalin. Using gas chromatography, we measure the anthracene concentration as $51 \pm 1\%$ in the top solution and $61 \pm 1\%$ in the bottom solution. The concentration difference is then $10 \pm 2\%$, an error of 20%, even though our chemical analyses are accurate to 2%. As a result, we might do better to use a differential refractometer to try to determine directly the concentration difference.

The second point about the diaphragm cell is the calibration constant β. This quantity is

$$\beta = \frac{A}{l} \left(\frac{1}{V_{\text{top}}} + \frac{1}{V_{\text{bottom}}} \right) \qquad \qquad . \ (5.5\text{-}2)$$

in which A is the area available for diffusion, l is the effective thickness of the diaphragm, and V_{top} and V_{bottom} are the volumes of the two cell compartments. We should note that A is the total area open for diffusion and so is not a strong function of the pore size in the diaphragm. As a rule, small pores are

Table 5.5-1. *Characteristics of the best methods of measuring diffusion coefficients*

	Nature of diffusion	Apparatus expense	Apparatus construction	Concentration difference required	Method of obtaining data	Overall value
The four best methods						
Diaphragm cell	Pseudo-steady state	Small	Easy	Large	Concentration at known time; requires chemical analysis	Excellent; simple equipment outweighs occasionally erratic results
Infinite couple	Unsteady in an infinite slab	Small	Easy	Large	Concentration vs. position at known time; requires chemical analysis	Excellent, but restricted to solids
Taylor dispersion	Decay of a pulse	Moderate	Easy	Average	Refractive index vs. time at known position	Excellent, especially for dilute solutions
Capillary method	Unsteady out of finite cell	Small	Easy	Average	Concentration vs. time; usually requires radioactive counter	Very good, but commonly used only with radioactive tracers
Three very accurate methods						
Gouy interferometer	Unsteady in an infinite cell	Large	Moderate	Small	Refractive-index gradient vs. position and time is photographed	Very good; excellent data at great effort
Rayleigh or Mach-Zehnder interferometer	Unsteady in an infinite cell	Large	Difficult	Small	Refractive index vs. position and time is photographed	Very good; best for concentration-dependent diffusion

Method	Condition				Measurement	Assessment
Bryngdahl interferometer	Unsteady in an infinite cell	Large	Difficult	Very small	Refractive-index difference vs. time is photographed	Very hard to use, but suitable for very dilute solutions
Other interesting methods						
Laser Doppler light scattering	Homogeneous solution	Large	Difficult	None	Scattered light measurements with photomultiplier tube	Good; most suitable for large solutes; results sometimes hard to interpret
Spinning disc	Dissolution of solid or liquid	Small	Easy	Large	Concentration vs. time; requires chemical analysis	Good; requires diffusion-controlled dissolution, a stringent restraint
Wedge interferometer	Unsteady in an infinite cell	Moderate	Easy	Large	Refractive index vs. time is photographed	Fair; much harder to use than most authors suggest
Steady-state methods	Steady diffusion across known length	Moderate	Moderate	Large	Small concentration changes require exceptional analysis	Fair; easy analysis does not compensate for very difficult experiments
Loschmidt method	Unsteady in a finite cell	Moderate	Moderate	Large	Concentration at known time; requires chemical analysis	Fair; historically useful, but now seems obsolete
Harned conductance method	Unsteady in a finite cell	Moderate	Easy	Average	Electrical resistance vs. time	Poor; restricted to electrolytes, and hard to use

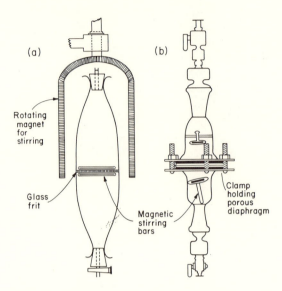

Fig. 5.5-1. Diaphragm cells. The cell on the left, which uses a porous glass frit as a diaphragm, is more accurate than that on the right, which uses filter paper as a diaphragm. However, the cell with the glass frit requires a much longer experiment.

preferred. Large pores may give a slightly larger area, but they often allow accidental mixing caused by flow through the diaphragm. Because A and l are, as a rule, not exactly known, β must be found by experiment. In liquids, this calibration is commonly made with KCl–water or urea–water (Stokes, 1950; Gosting & Akeley, 1952). Sucrose–water is less reliable because the solution often becomes contaminated by microorganisms. In gases, calibration depends on the method chosen to measure concentration.

The time required for diaphragm-cell measurements is determined by the value of β, and hence by the nature of the diaphragm. For accurate work the diaphragm should be a glass frit, and the experiments may take several days; for routine laboratory work the diaphragm can be a piece of filter paper, and the experiments may take as little as a few hours. For studies of membrane transport, a piece of membrane can be used in place of the filter paper. For studies in gases, the entire diaphragm can be replaced by a long, thin capillary tube, like the apparatus in Fig. 3.1-2.

The third point is that diffusion should always take place vertically. In other words, the diaphragm should lie in the horizontal plane. If the diaphragm is vertical, free convection can be generated, leading to spurious results (Toor, 1967). Interestingly, if the diaphragm is horizontal, then placing the more dense solution in the upper compartment may be done without fear of free convection. Many investigators routinely do this, feeling that they get superior results. At the same time, most investigators have done

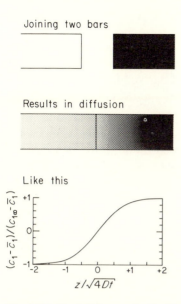

Fig. 5.5-2. The infinite couple. In this method, two solid bars of different compositions are joined together at zero time. The concentration profiles shown develop with time and are measured chemically.

away with the elaborate initial diffusion period suggested in early experiments. This period is significant only when the diaphragm volume is about one-sixth of the compartment volumes (Mills et al., 1968).

The final point about this method is its occasional unreliability. Every good experimentalist subjectively judges the quality of his experiments as he goes along. Most can correctly estimate an experiment's success even without detailed analysis. With the diaphragm cell, however, I have never been able to guess. Experiments I expect to be erratic often are, but experiments that I think are correct sometimes give answers that are in error by an order of magnitude. One of my students minimized such unpleasant surprises by carefully wrapping his cells in a particular brand of plastic bag purchased from a particular store in Cleveland, Ohio. For him, this worked. I have never found a similar trick.

Infinite couple

This experimental geometry, which is limited to solids, consists of two solid bars of differing compositions, as shown in Fig. 5.5-2 (Matano, 1933; Reed-Hill, 1973). To start an experiment, the two bars are joined together and quickly raised to the temperature at which the experiment is to be made. After a known time, the bars are quenched, and the composition is measured as a function of position. In the past, this analysis was made by grinding off small amounts of bar and determining the composition by a series of wet chemical tests; at present, the analysis is made more easily and quickly by an electron microprobe.

Because diffusion in solids is a slow process, the compositions at the ends of the solid bars away from the interface do not change with time. As a result, the concentration profile is that derived in Section 2.3:

$$\frac{c_1 - \bar{c}_1}{c_{1\infty} - \bar{c}_1} = \text{erf}\left(\frac{z}{\sqrt{4Dt}}\right) \tag{5.5-3}$$

in which $c_{1\infty}$ is the concentration at that end of the bar where $z = \infty$ and $\bar{c}_1$ $[= (c_{1\infty} + c_{1-\infty})/2]$ is the average concentration in the bars. The measured concentration profile is fit numerically using iterative nonlinear least squares to find the diffusion coefficient.

It must be remembered that diffusion in solids can be more complex than these paragraphs suggest. Some of this complexity stems from the different mechanisms by which diffusion in solids can occur. More subtle complexities arise from factors like residual stress in metal or the reference velocity in which diffusion is based. Such complexities dictate caution.

Taylor dispersion

The third of the four key methods of measuring diffusion is Taylor dispersion, illustrated schematically in Fig. 5.5-3 (Ouano, 1972; Maynard & Grushka, 1975; Pratt & Wakeham, 1977). This method, which is valuable for both gases and liquids, employs a long tube filled with solvent that slowly moves in laminar flow. A sharp pulse of solute is injected near one end of the tube. When this pulse comes out the other end, its shape is measured with a differential refractometer. Except for the refractometer, which can be purchased off the shelf, the apparatus is inexpensive and easy to build. This apparatus can be used routinely by those with very little training. It can be operated relatively easily at high temperature and pressure. It has the potential to give results accurate to better than 1%.

The concentration profile found in this apparatus is that for the decay of a pulse (see Section 4.2):

$$c_1 = \frac{M}{\pi R_0^2} \frac{e^{-z^2/4Et}}{\sqrt{4\pi Et}} \tag{5.5-4}$$

where M is the total solute injected, R_0 is the tube radius, v^0 is the average velocity of the flowing solvent, and E is a dispersion coefficient given by

$$E = \frac{(v^0 R_0)^2}{48D} \tag{5.5-5}$$

Because the refractive index varies linearly with the concentration, knowledge of the refractive-index profile can be used to find the concentration profile and the diffusion coefficient.

The fascinating aspect of this apparatus is the way in which the diffusion coefficient appears. Equation 5.5-4 has the same mathematical form as Eq. 2.4-14, but the dispersion coefficient E replaces the diffusion coefficient. So far, so good. However, E varies inversely with D. Consequently, a widely

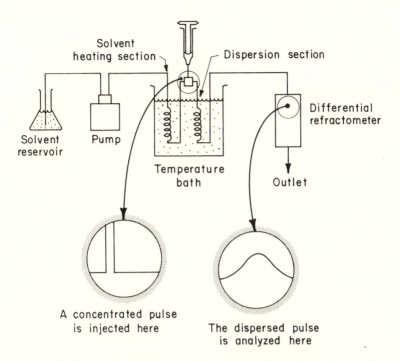

Fig. 5.5-3. The Taylor dispersion method. A sharp pulse is injected into a tube filled with flowing solvent. The dispersed pulse is measured at the tube's outlet. Interestingly, the pulse is dispersed more if the diffusion is slow.

spread pulse means a large E and a small D, as shown in Fig. 5.5-3. A very sharp pulse indicates small dispersion and hence fast diffusion.

Capillary method

This method is most suitable for measurements with radioactive tracers (Dunlop et al., 1972). It uses a small diffusion cell made of precision-bore capillary tubing perhaps 3 cm long and 0.05 cm in diameter. One end of this cell is sealed shut. After the cell is filled with a solution of known concentration, it is dropped into a large, stirred, thermostated solvent bath. At the end of the experiment, the cell is removed and the solute concentration within the cell is measured. The diffusion coefficient D can then be found from the equation

$$\frac{\bar{c}_1}{c_{10}} = \frac{8}{\pi^2} \sum_{n=1}^{\infty} \frac{1}{(2n-1)^2} e^{-\pi^2(2n-1)^2(Dt/4l^2)} \tag{5.5-6}$$

in which c_{10} and $\bar{c}_1$ are the average concentrations in the cell at times zero and t, respectively, and l is the length of the cell.

Three characteristics of this method deserve special mention. First, with careful technique it is accurate to better than 0.3%. The key caveat is "careful technique"; it is unusually easy to fool yourself with this equipment, obtaining reproducible inaccurate results. Second, the small size of the diffusion cell dictates careful chemical analysis of very small volumes of solution. In practice, this suggests using either radioactive tracers or some other microanalytical method. Finally, the power series in Eq. 5.5-6 converges rapidly. If you use reasonably long experiments, you can base your analysis on the first term in this series.

Very accurate methods

The four basic methods discussed earlier commonly give results accurate to a few percent, a suitable goal for most research. If higher accuracy is needed, we should turn to the three interferometers listed in Table 5.5-1. These instruments all depend on measuring an unsteady-state refractive profile in a transparent system, and so they are most useful for liquids. Their higher accuracy is purchased at a much greater cost in both equipment and effort.

The interferometers differ optically. The Gouy interferometer, shown schematically in Fig. 5.5-4(a), is the most highly developed and is accurate to better than 0.1% (Gosting, 1956). It is simple to build, easy to align, and very reliable. If one already has a comparator for measuring the interference fringes, this instrument is not particularly expensive. The Gouy method has been so highly developed that the extremely specialized jargon used in its operation may discourage newcomers. In fact, the experiments are simple to do; the hardest step is to understand the elaborate theory well enough to write the appropriate computer program. Average results with this instrument are at least equivalent to the best results obtained with any other device.

The Gouy interferometer measures the refractive-index gradient between two solutions that are diffusing into each other. The basic apparatus for measuring the gradient uses the lenses (L) to send parallel light rays from a light source (LS) through the diffusion cell (C). If this cell contains a refractive-index profile, then light passing through the center of the cell will be deflected to produce an interference pattern of black horizontal lines, as shown at the right in Fig. 5.5-4. The amount of this deflection at an intensity minimum between the lines, which is proportional to the refractive-index gradient, can be measured as a function of the "fringe number." This number, which is a convenient way of keeping track of where you are in the pattern, is defined by counting minima from the bottom of the pattern shown. The first minimum has fringe number 0; the second has fringe number 1, etc. This number can be related to a dimensionless position within the diffusion cell. Thus, the apparatus gives a refractive-index gradient versus cell position and time.

The Mach–Zehnder and Rayleigh interferometers are very solid alternatives to the Gouy interferometer (Longsworth, 1950; Creeth & Gosting, 1958). Although they are difficult to construct and adjust, they give information that is simpler to interpret. In the Mach–Zehnder apparatus, shown in

(a) Gouy interferometer

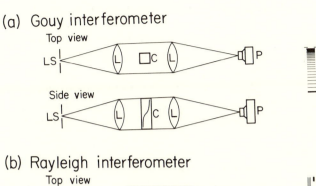

(b) Rayleigh interferometer

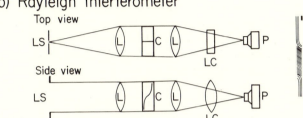

(c) Mach–Zehnder interferometer

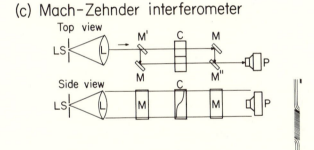

Fig. 5.5-4. Interferometers for accurate diffusion measurements. These three instruments are expensive to build and hard to operate, but they give very accurate results. Each produces interference fringes like those shown at the right of each schematic. LS, light source; L, collimating lens; C, diffusion cell; LC, cylindrical lens; M, mirror; M', M'', half-silvered mirrors.

Fig. 5.5-4(c), collimated light is split by the first half-silvered mirror M'. Half the light passes through each of the twin cells C and is recombined by the second half-silvered mirror M''. In the Rayleigh apparatus, these mirrors are replaced by a cylindrical lens, shown in Fig. 5.5-4(b).

Both instruments measure refractive index versus cell position. If both cells contain homogeneous solutions, the interference fringes are sets of parallel vertical lines; if one cell contains a refractive-index gradient caused by diffusion, the interference fringes look like those shown at the right of the figure. For both interferometers, these fringes can be used to calculate the diffusion coefficient.

The Bryngdahl interferometer uses the same cell as in other interferometric instruments, but it depends on more complicated optics (Bryngdahl, 1957). These optics include birefringent crystals, which shear the wave front to produce fringes at specific refractive-index differences. In some cases, these differences approximate the second derivative of the refractive-index profile. The instrument requires very small refractive-index differences; so it is the best instrument for studying diffusion of sparingly soluble solutes. However, the complexity of the optics restricts the utility of the instrument.

Other methods

The remaining methods for measuring diffusion are listed in Table 5.5-1 roughly in order of their value. The most powerful is the quasi-elastic light-scattering method or laser Doppler method (Berne & Pecora, 1976; Bloomfield, 1981). This method measures the change in light scattered by a single solute particle. Because large solutes scatter much more light than small ones, the technique is especially suitable for liquid solutions of proteins, viruses, or micelles. In some cases, the light scattered by these particles can be so sensitively analyzed that both translational and rotational diffusion can be calculated.

The chief disadvantage to quasi-elastic light scattering is the cost of the equipment. This cost is so high that the equipment tends to become the fiefdom of scientific entrepreneurs who are ever anxious to find new problems for this method. Such a situation produced some collaborations in which all participants did not know what they were doing. I currently view many quasi-elastic results skeptically. However, the method has potential.

The spinning-disc method is simple, the antithesis of light-scattering techniques (Levich, 1962; Chan et al., 1976). The apparatus is just a solid or liquid disc of solute slowly rotating in a solvent volume (see Fig. 3.4-3). The solute concentrations in the solvent are analyzed versus time. If the disc's dissolution is diffusion-controlled, these concentrations allow calculation of the diffusion coefficient from Example 3.4-3. If the disc's dissolution is not diffusion-controlled, we must choose another method.

The wedge interferometer is cheap and cute, a simple alternative to the expensive interferometers described earlier (Duda et al., 1969). It consists of two microscope slides separated at one edge with a coverslip. To start an experiment, one places drops of two different solutions next to each other on one slide. One then places the other slide and coverslip so that the drops are in contact in a wedge-shaped channel. When this wedge is put in a microscope, interference fringes indicate the concentration profile. Measuring the change of fringe position versus time allows calculation of the diffusion coefficient simply, cheaply, and approximately. Moreover, because only drops of solution are needed, one needs only very small amounts of solute.

The last three methods are less valuable, and so will be mentioned only briefly. The steady-state methods are like the diaphragm cell, but they replace the two well-stirred compartments with two flowing solutions (Rao &

Bennett, 1971). Such a replacement gives a situation in true steady state. However, the two solutions must flow at exactly the same rate; so expensive pumps and valves are needed. The experiments can become horrendously difficult and can consume huge amounts of solution. Nonetheless, the analysis is simple. The Loschmidt apparatus consists of two tubes, containing different solutions, that are connected at time zero and disconnected later (Dunlop et al., 1972). Because the apparatus is inexpensive, it is often used in student labs; because it requires careful technique, it almost always gives bad results in these labs. The Harned conductance method is similar to the Loschmidt method, but one does not disconnect the cell halves (Harned & Nuttall, 1949). Instead, the cell concentrations are electrically measured at specific positions. The method sounds simple conceptually, but it is terrible to operate; so its mention constitutes a historical artifact.

Section 5.6. A final perspective

This chapter shows that diffusion coefficients in gases and in liquids can often be accurately estimated, but that coefficients in solids and in polymers cannot. In gases, estimates are based on the Chapman–Enskog kinetic theory and are accurate to around 8%. In liquids, estimates are based on the Stokes–Einstein equation or its empirical parallels. These estimates, accurate to around 20%, can be supplemented by a good supply of experimental data. In solids and polymers, theories allow coefficients to be correlated but rarely predicted.

These common generalizations help to solve only the routine problems with which we are faced. Many problems remain. For example, we may want to know the rate at which hydrochloric acid diffuses into oil-bearing sandstone. We may need to estimate the drying speed of lacquer. We may seek the rate of flavor release from lemon pie filling. All these examples depend on diffusion; none can be accurately estimated with the common generalizations.

In some cases, diffusion coefficients can be adequately estimated by more carefully considering the chemistry. Specific cases, discussed in the next two chapters, includes electrolytes and wool dyeing. However, in most nonroutine problems the detailed chemistry is not known, and experiments are essential. The primer on experiments given in this chapter should be your initiation.

References

American Society for Metals (1973). *Diffusion*. Metals Park: ASM.
Barrer, R. M. (1941). *Diffusion in and through Solids*. New York: Macmillan.
Batchelor, G. K. (1972). *Journal of Fluid Mechanics*, **52**, 245; (1976), **71**, 1.
Berne, B. J., & Pecora, R. (1976). *Dynamic Light Scattering*. New York: Wiley.
Bloomfield, V. A. (1981). *Annual Review of Biophysics and Bioengineering*, **10**, 421.
Bryngdahl, O. (1957). *Acta Chemica Scandinavica*, **11**, 1017.

Chan, A. F., Evans, D. F., & Cussler, E. L. (1976). *American Institute of Chemical Engineers Journal,* **22,** 1006.

Chapman, S., & Cowling, T. G. (1970). *The Mathematical Theory of Non-Uniform Gases,* 3rd ed. Cambridge University Press.

Chen, S. H., Davis, H. T., & Evans, D. F. (1981). *Journal of Physical Chemistry,* **75,** 1422.

Choy, E. M., Evans, D. F., & Cussler, E. L. (1974). *Journal of the American Chemical Society,* **96,** 7085.

Clark, A. M., & Long, J. V. P. (1971). In: *Diffusion Processes, Vol. II,* ed. J. N. Sherwood et al. London: Gordon and Breach.

Cowie, J. M. G., & Cussler, E. L. (1967). *Journal of Chemical Physics,* **46,** 4886.

Creeth, J. M., & Gosting, L. J. (1958). *Journal of Physical Chemistry,* **62,** 58.

Cunningham, R. E., & Williams, R. J. J. (1980). *Diffusion in Gases and Porous Media.* New York: Plenum.

Cussler, E. L. (1976). *Multicomponent Diffusion.* Amsterdam: Elsevier.

Darken, L. S. (1948). *Transactions of the American Institute of Mining, Metallurgical and Petroleum Engineers,* **175,** 184.

deGennes, P.-G. (1979). *Scaling Concepts in Polymer Physics.* Ithaca: Cornell University Press.

Duda, J. L., Segelko, W. L., & Vrentas, J. S. (1969). *Journal of Physical Chemistry,* **73,** 141.

Duda, J. L., & Vrentas, J. S. (1982). *American Institute of Chemical Engineers Journal,* **28,** 279.

Dunlop, P. J., Steele, B. J., & Lane, J. E. (1972). In: *Physical Methods of Chemistry,* ed. A. Weissberger & B. W. Rossiter. New York: Wiley.

Einstein, A. (1905). *Annalen der Physik,* **17,** 549.

Farrington, G. C., & Briant, J. L. (1979). *Science,* **204,** 1371.

Ferry, J. D. (1980). *Viscoelastic Properties of Polymers.* New York: Wiley.

Franklin, W. M. (1975). In: *Diffusion in Solids: Recent Developments,* ed. A. S. Nowick & J. J. Burton. New York: Academic Press.

Frisch, H. L. (1980). *Polymer Engineering and Science,* **20,** 2.

Fuller, E. N., Schettler, P. D., & Giddings, J. C. (1966). *Industrial and Engineering Chemistry,* **58,** 19.

Glasstone, S., Laidler, K. J., & Eyring, H. (1941). *Theory of Rate Processes.* New York: McGraw-Hill.

Gosting, L. J. (1956). *Advances in Protein Chemistry,* **11,** 429.

Gosting, L. J., & Akeley, D. F. (1952). *Journal of the American Chemical Society,* **74,** 2058.

Harned, H. S., & Nuttall, R. L. (1949). *Journal of the American Chemical Society,* **71,** 1460.

Hartley, G. S., & Crank, J. (1949). *Transactions of the Faraday Society,* **45,** 801.

Hirschfelder, J., Curtiss, C. F., & Bird, R. B. (1954). *Molecular Theory of Gases and Liquids.* New York: Wiley.

Hiss, T. G., & Cussler, E. L. (1973). *American Institute of Chemical Engineers Journal,* **19,** 698.

Jeans, J. (1921). *Dynamic Theory of Gases,* 3rd ed. Cambridge University Press.

King, C. J., Hsueh, L., & Mao, K.-W. (1965). *Journal of Chemical and Engineering Data,* **10,** 348.

Kirkwood, J. G. (1967). *Macromolecules.* London: Gordon and Breach.

Kosanovich, G. M., & Cullinan, H. T. (1976). *Industrial and Engineering Chemistry Fundamentals,* **15,** 41.

Levich, V. (1962). *Physicochemical Hydrodynamics.* Englewood Cliffs, N.J.: Prentice-Hall.

Longsworth, L. G. (1950). *Review of Scientific Instruments,* **21,** 524.

Maynard, V. R., & Grushka, E. (1975). In: *Advances in Chromatography, Vol. 12,* ed. J. C. Giddings et al. New York: Dekker.

Matano, C. (1933). *Japanese Journal of Physics,* **8,** 109.

Mills, R., Woolf, L. A., & Watts, R. O. (1968). *American Institute of Chemical Engineers Journal,* **14,** 671.

Ouano, A. C. (1972). *Industrial and Engineering Chemistry Fundamentals,* **11,** 268.

Perrin, F. (1936). *Journal de Physique et le Radium,* **7,** 1.

Pratt, K. C., & Wakeham, W. A. (1977). *Journal of the Chemical Society, Faraday Transactions,* **73,** 977.

Rao, S. S., & Bennett, C. O. (1971). *American Institute of Chemical Engineers Journal,* **17,** 75.

Reed, C. C., & Anderson, J. L. (1980). *American Institute of Chemical Engineers Journal,* **26,** 816.

Reed-Hill, R. E. (1973). *Physical Metallurgy Principles,* 2nd ed. New York: Van Nostrand Reinhold.

Reid, R. C., Sherwood, T. K., & Prausnitz, J. M. (1977). *Properties of Gases and Liquids,* 3rd ed. New York: McGraw-Hill.

Russel, W. B. (1981). *Annual Review of Fluid Mechanics,* **13,** 425.

Scheibel, E. G. (1954). *Industrial and Engineering Chemistry,* **46,** 2007.

Sherwood, T. K., Pigford, R. L., & Wilke, C. R. (1975). *Mass Transfer.* New York: McGraw-Hill.

Shewmon, P. G. (1963). *Diffusion in Solids.* New York: McGraw-Hill.

Smithells, C. J. (1976). *Metals Reference Book, Vol. II.* London: Butterworth.

Stark, J. P. (1976). *Solid State Diffusion.* New York: Wiley.

Stokes, G. G. (1850). *Transactions of the Cambridge Philosophical Society,* **9,** 8.

Stokes, R. H. (1950). *Journal of the American Chemical Society,* **72,** 763, 2243; (1951) **73,** 3528.

Sutherland, W. (1905). *Philosophical Magazine,* **9,** 781.

Toor, H. L. (1967). *Industrial and Engineering Chemistry Fundamentals,* **6,** 454.

Tyrrell, H. J. V. (1961). *Diffusion and Heat Flow in Liquids.* London: Butterworth.

Vignes, A. (1966). *Industrial and Engineering Chemistry Fundamentals,* **5,** 189.

Vrentas, J. S., & Duda, J. L. (1977). *Journal of Polymer Science, Part A-2. Polymer Physics,* **15,** 403, 417, 441.

Vrentas, J. S., & Duda, J. L. (1980). *Journal of Applied Polymer Science,* **25,** 1297.

Wilke, C. R., & Chang, P. C. (1955). *American Institute of Chemical Engineers Journal,* **1,** 264.

6   SOLUTE–SOLUTE INTERACTIONS

In this chapter we turn to systems in which there are significant interactions between diffusing solute molecules. These interactions can strongly affect the apparent diffusion coefficients. In some cases these effects produce unusual averages of the diffusion coefficients of different solutes; in others, they suggest a strong dependence of diffusion on concentration; in still others, they result in diffusion that is thousands of times slower than expected.

Most of the interactions discussed in this chapter occur because of fast chemical reactions between solute molecules. In many cases, like the dyeing of wool, these reactions are explicit; the dye reacts with sites on the wool. In other cases, the reaction involved is implicit and is not immediately recognized. One easy example is the diffusion of acetic acid, where protons and acetate ions constantly react to form acetic acid molecules. Other less obvious examples include the diffusion of detergents and of ion pairs.

Systems with these fast chemical reactions are common. Such reactions can produce new solutes. For example, diffusion of sodium chloride almost always involves diffusion of sodium and chloride ions, not a single sodium chloride species. As a result, I have sometimes described these systems as solutions of "hidden solutes." It is these hidden solutes that are responsible for the array of effects observed.

In this chapter, Section 6.1 deals with diffusion of strong electrolytes, Section 6.2 discusses diffusion of associating solutes, and Section 6.3 covers diffusion in dyeing. In each section, the emphasis is on how the chemistry affects the diffusion coefficients.

6.1. Diffusion of strong electrolytes

Every high school chemistry student knows that when sodium chloride is dissolved in water, it is ionized. Sodium chloride in water does not diffuse as a single molecule; instead, the sodium ions and chloride ions move freely through the solution. The movement of the ions means that a 0.1-M sodium chloride solution passes an electric current 1 million times more easily than water does. The large ion size relative to electrons means that such a solution passes current ten thousand times less easily than a metal does.

The diffusion of sodium chloride can be accurately described by a single

Table 6.1-1. *Ionic diffusion coefficients in water at 25°C*

Cation$_i$	D_i	Anion$_i$	D_i
H^+	9.31	OH^-	5.28
Li^+	1.03	F^-	1.47
Na^+	1.33	Cl^-	2.03
K^+	1.96	Br^-	2.08
Rb^+	2.07	I^-	2.05
Cs^+	2.06	NO_3^-	1.90
Ag^+	1.65	CH_3COO^-	1.09
NH_4^+	1.96	$CH_3CH_2COO^-$	0.95
$N(C_4H_9)_4^+$	0.52	$B(C_6H_5)_4^-$	0.53
Ca^{2+}	0.79	SO_4^{2-}	1.06
Mg^{2+}	0.71	CO_3^{2-}	0.92
La^{3+}	0.62	$Fe(CN)_6^{3-}$	0.98

[a] Values at infinite dilution in 10^{-5} cm²/sec. Calculated from data of Robinson and Stokes (1960).

diffusion coefficient. Somehow this does not seem surprising, because we always refer to sodium chloride as if it were a single solute, and ignore the knowledge that it ionizes. We get away with this selective ignorance because the sodium and chloride ions diffuse at the same rate. If they did not do so, we could easily separate anions from cations.

Values of ionic diffusion coefficients are given in Table 6.1-1 (Robinson & Stokes, 1959). These data, which are hidden in the literature of electrochemistry, are obtained by a variety of experimental methods, including tracer diffusion determinations. The table shows that different ions have different diffusion coefficients. The proton and the hydroxyl ion are unusually fast; big, fat organic ions like tetrabutylammonium and tetraphenylboride are slow. Somewhat surprisingly, a potassium ion diffuses faster than a lithium ion does. This suggests that in aqueous solution, a potassium ion is smaller than a lithium ion. These sizes are unexpected from crystallographic measurements on the solid state that show the potassium ion is larger. The sizes in solution occur because the potassium ion is less strongly hydrated than the lithium (Franks, 1975).

Another interesting result in Table 6.1-1 is that sodium ion diffuses more slowly than chloride ion. In other words, the sodium ion does not have the same diffusion coefficient as the chloride ion. However, because sodium chloride diffuses with only one coefficient, the ionic diffusion coefficients must somehow be combined to give an average value. We shall now calculate this average, first for a simple 1-1 electrolyte like sodium chloride, and then for more complicated electrolytes. With these results as a basis, we shall then briefly discuss an old but now underused measurement: electrical conductance.

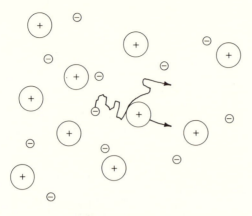

Fig. 6.1-1. Electrolyte diffusion. The two ions have the same charge and are present at the same local concentration. The larger cations (the positive ions) inherently move more slowly than the smaller anions (the negative ions). However, because of electroneutrality, both ions have the same net motion, and hence the same flux.

Basic arguments

Imagine a large, fat grandfather taking a small, rambunctious girl for a walk. The rate at which the two travel will be largely determined by the grandfather. He will move slowly, even ponderously, toward their goal. The girl may run back and forth, taking many more steps and so covering more distance, but her progress will be dominated by her elder.

In the same way, the diffusion of a large, fat cation and a small, quick anion will be dominated by the slower ion. The diffusion will proceed as does the walk, and the smaller ion may move around more. However, the two ions are tied together electrostatically, and so their overall progress will be the same and will tend to be dominated by the slower ion (Fig. 6.1-1).

To examine this analogy more exactly, we must first write a flux equation for ion diffusion. In this effort, we consider only dilute solutions, like those in Chapter 2, and so ignore problems like the complicated reference velocities of Chapter 3. The obvious choice of a flux equation is the simplest form of Fick's law, which for a sodium ion will be

$$-\mathbf{j}_{Na} = D_{Na}\nabla c_{Na} \tag{6.1-1}$$

However, we quickly realize that this choice is inadequate, for it suggests that an electric field will not affect diffusion.

To include this electrical effect, we return to the argument used to derive the Stokes–Einstein equation in Section 5.2: that the ion velocity is proportional to the sum of all the forces acting on the ion. In symbolic terms, this is (Robinson & Stokes, 1959)

$$\begin{pmatrix} \text{ion} \\ \text{velocity} \end{pmatrix} = \begin{pmatrix} \text{ion} \\ \text{mobility} \end{pmatrix} \begin{pmatrix} \text{chemical} \\ \text{forces} \end{pmatrix} + \begin{pmatrix} \text{electrical} \\ \text{forces} \end{pmatrix} \tag{6.1-2}$$

$$\mathbf{v}_i = -u_i(\nabla\mu_i + z_i \mathscr{F}\nabla\psi)$$

where u_i is the ion mobility, z_i is the ionic charge (equal to $+1$ for Na), $\mathfrak{F}$ is Faraday's constant, and ψ is the electrostatic potential.

Each of these quantities deserves some discussion. First, the mobility u_i is a physical property of the ion, a phenomenological coefficient that must be measured by experiment. This mobility is often taken to be $1/6\pi\eta R_0$, which, we recall, is a feature of the Stokes–Einstein equation. In fact, the use of this value simply restates our ignorance of mobility in terms of an effective ion radius, R_0.

Because the mobility is almost equivalent to the diffusion coefficient, it is something of a cultural artifact. It is included here because many papers dealing with electrolyte transport report their results in terms of mobilities, not in terms of diffusion coefficients. Faraday's constant is even more of a cultural artifact; it is a unit conversion factor explicitly included whenever this equation is written. The apparent suspicion is that no one can properly use electrostatic units without a warning.

The charge and potential in Eq. 6.1-2 make explicit the electrical effects connecting the ions. Including the charge seems sensible; note that if the ion has a negative charge, the direction of the electrical effect is reversed. The potential also looks sensible. It has two distinct parts. One part includes the effect of any potential applied to the system, for example, by electrodes attached to a battery. A second part is the potential generated by the different diffusion rates of different ions. For example, for sodium chloride, the potential includes the electrostatic interaction of the quicker chloride ions and the more sluggish sodium ions. It is thus the route by which we average ion diffusion coefficients.

To rewrite Eq. 6.1-2 as a flux relation, we take advantage of the fact that we are working in dilute solution, and so assume that the solution is ideal:

$$\nabla\mu_i = \frac{RT}{c_i}\nabla c_i \tag{6.1-3}$$

When this result is combined with Eq. 6.1-2, we get

$$-\mathbf{v}_i = \frac{[u_iRT]}{c_i}\left(\nabla c_i + c_iz_i\frac{\mathfrak{F}\nabla\psi}{RT}\right) \tag{6.1-4}$$

which is equivalent to the flux equation

$$-\mathbf{j}_i = -c_i\mathbf{v}_i$$

$$= [RTu_i]\left(\nabla c_i + c_iz_i\frac{\mathfrak{F}\nabla\psi}{RT}\right) \tag{6.1-5}$$

$$= [D_i]\left(\nabla c_i + c_iz_i\frac{\mathfrak{F}\nabla\psi}{RT}\right)$$

These relations, sometimes called the Nernst–Planck equations (Bard & Faulkner, 1980), could be written down directly as a definition for D_i. If this were done, then the restriction to dilute solutions in Eq. 6.1-3 and the implicit neglect of a reference velocity in the first line of Eq. 6.1-5 would be hidden in the final flux equation, lumped into the experimental coefficient D_i.

I find the derivation a sensible, reassuring rationalization, even though I know that it is arbitrary.

1-1 Electrolytes

We now want to describe the ion fluxes of a single strong 1-1 electrolyte. Such an electrolyte ionizes completely, producing equal numbers of cations and anions. Although the concentrations of anions and cations may vary through the solution, the concentrations and the concentration gradients of these species are equal everywhere because of electroneutrality:

$$c_1 = c_2$$
$$\nabla c_1 = \nabla c_2$$

(6.1-6)

where 1 and 2 refer, respectively, to cation and anion. Like the ion concentrations, the ion fluxes are also related:

$$\mathbf{j}_1 - \mathbf{j}_2 = |z|\mathbf{i}$$

(6.1-7)

where $|z|$ is the magnitude of the ionic charge and $\mathbf{i}$ is the current density in appropriate units. This current density is defined as positive when it goes from positive to negative.

To find the electrolyte flux, we first return to the basic flux equation for each ion:

$$-\mathbf{j}_1 = D_1(\nabla c_1 + |z|c_1 \mathfrak{F}\nabla\psi/RT)$$

(6.1-8)

$$-\mathbf{j}_2 = D_2(\nabla c_2 - |z|c_2 \mathfrak{F}\nabla\psi/RT)$$

(6.1-9)

These equations can be combined with Eq. 6.1-7 to find the current:

$$|z|\mathbf{i} = D_2\nabla c_2 - D_1\nabla c_1 - (D_1 c_1 + D_2 c_2)|z|\mathfrak{F}\nabla\psi/RT$$

(6.1-10)

But this equation now allows $\nabla\psi$ to be removed from the flux equations:

$$-\mathbf{j}_1 = \frac{2D_1 D_2}{D_1 + D_2} \nabla c_1 - \frac{D_1}{D_1 + D_2} (|z|\mathbf{i})$$

(6.1-11)

where we have used the fact that $c_1 = c_2$ to simplify the final expression. A similar equation for the anion flux $\mathbf{j}_2$ can be derived.

Two important limits of the flux $\mathbf{j}_1$ exist. First, when there is no current,

$$\mathbf{j}_1 = \mathbf{j}_2 = -D\nabla c_1 = -\left[\frac{2}{1/D_1 + 1/D_2}\right]\nabla c_1$$

(6.1-12)

The quantity in brackets is the average diffusion coefficient of the electrolyte. Because it is a harmonic average of the diffusion coefficients of the individual ions, it is dominated by the slower ion. However, there is only one diffusion coefficient for the two diffusing ions because the ions are electrostatically coupled.

The second interesting limit of Eq. 6.1-11 occurs when the solution is well mixed, so that no gradients of anion and cation exist. In this case,

$$\mathbf{j}_1 = [t_1](|z|\mathbf{i}) = \left[\frac{D_1}{D_1 + D_2}\right] (|z|\mathbf{i})$$

(6.1-13)

$$\mathbf{j}_2 = [t_2](-|z|\mathbf{i}) = \left[\frac{D_2}{D_1 + D_2}\right] (-|z|\mathbf{i})$$

(6.1-14)

where the t_i, equal to the quantities in brackets, are the transference numbers, i.e., the fractions of current transported by specific ions. Unlike the diffusion coefficient, these transference numbers are arithmetic averages of the ion diffusion coefficients. As a result, the transference numbers and the current in solution are both dominated by the faster ion.

Example 6.1-1: Diffusion of hydrogen chloride

What is the diffusion coefficient at 25°C for a very dilute solution of HCl in water? What is the transference number for the proton under these conditions?

Solution. From the data in Table 6.1-1, the ionic diffusion coefficients are $9.31 \cdot 10^{-5}$ cm²/sec for H^+ and $2.03 \cdot 10^{-5}$ cm²/sec for Cl^-. The electrolyte diffusion coefficient is given by Eq. 6.1-12:

$$D_{HCl} = \left[\frac{2}{1/D_{H^+} + 1/D_{Cl^-}} \right] = 3.3 \cdot 10^{-5} \text{ cm}^2/\text{sec}$$

The slow ion dominates. The result is only 1.5 times greater than the chloride's diffusion coefficient, but it is 3.5 times less than the proton's diffusion coefficient.

The transference number, t_{H^+}, can be found in a straightforward manner from Eq. 6.1-13:

$$t_{H^+} = \frac{D_{H^+}}{D_{H^+} + D_{Cl^-}}$$

The faster protons carry 82% of the current. The cause of this fast proton diffusion is discussed in Section 7.2.

Non-1-1 electrolytes

We now turn from the simple 1-1 electrolytes to more complicated electrolytes. Mathematical description of non-1-1 electrolytes is parallel to that developed earlier, but more complex algebraically. The basic flux equation is the same as Eq. 6.1-5:

$$-\mathbf{j}_i = D_i(\nabla c_i + c_i z_i \mathcal{F} \nabla \psi / RT) \tag{6.1-15}$$

The constraints on concentration and flux at zero current are

$$z_1 c_1 + z_2 c_2 = 0 \tag{6.1-16}$$

and

$$z_1 \mathbf{j}_1 + z_2 \mathbf{j}_2 = 0 \tag{6.1-17}$$

When the electrostatic potential is eliminated, the diffusion equation for ion 1 becomes (Harned & Owen, 1950)

$$-\mathbf{j}_1 = D\nabla c_1 = \left[\frac{D_1 D_2 (z_1^2 c_1 + z_2^2 c_2)}{D_1 z_1^2 c_1 + D_2 z_2^2 c_2} \right] \nabla c_1 \tag{6.1-18}$$

where the quantity in brackets is D, the diffusion coefficient of the electrolyte.

This equation can be somewhat misleading because of the unequal charge. For example, imagine that we are interested in the diffusion of very dilute solutions of sodium sulfate. If the sodium is ion 1, then its flux will be twice the total flux of sulfate. This is twice the flux of what we carelessly might regard as molecular sodium sulfate, even though such a molecule does not exist. Accordingly, when only one electrolyte is present, we may wish to rewrite this equation in terms of the total electrolyte flux j_T and the total electrolyte concentration c_T, defined as

$$j_T = j_1/|z_2| = j_2/|z_1| \qquad (6.1\text{-}19)$$
$$c_T = c_1/|z_2| = c_2/|z_1| \qquad (6.1\text{-}20)$$

Note that c_T is expressed in equivalents. The diffusion equation for a single non-1-1 electrolyte now becomes

$$-j_T = D\nabla c_T = \left[\frac{|z_1| + |z_2|}{|z_2|/D_1 + |z_1|/D_2}\right] \nabla c_T \qquad (6.1\text{-}21)$$

where the quantity in brackets is again the diffusion coefficient of the non-1-1 electrolyte.

This diffusion forms a curious contrast with the special case of a 1-1 electrolyte described by Eq. 6.1-12. Both equations involve a type of harmonic average of the ionic diffusion coefficients. Thus, we might expect that both cases are more strongly influenced by the slower ion. However, if this slower ion has a much larger charge than the faster ion, the faster ion may come to dominate the diffusion, because the harmonic average is weighted by the ion charge. The effect of this weighting can be more clearly shown by examples.

Example 6.1-2: Diffusion of lanthanum chloride
What is the diffusion coefficient of 0.001-M lanthanum chloride?

Solution. From Table 6.1-1, the diffusion coefficients of La^{3+} and Cl^- are $0.62 \cdot 10^{-5}$ cm^2/sec and $2.03 \cdot 10^{-5}$ cm^2/sec, respectively. In water, the average coefficient can be found either from Eq. 6.1-18 or from Eq. 6.1-21. From Eq. 6.1-21, taking La^{3+} as ion 1 and chloride as ion 2, we get

$$D = \frac{|z_1| + |z_2|}{|z_1|/D_2 + |z_2|/D_1}$$
$$= \left[\frac{|3| + |-1|}{|3|/2.03 \cdot 10^{-5} + |-1|/0.63 \cdot 10^{-5}}\right] \text{cm}^2/\text{sec}$$
$$= 1.29 \cdot 10^{-5} \text{ cm}^2/\text{sec}$$

From Eq. 6.1-18, because $c_1 = 0.001$-M and $c_2 = 0.003$-M, we can find the same result.

Example 6.1-3: Diffusion of lanthanum chloride in excess sodium chloride
How will the results of the previous example be changed if the lanthanum chloride diffuses through 1-M NaCl?

Solution. Answering this question requires the assumption that there are no ternary diffusion effects in this system. These effects may arise because the diffusion of sodium ion couples with the diffusion of chloride ion, which in turn affects the diffusion of La^{3+}. However, these effects vanish for any solute present in high dilution, as $LaCl_3$ is in this case (see Section 8.4).

Because of the added sodium chloride, we cannot use Eq. 6.1-21, which is valid only for a single non-1-1 electrolyte. We can use Eq. 6.1-18. If we again label lanthanum as ion 1 and chloride as ion 2, we recognize that c_1 equals 0.001-M, but c_2 is about 1-M. These unequal concentrations mean that Eq. 6.1-18 becomes

$$-\mathbf{j}_1 = \frac{D_1 D_2 (z_2^2 c_2)}{D_1 z_1^2 c_1 + D_2 z_2^2 c_2} \nabla c_1$$
$$= D_1 \nabla c_1$$

In other words, the diffusion of the lanthanum is $0.62 \cdot 10^{-5}$ cm²/sec, which is the same as the solitary ion. Thus, the diffusion of dilute $LaCl_3$ in concentrated $NaCl$ is dominated by the diffusion of the uncommon ion, La^{3+}.

Example 6.1-4: Protein diffusion

Proteins, those naturally occurring polymers of amino acids, have large numbers of side chains that end in amino (–NH₂) or carboxylic acid (–COOH) groups. At low pH, the amino groups become positively charged (–NH₃⁺); at high pH, the carboxylic acids become negatively charged (–COO⁻). Proteins will commonly have both positive and negative charges, the relative numbers of which vary with pH. At some intermediate pH, there will be equal numbers of positive and negative charges (i.e., there will be no net charge). This intermediate pH is called the isoelectric point. How would you expect protein diffusion to vary with pH and with added $NaCl$?

Solution. The answer to this important question is not completely known, in spite of research efforts over 50 years. We shall give only a qualitative answer, arguing largely by analogy with the previous example.

First, we must recognize that proteins themselves are big molecules, often with molecular weights in excess of 100,000. They have a large number of side chains, and hence a large potential charge. At the same time, the proteins' large molecular weight means that their molar concentration is very small. Moreover, because small electrolytes like sodium chloride are inevitably present as significant impurities, they are commonly and deliberately added in excess to protein solutions.

As a result, protein diffusion can vary widely. At high electrolyte concentration and at the isoelectric point, the protein has no net charge, and the electrostatic effects are minimal. The diffusion coefficient then has a value largely independent of changes in the solution. Under almost any other conditions, the diffusion coefficients are strongly concentration-dependent, because of increased protein charge or increased electrostatic coupling. This concentration dependence is exemplified by the results in Fig. 6.1-2.

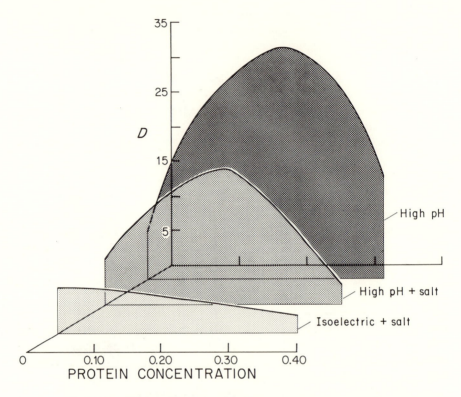

Fig. 6.1-2. Diffusion of bovine serum albumin. Proteins are naturally occurring polymers of amino acids. The amino acid side chains have a charge dependent on the ionic strength and the pH. This variable charge can produce a diffusion coefficient that is highly concentration-dependent, as shown for bovine serum albumin. [Data from Colton et al. (1981), with permission.]

Many scientists find these results frustrating. They are interested not in protein diffusion for itself but in what a protein diffusion coefficient reveals about the protein molecule. They are disturbed that the effects on diffusion of the molecular size and shape are obscured by electrostatics. They must be satisfied with having to guess the protein's properties from its diffusion in a concentrated salt solution of carefully controlled pH, instead of in pure water.

Diffusion versus conductance
     Although diffusion is a very common process, diffusion coefficients can be difficult to measure. This is true for most of the systems discussed in this book, including solutions of electrolytes. However, for electrolyte solutions, the electrical resistance and its reciprocal, the electrical conductivity, are very easy to measure with a Wheatstone bridge. Indeed, nothing in my experimental experience is as satisfying as a conductance experiment; I get

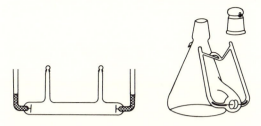

Fig. 6.1-3. Conductance cells. These cells are used to measure with extremely high accuracy the resistance of an electrolyte solution. This information is related to the diffusion coefficient of the electrolyte. As a result, a conductance experiment sometimes is a superior method of studying diffusion.

fantastically accurate results with embarrassingly little effort. Because diffusion and conductance give similar information about the system, it is well worth comparing the two processes in some detail.

The conductance of a single electrolyte in solution is most easily measured in cells like those shown in Fig. 6.1-3 (Evans & Matesich, 1973). The electrical resistance of the stirred solution is measured with a Wheatstone bridge and a rapidly oscillating AC field of fixed maximum voltage, so that the solution remains homogeneous throughout the experiment. The resistance is inversely proportional to the current through the cell, but the current, in turn, is proportional to the ion fluxes:

$$\text{(resistance)}^{-1} = K_{\text{cell}}\mathbf{i} = K_{\text{cell}}(z_1\mathbf{j}_1 + z_2\mathbf{j}_2) \tag{6.1-22}$$

The proportionality constant K_{cell} in this relation is a function of the electrode area, the electrode separation, and the cell shape. It is found by calibration of the cell, most commonly with a potassium chloride solution.

The ion fluxes in the cell are described by equations analogous to those used for ion diffusion. First, we assume that the ion flux is proportional to the ion concentration:

$$\mathbf{j}_i = c_i\mathbf{v}_i \tag{6.1-23}$$

We also assume that the ion velocity is proportional to the electrical force acting on the ion:

$$\mathbf{v}_i = -u_i z_i \mathfrak{F}\nabla\psi \tag{6.1-24}$$

where, as in Eq. 6.1-2, u_i is the ion mobility and ψ is the electrostatic potential acting on the ions. Because in this case the solution is homogeneous, the concentration gradient is zero. The only flux comes from the electrostatic potential applied by the electrodes.

We now can combine Eqs. 6.1-22 through 6.1-24 to find an expression for the resistance in terms of the ion mobilities:

$$\text{(resistance)}^{-1} = K_{\text{cell}}(z_1^2 c_1 u_1 + z_2^2 c_2 u_2)\mathfrak{F}\nabla\psi \tag{6.1-25}$$

The ion concentrations are related to the total concentration c_T by

$$c_T = c_1/|z_2| = c_2/|z_1| \tag{6.1-26}$$

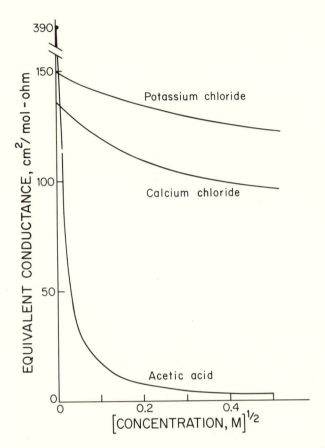

Fig. 6.1-4. Equivalent conductance versus concentration. Conductance varies with concentration, especially at high dilution. For strong electrolytes like KCl and CaCl₂, these variations are chemically interesting but practically unimportant. For weak electrolytes like acetic acid, the variation is larger (see Section 6.2).

Equations 6.1-25 and 6.1-26 can now be combined and simplified to define the most convenient measure of conductivity, the equivalent conductance:

$$\Lambda = |z_1|u_1 + |z_2|u_2$$

$$= \{(\text{resistance})[K_{\text{cell}}\mathcal{F}\nabla\psi]|z_1 z_2|c_T\}^{-1} \tag{6.1-27}$$

The quantity Λ is that most frequently reported in studies of conductance. It can be measured by determining each of the quantities in the braces. Because the electrochemical-potential gradient is fixed, the entire quantity in brackets can be treated as a cell constant.

The equivalent conductance Λ can be extremely accurately measured, often to accuracies of 0.01%. It is known to vary slightly with concentration, as shown in Fig. 6.1-4. This variation follows the equation

$$\Lambda = \Lambda_0 - S\sqrt{c_T} + Ec_T \ln c_T + Jc_T + J'c_T^{3/2} \tag{6.1-28}$$

where Λ_0, S, E, J, and J' are all constants. The limiting equivalent conductance Λ_0 is a property of the ions and is not well understood theoretically. The limiting slope S, first calculated by Onsager, is a function only of the charges on the ions and is thus characteristic of electrostatic interactions between the ions. The higher constants, E, J, and J', include more electrostatic interactions, ion–solute interactions, and the ion associations more commonly encountered with weak electrolytes.

In many practical problems, the ion transport is well described by assuming that Λ is a constant. After all, the concentration variations are less than 20% for aqueous solutions of most strong electrolytes. However, some solution chemists will attack this assertion with the swordlike insights of Onsager and Fuoss (1932), Fuoss and Onsager (1957), and Justice and Justice (1977). These chemists ignore the ion properties implicit in Λ_0 and instead extol those contained in E, J, and J'. If your purpose is knowledge of ion properties, listen to the chemists. If your purpose is knowledge of mass transfer, assume that Λ is a constant.

We now want to relate the equivalent conductance Λ to ion properties and, more specifically, to ion diffusion coefficients. First, because the ions migrate independently in a dilute-solution conductance experiment, we can define, from Eq. 6.1-27,

$$\Lambda = \lambda_1 + \lambda_2 \tag{6.1-29}$$

where

$$\lambda_i = |z_i| u_i \tag{6.1-30}$$

The λ_i, called the equivalent ionic conductances, cannot be found from measurements of Λ alone, but require other independent determinations, most commonly the transference numbers given in Eqs. 6.1-13 and 6.1-14. The λ_i depend not only on the ion mobility but also on the charge. More specifically, if two cations have the same size but not the same charge, they will have the same mobility, though not the same equivalent ionic conductance.

The equivalent ionic conductances are closely related to the ionic diffusion coefficients through the mobilities:

$$D_i = k_B T u_i$$

$$= \left[\frac{k_B T}{|z_i|} \right] \lambda_i \tag{6.1-31}$$

This looks trivial, but it has the potential to be extremely useful, for it suggests how diffusion can be found from conductance.

Surprisingly, Eq. 6.1-31 is not often used, even though it is so simple and valuable. Part of the reason for this neglect is probably that curse of science, the units in the equation. The λ_i are most commonly expressed in "conductance units," which are mercilessly $cm^2/mol\text{-}ohm$. These units are apparently chosen both as an experimental convenience and as a frustrating barrier to the uninitiated. As partial salve, we can write out the unit conversion factors to give, at 25°C,

$$D_i([=]\text{cm}^2\text{-sec}^{-1}) = \frac{2.662 \cdot 10^{-7}}{|z_i|} \lambda_i([=]\text{cm}^2\text{-mol}^{-1}\text{-ohm}^{-1}) \quad (6.1\text{-}32)$$

This relation was used to find the values in Table 6.1-1.

Equation 6.1-32 shows that conductance measurements might be a substitute for those of diffusion and other aspects of mass transfer. This would be appealing, because conductance is much, much easier to measure. Why not measure conductance and forget diffusion?

This idea has both merit and danger. The merit is the simplicity; the two methods do give closely related information. The danger is best illustrated by the case of a singly charged 1-1 electrolyte for which (Robinson & Stokes, 1960)

$$\Lambda = \lambda_1 + \lambda_2$$

$$= \frac{2}{k_B T} \left(\frac{D_1 + D_2}{2} \right) \quad (6.1\text{-}33)$$

and

$$D = \frac{2}{1/D_1 + 1/D_2} \quad (6.1\text{-}34)$$

The equivalent conductance depends on an arithmetic average of ion diffusion, but the diffusion coefficient is related to a harmonic average. The conductance is dominated by the larger D_i, by the faster ion. The diffusion is influenced most by the smaller D_i, by the slower ion.

These results mean that conductance and diffusion represent different averages over similar molecular processes. They will not be completely equivalent unless both ions have the same mobility (i.e., when $D_1 = D_2$ and each transference number equals one-half). In this case, the arithmetic and harmonic averages are equal, and diffusion and conductance give completely equivalent information. Interestingly, this is almost exactly true for KCl in water, where KCl conductance is almost completely equivalent to KCl diffusion. It is also true for tetrabutylammonium tetraphenylboride in organic solutions. Before you try a difficult diffusion experiment, think about an easy conductance run.

Example 6.1-5: Calcium chloride diffusion from conductance

Estimate the diffusion coefficient of $CaCl_2$ from conductance measurements. The equivalent conductance at infinite dilution is 59.5 for Ca^{2+} and 76.4 for chloride. The experimental value of the diffusion coefficient is about $1.32 \cdot 10^{-5}$ cm²/sec.

Solution. From Eq. 6.1-32 we can find the ionic diffusion coefficients

$$D_{Ca} = 0.79 \cdot 10^{-5} \text{ cm}^2/\text{sec}$$

$$D_{Cl} = 2.03 \cdot 10^{-5} \text{ cm}^2/\text{sec}$$

The diffusion coefficient can be found from these ionic values by using Eq. 6.1-23:

$$D_{CaCl_2} = \left[\frac{2 + 1}{(2/2.03) + (1/0.79)} \right] \cdot 10^{-5}$$

$$= 1.33 \cdot 10^{-5} \text{ cm}^2/\text{sec}$$

This result is accurate in very dilute solution. At higher concentrations, the diffusion coefficient drops to about 1.1 at 0.2-M and then rises slightly.

Section 6.2. Diffusion of associating solutes

We now switch from solutes that dissociate completely to form ions to solutes that associate to form aggregates. We again want to find the diffusion coefficient averaged over the various species present.

The analysis of these systems began in 1884, when Arrhenius (1884) suggested that materials like acetic acid partially dissociate in water. Many who study diffusion vaguely remember this variation but ignore it in their experiments. Interestingly, the diffusion of such solutes can lead to curious and dramatic results. These results have been scattered through different academic disciplines, and so have tended to be ignored. As an example, consider diffusion of potassium chloride across a thin membrane (Fig. 6.2-1) (Reusch & Cussler, 1973). The membrane consists of a chloroform solution of a macrocyclic polyether stabilized by a thin sheet of porous polymer. It separates a KCl solution from water. We expect the KCl flux across this membrane to be proportional to the KCl concentration; doubling this concentration should double the flux (see Example 2.2-1).

As shown, the flux of potassium chloride across this membrane is not proportional to the concentration of potassium chloride, but to this concentration squared. Doubling the KCl concentration increases the diffusion flux not by two but by four times. Why is this?

One possible reason is that the experiments are wrong. This was my first guess. I made an unfortunate student repeat these studies until he hoped never to see another crystal of potassium chloride. The experiments were reproducible. Another guess is that there is some sort of electrostatic effect like those in the previous section. (Electrostatic effects always cause trouble. Right?) Applying an electric field, however, does not affect the fluxes, so that electrostatics cannot be that important.

The correct explanation involves ion pairs. An ion pair consists of a cation and an anion, locked together electrostatically (Fuoss & Accasina, 1959), "held together with electrostatic glue." Within the organic membrane, most of the potassium chloride is present as ion pairs, and only a very small fraction exists as free potassium ions and free chloride ions. It is almost as if the picture of ionized electrolytes has been reversed, because potassium chloride diffusion now involves diffusion not of ions but of pseudo-molecules of potassium chloride.

What happens in these experiments is this: Potassium and chloride ions present in the aqueous solution near the membrane interface quickly dissolve in the membrane and then almost as quickly associate into an ion pair.

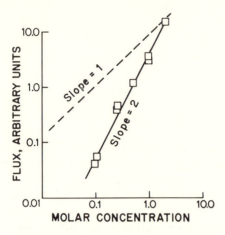

Fig. 6.2-1. Potassium chloride flux across an organic membrane. In these experiments, a concentrated solution of KCl diffuses across a polyether–chloroform membrane into pure water. The flux observed is not proportional to the salt's concentration but to this concentration squared. This effect occurs because potassium and chloride ions associate within the membrane to form ion pairs. [Data from Reusch & Cussler (1973).]

The concentration of ion pairs near the membrane surface is thus proportional to the product of potassium and chloride ions, or rather the square of the concentration of potassium chloride in the adjacent aqueous solution. The flux of potassium chloride across the membrane depends on the concentration difference of the ion pairs, and hence on the square of the aqueous potassium chloride concentration. This result is typical of those discussed in this section.

Weak electrolytes

Weak 1-1 electrolytes rapidly dimerize in solution. For example, an aqueous solution of acetic acid contains hydrated protons, acetate ions, and acetic acid molecules. The cations and anions of these weak electrolytes diffuse at the same rate because of electrostatic interactions. As such, these ions are like monomers of the same species. The electrolyte molecules, which are roughly equivalent to dimers, diffuse at a different rate. As the diffusion proceeds, the concentration changes, and so the fraction of solute present as monomer or dimer also changes. The result is a concentration-dependent diffusion process (Stokes, 1965; Dunn & Stokes, 1965).

We want to describe one-dimensional steady-state diffusion of a dimerizing solute. For this case, we imagine that the system contains a single solute, present as monomers at concentration c_1 and as dimers at concentration c_2. The steady-state continuity equations of these species are

$$0 = D_1 \frac{d^2 c_1}{dz^2} - 2r_2 \tag{6.2-1}$$

$$0 = D_2 \frac{d^2 c_2}{dz^2} + r_2 \tag{6.2-2}$$

where r_2 is the rate of production of the dimer. Although electrostatic coupling between the ions is not explicit, it can be shown that these equations are exact even if electrostatic coupling is included.

The concentrations c_1 and c_2 are subject to two constraints. First, they are related to the total solute concentration c_T:

$$c_T = c_1 + 2c_2 \tag{6.2-3}$$

This total concentration is expressed in equivalents. For example, for a 1-M acetic acid solution, it equals 2-M; but for a 1-M urea solution, it equals 1-M. Second, the dimerization will usually be more rapid than diffusion, so that monomer and dimer are everywhere in equilibrium:

$$c_2 = Kc_1^2 \tag{6.2-4}$$

where K is the equilibrium constant for this dimerization, identical with the *association* constant of a weak electrolyte.

To this point, we have discussed weak electrolytes without defining what "weak" means. Such a definition depends on the value of the constant K. This value, a strong function of the chemistry of the system, increases rapidly as the dielectric constant of the solvent is decreased. If the values of concentration and K are such that association is significant, then the electrolyte behaves as if it is weak. For example, acetic acid is a weak electrolyte in water and is essentially un-ionized in chloroform. Potassium chloride is fully ionized in water, but it behaves as a weak electrolyte in chloroform.

On this basis, we can describe the diffusion of a dimerizing solute. We multiply Eq. 6.2-2 by two, add it to Eq. 6.2-1, and integrate once to obtain

$$-J_T = D_1 \frac{dc_1}{dz} + 2D_2 \frac{dc_2}{dz} \tag{6.2-5}$$

where J_T is an integration constant. Physically, J_T corresponds to the total solute flux, again expressed in equivalents. The concentrations c_1 and c_2 can be found in terms of c_T from Eqs. 6.2-3 and 6.2-4; when the result is combined with Eq. 6.2-5, we obtain

$$-J_T = (D_1 + 4KD_2 c_1) \frac{dc_1}{dz} = D \frac{dc_T}{dz}$$

$$= \left(\frac{D_1 + D_2[\sqrt{1 + 8Kc_T} - 1]}{\sqrt{1 + 8Kc_T}} \right) \frac{dc_T}{dz} \tag{6.2-6}$$

The quantity in parentheses is the apparent diffusion coefficient D of the dimerizing solute. At very low c_T, this quantity equals the monomer diffusion coefficient D_1. At high c_T, it equals the dimer's value, D_2. Other values are shown in Fig. 6.2-2.

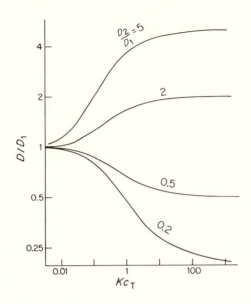

Fig. 6.2-2. The diffusion coefficient of a dimerizing solute. As a solute dimerizes, its average diffusion coefficient changes from that of the monomer to that of the dimer. The concentration c_T at which this occurs is roughly the reciprocal of the association constant K.

Example 6.2-1: Diffusion of acetic acid

What is the diffusion coefficient of the acetic acid molecule if the apparent diffusion coefficient of acetic acid is $1.80 \cdot 10^{-5}$ cm²/sec at 25°C and 1.0-M? The pK_a of acetic acid is 4.756.

Solution. The pK_a of a weak acid HA is defined as

$$pK_a = -\log_{10} \frac{[H^+][A^-]}{HA}$$

In this case, the $[H^+]$ and $[A^-]$ concentrations are equal. Comparing this with Eq. 6.2-4, we see that

$$K = 10^{pK_a} = 5.70 \cdot 10^4$$

If we insert this into Eq. 6.2-6, we find that the term containing D_2 dominates completely, and

$$D_2 \doteq D = 1.80 \cdot 10^{-5} \text{ cm}^2/\text{sec}$$

In passing, note that the diffusion coefficient of the fully ionized acid found from Eq. 6.1-12 and Table 6.1-1 is $1.95 \cdot 10^{-5}$ cm²/sec, a similar value.

Aggregating solutes

We now want to calculate the average diffusion coefficient for solutes that aggregate much more than the simple weak electrolytes discussed earlier (Weinheimer et al., 1981). Three cases of this aggregation are shown

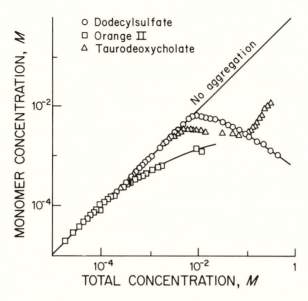

Fig. 6.2-3. Types of solute aggregation. The detergent sodium dodecylsulfate aggregates abruptly to form micelles, and the dye Orange II has isodesmic aggregates (see Fig. 6.2-4). The bile salt sodium taurodeoxycholate falls between these two limits. These results were obtained using ion-selective electrodes. [Data from Kale et al. (1980).]

in Fig. 6.2-3. The most dramatic case is the detergent sodium dodecylsulfate (SDS). Molecules of this detergent remain separate at low concentration, but then suddenly aggregate. The resulting aggregates, called "micelles," are most commonly visualized as an ionic hydrophilic skin surrounding an oily hydrophobic core [Fig. 6.2-4(a)]. In fact, detergents clean in this way; they capture oil-bearing particles in their cores.

In contrast, molecules of the dye Orange II aggregate gently, resulting in a slow and steady deviation from the unaggregated limit. Such aggregation results from a stacking of dye molecules, like that shown schematically in Fig. 6.2-4(b). When the ease of stacking is the same for all sizes in the stack, this aggregation is called "isodesmic." The third case involving the bile salt taurodeoxycholate is intermediate between the other two.

The two situations of micelle formation and isodesmic stacking represent two limiting forms of solute aggregation. These two limits are discussed in the following paragraphs.

Micelle formation

The diffusion coefficient measured in a detergent solution represents an average over the monomer and micelle present in solution (Weinheimer et al., 1981). Steady-state diffusion in such a system of monomer and micelle obeys the continuity equations

(a) <u>Long-chain surfactants</u>   (e.g., $CH_3(CH_2)_{11}SO_4^-$)

Many monomers
form one size
micelle

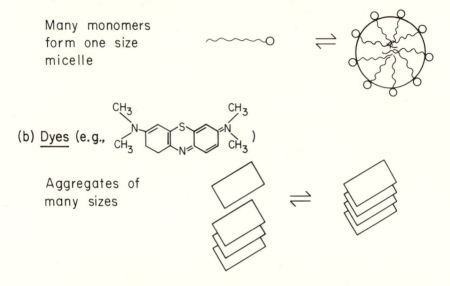

(b) <u>Dyes</u> (e.g.,

Aggregates of
many sizes

Fig. 6.2-4. Micelle formation and isodesmic aggregation. In the type of micelle formation discussed here, n monomers combine to form an n-mer. No other sizes are present. In isodesmic association, monomers add with equal facility to monomers or aggregates of any size.

$$0 = D_1 \frac{d^2c_1}{dx^2} - nr_m \qquad (6.2\text{-}7)$$

$$0 = D_m \frac{d^2c_m}{dx^2} + r_m \qquad (6.2\text{-}8)$$

where the subscripts 1 and m refer, respectively, to the monomer and the micelle, and r_m represents the rate of formation of micelles. Equation 6.2-8 is multiplied by n, added to Eq. 6.2-7, and integrated to give

$$-J_T = D_1 \frac{dc_1}{dx} + nD_m \frac{dc_m}{dx} \qquad (6.2\text{-}9)$$

The integration constant J_T is the total flux of solute.

Equation 6.2-9 is not useful, because it is written in terms of the unknown gradients of c_1 and c_m, rather than in terms of the known total-solute gradient c_T. To remove these unknowns, we assume that micelle formation is fast, so that

$$c_m = Kc_1^n \qquad (6.2\text{-}10)$$

where K is the equilibrium constant for the fast micelle-forming reaction. We also need the mass balance:

$$c_T = c_1 + nc_m \qquad (6.2\text{-}11)$$

We would like to combine Eqs. 6.2-9 through 6.2-11 to get the answer we want. After all, we did this easily in the case of solute dimerization. Instead of the quadratic equation for dimerization, however, we now have an nth-order equation for micelle formation. The exact order depends on the type of micelle, but typically it ranges from the 10th order to the 100th order.

We cannot analytically solve such an equation without approximation. We first recognize that detergent solutions have physical properties like conductance and surface tension that suddenly change at a critical concentration at which micelles start to form in significant numbers. The "critical micelle concentration" can thus be estimated experimentally. However, we should recognize that such a concept is consistent with the chemical equilibrium in Eq. 6.2-10 only if n is very large. If n is not large, then the solution's physical properties will change more gradually, and a critical micelle concentration will be more difficult to define experimentally.

Once this approximation is made, the results tumble into place. We expect the monomer concentration c_1 to approximately equal that at the critical micelle concentration; so, from Eq. 6.2-11,

$$c_m = \frac{1}{m} (c_T - c_{CMC}) \qquad (6.2\text{-}12)$$

where c_{CMC} is the critical micelle concentration. A more accurate estimate of c_1 can now be found from Eq. 6.2-10:

$$c_1 = \left[\frac{1}{nK} (c_T - c_{CMC}) \right]^{1/n} \qquad (6.2\text{-}13)$$

Inserting these results into Eq. 6.2-9, we find

$$-J_T = \left[D_M + \frac{D_1}{n} \frac{(nK)^{-1/n}}{(c_T - c_{CMC})^{1-(1/n)}} \right] \frac{dc_T}{dz} \qquad (6.2\text{-}14)$$

or, because n is large,

$$-J_T = D \frac{dc_T}{dz} = \left[D_m + \frac{D_1(nK)^{-1/n}}{n(c_T - c_{CMC})} \right] \frac{dc_T}{dz} \qquad (6.2\text{-}15)$$

which is the desired result. The quantity in square brackets is the average diffusion coefficient D found experimentally.

To my surprise, this analysis works for nonionic detergents. The apparent diffusion coefficient does vary inversely with $(c_T - c_{CMC})$, as shown in Fig. 6.2-5. The intercept on this plot agrees closely with the micelle's diffusion coefficient estimated in other ways. The slope is consistent with independent measurements of K and n.

However, this analysis does not work for ionic detergents at low ionic strength. For example, the diffusion coefficient of sodium dodecylsulfate increases significantly at concentrations above the critical micelle concentration, as shown in Fig. 6.2-6. This increase turns out to be of electrostatic origin, and it can be described by material like that in Section 6.1. At high ionic strength, these electrostatic effects are less important, and Eq. 6.2-15 is again verified.

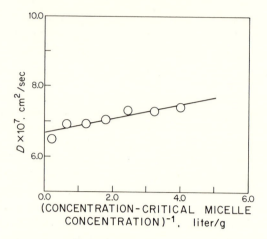

Fig. 6.2-5. Diffusion of the detergent Triton X-100 at 25°C. The variation with concentration is predicted by Eq. 6.2-19. The intercept is the micelle's diffusion coefficient, and the slope is related to the monomer's diffusion coefficient. [From Weinheimer et al. (1981), with permission.]

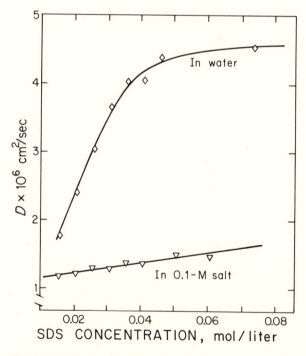

Fig. 6.2-6. Diffusion of sodium dodecylsulfate (SDS) at 25°C. The diffusion coefficients in this case increase as SDS concentration and solution viscosity rise. This increase is the result of aggregation and electrostatic interaction. [Data from Weinheimer et al. (1981).]

Isodesmic association

As the last topic in this section, we want to calculate the average diffusion coefficient for systems in which aggregation occurs one molecule at a time. The simplest case is called the isodesmic model. It assumes that

$$c_i = Kc_{i-1}c_1 \tag{6.2-16}$$

where K is an equilibrium constant that is independent of the size of the aggregate. Note that the equilibrium constant for forming dimers from two monomers is assumed to be the same as that for forming heptamers from hexamers and monomers (Stokes, 1965).

Equations 6.2-10 and 6.2-16 show why the isodesmic model and micelle formation represent two extreme limits of solute aggregation. In the isodesmic case, aggregates of any size form with equal facility, because all the steps are equal. In the micelle case, aggregates form only of that special micelle of n monomers; the equilibrium constants are zero for all but that special size.

To find the average diffusion coefficient of a solute associating isodesmically, we again start with the steady-state continuity equations:

$$0 = D_1 \frac{d^2c_1}{dz^2} - 2r_2 - r_3 - r_4 - \cdots \tag{6.2-17}$$

$$0 = D_2 \frac{d^2c_2}{dz^2} + r_2 - r_3 \tag{6.2-18}$$

$$0 = D_3 \frac{d^2c_3}{dz^2} + r_3 - r_4 \tag{6.2-19}$$

$$\vdots$$

Again, these equations can be added together to eliminate reaction terms:

$$0 = \sum_{i=1}^{\infty} iD_i \frac{d^2c_i}{dz^2} \tag{6.2-20}$$

Integrating this result gives

$$-J_T = \sum_{i=1}^{\infty} iD_i \frac{dc_i}{dz} \tag{6.2-21}$$

where J_T is again an integration constant physically equal to the total solute flux in both aggregated and monomer forms.

As earlier in this section, we now rewrite the unknown concentrations $\{c_i\}$ in terms of the known total concentration of solute. Doing this requires two constraints. One of these is that of isodesmic equilibria (Eq. 6.2-16). The other is a mass balance:

$$c_T = c_1 + 2c_2 + 3c_3 + \cdots$$

$$= \sum_{i=1}^{\infty} ic_i \tag{6.2-22}$$

When these constraints are combined, we find

$$c_T = \frac{c_1}{(1 - Kc_1)^2} \qquad (6.2\text{-}23)$$

This quadratic can be solved for c_1 as a function of c_T, and the result combined with Eqs. 6.2-16 and 6.2-21 gives the total flux J_T as a function of total solute concentration c_T. This solution is an algebraic mess. A more useful form is the power series

$$-J_T = D\frac{dc_T}{dz} = \{D_1 - Kc_T(4D_1 - 4D_2) + K^2c_T^2(15D_1 - 24D_2 + 9D_3)$$

$$- K^3c_T^3(56D_1 - 112D_2 + 72D_3 + 16D_4) + \cdots\}\frac{dc_T}{dz}$$

$$(6.2\text{-}24)$$

Note that the apparent diffusion coefficient D given in braces does not vary with concentration if the diffusion coefficients are all equal (i.e., if $D_1 = D_2 = D_3 = \ldots$).

Section 6.3. Diffusion in dyeing

The final section of this chapter is concerned with the diffusion of dyes into fibers, especially wool. As before, we focus on the diffusion coefficients basic to this process. Although dyeing is an ancient craft, its details remain largely empirical, without a strong scientific framework (Rattee & Breuer, 1974). As a result, our discussion will be largely qualitative.

When wool fibers are placed in a dyebath, they slowly pick up color. The amount m_1 of dye taken up by a fiber is found to be (Peters, 1968; McGregor, 1974)

$$m_1 = \sqrt{Dt}\,(R_0LHC_{10}) \qquad (6.3\text{-}1)$$

in which D is some sort of coefficient, t is the dyeing time, R_0 and L are the radius and length of the wool fibers, C_{10} is the dyebath concentration, and H is a partition coefficient including both physical and chemical factors. Note that this equation suggests that the average concentration in the fiber $\bar{c}_1$ is

$$\bar{c}_1 = \frac{m_1}{\pi R_0^2 L} \propto \left[\frac{Dt}{R_0^2}\right]^{1/2}\left(\frac{HC_{10}}{\pi}\right) \qquad (6.3\text{-}2)$$

Thus, for equal dyeing, the quantity Dt/R_0^2 should be constant. This means that to reach the same color, wool 25 μ in diameter must be dyed 60% longer than wool 20 μ in diameter.

These experimental facts are consistent with a diffusion process. Diffusion is slow, like dyeing. Diffusion laws predict an uptake proportional to the square root of time, as in Eq. 6.3-1. These laws also suggest that Dt/R_0^2 is a fundamental variable; indeed, the chapters of this book teem with this variable.

As a result, the coefficient D in Eq. 6.3-1 is sensibly defined as an apparent diffusion coefficient and is measured experimentally. It is a useful concept in correlating experience and in anticipating the effects of changes in process conditions. It is rarely reliable as a predictive tool, but that is not too surprising. The dyeing of a fleece depends on factors like the type of sheep, the weather in which the sheep is raised, and the amount that the wool is washed or chemically treated.

However, when we look at the diffusion coefficients reported for dyes in wool, we begin to wonder what these coefficients really are. To begin with, the diffusion coefficients are around 10^{-12} cm^2/sec. This seems much too small, because wool absorbs around 35% water, so that it should have characteristics of a concentrated aqueous solution or a gel. These low values are sometimes rationalized as a result of a thin outer layer or epicuticle surrounding the fibers. However, if such a layer were the chief resistance to dyeing, then the dye uptake should not have the time dependence of Eq. 6.3-1.

In addition to their small size, the diffusion coefficients in wool vary sharply with temperature. As in solids, this temperature dependence is conveniently correlated as

$$D = D_0 e^{-a/RT} \tag{6.3-3}$$

where D_0 and A are constants. For wool, a is commonly around 25 kcal. For diffusion in solids, it can be as high as this, but solids do not contain 35% water. For diffusion in water, a is usually around 2 kcal.

Finally, the diffusion coefficients of dyes in wool are larger for larger dyes than for smaller ones, and they vary significantly with dye concentration (Fig. 6.3-1). Thus, the apparent diffusion coefficient is much smaller, much more temperature-dependent, and much more concentration-dependent than would be expected from our experience to date.

The chief reason for the unusual behavior of the dye's diffusion coefficient is that the dye reacts with the wool fiber. Wool is largely keratin, a protein. The side chains of the protein include $-NH_2$ and $-S-S-$ linkages as residues of amino acids like arginine or cystine. The reaction with these side chains is most often electrostatic:

$$WNH_2 + H^+ + XSO_3^- \rightleftharpoons WNH_3^+ XSO_3^- \downarrow \tag{6.3-4}$$

It can also be covalent:

$$WNH_2 + XCl \rightleftharpoons WNHX + HCl \tag{6.3-5}$$

In both cases, W indicates the wool, and X symbolizes the dye.

Not surprisingly, the reactions between dye and wool are affected by many factors other than electrostatic interactions and the formation of covalent bonds. Factors often suggested include van der Waals forces, hydrogen and hydrophobic bonds, dye aggregation, and steric hindrance. This last reflects the fact that wool contains many small pores, filled with water and coated with reaction sites.

The effects of all these chemical factors on dye diffusion are not known. We can, however, draw a qualitative conclusion from the calculation in Example 2.3-3. In this example, we found that the apparent diffusion coefficient equaled the actual coefficient divided by one plus a reaction equilibrium constant. This equilibrium constant is large; so the apparent coefficient

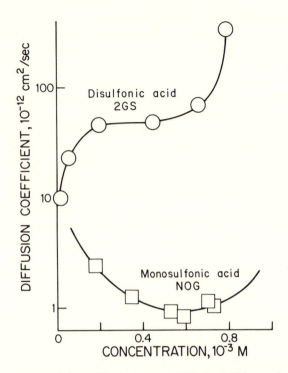

Fig. 6.3-1. Dye diffusion in wool. Diffusion in this case in unexpectedly slow and highly dependent on dye concentration. The reason is that the dye reacts rapidly with sites within the wool. Note that the larger disulfonic acid Naphthalene Orange (2GS) diffuses more rapidly than the monosulfonic acid Naphthalene Orange G (NOG). [Data from Medley (1964).]

will be small. The equilibrium constant varies widely with temperature, for it depends on chemical, not physical, factors. This constant can be a strong function of concentration. Thus, we conclude that the apparent diffusion coefficients that occur for the dyeing of wool are unusual because of fast chemical reaction.

Section 6.4. Conclusions

In this chapter we have been concerned with diffusion coefficients in systems containing strong solute–solute interactions. For strong electrolytes, we have good information about the ions themselves and can calculate the average diffusion coefficients involved. We find that these averages are harmonic, weighted with electrostatic charge. As a result, strong electrolyte diffusion is most influenced by the slow ions involved.

The rest of the chapter dealt with other types of solute–solute interactions, including those in weak electrolytes, micelles, and dyes. Diffusion coefficients for these systems are much less well understood than for strong electrolytes. In most cases the interactions produce concentration-dependent diffusion coefficients. This concentration dependence is often small and of little industrial significance. However, it provides information about the way the solute is acting in solution, and so can be significant chemically.

In using the material in this chapter, you should carefully remember your goals. For example, if you are developing separation processes, you need only remember that the diffusion coefficients you use are averages over the various species present in solution. You can ignore the details of this chapter. On the other hand, if your aim is a chemical understanding of processes like detergency, you need to consider how aggregation affects diffusion.

References

Arrhenius, S. A. (1884). *Göteborgs K. Vetenskaps-Och Vitterhets-Samhälles Bihang Till Handlingar,* **8,** No. 13–14.

Bard, A. J., & Faulkner, L. R. (1980). *Electrochemical Methods: Fundamentals and Applications.* New York: Wiley.

Colton, C. K., Smith, K. A., & deLeo, A. (1981). Paper presented at the American Institute of Chemical Engineers meeting in New Orleans, November 1981.

Dunn, L. A., & Stokes, R. H. (1965). *Australian Journal of Chemistry,* **18,** 285.

Evans, D. F., & Matesich, M. A. (1973). In: *Techniques of Electrochemistry, Vol. 2,* ed. E. Yeager & A. J. Salkind. New York: Wiley.

Franks, F., ed. (1975). *Water – A Comprehensive Treatise.* New York: Plenum.

Fuoss, R. M., & Accasina, F. (1959). *Electroconductance.* New York: Wiley Interscience.

Fuoss, R. M., & Onsager, L. (1957). *Journal of Physical Chemistry,* **61,** 668; (1962) **66,** 1722; (1963) **67,** 621; (1964) **68,** 1.

Harned, H. S., & Owen, B. B. (1950). *Physical Chemistry of Electrolyte Solutions,* 2nd ed. New York: Van Nostrand Reinhold.

Justice, J.-C., & Justice, M. C. (1977). *Discussions of the Faraday Society,* **64,** 265.

Kale, K., Cussler, E. L., & Evans, D. F. (1980). *Journal of Physical Chemistry,* **84,** 593.

McGregor, R. (1974). *Diffusion and Sorption in Fibers and Films,* Chapter 15. London: Academic Press.

Medley, J. A. (1964). *Transactions of the Faraday Society,* **60,** 1010.

Onsager, L., & Fuoss, R. M. (1932). *Journal of Physical Chemistry,* **36,** 2689.

Peters, R. H. (1968). In: *Diffusion in Polymers,* ed. J. Crank & G. S. Park. London: Academic Press.

Rattee, I., & Breuer, M. M. (1974). *Physical Chemistry of Dye Adsorption.* London: Academic Press.

Reusch, C. F., & Cussler, E. L. (1973). *American Institute of Chemical Engineers Journal,* **19,** 736.

Robinson, R. G., & Stokes, R. H. (1960). *Electrolyte Solutions.* London: Butterworth.

Stokes, R. H. (1965). *Journal of Physical Chemistry,* **69,** 4012.

Weinheimer, R. M., Evans, D. F., & Cussler, E. L. (1981). *Journal of Colloid and Interface Science,* **80,** 357.

7 SOLUTE–SOLVENT AND SOLUTE–BOUNDARY INTERACTIONS

In this chapter we want to explore the effects on diffusion caused by interactions of the solute with the solvent or with the system's boundaries. These effects can be dramatic. For example, near critical points, the diffusion coefficient precipitously plummets from a perfectly ordinary value to zero. In living cells, the rate of infection by viruses is sharply altered by virus interactions with cell walls.

The discussion of these problems involves a somewhat different strategy than that used earlier in this book. In Chapters 1–3 we treated the diffusion coefficient as an empirical parameter, an unknown constant that kept popping up in a variety of mathematical models. In more recent chapters we have focused on the values of these coefficients measured experimentally. In the simplest cases these values can be found from kinetic theory or from solute size; in more complicated cases these values require combining the effects of different ions or other related solutes. In all these cases, the goal is to use our past experience to estimate the diffusion coefficients, from which diffusion fluxes and the like can be calculated.

In this chapter we examine some of the physical phenomena that give strange and unexpected results. These results include fluxes that are larger or smaller than expected because of solute–solvent and solute–boundary interactions. In Section 7.1 we discuss the effect of solvation of the solute on diffusion. Such interaction is frequently postulated in water, and this leads to the idea of "hydration number," or the number of water molecules that react with a solute molecule. In Section 7.2 we turn to diffusion studied by means of radioactively labeled tracers. Such tracers are used routinely, both without trouble and without careful thought. In Section 7.3 we discuss diffusion near critical points in which solvent and solute interact extremely strongly. Finally, in Section 7.4 we turn to diffusion in composite materials, especially in small pores. Here the solute's diffusion can be altered by the pores' walls. These four sections thus survey the spectrum of solute–solvent and solute–boundary interactions, the understanding of which is the goal of this chapter.

Section 7.1. Hydration

The simplest type of solute–solvent interaction is a process in which solute and solvent combine to form a new species, which is that actually diffusing. In this section we discuss hydration as typical of this combination.

The idea of hydration is based on the following flux equation:

$$-\mathbf{j}_1 = D_0 \left(1 + \frac{\partial \ln \gamma_1}{\partial \ln c_1}\right) \nabla c_1$$

$$= \frac{k_B T}{6\pi\mu R_0} \left(1 + \frac{\partial \ln \gamma_1}{\partial \ln c_1}\right) \nabla c_1 \qquad (7.1\text{-}1)$$

in which D_0 is a new diffusion coefficient, μ is the solvent viscosity, R_0 is the solute radius, and γ_1 is an activity coefficient. This equation makes two implicit assumptions: that the solute's flux is proportional to chemical potential gradient and that the diffusion coefficient in dilute solution is given by the Stokes–Einstein equation (Hinton & Amis, 1971).

Hydration can affect this equation in two ways. First, the solute radius R_0 must be that of the hydrated species. This can be related to the true solute radius R_0' by the equation

$$\tfrac{4}{3}\pi R_0^3 = \tfrac{4}{3}\pi (R_0')^3 + n \left(\frac{\tilde{V}_{H_2O}}{\tilde{N}}\right) \qquad (7.1\text{-}2)$$

in which $\tilde{V}_{H_2O}$ is the molar volume of water and n is the "hydration number," the number of water molecules bound to a solute. If the diffusion coefficient at infinite dilution is known, R_0 can be calculated, R_0' can be estimated from crystallographic data, and n can be calculated. This kind of hydration *decreases* diffusion.

Hydration can also be calculated from the concentration dependence of diffusion by assuming that this concentration dependence is the result of hydration. Ideas like this were first used by Scatchard (1921) to rationalize the activity coefficient of sucrose. To do this, one assumes that the solute activity $c_1\gamma_1$ equals the solute's true mole fraction corrected for hydration:

$$c_1\gamma_1 = \frac{\text{number of hydrated solute molecules}}{\left(\begin{array}{c}\text{number of hydrated solute}\\\text{molecules}\end{array}\right) + \left(\begin{array}{c}\text{number of "free"}\\\text{water molecules}\end{array}\right)}$$

$$= \frac{\text{number of solute molecules}}{(1 - n)\left(\begin{array}{c}\text{number of solute}\\\text{molecules}\end{array}\right) + \left(\begin{array}{c}\text{total number of}\\\text{water molecules}\end{array}\right)}$$

$$= \frac{c_1}{(1 - n)c_1 + c_2} \qquad (7.1\text{-}3)$$

The two concentrations are related through the partial molar volumes:

$$c_1 \tilde{V}_1 + c_2 \tilde{V}_2 = 1 \qquad (7.1\text{-}4)$$

Combining Eqs. 7.1-1, 7.1-3, and 7.1-4, we obtain

Table 7.1-1. *Hydration numbers found by various methods*

Ion	Observed diffusion coefficient at infinite dilution[a]	Hydration numbers from diffusion at infinite dilution	Hydration numbers from diffusion's concentration dependence[b]	Hydration numbers from activity coefficients[c]	Hydration numbers from transference methods[c]
H^+	9.33	-1.3	—	4	1
Li^+	1.03	1.3	2.8	4	14
Na^+	1.34	0.5	1.2	3	8
K^+	1.96	-0.1	0.9	1	5
Cs^+	2.06	-0.5	0.5	0	5
Cl^-	2.03	-0.7	0	1	4
Br^-	2.08	-0.9	0.2	1	5
I^-	2.04	-1.2	0.7	2	2

[a] $\times 10^{-5}$ cm²/sec.
[b] Data of Robinson & Stokes (1959).
[c] Data of Hinton & Amis (1971).

$$D = D_0 \left[1 - \frac{(1 - n - \bar{V}_1/\bar{V}_2)c_1}{1/\bar{V}_2 + (1 - n - \bar{V}_1/\bar{V}_2)c_1} \right] \qquad (7.1\text{-}5)$$

For dilute solutions, c_1 is small; for ideal gases or solutions of constant density, $\bar{V}_1/\bar{V}_2$ is unity, and Eq. 7.1-5 becomes

$$D = D_0[1 + n\bar{V}_2 c_1] \qquad (7.1\text{-}6)$$

This result is often decorated with viscosity and electrostatic corrections. However, the basic message remains; hydration tends to *increase* diffusion.

These ideas are frequently qualitatively useful, but they are rarely quantitatively applicable. The data in Table 7.1-1 illustrate this by comparing hydration numbers found from diffusion, from activity coefficients, and from transference methods. Qualitatively, these values supply insights. For example, the diffusion of lithium is slower than that of sodium, which is slower than that of potassium, etc. This suggests that the radii of the diffusing solutes are in the order $Li^+ > Na^+ > K^+ > Cs^+$, exactly the reverse of the ionic radii found in the solid state. Such inverted behavior seems to be the result of hydration.

However, the hydration numbers make little quantitative sense. The values found from Eq. 7.1-2 are shown in the third column of Table 7.1-1. Although these values are often negative, we could force them to be positive by replacing the factor 6π in the Stokes–Einstein equation with some other theoretically rationalized value. The values calculated from Eq. 7.1-6, shown in the fourth column, do have the courtesy to remain positive, but they are far from being integers. I am always unsure how a cesium ion can react with half a molecule of water.

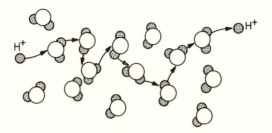

Fig. 7.1-1. Proton diffusion in water. Proton diffusion occurs by the chain reaction shown between water molecules. Such a jump mechanism also exists in alcohol, but not in alcohol–water mixtures.

In addition, the hydration numbers found from diffusion show little relation with those calculated from values from the other types of experiments shown in Table 7.1-1. These other values also suggest that the hydration, and hence the solute radii, are in the sequence $Li^+ > Na^+ > K^+ > Cs^+$. The hydration numbers, however, differ from those based on diffusion. Graciously, you may conclude that the field of hydration is in a state of confusion. More ruthlessly, you may decide that these ideas do not work.

One serious effort to provide an alternative to these ideas is the concept of "water structure" (Franks, 1975). This concept suggests that water has a fair degree of tetrahedral bonding, which structures it like ice. When ions are dissolved in water, they either destroy or enhance the icelike structure. If the ions destroy the structure, they are immediately surrounded by water of a reduced apparent viscosity, equivalent to a higher local temperature. Such "structure-breaking" ions include cesium and iodide, both of which diffuse more rapidly than expected.

In contrast, lithium, fluoride, and the tetraalkylammonium ions apparently enhance water structure. These ions are immediately surrounded by water that is more organized than water in the bulk. Such more organized water has a higher viscosity, equivalent to a lower temperature. Such "structure-making" ions diffuse more slowly than expected. Ideas of water structure seem to me an attractive qualitative alternative to those of hydration.

One final point concerns the diffusion of protons. From Table 7.1-1 we see that the diffusion coefficient of H^+ is five times greater than that of any other ion. This speed is inconsistent with the ion's size, which would suggest a more normal value. The reason for this behavior is that proton transport occurs by a different mechanism (Franks, 1975). In this mechanism, shown schematically in Fig. 7.1-1, a proton does not move through water as an intact entity. Instead, it reacts with a water molecule, forcing a proton off the other side. This newly generated molecule reacts again to produce a third proton; this third proton continues the chain reaction.

The proton jump mechanism, which I find emotionally appealing, can exist in solvents other than water, but to nowhere near the same degree. For example, in pure water and in pure methanol, proton diffusion seems anomalously fast. However, at about 95 mole percent methanol, it goes through a minimum (Shedlovsky and Kay, 1956). Apparently, at this concentration, the protons are tightly bound to water and diffuse as an intact solute species. Such arguments are ethereal, but fun.

Section 7.2. Tracer diffusion

Imagine we want to study the diffusion of steroids like progesterone through human blood. The amounts of these steroids will be very small, making direct chemical analysis difficult. As a result, we synthesize steroids that contain carbon 14 as a radioactive label. We then analyze for steroid concentration, using a liquid scintillation counter, and calculate diffusion coefficients from the appropriate set of concentration measurements.

This measurement of tracer diffusion in dilute solution is a good strategy. The use of radioactive tracers provides a near unique opportunity for a specific chemical analysis in highly dilute solution. Such analysis is especially important in biological systems, where complex chemistry may compromise analysis. Moreover, in dilute solution, the diffusion coefficients found with radioactive tracers are almost always indistinguishable from those measured in other ways. Exceptions occur in those systems in which the solute moves by a jump mechanism like that for protons (see Section 7.1) or in which the solute's molecular weight is significantly altered by the isotopic mass.

In concentrated solution, tracer diffusion is a much more complex process, and it does not provide coefficients identical with those in the binary system. This is illustrated by the data in Fig. 7.2-1 (Mills, 1965). In this figure we see that the diffusion coefficients using different radioactive isotopes can differ from each other and from the binary diffusion coefficient. On reflection, we realize that this is not surprising; the diffusion of radioactively tagged benzene in untagged benzene is obviously a different process than the diffusion of tagged cyclohexane in benzene.

Explaining these differences requires more careful definitions (Albright & Mills, 1965; Curran et al., 1967). *Binary diffusion* occurs with two chemically distinct species. In contrast, *intradiffusion* occurs with three distinguishable species. One of these species is chemically different. The other two species are very similar, for they have the same chemical formula, the same boiling point, the same viscosity, and so forth. They differ only in their isotopic composition or their optical rotation. Nonetheless, this means that intradiffusion involves three species.

There are two important special cases of intradiffusion. The first, *tracer diffusion,* is the limit when the concentration of one similar species is small. This is the usual situation when one uses radioactive isotopes, for high concentrations of radioactive material are expensive, risky, and unnecessary. The second special case, *self-diffusion,* occurs when the system contains a radioactively tagged solute in an untagged but otherwise chemically

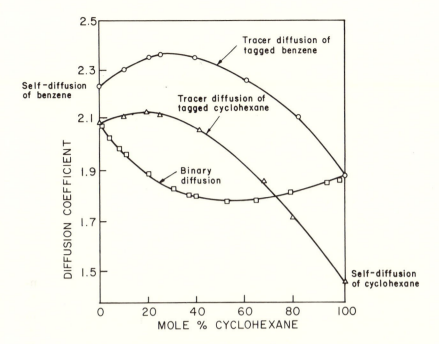

Fig. 7.2-1. Binary and tracer diffusion at 25°C. The tracer diffusion coefficient equals the binary coefficient only in certain special cases. All coefficients are $\times 10^{-5}$ cm^2/sec. [From Mills (1965), with permission.]

identical solvent. This system may also contain traces of other solutes, and so still may have more than two components. These different definitions are identified in Fig. 7.2-1.

The complete relations between binary diffusion and intradiffusion are known for the case of dilute gases described by the Chapman–Enskog theory (Bremer & Cussler, 1970). Imagine that this system consists of species 1 (the radioactively tagged solute), species 2 (the untagged, chemically identical solute), and species 3 (the chemically different solvent). Diffusion in this system is described in terms of solute–solvent collisions and solute–solute collisions. Solute–solvent collisions are characterized by collision diameters σ_{13} and ε_{13}. Solute–solute collisions are described by σ_{12} and ε_{12}.

With these diameters and energies, the binary diffusion coefficient D can be shown to be a function only of solute–solvent collisions:

$$D = D(\sigma_{23}, \varepsilon_{23}) = D(\sigma_{13}, \varepsilon_{13}) \qquad (7.2\text{-}1)$$

On the other hand, the intradiffusion coefficient D^* is a weighted harmonic average of solute–solvent and solute–solute collisions:

$$D^* = \cfrac{1}{\cfrac{y_3}{D(\sigma_{23}, \varepsilon_{23})} + \cfrac{y_1 + y_2}{D(\sigma_{12}, \varepsilon_{12})}} \qquad (7.2\text{-}2)$$

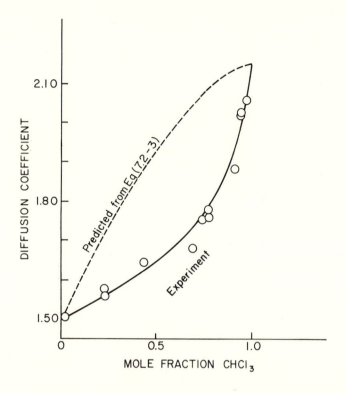

Fig. 7.2-2. Binary diffusion predicted from tracer diffusion. In general, binary diffusion cannot be predicted from tracer diffusion and activity data using empirical relations like Eq. 7.2-3. The data, for chloroform–carbon tetrachloride at 25°C, are $\times 10^{-5}$ cm²/sec. [From Kelly et al. (1971), with permission.]

Note that when $(y_1 + y_2)$ is nonzero, D^* is not equal to D. In the limit of infinite dilution, both y_1 and y_2 approach zero, and D^* equals D.

Many investigators have tried to discover empirical connections between binary diffusion and intradiffusion. The most common is the assertion that

$$D = D^* \left(1 + \frac{\partial \ln \gamma_1}{\partial \ln c_1}\right) \tag{7.2-3}$$

in which D is the binary diffusion coefficient, D^* is the intradiffusion coefficient measured with a radioactive tracer, and the quantity in parentheses is the increasingly familiar activity correction for diffusion. This empirical assertion is often buttressed by theoretical arguments, especially those based on the irreversible thermodynamics described in Section 8.2. Equation 7.2-3 does not always work experimentally, as shown by the results in Fig. 7.2-2.

Why Eq. 7.2-3 sometimes fails is illustrated by the case of dilute gases. Binary diffusion involves only solute–solvent interactions. Intradiffusion and tracer diffusion are the result not only of solute–solvent interactions but

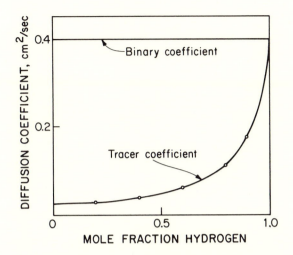

Fig. 7.2-3. Binary versus tracer diffusion of hydrogen gas and benzene vapor. The benzene is the species tagged with radioactivity. The differences between the binary and tracer values are unusually large in this case.

also of solute–solute interactions. Thus, D^* contains different information than D, information characteristic of dynamic collisions as well as equilibrium activities. This difference means that D^* cannot be found only from D and activity coefficients.

Example 7.2-1: Tracer and binary diffusion of hydrogen and benzene

Find the tracer diffusion coefficient of ^{14}C-tagged benzene in gas mixtures of hydrogen and benzene. At 25°C, the binary diffusion coefficient is 0.40 cm²/sec, and the self-diffusion coefficient of benzene is 0.03 cm²/sec.

Solution. To be consistent with the preceding development, let species 1, 2, and 3 be radioactively tagged benzene, untagged benzene, and hydrogen, respectively. Then, from Eq. 7.2-1, we see that the binary coefficient is

$$D = 0.40 \text{ cm}^2/\text{sec}$$

This coefficient is independent of concentration. The tracer diffusion coefficient is found from Eq. 7.2-2:

$$D^* = \cfrac{1}{\cfrac{y_3}{0.40} + \cfrac{1 - y_3}{0.03}}$$

This result is shown versus hydrogen concentration in Fig. 7.2-3. In this case, the binary and tracer values differ by an unusually large amount, a consequence of the exceptional mobility of hydrogen.

Section 7.3. Diffusion near critical or consolute points

Two limiting types of solute–solvent interactions are possible in solutions. In the first, discussed in Section 7.1, solute and solvent interact to form a new species. In thermodynamic terms, the solvated solute has a lower free energy; in human terms, solvent and solute like each other.

The second type of solute–solvent interaction occurs near critical or consolute points. Here, solute and solvent are on the verge of a phase separation. In this situation, solute and solvent are not randomly distributed, but tend to form small clusters of molecules of one species. In human terms, solute and solvent are chemical racists; they do not like each other and prefer their own kind.

In this section we discuss diffusion under these near critical conditions. Consolute points in liquids are much easier to study than critical points in gases; so we emphasize these in our discussion. Such points are shown schematically as the ×'s in the inset in Fig. 7.3-1. They represent the temperature and composition where two liquids become miscible in all proportions. Where this miscibility occurs with increasing temperature, we observe an upper consolute point. Where it requires a decrease in temperature, we find a lower consolute point.

Near these points, the binary diffusion coefficient D drops precipitously to zero, as shown in Fig. 7.3-1. This drop has been explained in three different ways. Because all three ways remain current in the literature, they merit a review here.

The first explanation for the variation in Fig. 7.3-1 rests on modification of relations like the Stokes–Einstein equation. This equation depends on the assumption that diffusion flux is proportional to chemical potential gradient, that

$$-\mathbf{j}_1 = \frac{D_0 c_1}{RT} \nabla \mu_1 \tag{7.3-1}$$

The chemical potential per mole can be written in terms of the activity $\gamma_1 c_1$:

$$\mu_1 = \mu_1^0 + RT \ln \gamma_1 c_1 \tag{7.3-2}$$

We easily see that

$$-\mathbf{j}_1 = \left[\frac{D_0 c_1}{RT} \left(\frac{\partial \mu_1}{\partial c_1} \right) \right] \nabla c_1 = \left[D_0 \left(1 + \frac{\partial \ln \gamma_1}{\partial \ln c_1} \right) \right] \nabla c_1 \tag{7.3-3}$$

where the quantity in square brackets is the diffusion coefficient D. To simplify the algebra, we further assume that the molar concentration c is constant. Removing this assumption has no effect on the conclusions. Because $c_1 = cx_1$,

$$D = D_0 \left(1 + \frac{\partial \ln \gamma_1}{\partial \ln x_1} \right) \tag{7.3-4}$$

This type of derivation is used commonly, almost without question.

This result can explain some of the features of the data in Fig. 7.3-1. At the consolute point, $\partial \ln \gamma_1 / \partial \ln x_1 = -1$ (King, 1969), but D_0 is approximately

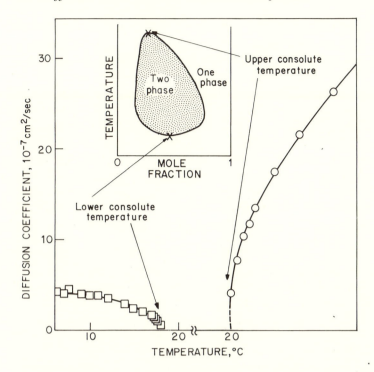

Fig. 7.3-1. Diffusion near consolute temperatures. The squares are for triethyl-amine–water, and the circles represent hexane–nitrobenzene. At the consolute or critical-solution point, the binary diffusion coefficient is zero. [Data from Claersson & Sundelöf (1957) and Haase & Siry (1968).]

constant (Harman et al., 1972). Thus, Eq. 7.3-3 correctly predicts that diffusion at the consolute point is zero.

However, Eq. 7.3-4 does not correctly predict all aspects of diffusion near the consolute point. To illustrate this, consider the system hexane–nitrobenzene, for which results are given in Fig. 7.3-1. The chemical potential of this system is found by experiment to fit the equation

$$\mu_1 = \mu_1^0 + RT \ln x_1 + \omega x_2^2 \tag{7.3-5}$$

This type of chemical potential is sometimes called a "regular solution." At the consolute point of such a solution, $x_1 = x_2 = 0.5$ and $\omega = 2RT_C$, where T_C is the consolute temperature (Prigogine & Defay, 1954). As a result, we find that at the consolute composition,

$$\frac{x_1}{RT}\frac{\partial \mu_1}{\partial c_1} = 1 + \frac{\partial \ln \gamma_1}{\partial \ln x_1} = 1 - \frac{4x_1 x_2 T_C}{T} = 1 - \frac{T_C}{T} \tag{7.3-6}$$

Combining this with Eq. 7.3-4, we obtain

$$D = D_0 \left[\frac{T - T_C}{T}\right] \tag{7.3-7}$$

The temperature variation in brackets is much greater than the temperature variation of D_0 near the consolute point. However, this linear variation is not observed in Fig. 7.3-1, so that Eq. 7.3-4 and Eq. 7.3-7 are inconsistent with experiment.

The second explanation of the results in Fig. 7.3-1 assumes that Fick's law is not an adequate description of diffusion in this region (Anisimov & Perelman, 1966), and so must be replaced by

$$D = D_0[1 + Z\nabla^2 x_1 + \cdots] \tag{7.3-8}$$

where Z is a constant. In other words, the linear form of Fick's law must now include higher terms, just as the flow of non-Newtonian fluids can require higher terms than that in Newton's law of viscosity. Although the details cannot be concisely given, this approach predicts that the diffusion should approach zero at a consolute point; it also can predict the correct temperature dependence.

However, Eq. 7.3-8 does not predict the concentration profiles in a free-diffusion experiment (Brunel & Breuer, 1971). If Fick's law is correct, the concentration gradient in such an experiment should vary with the square root of time. If Eq. 7.3-8 is correct, the variation should be with the fourth root of time. Experiments near the consolute point conclusively show variation with the square root of time, as shown in Fig. 7.3-2. Thus, Eq. 7.3-8 is not valid over the current experimental range.

The third explanation of the data in Fig. 7.3-1 is currently the most successful. It assumes that long-range fluctuations dominate behavior near the consolute point. When fluctuations of concentration and of fluid velocity couple, diffusion occurs. Under ordinary conditions, the concentration fluctuations are dominated by motion of single molecules; but near the critical point, these fluctuations include motion of clusters of molecules. The velocity fluctuations exist even when the average fluid velocity is zero. The result is like a turbulent "eddy diffusion coefficient," but without flow.

When the details of these coupled fluctuations are considered, the diffusion coefficient is found to be

$$D = \frac{k_B T}{2\pi\mu\xi} \tag{7.3-9}$$

where the correlation length ξ is approximately the average size of a cluster. This approach retains the same temperature and viscosity dependence as the Stokes–Einstein equation. The factor 2π in place of 6π is not a major change. However, both the diffusion coefficient D and the length ξ vary dramatically with the thermodynamic factor $(1 + \partial \ln \gamma_1/\partial \ln x_1)$.

The calculation of ξ as a function of temperature and composition can proceed in two different ways. The best way is to depend on scaling laws developed for phase transitions that in turn are based most frequently on the Ising model (Stanley, 1971). Such calculations give the temperature dependence at the critical composition:

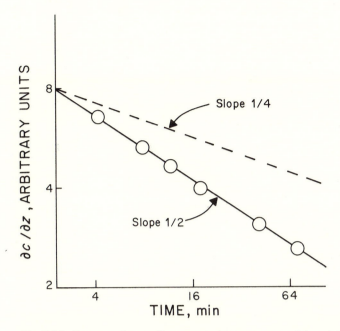

Fig. 7.3-2. Concentration gradient versus time near a consolute point. If Fick's law is valid, the data should fall along a line of slope one-half. If a new diffusion law is involved, the data should fall along a line of one-fourth. [Data from Brunel & Breuer (1971).]

$$D \propto \left(\frac{T}{T_C} - 1\right)^{0.62} \tag{7.3-10}$$

Unfortunately, these calculations are not so complete as to give the concentration dependence of ξ and D. The alternative, less accurate route in the calculation of ξ is to use simple models of the chemical potential like that developed by Kirkwood and Buff (1951). These models lead to an equation at the consolute temperature:

$$D = D_0 \left[1 + \frac{A}{x_1 x_2}\left(\frac{\partial \ln x_1}{\partial \ln \gamma_1 x_1} - 1\right)\right]^{-1/2} \tag{7.3-11}$$

in which A is a constant of the order of one-half. They also lead to an equation at the consolute concentration:

$$D \propto \left(\frac{T}{T_C} - 1\right)^{1/2} \tag{7.3-12}$$

Using the Kirkwood–Buff model gives a slightly smaller exponent than the Ising-based models (Cussler, 1980).

The predictions based on coupled fluctuations are compared with those of the more traditional results in Fig. 7.3-3. This figure includes data on four

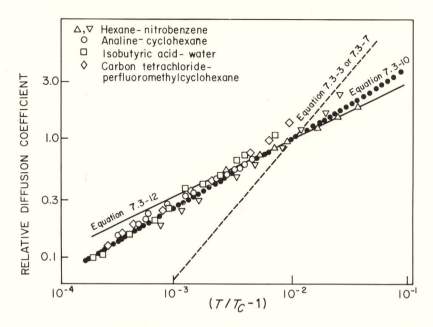

Fig. 7.3-3. Diffusion versus temperature near a consolute point. The classic theories shown by the broken line are much less successful than the predictions of scaling laws (dotted line) or of cluster diffusion (solid line). [From Cussler (1980), with permission.]

different systems obtained in five different laboratories using four different experimental methods. The data all appear consistent. They fall very close to the predictions of Eq. 7.3-10, which is based on the coupled fluctuations as described by scaling laws. They are in reasonable agreement with the predictions of Eq. 7.3-12, which uses simple statistical models for chemical potential. These results support the explanation of diffusion near the critical point in terms of coupled fluctuations of concentration and velocity.

Example 7.3-1: Diffusion through a consolute point

Imagine a diaphragm cell of two well-stirred compartments (see Example 2.2-4). One compartment contains water, and the other contains triethylamine. Diffusion occurs across the diaphragm between the two compartments. However, this experiment will be made at the consolute temperature of 18.6°C. As a result, somewhere within the diaphragm, the concentration must be that at the consolute point, and the diffusion coefficient at that point will approach zero, as shown in Fig. 7.3-1. What will happen in this experiment?

Solution. If we make a mass balance on a thin slice of the diaphragm, we find that

$$0 = \frac{-dn_1}{dz} = \frac{-dj_1}{dz}$$

This means that there will be a steady-state flux across the diaphragm. When we combine this result with Fick's law, we find

$$-j_i = D(c_1) \frac{dc_1}{dz} = \text{constant}$$

At the consolute concentrations, $D(c_1)$ approaches zero; so dc_1/dz must approach infinity. Thus, in this experiment, the flux behaves normally, but the concentration gradient reflects the unusual properties of the diffusion coefficient (see Problem 7.6).

Section 7.4. Solute–boundary interactions

When a solute diffuses through small pores, its speed is affected by the size and shape of the pores. For example, a solute will diffuse faster through a large straight pore than through a small crooked one. In this section we explore these effects in more detail. We first discuss diffusion in porous media, and then switch to more idealized situations in which more exact results are possible.

Diffusion in porous media

Imagine a solute diffusing through the fluid-filled pores of the porous solid shown schematically in Fig. 7.4-1(a). In this case, diffusion takes place only through the cramped and tortuous pores of the composite. Because the pores are not straight, the diffusion effectively takes place over a longer distance than it would in a homogeneous material. Because the solid is impermeable, diffusion occurs over a smaller cross-sectional area than that available in a homogeneous material.

The effects of longer pores and smaller areas are often lumped together in the definition of a new, effective diffusion coefficient D_{eff} (Geankoplis, 1972):

$$D_{\text{eff}} = \frac{\varepsilon}{a^2} D \tag{7.4-1}$$

D is the diffusion coefficient within the pores, in which ε is the void fraction, and a is the actual pore length per distance in the direction of diffusion. In other cases, this effective diffusion coefficient is less explicitly defined (Aris, 1975):

$$D_{\text{eff}} = \frac{D}{\tau} \tag{7.4-2}$$

where τ is the tortuosity that attempts to account for the reduced area of longer pores. Tortuosities usually range between 2 and 6, averaging about 3. These values can be rationalized because solutes diffuse in three directions instead of one; so they diffuse about three times as far. Such rationalization is suspect. I have measured tortuosities as high as 10, which I find hard to

(a) Tortuous diffusion in a porous solid

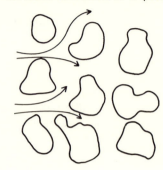

(b) Diffusion through a spatially periodic composite

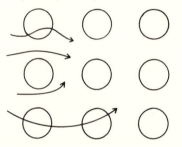

Fig. 7.4-1. Diffusion in porous media. This diffusion can be altered by a tortuous path, as shown in (a), or by two phases in which the diffusion coefficients are different, as shown in (b).

justify on geometrical arguments alone. Moreover, the tortuosity measured by diffusion may bear little relation to the tortuosity found for fluid flow.

Diffusion in a periodic composite

One special case of diffusion in a porous solid occurs when the solid consists of periodically spaced spheres like those shown in Fig. 7.4-1(b). In this case, we assume that diffusion can take place both in the interstitial region between the spheres and through the spheres themselves. The effective diffusion coefficient D_{eff} can be calculated exactly (Maxwell, 1873):

$$\frac{D_{\text{eff}}}{D} = \frac{\dfrac{2}{D_s} + \dfrac{1}{D} - 2\phi_s \left(\dfrac{1}{D_s} - \dfrac{1}{D} \right)}{\dfrac{2}{D_s} + \dfrac{1}{D} + \phi_s \left(\dfrac{1}{D_s} - \dfrac{1}{D} \right)} \tag{7.4-3}$$

in which D is the diffusion coefficient in the interstitial pores, D_s is the diffusion coefficient through the spheres, and ϕ_s is the volume fraction of the spheres in the composite material.

This is a fascinating result. It says that diffusion does not depend on the size of the spheres in the composite but only on their volume fraction. It does not matter if the spheres are birdshot or basketballs – the diffusion is the same if the volume fraction is the same.

A second interesting consequence of Eq. 7.4-3 is that the properties of the continuous phase dominate the diffusion process. To demonstrate this, we imagine that the spheres are impenetrable, so that D_s is zero. Then Eq. 7.4-3 becomes

$$\frac{D_{\text{eff}}}{D} = \frac{2(1 - \phi_s)}{2 + \phi_s} \tag{7.4-4}$$

If ϕ_s is ½, then $D_{\text{eff}}/D = 2/5$. The diffusion is 40% of what it would be without the spheres.

Now consider the other limit in which diffusion through the spheres is extremely rapid, so that $D_s \to \infty$. In this case,

$$\frac{D_{\text{eff}}}{D} = \frac{1 + 2\phi_s}{1 - \phi_s} \tag{7.4-5}$$

If ϕ_s is still ½, then D_{eff}/D is 4. Thus, changing the diffusion coefficient in the spheres from zero to infinity changes the effective diffusion coefficient only by a factor of 10.

I must emphasize that Eq. 7.4-3 is limited to periodic composites with one continuous phase and one discontinuous phase. If we want to calculate the diffusion of slaked lime through water-saturated brick, neither water nor brick is discontinuous, and Eq. 7.4-3 does not apply. In that case, we must return to empirically determined tortuosities.

Gas diffusion at low pressure: Knudsen diffusion

The preceding discussion began with the concept of tortuosity, an approximation that is applicable to any porous medium. The discussion then switched to periodic arrays of spheres as a model of a composite medium. We now continue this discussion for more specific geometry, a straight cylindrical pore.

The basic concern here is the diffusion of individual solute molecules in small pores. If the pores are very small, a particular molecule will collide most often with pore walls, as suggested by Fig. 7.4-2(a). It will not collide frequently with other molecules.

This type of transport is called Knudsen diffusion. It is important whenever the distance between molecular collisions is greater than the pore diameter. This ratio of distances is dignified as a dimensionless group, the Knudsen number Kn (Cunningham & Williams, 1980):

$$\text{Kn} = \frac{l}{d} \tag{7.4-6}$$

in which l is now the mean free path and d is the pore diameter. If the Knudsen number is small, diffusion has the same characteristics as it does outside of the pores, and it is analyzed with the effective coefficients and

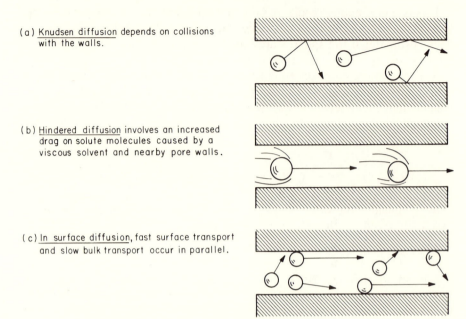

(a) Knudsen diffusion depends on collisions with the walls.

(b) Hindered diffusion involves an increased drag on solute molecules caused by a viscous solvent and nearby pore walls.

(c) In surface diffusion, fast surface transport and slow bulk transport occur in parallel.

Fig. 7.4-2. Diffusion in a pore. Diffusion here can be affected by collisions with the wall, by viscous drag on the wall, or by adsorption on the wall.

tortuosities given earlier. If the Knudsen number is large, diffusion is dominated by collisions with the boundaries; this requires a different description.

For liquids, the mean free path is commonly a few angstroms; so the Knudsen number is almost always small, and Knudsen diffusion is not important.

In gases, the mean free path can be estimated from the following (Cunningham & Williams, 1980):

$$l = \frac{k_B T}{\sqrt{2}\,\pi\sigma_{ii}^2 p} \tag{7.4-7}$$

in which σ_{ii} is the collision diameter of the diffusing species. This mean free path can be large. For example, for air at room temperature and pressure, it is over 600 Å; for hydrogen at 300°C and 1 atm, it is over 2,000 Å. Because pores smaller than these values often exist, for example, in porous catalysts, Knudsen diffusion can be a significant effect in gases.

When the mean free path and the Knudsen number are large, the flux is given by an alternative form of flux equation:

$$n_i = -D_{Kn}\frac{dc_i}{dx} \tag{7.4-8}$$

The Knudsen diffusion coefficient D_{Kn} can be found for dilute gases from (Evans et al., 1961)

$$D_{Kn} = 4{,}850d\,\sqrt{T/\bar{M}} \tag{7.4-9}$$

in which d is the pore diameter in centimeters and, as before, T and $\tilde{M}$ are temperature and molecular weight. Unlike gas diffusion out of the pores, the diffusion coefficient is independent of pressure.

Diffusion in liquid-filled pores: the Rankin equation

When a solute diffuses in gas-filled pores, the solute molecules can frequently collide with the pore walls. When solute molecules diffuse in liquid-filled pores, these direct encounters with the pore walls are infrequent. This is because molecules in liquids are so close together that they rarely move more than a few angstroms before hitting each other.

Solute diffusion liquid-filled pores can be profoundly affected by the different mechanism shown in Fig. 7.4-2(b). This mechanism is modeled as a viscous drag caused by the adjacent walls. The results of a spherical particle diffusing in a right-circular cylindrical pore are summarized as the Rankin equation, calculated in analytical form by Gajdos and Brenner (1978) as

$$\frac{D}{D_0} = 1 + \frac{9}{8} \lambda \ln \lambda - 1.54\lambda + O(\lambda^2) \tag{7.4-10}$$

in which λ $(= 2R_0/d)$ is the solute diameter divided by the pore diameter. This result is accurate to around 2% when λ is less than 0.2 (Quinn et al., 1972).

Surface diffusion

The final type of solute–boundary interaction considered here is surface diffusion. This term is often used as an excuse for many unexpected observations (Cunningham & Williams, 1980). Although these observations certainly are correct, the interpretation as surface diffusion remains, for me, incomplete and unsatisfying.

The basic physical picture, given in Fig. 7.4-2(c), postulates two physical processes occurring in parallel. The first of these processes, taking place within the pore, is the usual form of diffusion. This diffusion is relatively slow. The second process, on the walls of the pores, is the relatively fast surface diffusion. Such a process includes rapid solute adsorption, rapid transport while adsorbed, and rapid solute desorption.

This physical picture is used to rationalize diffusion coefficients in porous solids that occur more rapidly than expected. It explains diffusion that decreases with temperature as a decreased adsorption and hence decreased surface transport. However, more complete studies of surface diffusion have been frustrated by experimental problems. For example, at a liquid–liquid interface, one must study simultaneously surface diffusion along the interface and bulk diffusion in each liquid phase. Such complexity has inhibited understanding.

Example 7.4-1: Diffusion in a porous catalyst

Imagine a catalyst sphere to be used for the dehydrogenation reaction

$$C_2H_6 \rightarrow C_2H_4 + H_2$$

At 300°C and 1 atm, the effective diffusion coefficient of ethane in a 0.5-cm sphere is 0.17 cm²/sec. What is the tortuosity?

Solution. The chemical reaction produces a ternary mixture of ethane, ethylene, and hydrogen. Such a mixture may require consideration of the multicomponent diffusion equations in Chapter 8. However, if conversion is low, the diffusion coefficient can be estimated with the same precision as for a mixture of ethane and ethylene. From Eq. 5.1-1, we find

$$D = \frac{1.86 \cdot 10^{-3} T^{3/2} (1/\tilde{M}_1 + 1/\tilde{M}_2)^{1/2}}{p\sigma_{12}^2 \Omega}$$

$$= \frac{1.86 \cdot 10^{-3}(573)^{3/2}(1/28 + 1/26)^{1/2}}{(1)[(4.23 + 4.16)/2]^2(0.99)}$$

$$= 0.40 \text{ cm}^2/\text{sec}$$

The tortuosity is then

$$\tau = \frac{D}{D_{\text{eff}}} = \frac{0.40 \text{ cm}^2/\text{sec}}{0.17 \text{ cm}^2/\text{sec}} = 2.3$$

This value is typical.

Example 7.4-2: Effective diffusion in an inhomogeneous gel

The diffusion coefficient of KCl through a protein gel is $6 \cdot 10^{-7}$ cm²/sec. However, the gel is not homogeneous, because it contains water droplets about 10^{-2} cm in diameter that are separated by only $2 \cdot 10^{-2}$ cm. The diffusion in these water droplets is about $2 \cdot 10^{-5}$ cm²/sec. What is the diffusion in the homogeneous gel?

Solution. The volume fraction of water droplets can be found by considering a unit cell, $2 \cdot 10^{-2}$ cm on a side s, drawn around each 10^{-2}-cm droplet:

$$\phi_D = \frac{\frac{4}{3}\pi r^3}{s^3} = \frac{\frac{4}{3}\pi \left(\frac{10^{-2} \text{ cm}}{2}\right)^3}{(2 \cdot 10^{-2} \text{ cm})} = 0.065$$

The diffusion in the gel is then found from Eq. 7.4-3:

$$\frac{D_{\text{eff}}}{D} = \frac{\dfrac{2}{D_s} + \dfrac{1}{D} - 2\phi_s \left(\dfrac{1}{D_s} - \dfrac{1}{D}\right)}{\dfrac{2}{D_s} + \dfrac{1}{D} + \phi s \left(\dfrac{1}{D_s} - \dfrac{1}{D}\right)}$$

$$\frac{6 \cdot 10^{-7} \text{ cm}^2/\text{sec}}{D} = \frac{\dfrac{2}{2 \cdot 10^{-5} \text{ cm}^2/\text{sec}} + \dfrac{1}{D} - 2(0.065)\left(\dfrac{1}{2 \cdot 10^{-5} \text{ cm}^2/\text{sec}} - \dfrac{1}{D}\right)}{\dfrac{2}{2 \cdot 10^{-5} \text{ cm}^2/\text{sec}} + \dfrac{1}{D} + 0.065\left(\dfrac{1}{2 \cdot 10^{-5} \text{ cm}^2/\text{sec}} - \dfrac{1}{D}\right)}$$

Solving, we find that D equals about $5 \cdot 10^{-7}$ cm²/sec.

Example 7.4-3: Diffusion of hydrogen in small pores

Find the steady diffusion flux at 100°C and 1 atm for hydrogen diffusing into nitrogen through a plug effectively 0.6 cm thick with 130-Å pores. Then estimate the flux through $18.3 \cdot 10^{-4}$-cm pores.

Solution. The mean free path for hydrogen can be found from Eq. 7.4-7:

$$
\begin{aligned}
\lambda &= \frac{k_B T}{\sqrt{2}\,\pi\sigma_{11}^2 p} \\
&= \frac{(1.38 \cdot 10^{-16}\ \text{g-cm}^2/\text{sec}^2\text{-}°\text{K})(373°\text{K})}{\sqrt{2}\,\pi(2.83 \cdot 10^{-8}\ \text{cm})^2(1.01 \cdot 10^6\ \text{g/cm-sec}^2)} \\
&= 1{,}400\ \text{Å}
\end{aligned}
$$

This mean free path is greater than the pore diameter; so the Knudsen number is large. Thus, diffusion takes place in the Knudsen regime. For steady-state transport, Eq. 7.4-8 can be integrated to give

$$
\begin{aligned}
n_1 &= [D_{\text{Kn}}]\frac{\Delta c_1}{l} \\
&= \left[4{,}850(130 \cdot 10^{-8})\sqrt{\frac{373}{2}}\right]\ \text{cm}^2/\text{sec}\left\{\frac{\left(\dfrac{1\ \text{g-mol}}{22.4 \cdot 10^3\ \text{cm}^3}\right)\dfrac{273°\text{K}}{373°\text{K}}}{0.6\ \text{cm}}\right\} \\
&= 0.37 \cdot 10^{-5}\ \text{g-mol/cm}^2\text{-sec}
\end{aligned}
$$

There are two interesting features of this result. First, hydrogen molecules spend their time colliding with pore walls, not with nitrogen molecules. Consequently, the properties of nitrogen do not appear in the calculation. Second, we have assumed that the pores are as long as the plug is thick; so the pores are implicitly taken to be straight. Any tortuosity would reduce the flux.

For the $18.3 \cdot 10^{-4}$-cm pores, the mean free path is much less than the pore diameter, because now the Knudsen number is small. In this case, the flux equation contains the usual diffusion coefficient, calculated from Eq. 5.1-1:

$$
\begin{aligned}
n_1 &= j_1 = [D]\frac{\Delta c_1}{l} \\
&= \frac{1.86 \cdot 10^{-3}(373)^{3/2}\left(\dfrac{1}{2.01}+\dfrac{1}{28.0}\right)^{1/2}}{1\ \text{atm}\left(\dfrac{2.92+3.68}{2}\right)^2(0.80)}\left\{\frac{\dfrac{1\ \text{mol}}{22.4 \cdot 10^3\ \text{cm}^3}\left(\dfrac{273°\text{K}}{373°\text{K}}\right)}{0.6\ \text{cm}}\right\} \\
&= 6.1 \cdot 10^{-5}\ \text{g-mol/cm}^2\text{-sec}
\end{aligned}
$$

This flux is substantially greater than that in the Knudsen limit.

Example 7.4-4: Pores in cell walls

Some experiments on living cells suggest that there are pores 30 Å in diameter in the cell wall. Estimate the diffusion coefficient at 37°C for a solute 5 Å in diameter through such a pore.

Solution. To find the solute's diffusion coefficient in bulk solution, we use the Stokes–Einstein equation. Combining this with Eq. 7.4-20, we find

$$D = \frac{k_B T}{6\pi\mu R_0} \left[1 + \frac{9}{8}\left(\frac{2R_0}{d}\right) \ln\left(\frac{2R_0}{d}\right) - 1.54\left(\frac{2R_0}{d}\right) + \cdots \right]$$

$$= \frac{1.38 \cdot 10^{-16} \text{ g-cm}^2/\text{sec}^2\text{-}°\text{K})(3.10°\text{K})}{6\pi(0.01 \text{ g/cm-sec})(2.5 \cdot 10^{-8} \text{ cm})}$$

$$\cdot \left[1 + \frac{9}{8}\left(\frac{5}{30}\right) \ln\left(\frac{5}{30}\right) - 1.54\left(\frac{5}{30}\right) + \cdots \right]$$

$$= (9.1 \cdot 10^{-6} \text{ cm}^2/\text{sec})(1 - 0.33 - 0.26 + \cdots)$$

$$= 3.7 \cdot 10^{-6} \text{ cm}^2/\text{sec}$$

Note that we have implicitly assumed that the pore is filled with water by using the viscosity of water in this estimate.

Section 7.5. A final perspective

At the start of this book we argued that the simplest way to look at diffusion was as a dilute solution of a particular solute moving through a homogeneous solvent. Such an argument led to the idea of a diffusion coefficient, a particular property of solute and solvent.

In this chapter we have discussed the effects on the diffusion coefficient of the solute's interaction with other parts of the system. Sometimes the solute combines with solvent molecules, but near consolute points it avoids them. In a porous medium, the solute's diffusion may be slowed or accelerated; it may collide with pore walls during Knudsen diffusion, or be adsorbed in surface diffusion. In every case, the changes in diffusion can be major.

In describing these effects, scientists have used elaborate methods. For example, the mathematics leading to the Rankin equation for hindered diffusion pose, for me, a truly formidable exercise in fluid mechanics. Obtaining the results of diffusion in composite media required the genius of Clerk Maxwell.

However, these descriptions are limited by the approximate models of the diffusion process. For example, the ideas of hydration are certainly inexact. Even the Rankin equation depends on the model of a rigid solute sphere in a solvent continuum.

As a result, I believe that the results in this chapter are best applied when using your scientific judgment. I do not think any of the ideas are gospel. Instead, they are approximations, subject to corrections found in future research. I wish you luck in finding these corrections.

References

Albright, J. G., & Mills, R. (1965). *Journal of Physical Chemistry*, **69**, 3120.

Anisimov, S. I., & Perelman, T. L. (1966). *International Journal of Heat and Mass Transfer*, **9**, 1279.

Aris, R. (1975). *Mathematical Theory of Diffusion and Reaction in Permeable Catalysis*, p. 25. Oxford: Clarendon Press.

Bremer, M. F., & Cussler, E. L. (1970). *American Institute of Chemical Engineers Journal*, **16**, 832.

Brunel, M. E., & Breuer, M. M. (1971). *Diffusion Processes, Vol. 1*, ed. J. N. Sherwood et al., p. 119. London: Gordon and Breach.

Claersson, S., & Sundelöf, L. O. (1957). *Journal de Chimie Physique*, **54**, 914.

Cunningham, R. E., & Williams, R. J. J. (1980). *Diffusion in Gases and Porous Media*. New York: Plenum Press.

Curran, P. F., Taylor, A. E., & Soloman, A. K. (1967). *Biophysical Journal*, **7**, 879.

Cussler, E. L. (1980). *American Institute of Chemical Engineers Journal*, **26**, 43.

Evans, R. B., Watson, G. M., & Mason, E. A. (1961). *Journal of Chemical Physics*, **33**, 2076.

Franks, F., ed. (1975). *Water – A Comprehensive Treatise*. New York: Plenum Press.

Gajdos, L. J., & Brenner, H. (1978). *Separation Science and Technology*, **13**, 215.

Geankoplis, C. (1972). *Mass Transfer Phenomena*, pp. 149ff. New York: Holt, Rinehart and Winston.

Haase, R., & Siry, M. (1968). *Zeitschrift für Physicalische Chemie (Neue Folge)*, **57**, 56.

Harman, H., Hohersil, C., & Richtering, H. (1972). *Berichte der Bunsen Gesellschaft für Physikalische Chemie*, **76**, 249.

Hinton, J. F., & Amis, E. S. (1971). *Chemical Reviews*, **71**, 621.

Kelly, C. M., Wirth, G. B., & Anderson, D. K. (1971). *Journal of Physical Chemistry*, **75**, 3293.

King, M. B. (1969). *Phase Equilibrium in Mixtures*. New York: Pergamon Press.

Kirkwood, J. G., & Buff, F. P. (1951). *Journal of Chemical Physics*, **19**, 774.

Maxwell, J. C. (1873). *A Treatise on Electricity and Magnetism, Vol. 1*, p. 365. Oxford: Clarendon Press.

Mills, R. (1965). *Journal of Physical Chemistry*, **69**, 3116.

Prigogine, I., & Defay, R. (1954). *Treatise on Thermodynamics*. London: Longman.

Quinn, J. A., Anderson, J. L., Ho, W. S., & Petzny, W. J. (1972). *Biophysical Journal*, **12**, 990.

Robinson, R. G., & Stokes, R. H. (1959). *Electrolyte Solutions*, Chapter 11. London: Butterworth.

Scatchard, G. (1921). *Journal of the American Chemical Society*, **43**, 2406.

Shedlovsky, T., & Kay, R. L. (1956). *Journal of Physical Chemistry*, **60**, 151.

Stanley, H. E. (1971). *Introduction to Phase Transitions and Critical Phenomena*. New York: Oxford University Press.

8 MULTICOMPONENT DIFFUSION

Throughout this book we have routinely assumed that diffusion takes place in binary systems. We have described these systems as containing a solute and a solvent, although such specific labels are arbitrary. We often have further assumed that the solute is present at low concentration, so that the solutions are always dilute. Such dilute systems can be analyzed much more easily than concentrated ones.

In addition to these binary systems, other diffusion processes include the transport of many solutes. One group of these processes occurs in the human body. Simultaneous diffusion of oxygen, sugars, and proteins takes place in the blood. Mass transfer of bile salts, fats, and amino acids occurs in the small intestine. Sodium and potassium ions cross many cell membranes by means of active transport. All these physiological processes involve simultaneous diffusion of many solutes.

This chapter describes diffusion for these and other multicomponent systems. The current formalism of multicomponent diffusion, however, is of limited value. The more elaborate flux equations and the slick methods used to solve them are often unnecessary for an accurate description. There are two reasons for this. First, multicomponent effects are minor in dilute solutions, and most solutions are dilute. For example, the diffusion of sugars in blood is accurately described with the binary form of Fick's law. Second, some multicomponent effects are often more lucid if described without the cumbersome equations splattered through this chapter. For example, the diffusion of oxygen and carbon dioxide in blood is better described by considering explicitly the chemical reactions with hemoglobin.

Nonetheless, some concentrated systems are best described using multicomponent diffusion equations. Examples of these systems, which commonly involve unusual chemical interactions, are listed in Table 8.0-1. They are best described using the equations derived in Section 8.1. These equations can be rationalized using the theory of irreversible thermodynamics, a synopsis of which is given in Section 8.2. In most cases, the solution to multicomponent diffusion problems is automatically available if the binary solution is available; the reasons for this are given in Section 8.3. Finally, some values of ternary diffusion coefficients are given in Section 8.4 as an indication of the magnitude of the effects involved.

Table 8.0-1. *Systems with large multicomponent effects*

Type of system	Examples
Solutes of very different sizes	Hydrogen–methane–argon (Arnold & Toor, 1967) Polystyrene–cyclohexane–toluene (Cussler & Lightfoot, 1965)
Solutes in highly nonideal solutions	Mannitol–sucrose–water (Ellerton & Dunlop, 1967) Acetic acid–chloroform–water (Vitagliano et al., 1969)
Concentrated electrolytes	Sodium sulfate–sulfuric acid–water (Wendt, 1962) Sodium chloride–polyacrylic acid–water (Vermeulen et al., 1967)
Concentrated alloys	Zinc–cadmium–silver (Carlson et al., 1972)
Membranes with mobile carriers	Sodium chloride–hydrochloric acid–monensin as a mobile carrier (Cussler et al., 1971) Oxygen–carbon dioxide–hemoglobin as a mobile carrier (Keller & Friedlander, 1966)

Section 8.1. The flux equations for multicomponent diffusion

Binary diffusion is often most simply described by Fick's law relative to the volume average velocity $\mathbf{v}^0$:

$$-\mathbf{j}_i = c_i(\mathbf{v}^0 - \mathbf{v}_i) = D\nabla c_i \qquad (8.1\text{-}1)$$

In most cases, multicomponent diffusion is most simply described by generalizing this equation to an *n*-component system (Onsager, 1945):

$$-\mathbf{j}_i = c_i(\mathbf{v}^0 - \mathbf{v}_i) = \sum_{j=1}^{n-1} D_{ij}\,\nabla c_j \qquad (8.1\text{-}2)$$

in which the D_{ij} are multicomponent diffusion coefficients. The relation between these coefficients and the binary values is not known, except for the dilute-gas limit, given for ternary diffusion in Table 8.1-1 (Cussler, 1976). In general, the diffusion coefficients are not symmetric (i.e., $D_{ij} \neq D_{ji}$). The diagonal terms (the D_{ii}) are called the "main-term" diffusion coefficients, because they commonly are large and similar in magnitude to binary values. The off-diagonal terms (the $D_{ij,i \neq j}$), called the "cross-term" diffusion coefficients, are often 10% or less of the main terms. Each cross term gives a measure of the flux of one solute that is engendered by the concentration gradient of a second solute.

For an *n*-component system, this equation contains $(n - 1)^2$ diffusion coefficients. This implies that one component must be arbitrarily designated as the solvent *n*. Because of the Onsager reciprocal relations discussed in

Table 8.1-1. *Ternary diffusion coeffi-cients: known functions of binary val-ues for ideal gases*

$$D_{11} = \left[\frac{\dfrac{y_1}{\mathcal{D}_{12}} + \dfrac{y_2 + y_3}{\mathcal{D}_{23}}}{\dfrac{y_1}{\mathcal{D}_{12}\mathcal{D}_{13}} + \dfrac{y_2}{\mathcal{D}_{12}\mathcal{D}_{23}} + \dfrac{y_3}{\mathcal{D}_{13}\mathcal{D}_{23}}} \right]$$

$$D_{12} = \left[\frac{y_1 \left(\dfrac{1}{\mathcal{D}_{13}} - \dfrac{1}{\mathcal{D}_{12}} \right)}{\dfrac{y_1}{\mathcal{D}_{12}\mathcal{D}_{13}} + \dfrac{y_2}{\mathcal{D}_{12}\mathcal{D}_{23}} + \dfrac{y_3}{\mathcal{D}_{13}\mathcal{D}_{23}}} \right]$$

$$D_{21} = \left[\frac{y_2 \left(\dfrac{1}{\mathcal{D}_{23}} - \dfrac{1}{\mathcal{D}_{12}} \right)}{\dfrac{y_1}{\mathcal{D}_{12}\mathcal{D}_{13}} + \dfrac{y_2}{\mathcal{D}_{12}\mathcal{D}_{23}} + \dfrac{y_3}{\mathcal{D}_{13}\mathcal{D}_{23}}} \right]$$

$$D_{22} = \left[\frac{\dfrac{y_1 + y_3}{\mathcal{D}_{13}} + \dfrac{y_2}{\mathcal{D}_{12}}}{\dfrac{y_1}{\mathcal{D}_{12}\mathcal{D}_{13}} + \dfrac{y_2}{\mathcal{D}_{12}\mathcal{D}_{23}} + \dfrac{y_3}{\mathcal{D}_{13}\mathcal{D}_{23}}} \right]$$

Section 8.2, the coefficients are not all independent, but instead are subject to certain restraints (Woolf et al., 1962):

$$\sum_{j=1}^{n-1} \sum_{l=1}^{n-1} \left(\frac{\partial \mu_l}{\partial c_i} \right)_{c_{k \neq i,n}} \alpha_{lj} D_{jk} = \sum_{j=1}^{n-1} \sum_{l=1}^{n-1} \left(\frac{\partial \mu_l}{\partial c_k} \right)_{c_{i \neq k,n}} \alpha_{lj} D_{ji} \tag{8.1-3}$$

where

$$\alpha_{lj} = \left(\delta_{lj} + \frac{c_j \bar{V}_l}{c_n \bar{V}_n} \right) \tag{8.1-4}$$

These restraints reduce the number of diffusion coefficients required to de-scribe diffusion to $(\frac{1}{2})[n(n-1)]$ for an n-component system (deGroot & Mazur, 1962). However, because applications of these restraints require detailed thermodynamic information that is rarely available, the restraints are frequently impossible to apply, and by default the system is treated as having $(n-1)^2$ independent diffusion coefficients.

Equation 8.1-2 is the most useful form of the multicomponent flux equa-tions. Because of an excess of theoretical zeal, many who work in this area have nurtured a glut of alternatives. These zealots most commonly use different driving forces or reference velocities. Unfortunately, most of their answers are of very limited value. The exception is for some metal alloys (see, e.g., Dayananda, 1968).

The one good alternative to Eq. 8.1-2 is the Stefan–Maxwell equation for dilute gases (Curtiss & Hirschfelder, 1949):

$$\nabla y_i = \sum_{j=1}^{n-1} \frac{y_i y_j}{\mathcal{D}_{ij}} (\mathbf{v}_j - \mathbf{v}_i) \tag{8.1-5}$$

This equation has two major advantages over Eq. 8.1-2. First, these diffusion coefficients are the binary values found from binary experiments or calculated from the Chapman–Enskog theory given in Section 5.1. Second, the Stefan–Maxwell equations do not require designating one species as a solvent, which is sometimes an inconvenience when using Eq. 8.1-2.

At the same time, the Stefan–Maxwell form has serious disadvantages. Most important, it is limited to dilute gases. Although a parallel form can be used for liquids (Lamm, 1944; Lightfoot et al., 1962), the new diffusion coefficients in this parallel form no longer equal the binary values. In addition, the Stefan–Maxwell form is difficult to combine with mass balances, without designating one of the species as a solvent. Thus, in practice, both advantages of this form are often lost. As a result, Eq. 8.1-2 remains the most useful form of flux equation. We next examine the origins of these equations more carefully using irreversible thermodynamics.

Section 8.2. Irreversible thermodynamics

The multicomponent flux equations given in Eq. 8.1-2 are empirical generalizations of Fick's law that define a set of multicomponent diffusion coefficients. Because such definitions are initially intimidating, many have felt the urge to rationalize the origin of these equations and buttress this rationale with "more fundamental principles." This emotional need is often met with derivations based on irreversible thermodynamics (Fitts, 1962; Katchalsky & Curran, 1967; Haase, 1969).

Because the derivation of irreversible thermodynamics is straightforward, it seems on initial reading to be extremely valuable. After all those years of laboring under the restraint of equilibrium, the treatment of departures from equilibrium seems like a new freedom. Eventually one realizes that although irreversible thermodynamics does give the proper form of the flux equations and clarifies the number of truly independent coefficients, this information is of little value, because it is already known from experiment. Irreversible thermodynamics tells us nothing about the nature and magnitude of the coefficients in the multicomponent equations, nor the resulting size and nature of the multicomponent effects. These are the topics in which we are interested. As a result, irreversible thermodynamics has enjoyed an overoptimistic vogue, first in chemical physics, then in engineering, and currently in biophysics. Subsequently it has been deemphasized as its limitations have become recognized. Because irreversible thermodynamics is of limited utility in describing multicomponent diffusion, only the barest outline will be given here.

The entropy production equation

Three basic postulates are involved in the derivation of Eq. 8.1-2 (Fitts, 1962). The first postulate states that thermodynamic variables such as entropy, chemical potential, and temperature can in fact be correctly defined in a *differential volume* of a system that is not at equilibrium. This is an excellent approximation, except for systems that are very far from equilibrium, such as explosions. In the simple derivation given here, we assume a system of constant density, temperature, and pressure, with no net flow or chemical reaction. More complete equations without these assumptions are derived elsewhere (e.g., Haase, 1969).

The mass balance for each species in this type of system is given by

$$\frac{\partial c_i}{\partial t} = -\nabla \cdot \mathbf{n}_i = -\nabla \cdot \mathbf{j}_i \tag{8.2-1}$$

In this continuity equation, we use the fact that at no net flow and constant density, $\mathbf{n}_i$ equals $\mathbf{j}_i$, the flux relative to the volume or mass average velocity. We also imply that the concentration is expressed in mass per unit volume. The left-hand side of this equation represents solute accumulation, and the right-hand side represents the solute diffusing in minus that diffusing out. The energy equation is similar:

$$\rho \frac{\partial \hat{H}}{\partial t} = -\nabla \cdot \mathbf{q} - \nabla \cdot \sum_{i=1}^{n} \bar{H}_i \mathbf{j}_i \tag{8.2-2}$$

The left-hand side of this relation is the accumulation; the first term on the right-hand side is the energy conducted in minus that conducted out; and the second term is the energy diffusing in minus that diffusing out. Because we are assuming an isothermal system, $\mathbf{q}$ is presumably zero; we include it here so that the equation will look more familiar.

By parallel arguments, we can write a similar equation for entropy:

$$\rho \frac{\partial \hat{S}}{\partial t} = -\nabla \cdot \mathbf{J}_s + \sigma \tag{8.2-3}$$

By analogy, the term on the left must be the entropy accumulation, and the first term on the right must be entropy in minus entropy out by both convection and diffusion. The second term on the right, σ, gives the entropy produced in the process. This entropy production, which must be positive, is the quantitative measure of irreversibility in the system and represents a novel contribution of irreversible thermodynamics.

To find the entropy production, we first recognize that in this isothermal system,

$$d\hat{G} = d\hat{H} - T\,d\hat{S} = \frac{1}{\rho} \sum_{i=1}^{n} \mu_i \, dc_i \tag{8.2-4}$$

in which μ_i is the partial Gibbs free energy per unit mass, not the usual form of chemical potential. This equation suggests that

$$\rho T \frac{\partial \hat{S}}{\partial t} = \rho \frac{\partial \hat{H}}{\partial t} - \sum_{i=1}^{n} \mu_i \frac{\partial c_i}{\partial t} \tag{8.2-5}$$

Combining with Eqs. 8.2-1 and 8.2-2,

$$\rho T \frac{\partial \hat{S}}{\partial t} = -\nabla \cdot \mathbf{q} - \nabla \cdot \sum_{i=1}^{n} \bar{H}_i \mathbf{j}_i - \sum_{i=1}^{n} \mu_i (\nabla \cdot \mathbf{j}_i) \tag{8.2-6}$$

However,

$$\mu_i(\nabla \cdot \mathbf{j}_i) = \nabla \cdot (\bar{H}_i - T\tilde{S}_i)\mathbf{j}_i - (\mathbf{j}_i \cdot \nabla\mu_i) \tag{8.2-7}$$

Combining Eqs. 8.2-6 and 8.2-7,

$$\rho \frac{\partial \hat{S}}{\partial t} = -\nabla \cdot \left[\frac{\mathbf{q}}{T} + \sum_{i=1}^{n} \tilde{S}_i \mathbf{j}_i \right] - \frac{1}{T} \sum_{i=1}^{n} \mathbf{j}_i \cdot \nabla\mu_i \tag{8.2-8}$$

By comparison with the entropy balance, Eq. 8.2-3, we see that the entropy flux is

$$\mathbf{J}_s = \frac{\mathbf{q}}{T} + \sum_{i=1}^{n} \tilde{S}_i \mathbf{j}_i \tag{8.2-9}$$

The first and second terms on the right-hand side are the entropy flux by conduction and by diffusion, respectively.

The entropy production can also be found by comparing Eqs. 8.2-3 and 8.2-8:

$$\sigma = -\frac{1}{T} \sum_{i=1}^{n} \mathbf{j}_i \cdot \nabla\mu_i \tag{8.2-10}$$

The terms in this equation have units of energy per volume per time per temperature. Not all the fluxes and gradients in Eq. 8.2-1 are independent, because

$$\sum_{i=1}^{n} \mathbf{j}_i = 0 \tag{8.2-11}$$

and, because the pressure and temperature are constant,

$$\sum_{i=1}^{n} c_i \nabla\mu_i = 0 \tag{8.2-12}$$

Using these restraints, we can rewrite Eq. 8.2-10 in terms of $n - 1$ fluxes and gradients relative to any reference velocity. In particular, for the volume average velocity, we can show that (deGroot & Mazur, 1962)

$$\sigma = -\frac{1}{T} \sum_{i=1}^{n-1} \mathbf{j}_i \cdot \mathbf{X}_i \tag{8.2-13}$$

where

$$\mathbf{X}_i = \sum_{j=1}^{n-1} \alpha_{ij} \nabla\mu_j \tag{8.2-14}$$

where α_{ij} is given by Eq. 8.1-14. The $\mathbf{X}_i$ are the more general forces describing diffusion relative to the volume average velocity; a wide variety of alternative forces are discussed elsewhere.

The linear laws

The second postulate in the derivation of irreversible thermodynamics (Fitts, 1962) states that a linear relation exists between the forces and fluxes in Eq. 8.2-13:

$$-\mathbf{j}_i = \sum_{j=1}^{n-1} L_{ij} \mathbf{X}_j \tag{8.2-15}$$

where the L_{ij} have the mind-bending name of "Onsager phenomenological coefficients." These L_{ij} are strong functions of concentration, especially in dilute solution, where they approach zero as $c_i \to 0$. The linear law can be derived mathematically by use of a Taylor series in which all but the first terms are neglected, but because I am unsure when this neglect is justified, I prefer to regard the linear relation as a postulate.

The Onsager relations

The third and final postulate is that the L_{ij} are symmetric; that is,

$$L_{ij} = L_{ji} \tag{8.2-16}$$

These symmetry conditions, called the Onsager reciprocal relations (Onsager, 1931), can be derived by means of perturbation theory if "microscopic reversibility" is valid. The physical significance of microscopic reversibility is best visualized for a binary collision in which two molecules start in some initial positions, collide, and wind up in some new positions. If the velocities of these molecules are reversed and if microscopic reversibility is valid, the two molecules will move backward, retracing their paths through the collision to regain their original initial positions, just like a movie running backward. Those unfamiliar with the temperament of molecules running backward may be mollified by recalling that the symmetry suggested by Eq. 8.2-16 has been verified experimentally. Thus, we can accept Eq. 8.2-16 as a theoretical result or as an experimentally verified postulate.

The flux equations

Using these three postulates, we can easily complete the derivation of the multicomponent flux equations from irreversible thermodynamics. We first rewrite Eq. 8.2-15 in terms of concentration gradients. Because the $\bar{V}_i$ are partial extensive quantities,

$$\sum_{i=1}^{n} \bar{V}_i \nabla c_i = 0 \tag{8.2-17}$$

Those less well versed in thermodynamics can get the same result by assuming that the partial molar volumes are constant. As a result, only $n - 1$ concentration gradients are independent:

$$\nabla \mu_i = \sum_{j=1}^{n-1} \left(\frac{\partial \mu_i}{\partial c_j}\right)_{c_{k \neq j, n}} \nabla c_j \tag{8.2-18}$$

Note that the concentrations that are held constant in this differentiation differ from those that are commonly held constant in partial differentiation. If we combine Eqs. 8.2-14, 8.2-15, and 8.2-18, we obtain

$$-\mathbf{j}_i = \sum_{j=1}^{n-1} D_{ij} \nabla c_j \tag{8.2-19}$$

where

$$D_{ij} = \sum_{k=1}^{n-1} \sum_{l=1}^{n-1} L_{ik} \alpha_{kl} \left(\frac{\partial \mu_l}{\partial c_j}\right)_{c_{m \neq j, n}} \tag{8.2-20}$$

where the α_{kl} are those given by Eq. 8.1-4. Thus, by starting our argument with conservation equations plus an equation for entropy production, we have derived multicomponent diffusion equations using only three postulates.

We still know nothing from this theory about the diffusion coefficients D_{ij}; we must evaluate these from experiment. Finding these coefficients commonly requires solving the flux equations with the techniques developed in the next section.

Section 8.3. Solving the multicomponent flux equations

In general, solving the multicomponent diffusion problems is not necessary if the analogous binary problem has already been solved (Toor, 1964; Stewart & Prober, 1964). We can mathematically convert the multicomponent problem into a binary problem, look up the binary solution, and then convert this solution back into the multicomponent one. In other words, multicomponent problems usually can be solved using a cookbook approach; little additional work is needed. Some use this cookbook to convert fairly comprehensible binary problems into multicomponent goulash that is harder to understand than necessary.

In this section we first give the results for ternary diffusion and then for the general approach. By starting with the ternary results, we hope to help those who need to solve simple problems. They should not have to dig through the algebra unless they decide to do so.

The ternary solutions
A binary diffusion problem has a solution that can be written as

$$\Delta c_1 = \Delta c_{10} F(D) \tag{8.3-1}$$

In this, Δc_1 is a concentration difference that generally varies with position and time; Δc_{10} is some reference concentration difference containing initial and boundary conditions; and $F(D)$ is the explicit function of position and time. For example, for the diaphragm cell, the binary solution is (see Example 2.2-4)

$$(c_{1B} - c_{1A}) = (c_{1B}^0 - c_{1A}^0)e^{-\beta D t} \tag{8.3-2}$$

where c_{1i}^0 and c_{1i} are the concentrations in the diaphragm-cell compartment i at times zero to t, respectively, β is the cell calibration constant, and D is the diffusion coefficient. By comparison of Eqs. 8.3-1 and 8.3-2, we see that Δc_1 is $c_{1B} - c_{1A}$, Δc_{10} is $c_{1B}^0 - c_{1A}^0$, and $F(D)$ is $e^{-\beta D t}$.

Every binary diffusion problem has an analogous ternary diffusion problem that is described by similar differential equations and similar initial and boundary conditions. The differential equations differ only in the form of Fick's law that is used. The conditions also parallel. For example, in a binary problem the solute concentration may be fixed at a particular boundary; so in the corresponding ternary problem, solute concentrations will also be fixed

Table 8.3-1. *Factors for solution of ternary diffusion problems*[a]

Eigenvalues

$$\sigma_1 = \tfrac{1}{2}[D_{11} + D_{22} + \sqrt{(D_{11} - D_{22})^2 + 4D_{12}D_{21}}]$$

$$\sigma_2 = \tfrac{1}{2}[D_{11} + D_{22} - \sqrt{(D_{11} - D_{22})^2 + 4D_{12}D_{21}}]$$

Weighting factors

$$P_{11} = \left(\frac{D_{11} - \sigma_2}{\sigma_1 - \sigma_2}\right)\Delta c_{10} + \left(\frac{D_{12}}{\sigma_1 - \sigma_2}\right)\Delta c_{20}$$

$$P_{12} = \left(\frac{D_{11} - \sigma_1}{\sigma_2 - \sigma_1}\right)\Delta c_{10} + \left(\frac{D_{12}}{\sigma_2 - \sigma_1}\right)\Delta c_{20}$$

$$P_{21} = \left(\frac{D_{21}}{\sigma_1 - \sigma_2}\right)\Delta c_{20} + \left(\frac{D_{22} - \sigma_2}{\sigma_1 - \sigma_2}\right)\Delta c_{10}$$

$$P_{22} = \left(\frac{D_{21}}{\sigma_2 - \sigma_1}\right)\Delta c_{20} + \left(\frac{D_{22} - \sigma_1}{\sigma_2 - \sigma_1}\right)\Delta c_{10}$$

---

[a] For further definitions, see Eqs. 8.1-2, 8.3-1, 8.3-3, and 8.3-4.

at the corresponding boundary. When this is true, the ternary diffusion problem has the solutions

$$\Delta c_1 = P_{11}F(\sigma_1) + P_{12}F(\sigma_2) \tag{8.3-3}$$

and

$$\Delta c_2 = P_{21}F(\sigma_1) + P_{22}F(\sigma_2) \tag{8.3-4}$$

in which the concentration differences Δc_1 and Δc_{10} are the dependent and independent values in the binary problem, $F(D)$ is again the solution to the binary problem, and the values of σ_i and P_{ij} are given in Table 8.3-1 (Cussler, 1976). The σ_i are the eigenvalues of the diffusion-coefficient matrix and hence are a type of pseudo-binary diffusion coefficient.

The calculation of the ternary diffusion profile is now routine. For example, the result for solute 1 in the diaphragm cell will be

$$c_{1B} - c_{1A} = \frac{(D_{11} - \sigma_2)(c_{1B}^0 - c_{1A}^0) + D_{12}(c_{2B}^0 - c_{2A}^0)}{\sigma_1 - \sigma_2} e^{-\sigma_1 \beta t}$$
$$+ \frac{(D_{11} - \sigma_1)(c_{1B}^0 - c_{1A}^0) + D_{12}(c_{2B}^0 - c_{2A}^0)}{\sigma_2 - \sigma_1} e^{-\sigma_2 \beta t} \tag{8.3-5}$$

The results for the second solute can be found from Eq. 8.3-4, or by rotating the indices in Eq. 8.3-5.

Example 8.3-1: Fluxes for ternary free diffusion

Find the fluxes and the concentration profiles in a dilute ternary free-diffusion experiment. In such an experiment, one ternary solution is suddenly brought into contact with a different composition of the same ternary solution. Find the flux and the concentrations versus position and time at small times.

Solution. When the two solutions come in contact for only a short time, they are effectively infinitely thick. The binary solution of this problem is (see Eq. 2.3-15)

$$\frac{c_1 - c_{10}}{c_{1\infty} - c_{10}} = \text{erf} \frac{z}{\sqrt{4Dt}}$$

in which c_{10} and $c_{1\infty}$ are, respectively, the concentrations where the solutions are contacted (at $x = 0$) and far into one solution (at $x = \infty$), x and t are the position and time, and D is the binary diffusion coefficient. By comparison with Eq. 8.3-1, we see that Δc_1 is $c_1 - c_{10}$, Δc_{10} is $c_{1\infty} - c_{10}$, and $F(D)$ equals the error function of $z/\sqrt{4Dt}$. As a result, the concentration profile for solute 1 will be

$$c_1 - c_{10} = \left[\frac{(D_{11} - \sigma_2)(c_{1\infty} - c_{10}) + D_{12}(c_{2\infty} - c_{20})}{\sigma_1 - \sigma_2}\right] \text{erf} \frac{z}{\sqrt{4\sigma_1 t}}$$

$$+ \left[\frac{(D_{11} - \sigma_1)(c_{1\infty} - c_{10}) + D_{12}(c_{2\infty} - c_{20})}{\sigma_2 - \sigma_1}\right] \text{erf} \frac{z}{\sqrt{4\sigma_2 t}}$$

The close similarity between this result and that for the diaphragm cell is obvious.

The fluxes can be found in the same manner as the concentration profile. Because the solutions are dilute, there is negligible convection induced by diffusion; so

$$-n_1 \doteq -j_1 = D_{11} \frac{\partial c_1}{\partial z} + D_{12} \frac{\partial c_2}{\partial z}$$

Combining this with Eqs. 2.3-17, 8.3-3, and 8.3-4,

$$-j_1 = (D_{11}P_{11} + D_{12}P_{21}) \frac{e^{-z^2/4\sigma_1 t}}{\sqrt{\pi \sigma_1 t}}$$

$$+ (D_{11}P_{12} + D_{12}P_{22}) \frac{e^{-z^2/4\sigma_2 t}}{\sqrt{\pi \sigma_2 t}}$$

where again the P_{ij} are given in Table 8.3-1. These results are complex algebraically but straightforward conceptually.

The general solution

We now turn from the detail of ternary diffusion to the more general solution of the multicomponent problems. The general solution of these equations is most easily presented in terms of linear algebra, a notation that is not used elsewhere in this book. In this presentation, we consider the species concentrations as a vector of $\underline{c}$, and the multicomponent diffusion coefficients as a matrix $\underline{\underline{D}}$.

In matrix notation, the multicomponent flux equations are

$$-\mathbf{j} = \underline{\underline{D}} \cdot \nabla \underline{c} \tag{8.3-6}$$

The continuity equations for this case are

$$\frac{\partial \underline{c}}{\partial t} + (\nabla \cdot \mathbf{v}^0 \underline{c}) = -\nabla \cdot \mathbf{j} \tag{8.3-7}$$

These are subject to the initial and boundary conditions

$$\Delta \underline{c}(x, y, z, t = 0) = \Delta \underline{c_0} \tag{8.3-8}$$

$$\Delta \underline{c}(B, t) = 0 \tag{8.3-9}$$

$$\frac{\partial \underline{c}}{\partial z}(b, t) = 0 \tag{8.3-10}$$

where B and b represent two boundaries of the system. Note that the boundary conditions on all concentrations must have the same functional form. This is a serious restriction only for the case of simultaneous diffusion and chemical reaction.

We now assume that there exists a nonsingular matrix $\underline{t}$ that can diagonalize $\underline{\underline{D}}$ (Cullinan, 1965):

$$\underline{t}^{-1} \cdot \underline{\underline{D}} \cdot \underline{t} = \underline{\underline{\sigma}} = \begin{bmatrix} \sigma_1 & 0 & 0 & \cdots \\ 0 & \sigma_2 & 0 & \cdots \\ 0 & 0 & \sigma_3 & \cdots \\ \vdots & \vdots & \vdots & \end{bmatrix} \tag{8.3-11}$$

where $\underline{t}^{-1}$ is the inverse of $\underline{t}$, and $\underline{\underline{\sigma}}$ is the diagonal matrix of the eigenvalues of the diffusion coefficient matrix $\underline{\underline{D}}$. The assumption that $\underline{\underline{D}}$ can be put into diagonal form is not necessary for a general mathematical solution, but because this assumption is valid for all cases encountered in practice, it is used here. For the case of ternary diffusion,

$$\underline{t} = \begin{bmatrix} t_{11} & t_{12} \\ t_{21} & t_{22} \end{bmatrix} = \frac{\begin{bmatrix} 1 & \dfrac{D_{12}}{D_{22} - \sigma_1} \\ \dfrac{D_{22} - \sigma_2}{D_{12}} & 1 \end{bmatrix}}{\det(\underline{t})} = \frac{\begin{bmatrix} 1 & \dfrac{D_{11} - \sigma_1}{D_{21}} \\ \dfrac{D_{21}}{D_{11} - \sigma_2} & 1 \end{bmatrix}}{\det(\underline{t})} \tag{8.3-12}$$

Correspondingly,

$$\underline{t}^{-1} = \begin{bmatrix} t_{11}^{-1} & t_{12}^{-1} \\ t_{21}^{-1} & t_{22}^{-1} \end{bmatrix} = \begin{bmatrix} 1 & \dfrac{D_{12}}{\sigma_1 - D_{22}} \\ \dfrac{\sigma_2 - D_{22}}{D_{12}} & 1 \end{bmatrix} = \begin{bmatrix} 1 & \dfrac{\sigma_1 - D_{11}}{D_{21}} \\ \dfrac{D_{21}}{\sigma_2 - D_{11}} & 1 \end{bmatrix} \tag{8.3-13}$$

where

$$\det(\underline{t}) = \frac{\sigma_1 - \sigma_2}{\sigma_1 - D_{22}} = \frac{\sigma_2 - \sigma_1}{\sigma_2 - D_{11}} \tag{8.3-14}$$

Remember that the product of $\underline{t}$ and its inverse $\underline{t}^{-1}$ is the unit matrix.

We now use this new matrix $\underline{t}$ to define a new combined concentration $\underline{\Psi}$

$$\underline{c} = \underline{t} \cdot \underline{\Psi} \tag{8.3-15}$$

We then combine Eqs. 8.3-6, 8.3-7, and 8.3-15 and premultiply the equation by $\underline{t}^{-1}$ to obtain

$$\frac{\partial \underline{\Psi}}{\partial t} + \nabla \cdot \mathbf{v}^0 \underline{\Psi} = \underline{\underline{\sigma}} \cdot \nabla^2 \underline{\Psi} \tag{8.3-16}$$

which represents a set of scalar equations

$$\frac{\partial \Psi_i}{\partial t} + \nabla \cdot \mathbf{v}^0 \Psi_i = \sigma_i \cdot \nabla^2 \Psi_i \tag{8.3-17}$$

In this operation, we have made the good assumption that $\underline{\underline{D}}$ and hence both $\underline{\underline{t}}$ and $\underline{\underline{\sigma}}$ are not functions of composition.

The initial and boundary conditions can also be written in terms of the new combined concentration $\underline{\Psi}$:

$$\Delta \underline{\Psi}(x, y, z, 0) = \Delta \underline{\Psi}_0 = \underline{\underline{t}}^{-1} \cdot \Delta \underline{c}_0 \tag{8.3-18}$$

$$\Delta \underline{\Psi}(B, t) = 0 \tag{8.3-19}$$

$$\frac{\partial \underline{\Psi}}{\partial z} (b, t) = 0 \tag{8.3-20}$$

Thus, a set of coupled differential equations has been separated into uncoupled equations written in terms of the new concentration $\underline{\Psi}$.

Equations 8.3-16 through 8.3-20 have exactly the same form as the associated *binary* diffusion problem:

$$\frac{\partial c_1}{\partial t} + \nabla \cdot \mathbf{v}^0 c_1 = D \nabla^2 c_1 \tag{8.3-21}$$

which has the same initial and boundary conditions for each species as those given in Eqs. 8.3-8 through 8.3-10. If this binary problem has the solution

$$\Delta c_1 = F(D) \Delta c_{10} \tag{8.3-22}$$

then Eqs. 8.3-16 through 8.3-20 must have the solution

$$\Delta \Psi_i = F(\sigma_i) \Delta \Psi_{i0} \tag{8.3-23}$$

where the eigenvalue σ_i is substituted everywhere that the binary diffusion coefficient occurs in the binary solution. If we rewrite our solution in terms of the actual concentrations, we find that

$$\Delta \underline{c} = \underline{\underline{t}} \cdot \underline{\underline{F}}(\underline{\underline{\sigma}}) \cdot \underline{\underline{t}}^{-1} \cdot \Delta \underline{c}_0 \tag{8.3-24}$$

Thus, we know the concentration profiles in the multicomponent system in terms of its binary analogue. The results for the ternary case are given in Table 8.3-1.

Many find this derivation difficult to grasp, even after they apparently understand every step. Their trouble usually stems from a mathematical, not physical, problem. They do not see why the derivation is more than a trick, a slick invention. The reason is that Eq. 8.3-17 and its associated conditions are shown to be mathematically the same as the binary solution. If we change the symbol Ψ_i to c_1, Eq. 8.3-17 and Eq. 8.3-21 are exactly the same. The physical circumstances in the multicomponent problem may be more elaborate, but the identity of the differential equations signals that the mathematical solutions are identical.

Example 8.3-2: Steady-state multicomponent diffusion across a thin film

In steady-state binary diffusion, we found that the solute's concentration varied linearly across a thin film. Will solute concentrations vary linearly in the multicomponent case? What will the flux be?

Solution. By comparison with Eq. 2.2-9, we see that

$$(c_1 - c_{10}) = \left(\frac{z}{l}\right)(c_{1l} - c_{10})$$

By comparing this with Eq. 8.3-22, we see that $F(D)$ equals (z/l). From Eq. 8.3-24, for the multicomponent case,

$$\Delta \underline{c} = \left(\underline{t} \cdot \frac{z}{l} \underline{\delta} \cdot \underline{t}^{-1}\right) \cdot \Delta \underline{c}_0 = \left(\frac{z}{l}\right) \Delta c_0$$

Thus, the concentration profile of each solute remains linear. The flux is

$$
\begin{aligned}
-\underline{j} &= \underline{D} \cdot \Delta \underline{c} \\
&= \underline{D} \cdot \frac{\Delta \underline{c}_0}{l}
\end{aligned}
$$

or

$$j_i = \sum_{j=1}^{n} \frac{D_{ij}}{l}(c_{j0} - c_{jl})$$

Note that a solute's flux can be in the opposite direction to that expected if other gradients exist in the system.

Section 8.4. Ternary diffusion coefficients

In this section we report a variety of values for ternary diffusion coefficients. These coefficients support the generalizations given at the beginning of this chapter that said that multicomponent effects were significant when the system was concentrated and contained interacting species. These interactions can originate from chemical reactions, from electrostatic coupling, or from major differences in molecular weights.

Typical diffusion coefficients for gases are shown in Table 8.4.1. These values are not experimental, but are calculated from the Chapman–Enskog theory (see Section 5.1) and from Table 8.1-1. The first two rows in the table show how the values of D_{12} and D_{21} are larger as the solution becomes concentrated. The second and third rows refer to the same solution, but with a different species chosen as the solute. The difference in the diffusion coefficients illustrates how ternary diffusion coefficients can be difficult to interpret. The final three rows are other characteristic situations.

Ternary diffusion coefficients in liquids and solids cannot be found from binary values, but only from experiments. When experiments are not avail-

Table 8.4-1. *Ternary diffusion coefficients in gases at* $25°C^a$

System	D_{11}	D_{12}	D_{21}	D_{22}
Hydrogen ($x_1 = 0.05$) Methane ($x_2 = 0.05$) Argon ($x_3 = 0.90$)	0.78	−0.00	0.03	0.22
Hydrogen ($x_1 = 0.2$) Methane ($x_2 = 0.2$) Argon ($x_3 = 0.6$)	0.76	−0.01	0.12	0.25
Argon ($x_1 = 0.6$) Methane ($x_2 = 0.2$) Hydrogen ($x_3 = 0.2$)	0.64	−0.39	−0.12	0.37
Carbon dioxide ($x_1 = 0.2$) Oxygen ($x_2 = 0.2$) Nitrogen ($x_3 = 0.6$)	0.15	−0.00	−0.01	0.19
Hydrogen ($x_1 = 0.2$) Ethylene ($x_2 = 0.2$) Ethane ($x_3 = 0.6$)	0.56	0.00	0.11	0.13
Benzene ($x_1 = 0.2$) Cyclohexane ($x_2 = 0.2$) Hexane ($x_3 = 0.6$)	0.028	0.000	0.001	0.026

^{a} All coefficients have units of cm^2/sec and are calculated from the equations in Table 8.1-1.

able, which is usually the case, one can make estimates by assuming that the Onsager phenomenological coefficients are a diagonal matrix; that is,

$$L_{ij,i \neq j} = 0 \qquad (8.4-1)$$

In addition, we can assume that the main-term coefficients are related to the binary values given by

$$L_{ii} = \left(\frac{D_i c_i}{RT}\right) \qquad (8.4-2)$$

where D_i is the coefficient of species i in the solvent. These assumptions can be combined with Eq. 8.1-4 and Eq. 8.2-20 to give

$$D_{ij} = \sum_{l=1}^{n-1} \left(\frac{D_i c_i}{RT}\right)\left(\delta_{il} + \frac{c_l \bar{V}_i}{c_n \bar{V}_n}\right)\left(\frac{\partial \mu_l}{\partial c_j}\right)_{c_{k \neq j,n}} \qquad (8.4-3)$$

This is equivalent to saying that ternary effects result from activity coefficients. I routinely use this equation for making initial estimates.

Experimental values of ternary diffusion coefficients characteristic of liquids are shown in Table 8.4.2. In cases like KCl–NaCl–water, KCl–sucrose–water, and toluene–chlorobenzene–bromobenzene, the cross-term diffusion coefficients are small, less than 10% of the main diffusion coefficients. In these cases, we can safely treat the diffusion as a binary process.

The cross-term diffusion coefficients are much more significant for inter-

Table 8.4-2. *Ternary diffusion coefficients in liquids at 25°C[a]*

System	D_{11}	D_{12}	D_{21}	D_{22}
1.5-M KCl (1) 1.5-M NaCl (2) H_2O (3)[b]	1.80	0.33	0.10	1.39
0.10-M HBr (1) 0.25-M KBr (2) H_2O (3)[c]	5.75	0.05	−2.20	1.85
1-M H_2SO_4 (1) 1-M Na_2SO_4 (2) H_2O (3)[d]	2.61	−0.04	−0.51	0.91
0.06-g/cm³ KCl (1) 0.03-g/cm³ sucrose (2) H_2O (3)[e]	1.78	0.02	0.07	0.50
2-M urea (1) 0-M ^{14}C-tagged urea (2) H_2O (3)[f]	1.24	0.01	0	1.23
32 mol% hexadecane (1) 35 mol% dodecane (2) 33 mol% hexane (3)[g]	1.03	0.23	0.27	0.97
25 mol% toluene (1) 50 mol% chlorobenzene (2) 25 mol% bromobenzene (3)[h]	1.85	−0.06	−0.05	1.80
0.326-g/cm³ benzene (1) 0.265-g/cm³ propanol (2) Carbon tetrachloride (3)[i]	1.64	0.78	0.17	1.33
5 wt% cyclohexane (1) 5 wt% polystyrene (2) 90 wt% toluene (3)[j]	2.03	−0.09	−0.02	0.09

[a] All values × 10^{-5} cm²/sec.
[b] P. J. Dunlop, *J. Phys. Chem.*, **63**, 612 (1959).
[c] A. Reojin, *J. Phys. Chem.*, **76**, 3419 (1972).
[d] R. P. Wendt, *J. Phys. Chem.*, **66**, 1279 (1962).
[e] E. L. Cussler and P. J. Dunlop, *J. Phys. Chem.*, **70**, 1880 (1966).
[f] J. G. Albright and R. Mills, *J. Phys. Chem.*, **69**, 3120 (1966).
[g] T. K. Kett and D. K. Anderson, *J. Phys. Chem.*, **73**, 1268 (1969).
[h] J. K. Burchard and H. L. Toor, *J. Phys. Chem.*, **66**, 2015 (1962).
[i] R. A. Graff and T. B. Drew, *IEC Fund.*, **7**, 490 (1968) (data at 20°C).
[j] E. L. Cussler and E. N. Lightfoot, *J. Phys. Chem.*, **69**, 1135 (1965).

acting solutes. In cases like HBr–KBr–water and H_2SO_4–Na_2SO_4–water, this interaction is ionic; in other cases it may involve hydrogen-bond formation. Cross-term diffusion coefficients and the resulting ternary effects should be especially large in partially miscible systems, where few measurements have been made.

The ternary diffusion coefficients in metals shown in Table 8.4-3 are the largest cross-term diffusion coefficients. As a result, the flux of one compo-

Table 8.4-3. *Ternary interdiffusion coefficients in solids*[a]

Ternary system 1-2-3	Composition (mol%) and structure at temperature studied	Temperature (°C)	D_{11}^3	D_{12}^3	D_{21}^3	D_{22}^3
C–Si–Fe[b]	0.46C–1.97Si–97.57Fe (fcc)	1,050	4.8×10^{-7}	0.3×10^{-7}	≈ 0	2.3×10^{-9}
Al–Ni–Fe	47Al–18Ni–35Fe (bcc)[c]	1,004	4.4×10^{-11}	-0.2×10^{-11}	0.3×10^{-11}	1.6×10^{-11}
	43Al–8.5Ni–48.5Fe (bcc)[c]	1,004	16.4×10^{-11}	-5.9×10^{-11}	0.3×10^{-11}	2.5×10^{-11}
	8Al–44.5Ni–48.5Fe (fcc)[d]	1,000	23.3×10^{-12}	-9.2×10^{-12}	-9.9×10^{-12}	16.4×10^{-12}
Cr–Ni–Co[e]	9.5Cr–20.4Ni–70Co (fcc)	1,300	0.6×10^{-9}	-0.13×10^{-11}	-0.12×10^{-10}	0.23×10^{-9}
	8.8Cr–40Ni–51Co		1.09×10^{-9}	-0.6×10^{-11}	-2.6×10^{-10}	0.47×10^{-9}
	9.2Cr–79Ni–11.8Co (fcc)		1.25×10^{-9}	-2.5×10^{-11}	-5.1×10^{-10}	0.74×10^{-9}
Zn–Cd–Ag[f]	13.1Zn–3.5Cd–83.4Ag	600	1.2×10^{-10}	1.3×10^{-10}	0.13×10^{-10}	1.2×10^{-10}
	18.1Zn–4.4Cd–77.5Ag		3.3×10^{-10}	2.4×10^{-10}	0.6×10^{-10}	2.1×10^{-10}
	16.4Zn–8.5Cd–75.1Ag		4.4×10^{-10}	4.5×10^{-10}	1.4×10^{-10}	5.5×10^{-10}
Zn–Mn–Cu[g]	10.3Zn–1.8Mn–87.9Cu (fcc)	850	1.82×10^{-9}	0.11×10^{-9}	-0.02×10^{-9}	1.46×10^{-9}
Zn–Ni–Cu[h]	19Zn–43Ni–38Cu (fcc)	775	5.1×10^{-11}	-0.8×10^{-11}	-1.7×10^{-11}	1.2×10^{-11}
V–Zr–Ti[i]	9V–9Zr–82Ti	800	2.3×10^{-10}	0.1×10^{-10}	1.0×10^{-10}	4.4×10^{-10}
	17.5V–19.5Zr–63Ti		2.9×10^{-10}	1.5×10^{-10}	0.7×10^{-10}	1.8×10^{-10}
	37.5V–7.5Zr–55Ti		0.16×10^{-10}	0.18×10^{-10}	0.1×10^{-10}	0.23×10^{-10}
	5.0V–77.5Zr–17.5Ti (bcc)		12.4×10^{-10}	2.6×10^{-10}	-0.8×10^{-10}	2.8×10^{-10}

Table 8.4-3. (cont.)

Ternary system 1-2-3	Composition (mol%) and structure at temperature studied	Temperature (°C)	D^3_{11}	D^3_{12}	D^3_{21}	D^3_{22}
Cu–Ag–Au[j]	13.1Cu–34.0Ag–52.9Au	725	1.0×10^{-10}	0.1×10^{-10}	1.7×10^{-10}	1.3×10^{-10}
	60.3Cu–12.9Ag–26.8Au (fcc)		2.3×10^{-10}	1.1×10^{-10}	1.8×10^{-10}	3.1×10^{-10}
Co–Ni–Fe[k]	10.3Co–31.4Ni–58.3Fe	1,315	4×10^{-10}	0.9×10^{-10}	3×10^{-10}	7.1×10^{-10}
	35.5Co–35.4Ni–29.1Fe		6.5×10^{-10}	2.7×10^{-10}	3.2×10^{-10}	7.3×10^{-10}
	31.1Co–65.6Ni–3.3Fe (fcc)		6.1×10^{-10}	0.2×10^{-10}	4.0×10^{-10}	8.8×10^{-10}

[a] All diffusion coefficients are in $cm^2 \cdot sec^{-1}$ and are based on a solvent-fixed reference frame.
[b] J. Kirkaldy, *Cand. J. Phys.*, **35**, 435 (1957).
[c] T. D. Moyer and M. A. Dayananda, *Met. Trans.*, **7A**, 1035 (1976).
[d] G. H. Cheng and M. A. Dayananda, *Met. Trans.*, **10A**, 1415 (1979).
[e] A. G. Guy and V. Leroy, *The Electron Microprobe*, ed. McKinley, Heinrich, and Wittry. New York (1966).
[f] P. T. Carlson, M. A. Dayananda, and R. E. Grace, *Met. Trans.*, **3**, 819 (1972).
[g] M. A. Dayananda and R. E. Grace, *Trans. Met. Soc. AIME*, **233**, 1287 (1965).
[h] R. D. Sisson, Jr., and M. A. Dayananda, *Met. Trans.*, **8A**, 1849 (1977).
[i] A. Brunsch and S. Steel, *Zeitschrift für Metallkunde*, **65**, 765 (1974).
[j] T. O. Ziebold and R. E. Ogilvie, *Trans. Met. Soc. AIME*, **239**, 942 (1967).
[k] A. Vignes and J. P. Sabatier, *Trans. Met. Soc. AIME*, **245**, 1795 (1969).

nent in an alloy can frequently be against its concentration gradient, from low concentration into higher concentration. These effects are especially interesting when they are superimposed on the elaborate phase diagrams characteristic of alloys, because they can lead to local phase separations that dramatically alter the material's properties. As in gases and liquids, the methods of estimating ternary diffusion coefficients are risky. One must either rely on relations like Eq. 8.4-3 or undertake the difficult experiments involved. As a result, many avoid ternary diffusion even when they suspect it is important.

Section 8.5. Conclusions

Diffusion frequently occurs in multicomponent systems. When these systems are dilute, the diffusion of each solute can be treated with a binary form of Fick's law. In concentrated solutions, the fluxes and concentration profiles deviate significantly from binary expectations only in exceptional cases. These exceptions include mixed gases containing hydrogen, mixed weak electrolytes, partially miscible species, and some alloys.

When multicomponent diffusion is significant, it is best described with a generalized form of Fick's law containing $(n - 1)^2$ diffusion coefficients in an n-component system. This form of diffusion equation can be rationalized using irreversible thermodynamics. Concentration profiles in these multicomponent cases can be directly inferred from the binary results. However, multicomponent diffusion coefficients are difficult to estimate, and experimental values are fragmentary. As a result, you should make very sure that you need the more complicated theory before you attempt to use it.

References

Arnold, K. R., & Toor, H. L. (1967). *American Institute of Chemical Engineers Journal,* **13,** 909.

Carlson, P. T., Dayananda, M. A., & Grace, R. E. (1972). *Metallurgical Transactions,* **3,** 819.

Cullinan, H. T. (1965). *Industrial and Engineering Chemistry Fundamentals,* **4,** 133.

Cussler, E. L. (1976). *Multicomponent Diffusion.* Amsterdam: Elsevier.

Cussler, E. L., Evans, D. F., & Matesich, M. A. (1971). *Science,* **172,** 377.

Cussler, E. L., & Lightfoot, E. N. (1965). *Journal of Physical Chemistry,* **69,** 1135.

Curtiss, C. F., & Hirschfelder, J. O. (1949). *Journal of Chemical Physics,* **17,** 550.

Dayananda, M. A. (1968). *Transactions of the Metallurgical Society of AIME,* **242,** 1369.

deGroot, S. R., & Mazur, P. (1962). *Non-Equilibrium Thermodynamics.* Amsterdam: North Holland.

Ellerton, H. D., & Dunlop, P. J. (1967). *Journal of Physical Chemistry,* **71,** 1291, 1538.

Fitts, D. D. (1962). *Non-Equilibrium Thermodynamics.* New York: McGraw-Hill.

Haase, R. (1969). *Thermodynamics of Irreversible Processes.* London: Addison-Wesley.

Katchalsky, A., & Curran, P. F. (1967). *Non-Equilibrium Thermodynamics in Biophysics.* Cambridge: Harvard University Press.

Keller, K. H., & Friedlander, S. K. (1966). *Journal of General Physiology,* **49,** 663.

Lamm, O. (1944). *Arkiv for Kemi, Minerologi och Geologi,* **2,** 1813.

Lightfoot, E. N., Cussler, E. L., & Rettig, R. L. (1962). *American Institute of Chemical Engineers Journal,* **8,** 708.

Onsager, L. (1931). *Physical Review,* **37,** 405; (1931) **38,** 2265.

Onsager, L. (1945). *New York Academy of Sciences Annals,* **46,** 241.

Stewart, W. E., & Prober, R. (1964). *Industrial and Engineering Chemistry Fundamentals,* **3,** 224.

Toor, H. L. (1964). *American Institute of Chemical Engineers Journal,* **10,** 448, 460.

Vermeulen, T., Clazie, R. N., & Klein, G. (1967). OSW RD progress report No. 326.

Vitagliano, V., Laurentino, R., & Constantino, L. (1969). *Journal of Physical Chemistry,* **73,** 2456.

Wendt, R. P. (1962). *Journal of Physical Chemistry,* **66,** 1279.

Woolf, L. A., Miller, D. G., & Gosting, L. J. (1962). *Journal of the American Chemical Society,* **84,** 317.

PART III

Mass transfer

9   FUNDAMENTALS OF MASS TRANSFER

Diffusion is the process by which molecules, ions, or other small particles spontaneously mix, moving from regions of relatively high concentration into regions of lower concentration. This process can be analyzed in two ways. First, it can be described with Fick's law and a diffusion coefficient, a fundamental and scientific description used in the first two parts of this book. Second, it can be explained in terms of a mass transfer coefficient, an approximate engineering idea that often gives a simpler description. It is this simpler idea that is emphasized in this part of this book.

Analyzing diffusion with mass transfer coefficients requires assuming that changes in concentration are limited to that small part of the system's volume near its boundaries. For example, in the absorption of one gas into a liquid, we assume that all gases and liquids are well mixed, except near the gas–liquid interface. In the leaching of metal by pouring acid over ore, we assume that the acid is homogeneous, except in a thin layer next to the solid ore particles. In studies of digestion, we assume that the contents of the small intestine are well mixed, except near the villi at the intestine's wall. Such an analysis is sometimes called a "lumped-parameter model" to distinguish it from the "distributed-parameter model" using diffusion coefficients. Both models are much simpler for dilute solutions.

If you are beginning a study of diffusion, you may have trouble deciding whether to organize your results as mass transfer coefficients or as diffusion coefficients. I have this trouble, too. The cliché is that you should use the mass transfer coefficient approach if the diffusion occurs across an interface, but this cliché has many exceptions. Instead of depending on the cliché, I believe you should always try both approaches to see which is better for your own needs. In my own work, I have found that I often switch from one to the other as the work proceeds and my objectives evolve.

This chapter discusses mass transfer coefficients for dilute solutions; extensions to concentrated solutions are deferred to Section 11.5. In Section 9.1 we give a basic definition for a mass transfer coefficient and show how this coefficient can be used experimentally. In Section 9.2 we present other common definitions that represent a thicket of prickly alternatives rivaled only by standard states for chemical potentials. These various definitions are why mass transfer often has a reputation with students of being a difficult subject. In Section 9.3 we list existing correlations of mass transfer coeffi-

215

cients, and in Section 9.4 we explain how these correlations can be developed with dimensional analysis. Finally, in Section 9.5 we discuss processes involving diffusion across interfaces, a topic that leads to overall mass transfer coefficients found as averages of more local processes. This last idea is commonly called mass transfer resistances in series.

Section 9.1. A definition of mass transfer coefficients

The definition of mass transfer is based on empirical arguments like those used in developing Fick's law in Chapter 2. Imagine we are interested in the transfer of mass from some interface into a well-mixed solution. We expect that the amount transferred is proportional to the concentration difference and the interfacial area:

$$\begin{pmatrix} \text{amount of mass} \\ \text{transferred} \end{pmatrix} = k \begin{pmatrix} \text{interfacial} \\ \text{area} \end{pmatrix} \begin{pmatrix} \text{concentration} \\ \text{difference} \end{pmatrix} \qquad (9.1\text{-}1)$$

where the proportionality is summarized by k, called a mass transfer coefficient. If we divide both sides of this equation by the area, we can write the equation in more familiar symbols:

$$N_1 = k(c_{1i} - c_1) \qquad (9.1\text{-}2)$$

where N_1 is the flux at the interface and c_{1i} and c_1 are the concentrations at the interface and in the bulk solution, respectively.

This result makes practical sense. It says that if the concentration difference is doubled, the flux will double. It also suggests that if the area is doubled, the total amount of mass transferred will double, but the flux per area will not change. In other words, this definition suggests an easy way of organizing our thinking around a simple constant, the mass transfer coefficient k.

Unfortunately, this simple scheme conceals a variety of approximations and ambiguities. Before introducing these complexities, we shall go over some easy examples. These examples are important. Study them carefully before you go on to the harder material that follows.

Example 9.1-1: Humidification

Imagine that water is evaporating into initially dry air in the closed vessel shown schematically in Fig. 9.1-1(a). The vessel is isothermal at 25°C; so the water's vapor pressure is 23.8 mm Hg. This vessel has 0.8 liter of water with 150 cm² of surface area in a total volume of 19.2 liters. After 3 min, the air is 0.05% saturated. What is the mass transfer coefficient? How long will it take to reach 90% saturation?

Solution. The flux at 3 min can be found directly from the values given:

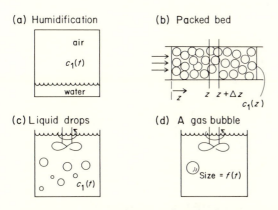

(a) Humidification

air

$c_1(t)$

water

(b) Packed bed

z $z+\Delta z$

$c_1(z)$

(c) Liquid drops

$c_1(t)$

(d) A gas bubble

Size = $f(t)$

Fig. 9.1-1. Four easy examples. We analyze each of the physical situations shown in terms of mass transfer coefficients. In (a), we assume that the air is at constant humidity, except near the air–water interface. In (b), we assume that water flowing through the packed bed is well mixed, except very close to the solid spheres. In (c) and (d), we assume that the liquid solution, which is the continuous phase, is at constant composition, except near the droplet or bubble surfaces.

$$N_1 = \frac{\left(\begin{array}{c}\text{vapor}\\\text{concentration}\end{array}\right)\left(\begin{array}{c}\text{air}\\\text{volume}\end{array}\right)}{\left(\begin{array}{c}\text{liquid}\\\text{area}\end{array}\right)(\text{time})}$$

$$= \frac{5 \cdot 10^{-4}\left(\frac{23.8}{760}\right)\left(\frac{1\text{ mol}}{22.4\text{ liters}}\right)\left(\frac{273}{298}\right)(18.4\text{ liters})}{(150\text{ cm}^2)(180\text{ sec})}$$

$$= 4.4 \cdot 10^{-8}\text{ mol/cm}^2\text{-sec}$$

The concentration difference is that at the water's surface minus that in the bulk solution. That at the water's surface is the value at saturation; that in bulk at short times is essentially zero. Thus, from Eq. 9.1-2, we have

$$4.4 \cdot 10^{-8}\text{ mol/cm}^2\text{-sec} = k\left(\frac{23.8}{760}\frac{1\text{ mol}}{22.4 \cdot 10^3\text{ cm}^3}\frac{273}{298} - 0\right)$$

$$k = 3.4 \cdot 10^{-2}\text{ cm/sec}$$

This value is lower than that commonly found for transfer in gases.

The time required for 90% saturation can be found from a mass balance:

$$\left(\begin{array}{c}\text{accumulation}\\\text{in gas phase}\end{array}\right) = \left(\begin{array}{c}\text{evaporation}\\\text{rate}\end{array}\right)$$

$$\frac{d}{dt}Vc_1 = AN_1$$

$$= Ak[c_1(\text{sat}) - c_1]$$

The air is initially dry; so

$$t = 0, \quad c_1 = 0$$

We use this condition to integrate the mass balance:

$$\frac{c_1}{c_1(\text{sat})} = 1 - e^{-(kA/V)t}$$

Rearranging the equation and inserting the values given, we find

$$t = -\frac{V}{kA} \ln \left(1 - \frac{c_1}{c_1(\text{sat})}\right)$$

$$= -\frac{18.4 \cdot 10^3 \text{ cm}^3}{(3.4 \cdot 10^{-2} \text{ cm/sec}) \cdot (150 \text{ cm}^2)} \ln(1 - 0.9)$$

$$= 8.3 \cdot 10^3 \text{ sec} = 2.3 \text{ hr}$$

It takes over 2 hr to saturate the air this much.

Example 9.1-2: Mass transfer in a packed bed

Imagine that 0.2-cm-diameter spheres of benzoic acid are packed into a bed like that shown schematically in Fig. 9.1-1(b). The spheres have 23 cm^2 surface per 1 cm^3 of bed. Pure water flowing at a superficial velocity of 5 cm/sec into the bed is 62% saturated with benzoic acid after it has passed through 100 cm of bed. What is the mass transfer coefficient?

Solution. The answer to this problem depends on the concentration difference used in the definition of the mass transfer coefficient. In every definition, we choose this difference as the value at the sphere's surface minus that in the solution. However, we can define different mass transfer coefficients by choosing the concentration difference at various positions in the bed. For example, we can choose the concentration difference at the bed's entrance and so obtain

$$N_1 = k[c_1(\text{sat}) - 0]$$

$$\frac{0.62c_1(\text{sat})(5 \text{ cm/sec})A}{(23 \text{ cm}^2/\text{cm}^3)(100 \text{ cm})A} = kc_1(\text{sat})$$

where A is the bed's cross section. Thus,

$$k = 1.3 \cdot 10^{-3} \text{ cm/sec}$$

This definition for the mass transfer coefficient is infrequently used.

Alternatively, we can choose as our concentration difference that at a position z in the bed and write a mass balance on a differential volume $A\Delta z$ at this position:

$$(\text{accumulation}) = \left(\begin{array}{c}\text{flow in}\\\text{minus flow out}\end{array}\right) + \left(\begin{array}{c}\text{amount of}\\\text{dissolution}\end{array}\right)$$

$$0 = A\left(c_1 v^0\Big|_z - c_1 v^0\Big|_{z+\Delta z}\right) + (A\Delta z)aN_1$$

Substituting for N_1 from Eq. 9.1-2, dividing by $A\Delta z$, and taking the limit as Δz goes to zero, we find

$$\frac{dc_1}{dz} = \frac{ka}{v^0}[c_1(\text{sat}) - c_1]$$

This is subject to the initial condition that

$$z = 0, \quad c_1 = 0$$

Integrating, we obtain an exponential of the same form as in the first example:

$$\frac{c_1}{c_1(\text{sat})} = 1 - e^{-(ka/v^0)z}$$

Rearranging the equation and inserting the values given, we find

$$k = -\left(\frac{v^0}{az}\right) \ln\left(1 - \frac{c_1}{c_1(\text{sat})}\right)$$

$$= -\frac{5 \text{ cm/sec}}{(23 \text{ cm}^2/\text{cm}^3)(100 \text{ cm})} \ln(1 - 0.62)$$

$$= 2.1 \cdot 10^{-3} \text{ cm/sec}$$

This value is typical of those found in liquids. This type of mass transfer coefficient definition is preferable to that used first, a point explored further in Section 9.2.

A tangential point worth discussing is the specific chemical system of benzoic acid dissolving in water. This system is academically ubiquitous, showing up again and again in problems of mass transfer. Indeed, if you read the literature, you can get the impression that it is the only system where mass transfer is important, which is not true. Why is it used so much?

Benzoic acid is studied thoroughly for three distinct reasons. First, its concentration is relatively easily measured, for the amount present can be determined by titration with base, by UV spectrophotometry of the benzene ring, or by radioactively tagging either the carbon or the hydrogen. Second, the dissolution of benzoic acid is accurately described by one mass transfer coefficient. This is not true of all dissolutions. For example, the dissolution of aspirin is essentially independent of events in solution (see Section 13.3). Third, and most subtle, benzoic acid is solid, so mass transfer takes place across a solid–fluid interface. Such interfaces are an exception in mass transfer problems; fluid–fluid interfaces are much more common. However, solid–fluid interfaces are the rule for heat transfer, the intellectual precursor of mass transfer. Experiments with benzoic acid dissolving in water can be compared directly with heat transfer experiments. These three reasons make this chemical system popular.

Example 9.1-3: Mass transfer in an emulsion

Bromine is being rapidly dissolved in water, as shown schematically in Fig. 9.1-1(c). Its concentration is about half saturated in 3 min. What is the mass transfer coefficient?

Solution. Again, we begin with a mass balance:

$$\frac{d}{dt} Vc_1 = AN_1 = Ak[c_1(\text{sat}) - c_1]$$

$$\frac{dc_1}{dt} = ka[c_1(\text{sat}) - c_1]$$

where $a\ (= A/V)$ is the surface area of the bromine droplets divided by the volume of aqueous solution. If the water initially contains no bromine,

$$t = 0, \quad c_1 = 0$$

Using this in our integration, we find

$$\frac{c_1}{c_1(\text{sat})} = 1 - e^{-kat}$$

Rearranging,

$$ka = -\frac{1}{t} \ln \left(1 - \frac{c_1}{c_1(\text{sat})} \right)$$

$$= -\frac{1}{3 \text{ min}} \ln(1 - 0.5)$$

$$= 3.9 \cdot 10^{-3} \text{ sec}^{-1}$$

This is as far as we can go; we cannot find the mass transfer coefficient, only its product with a.

Such a product occurs often and is a fixture of many mass transfer correlations. The quantity ka is very similar to the rate constant of a first-order reversible reaction with an equilibrium constant equal to unity. This particular problem is similar to the calculation of a half-life for radioactive decay. Such a parallel is worth thinking through, and will become a very useful concept in Chapter 13.

Example 9.1-4: Mass transfer from an oxygen bubble

A bubble of oxygen originally 0.1 cm in diameter is injected into excess stirred water, as shown schematically in Fig. 9.1-1(d). After 7 min, the bubble is 0.054 cm in diameter. What is the mass transfer coefficient?

Solution. This time, we write a mass balance not on the surrounding solution but on the bubble itself:

$$\frac{d}{dt} \left(c_1 \frac{4}{3} \pi r^3 \right) = A N_1$$

$$= -4\pi r^2 k[c_1(\text{sat}) - 0]$$

This equation is tricky; c_1 refers to the oxygen concentration in the bubble, 1 mol/22.4 liters at standard conditions; but $c_1(\text{sat})$ refers to the oxygen concentration at saturation in water, about $1.5 \cdot 10^{-3}$ mol/liter under similar conditions. Thus,

$$\frac{dr}{dt} = -k \frac{c_1(\text{sat})}{c_1}$$

$$= -0.034k$$

This is subject to the condition

$t = 0$, $r = 0.05$ cm

so integration gives

$r = 0.05$ cm $- 0.034kt$

Inserting the numerical values given, we find

0.027 cm = 0.05 cm $- 0.034k(420$ sec)

$k = 1.6 \cdot 10^{-3}$ cm/sec

Remember that this coefficient is defined in terms of the concentration in the liquid; it would be numerically different if it were defined in terms of the gas-phase concentration.

Section 9.2. Other definitions of mass transfer coefficients

We now want to return to some of the problems we glossed over in the simple definition of a mass transfer coefficient given in the previous section. We introduced this definition with the implication that it provides a simple way of analyzing complex problems. We implied that the mass transfer coefficient will be like the density or the viscosity, a physical quantity that is well defined for a specific situation.

In fact, the mass transfer coefficient is often an ambiguous concept, reflecting nuances of its basic definition. To begin our discussion of these nuances, we first compare the mass transfer coefficient with the other rate constants given in Table 9.2-1. The mass transfer coefficient seems a curious contrast, a combination of diffusion and dispersion. Because it involves a concentration difference, it has different dimensions than the diffusion and dispersion coefficients. It is a rate constant for an interfacial physical reaction, most similar to the rate constant of an interfacial chemical reaction.

Unfortunately, the definition of the mass transfer coefficient in Table 9.2-1 is not so well accepted that the coefficient's dimensions are always the same. This is not true for the other processes in this table. For example, the dimensions of the diffusion coefficient are always taken as L^2/t. If the concentration is expressed in terms of mole fraction or partial pressure, then appropriate unit conversions are made to ensure that the diffusion coefficient keeps the same dimensions.

This is not the case for mass transfer coefficients, where a variety of definitions are accepted. Four of the more common of these are shown in Table 9.2-2 (Treybal, 1980). This variety is largely an experimental artifact, arising because the concentration can be measured in so many different units, including partial pressure, mole and mass fractions, and molarity.

In this book, we assume that if the flux is in moles per area per time, then the concentration difference is expressed as moles per volume. If the flux is in mass per area per time, then the concentration is given as mass per volume. Thus, our mass transfer coefficient has dimensions of velocity. This definition recalls the analyses common 50 years ago, when understandable

Table 9.2-1. *Mass transfer coefficient compared with other rate coefficients*

Effect	Basic equation	Rate	Force	Coefficient
Mass transfer	$N_1 = k\Delta c_1$	Flux per area relative to an interface	Difference of concentration	The mass transfer coefficient k ($[=]L/t$) is a function of flow
Diffusion	$-\mathbf{j}_i = D\nabla c_1$	Flux per area relative to the volume average velocity	Gradient of concentration	The diffusion coefficient D ($[=]L^2/t$) is a physical property independent of flow
Dispersion	$-\overline{c_1'\mathbf{v}_1'} = E\nabla\bar{c}_1$	Flux per area relative to the mass average velocity	Gradient of time average concentration	The dispersion coefficient E ($[=]L^2/t$) depends on the flow
Homogeneous chemical reaction	$r_1 = \kappa_1 c_1$	Rate per volume	Concentration	The rate constant κ_1 ($[=]1/t$) is a physical property independent of flow
Heterogeneous chemical reaction	$r_1 = \kappa_1 c_1$	Flux per interfacial area	Concentration	The rate constant κ_1 ($[=]L/t$) is a surface property usually defined in terms of a bulk concentration

Table 9.2-2. *Common definitions of mass transfer coefficients*[a]

Basic equation	Typical units of k[b]	Remarks
$N_1 = k\Delta c_1$	cm/sec	Common in the older literature; used here because of its simple physical significance
$N_1 = k'\Delta p_1$	mol/cm²-sec-atm	Common for gas absorption; equivalent forms occur in biological problems (McCabe & Smith, 1975; Sherwood et al., 1975)
$N_1 = k''\Delta x_1$	mol/cm²-sec	Preferred for some theoretical calculations, especially in gases (Bennett & Myers, 1974)
$N_1 = k'''\Delta c_1 + c_1 v^0$	cm/sec	Used in an effort to include diffusion-induced convection (cf. k in Eq. 11.5-2 et seq.) (Bird et al., 1960)

[a] In this table, N_1 is defined as mol/$L^2 t$, and c_1 as mol/L^3. Parallel definitions where N_1 is in terms of $M/L^2 t$ and c_1 is M/L^3 are easily developed. Definitions mixing moles and mass are infrequently used.

[b] For a gas of constant molar concentration c, $k = RTk' = k''c$. For a dilute liquid solution $k = (\bar{M}_2 H/\rho)k' = (\bar{M}_2/\rho)k''$, where $\bar{M}_2$ is the molecular weight of the solvent, H is Henry's law constant of the solute, and ρ is the solution density.

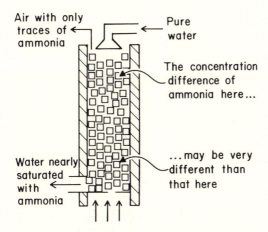

Fig. 9.2-1. Ammonia scrubbing. In this example, ammonia is separated by washing a gas mixture with water. As explained in the text, the example illustrates ambiguities in the definition of mass transfer coefficients. The ambiguities occur because the concentration difference causing the mass transfer changes and because the interfacial area between gas and liquid is unknown.

definitions of mass transfer were still in vogue. It is simpler than the approaches in most recent books, where elaborate definitions are flourishing. Although these can be superior in specific situations, I believe that they are harder to learn than the ideas used here.

The basic definition of mass transfer coefficient given in Eq. 9.1-2 or Table 9.2-1 is ambiguous, for the concentration difference involved is incompletely defined. To explore the ambiguity more carefully, consider the packed tower shown schematically in Fig. 9.2-1. This tower is basically a piece of pipe standing on its end and filled with crushed inert material like broken glass. Water trickles down through the column and absorbs the ammonia. In this tower, ammonia is scrubbed out of a gas mixture with water. The flux of ammonia into the water is proportional to the ammonia concentration at the air–water interface minus the ammonia concentration in the bulk water. The proportionality constant is the mass transfer coefficient. Unfortunately, the concentration difference between interface and bulk is not constant, but can vary along the height of the column. Which value of concentration difference should we use?

In this book, we always choose to use the local concentration difference at a particular position in the column. Such a choice is often called a "local mass transfer coefficient" to distinguish it from an "average mass transfer coefficient" (e.g., Perry & Chilton, 1973). Use of a local coefficient means that we often must make a few extra mathematical calculations. However, the local coefficient is more nearly constant, a smooth function of changes in other process variables. This definition was implicitly used in Examples

9.1-1, 9.1-3, and 9.1-4 in the previous section. It was used in parallel with a type of average coefficient in Example 9.1-2.

Still another potential source of ambiguity in the definition of the mass transfer coefficient is the interfacial area. As an example, we again consider the packed tower in Fig. 9.2-1. The surface area between water and gas is usually experimentally unknown, so that the flux per area is unknown as well. Thus, the mass transfer coefficient cannot be easily found. This problem is dodged by lumping the area into the mass transfer coefficient and experimentally determining the product of the two. We just measure the flux per column volume. This may seem like cheating, but it works like a charm.

I find these points difficult, hard to understand without careful thought. To spur this thought, try the harder examples that follow.

Example 9.2-1: Converting units of mass transfer coefficients

A packed tower is being used to study ammonia scrubbing with 25°C water. The mass transfer coefficients reported for this tower are 1.18 lb-mol NH_3/hr-ft^2 for the liquid and 1.09 lb-mol NH_3/hr-ft^2-atm for the gas. What are these coefficients in centimeters per second?

Solution. From Table 9.2-2 we see that the units of the liquid-phase coefficient correspond to k''. Thus,

$$k = \frac{\tilde{M}_2}{\rho} k''$$

$$= \left(\frac{18 \text{ lb/mol}}{62.4 \text{ lb/ft}^3}\right)\left(\frac{1.18 \text{ lb-mol } NH_3}{ft^2\text{-hr}}\right)\left(\frac{30.5 \text{ cm}}{ft}\right)\left(\frac{hr}{3{,}600 \text{ sec}}\right)$$

$$= 2.9 \cdot 10^{-3} \text{ cm/sec}$$

For the gas phase, we see from Table 9.2-2 that the coefficient has the units of k'. Thus,

$$k = RTk'$$

$$= \left(\frac{1.314 \text{ atm-ft}^3}{\text{lb-mol-}°K}\right)\left(\frac{1.09 \text{ lb-mol}}{hr\text{-}ft^2\text{-atm}}\right)\left(\frac{30.5 \text{ cm}}{ft}\right)\left(\frac{hr}{3{,}600 \text{ sec}}\right)(298°K)$$

$$= 3.6 \text{ cm/sec}$$

These conversions take time and thought, but are not difficult.

Example 9.2-2: Averaging a mass transfer coefficient

Imagine two porous solids whose pores contain different concentrations of a particular dilute solution. If these solids are placed together, the flux N_1 from one to the other will be (see Section 2.3)

$$N_1 = \sqrt{D/\pi t}\ \Delta c_1$$

By comparison with Eq. 9.1-2, we see that the local mass transfer coefficient is

$$k = \sqrt{D/\pi t}$$

Note that this coefficient is initially infinite.

We want to correlate our results not in terms of this local value but in terms of a total experimental time t_0. This implies an average coefficient $\bar{k}$, defined by

$$\bar{N}_1 = \bar{k}\Delta c_1$$

where $\bar{N}_1$ is the total solute transferred per area divided by t_0. How is $\bar{k}$ related to k?

Solution. From the problem statement, we see that

$$\bar{N}_1 = \frac{\int_0^{t_0} N_1 \, dt}{\int_0^{t_0} dt} = \frac{\int_0^{t_0} \sqrt{D/\pi t} \, \Delta c_1 \, dt}{t_0} = 2\sqrt{D/\pi t_0} \, \Delta c_1$$

Thus,

$$\bar{k} = 2\sqrt{D/\pi t_0}$$

which is twice the value of k evaluated at t_0. Note that "local" refers here to a particular time rather than a particular position.

Example 9.2-3: Log mean mass transfer coefficients

Consider again the packed bed of benzoic acid spheres shown in Fig. 9.1-1(b) that was basic to Example 9.1-2. Mass transfer coefficients in a bed like this are sometimes reported in terms of a log mean driving force:

$$\bar{N}_1 = k_{\log} \left(\frac{\Delta c_{1,\text{inlet}} - \Delta c_{1,\text{outlet}}}{\ln\left(\dfrac{\Delta c_{1,\text{inlet}}}{\Delta c_{1,\text{outlet}}}\right)} \right)$$

For this specific case, $\bar{N}_1$ is the total benzoic acid leaving the bed divided by the total surface area in the bed. The bed is fed with pure water, and the benzoic acid concentration at the sphere surfaces is at saturation; that is, it equals $c_1(\text{sat})$. Thus,

$$\bar{N}_1 = k_{\log} \frac{[c_1(\text{sat}) - 0] - [c_1(\text{sat}) - c_1(\text{out})]}{\ln\left(\dfrac{c_1(\text{sat}) - 0}{c_1(\text{sat}) - c_1(\text{out})}\right)}$$

Show how $k_{\log}$ is related to the local coefficient k used in the earlier problem.

Solution. By integrating a mass balance on a differential length of bed, we showed in Example 9.1-2 that for a bed of length L,

$$\frac{c_1(\text{out})}{c_1(\text{sat})} = 1 - e^{-(ka/v^0)L}$$

Rearranging, we find

$$\frac{c_1(\text{sat}) - c_1(\text{out})}{c_1(\text{sat}) - 0} = e^{-(ka/v^0)L}$$

Taking the logarithm of both sides and rearranging,

$$v^0 = \frac{kaL}{\ln \left(\dfrac{c_1(\text{sat}) - 0}{c_1(\text{sat}) - c_1(\text{out})}\right)}$$

Multiplying both sides by $c_1(\text{out})$,

$$c_1(\text{out})v^0 = kaL \left(\frac{[c_1(\text{sat}) - 0] - [c_1(\text{sat}) - c_1(\text{out})]}{\ln \left(\dfrac{c_1(\text{sat}) - 0}{c_1(\text{sat}) - c_1(\text{out})}\right)}\right)$$

By definition,

$$N_1 = \frac{c_1(\text{out})v^0 A}{a(AL)}$$

where A is the bed's cross section and Az is its volume. Thus,

$$N_1 = k \left(\frac{[c_1(\text{sat}) - 0] - [c_1(\text{sat}) - c_1(\text{out})]}{\ln \left(\dfrac{c_1(\text{sat}) - 0}{c_1(\text{sat}) - c_1(\text{out})}\right)}\right)$$

and

$$k_{\log} = k$$

The coefficients are identical.

Many argue that the log mean mass transfer coefficient is superior to the local value used most in this book. Their reasons are that the coefficients are the same or (at worst) closely related and that $k_{\log}$ is macroscopic and hence easier to measure. After all, these critics assert, you implicitly repeat this derivation every time you make a mass balance. Why bother? Why not use $k_{\log}$ and be done with it?

This argument has merit, but it makes me uneasy. I find that I need to think through the approximations of mass transfer coefficients every time I use them and that this review is easily accomplished by making a mass balance and integrating. I find that most students share this need. My advice is to avoid log mean coefficients until your calculations are routine.

Section 9.3. Correlations of mass transfer coefficients

In the previous two sections we have presented definitions of mass transfer coefficients and have shown how these coefficients can be found from experiment. Thus, we have a method for analyzing the results of mass transfer experiments. This method can be more convenient than diffusion when the experiments involve mass transfer across interfaces. Experiments of this sort include liquid–liquid extraction, gas absorption, and distillation.

However, we often want to predict how one of these complex situations will behave. We do not want to correlate experiments; we want to avoid experiments as much as possible. This avoidance is like that in our studies of diffusion, where we often looked up diffusion coefficients so that we could calculate a flux or a concentration profile. We wanted to use someone else's measurements rather than painfully make our own.

Table 9.3-1. *Significance of common dimensionless groups*

Group[a]	Physical meaning	Used in
Sherwood number $\dfrac{kl}{D}$	$\dfrac{\text{mass transfer velocity}}{\text{diffusion velocity}}$	Usual dependent variable
Stanton number $\dfrac{k}{v^0}$	$\dfrac{\text{mass transfer velocity}}{\text{flow velocity}}$	Occasional dependent variable
Schmidt number $\dfrac{\nu}{D}$	$\dfrac{\text{diffusivity of momentum}}{\text{diffusivity of mass}}$	Correlations of gas and liquid data
Lewis number $\dfrac{\alpha}{D}$	$\dfrac{\text{diffusivity of energy}}{\text{diffusivity of mass}}$	Simultaneous heat and mass transfer
Prandtl number $\dfrac{\nu}{\alpha}$	$\dfrac{\text{diffusivity of momentum}}{\text{diffusivity of energy}}$	Heat transfer; included here for completeness
Reynolds number $\dfrac{lv^0}{\nu}$	$\dfrac{\text{inertial forces}}{\text{viscous forces}}$ or $\dfrac{\text{flow velocity}}{\text{"momentum velocity"}}$	Forced convection
Grashöf number $\dfrac{l^3 g \Delta\rho/\rho}{\nu^2}$	$\dfrac{\text{buoyancy forces}}{\text{viscous forces}}$	Free convection
Péclet number $\dfrac{v^0 l}{D}$	$\dfrac{\text{flow velocity}}{\text{diffusion velocity}}$	Correlations of gas or liquid data
Second Damköhler number $\dfrac{\kappa l^2}{D}$ or (Thiele modulus)^2	$\dfrac{\text{reaction velocity}}{\text{diffusion velocity}}$	Correlations involving reactions (see Chapters 13–14)

[a] The symbols and their dimensions are as follows:

D diffusion coefficient (L^2/t)
g acceleration due to gravity (L/t^2)
k mass transfer coefficient (L/t)
l characteristic length (L)
v^0 fluid velocity (L/t)
α thermal diffusivity (L^2/t)
κ first-order reaction rate constant (t^{-1})
ν kinematic viscosity (L^2/t)
$\Delta\rho/\rho$ fractional density change

Dimensionless numbers

In the same way, we want to look up mass transfer coefficients whenever possible. These coefficients are rarely reported as individual values, but as correlations of dimensionless numbers. These numbers are often named, and they are major weapons that engineers use to confuse scientists. These weapons are effective because the names sound so scientific, like close relatives of nineteenth-century organic chemists.

The characteristics of the common dimensionless groups frequently used in mass transfer correlations are given in Table 9.3-1. Sherwood and Stanton numbers involve the mass transfer coefficient itself. The Schmidt, Lewis, and Prandtl numbers involve different kinds of diffusion, and the Reynolds, Grashöf, and Péclet numbers describe flow. The second Damköhler number, which certainly is the most imposing name, is one of many groups used for diffusion with chemical reaction.

A key point about each of these groups is that its exact definition implies a specific physical system. For example, the characteristic length l in the Sherwood number kl/D will be the membrane thickness for membrane transport, but the sphere diameter for a dissolving sphere. A good analogy is the dimensionless group "efficiency." An efficiency of 30% has very different implications for a turbine and for a running deer. In the same way, a Sherwood number of 2 means different things for a membrane and for a dissolving sphere. This flexibility is central to the correlations that follow.

Frequently used correlations

Some of the more common correlations of mass transfer coefficients are shown in Table 9.3-2. The accuracy of these correlations is typically of the order of 10%, but larger errors are not uncommon. Raw data can look more like the result of a shotgun blast than any sort of coherent experiment, because the data generally include wide ranges of chemical and physical properties. For example, the Reynolds number, that characteristic parameter of forced convection, can vary as much as a million times. The Schmidt number, the ratio (ν/D), is about 1 for gases but about 1,000 for liquids. Over a more moderate range, experimental data can be very reliable.

Many of the correlations in Table 9.3-2 have the same general form. They typically involve a Sherwood number, which contains the mass transfer coefficient, the quantity of interest. This Sherwood number varies with Schmidt number, a characteristic of diffusion, and with some other dimensionless group characteristic of the flow.

Variations of Sherwood number with Schmidt number are often similar even in widely different physical situations. In many cases the Sherwood number varies with the cube root of the Schmidt number. This narrow range facilitates estimates of k as a function of D.

The variation of Sherwood number with flow is more complex, because the flow has two different physical origins. In most cases the flow is caused by external stirring or pumping. For example, the liquids used in extraction are rapidly stirred; the gas in ammonia scrubbing is pumped through the packed tower; the blood in the artificial kidney is pumped by the heart through the dialysis unit. This type of externally driven flow is called "forced convection."

In other cases the fluid velocity is a result of the mass transfer itself. The mass transfer causes density gradients in the surrounding solution; these, in turn, cause flow. This type of internally generated flow is called "free convection." For example, the dispersal of pollutants and the dissolution of drugs are often accelerated by free convection.

Forced convection is characterized in Table 9.3-2 by a Reynolds number. The correlations shown are based on Reynolds numbers ranging from 10^{-2} to 10^7. In some cases, two correlations are necessary to describe the observed behavior. One of these correlations is accurate for turbulent flow, in which the fluid is flowing sufficiently fast that it undergoes random fluctuations around its average velocity. Anyone who has watched a river flow into an area of rapids will recognize this type of flow. The second correlation is accurate for laminar or streamline flow, in which the fluid flows smoothly and evenly. This type flow is characteristic of blood in a small artery.

Free convection, which occurs when the mass transfer process itself generates a flow, is characterized by a Grashöf number. The key part of this dimensionless group is the density change caused by the concentration difference between the surface and the bulk solution. This concentration difference can change in subtle ways. For example, in studies of drug dissolution, the concentration difference for dissolution in a pure solvent is the concentration at saturation minus the concentration in the pure solvent. However, if this dissolution takes place in a small volume of solvent, the solvent contains more and more drug as the experiment proceeds. The concentration difference between the solid drug and the surrounding solution will become smaller and smaller, and the flow caused by free convection and the mass transfer coefficient k will both drop.

These general remarks are buttressed by the following numerical examples.

Example 9.3-1: Dissolution rate of a spinning disc

A solid disc of benzoic acid 2.5 cm in diameter is spinning at 20 rpm and 25°C. How fast will it dissolve in a large volume of water? How fast will it dissolve in a large volume of air? The diffusion coefficients are $1.00 \cdot 10^{-5}$ cm²/sec in water and 0.233 cm²/sec in air. The solubility of benzoic acid in water is 0.003 g/cm³; its equilibrium vapor pressure is 0.30 mm Hg.

Solution. Before starting this problem, try to guess the answer. Will the mass transfer be higher in water or in air?

In each case, the dissolution rate is

$$N_1 = kc_1(\text{sat})$$

where $c_1(\text{sat})$ is the concentration at equilibrium. We can find k from Table 9.3-2:

$$k = 0.62D \left(\frac{\omega}{\nu}\right)^{1/2} \left(\frac{\nu}{D}\right)^{1/3}$$

For water, the mass transfer coefficient is

$$k = 0.62(1.00 \cdot 10^{-5} \text{ cm}^2/\text{sec}) \left(\frac{(20/60)(2\pi/\text{sec})}{0.01 \text{ cm}^2/\text{sec}}\right)^{1/2} \left(\frac{0.01 \text{ cm}^2/\text{sec}}{1.00 \cdot 10^{-5} \text{ cm}^2/\text{sec}}\right)^{1/3}$$

$$= 0.90 \cdot 10^{-3} \text{ cm/sec}$$

Table 9.3-2. *A selection of mass transfer correlations*[a]

Physical situation	Basic equation[b]	Key variables	Remarks
Solid interfaces			
Membrane	$\dfrac{kl}{D} = 1$	l = membrane thickness	Often applied even where membrane is hypothetical
Laminar flow along flat plate	$\dfrac{kz}{D} = 0.323 \left(\dfrac{zv^0}{v}\right)^{1/2} \left(\dfrac{v}{D}\right)^{1/3}$	z = distance from start of plate v^0 = bulk velocity	Solid theoretical foundation, which is unusual
Turbulent flow through horizontal slit	$\dfrac{kd}{D} = 0.026 \left(\dfrac{dv^0}{v}\right)^{0.8} \left(\dfrac{v}{D}\right)^{1/3}$	v^0 = average velocity in slit d = $(2/\pi)$ (slit width)	Mass transfer here is identical with that in a pipe of equal wetted perimeter
Turbulent flow through circular pipe	$\dfrac{kd}{D} = 0.026 \left(\dfrac{dv^0}{v}\right)^{0.8} \left(\dfrac{v}{D}\right)^{1/3}$	v^0 = average velocity in pipe d = pipe diameter	Same as slit, because only wall region is involved
Laminar flow through circular pipe[c]	$\dfrac{kd}{D} = 1.86 \left(\dfrac{dv^0}{D}\right)^{0.8}$	d = pipe diameter L = pipe length v^0 = average velocity in pipe	Not reliable when (dv/D) < 10 because of free convection
Forced convection around a solid sphere	$\dfrac{kd}{D} = 2.0 + 0.6 \left(\dfrac{dv^0}{v}\right)^{1/2} \left(\dfrac{v}{D}\right)^{1/3}$	d = sphere diameter v^0 = velocity of sphere	Very difficult to reach $(kd/D) = 2$ experimentally; no sudden laminar–turbulent transition
Free convection around a solid sphere	$\dfrac{kd}{D} = 2.0 + 0.6 \left(\dfrac{d^3[\Delta\rho]g}{\rho v^2}\right)^{1/4} \left(\dfrac{v}{D}\right)^{1/3}$	d = sphere diameter g = gravitational acceleration	For a 1-cm sphere in water, free convection is important when $\Delta\rho = 10^{-9}$ g/cm^3

Situation	Correlation	Key variables	Remarks
Spinning disc	$\dfrac{kd}{D} = 0.62 \left(\dfrac{d^2\omega}{\nu}\right)^{1/2} \left(\dfrac{\nu}{D}\right)^{1/3}$	d = disc diameter ω = disc rotation (radians/time)	Valid for Reynolds numbers between 100 and 20,000
Flow normal to capillary bed	$\dfrac{kd}{D} = f\left(\dfrac{dv^0}{\nu}, \dfrac{\nu}{D}\right)$	d = tube diameter v^0 = average velocity	Large number of correlations with different exponents found by analogy with heat transfer
Packed beds	$\dfrac{k}{v^0} = 1.17 \left(\dfrac{dv^0}{\nu}\right)^{-0.42} \left(\dfrac{\nu}{D}\right)^{-0.67}$	d = particle diameter v^0 = superficial velocity	The superficial velocity is that which would exist without packing
Fluid–fluid interfaces Drops or bubbles in stirred solution	$\dfrac{kL}{D} = 0.13 \left(\dfrac{L^4(P/V)}{\rho\nu^3}\right)^{1/4} \left(\dfrac{\nu}{D}\right)^{1/3}$	L = stirrer length P/V = power per volume	Correlations versus power per volume are common for dispersions
Large drops in unstirred solution	$\dfrac{kd}{D} = 0.42 \left(\dfrac{d^3\Delta\rho g}{\rho\nu^2}\right)^{1/3} \left(\dfrac{\nu}{D}\right)^{1/2}$	d = bubble diameter $\Delta\rho$ = density difference between bubble and surrounding fluid	"Large" is defined as ~0.3-cm diameter
Small drops of pure solute in unstirred solution	$\dfrac{kd}{D} = 1.13 \left(\dfrac{dv^0}{D}\right)^{0.8}$	d = bubble diameter v^0 = bubble velocity	These behave like rigid spheres
Falling films	$\dfrac{kz}{D} = 0.69 \left(\dfrac{zv^0}{D}\right)^{1/2}$	z = position along film v^0 = average film velocity	Frequently embroidered and embellished

[a] The symbols used include the following: ρ is the fluid density; ν is the kinematic viscosity; D is the diffusion coefficient of the material being transferred; k is the local mass transfer coefficient. Other symbols are defined for the specific situation.

[b] The dimensionless groups are defined as follows: (dv/ν) and $(d^2\omega/\nu)$ are the Reynolds number; $(d^2\Delta\rho g/\rho\nu^2)$ is the Grashöf number; kd/D is the Sherwood number; ν/D is the Schmidt number; k/ν is the Stanton number.

[c] The mass transfer coefficient given here is the value averaged over the length.

Source: Data from Calderbank (1967), McCabe and Smith (1975), Schlichting (1979), Sherwood et al. (1975), and Treybal (1980).

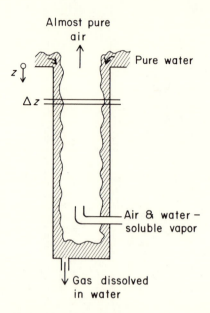

Fig. 9.3-1. Gas scrubbing in a wetted-wall column. A large amount of a water-soluble gas is being dissolved in a falling film of water. The problem is to calculate the length of the column necessary to reach a liquid concentration equal to 10% saturation.

Thus, the flux is

$$N_1 = (0.90 \cdot 10^{-3} \text{ cm/sec})(0.003 \text{ g/cm}^3)$$
$$= 2.7 \cdot 10^{-6} \text{ g/cm}^2\text{-sec}$$

For air, the values are very different:

$$k = 0.62(0.233 \text{ cm}^2/\text{sec}) \left(\frac{(20/60)(2\pi/\text{sec})}{0.15 \text{ cm}^2/\text{sec}}\right)^{1/2} \left(\frac{0.15 \text{ cm}^2/\text{sec}}{0.233 \text{ cm}^2/\text{sec}}\right)^{1/3}$$

$$= 0.47 \text{ cm/sec}$$

which is much larger than before. However, the flux is

$$N_1 = (0.47 \text{ cm/sec}) \left[\left(\frac{0.3 \text{ mm Hg}}{760 \text{ mm Hg}}\right)\left(\frac{1 \text{ mol}}{22.4 \cdot 10^3 \text{ cm}^3}\right)\left(\frac{273}{298}\right)\left(\frac{122 \text{ g}}{\text{mol}}\right)\right]$$

$$= 0.9 \cdot 10^{-6} \text{ g/cm}^2\text{-sec}$$

The flux in air is about one-third of that in water, even though the mass transfer coefficient in air is about 500 times larger than that in water. Did you guess this?

Example 9.3-2: Gas scrubbing with a wetted-wall column

Air containing a water-soluble vapor is flowing up and water is flowing down in the experimental column shown in Fig. 9.3-1. The water flow in the 0.07-cm-thick film is 3 cm/sec, the column diameter is 10 cm, and the air is essentially well mixed right up to the interface. The diffusion coefficient in water of the absorbed vapor is $1.8 \cdot 10^{-5}$ cm²/sec. How long a column is needed to reach a gas concentration in water that is 10% of saturation?

Solution. The first step is to write a mass balance on the water in a differential column height Δz:

(accumulation) = (flow in minus flow out) + (absorption)

$$0 = [\pi dl v^0 c_1]_z - [\pi dl v^0 c_1]_{z+\Delta z}$$
$$+ \pi d \Delta z k [c_1(\text{sat}) - c_1]$$

in which d is the column diameter, l is the film thickness, v is the flow, and c_1 is the ammonia concentration in the water. This balance leads to

$$0 = -l v^0 \frac{dc_1}{dz} + k[c_1(\text{sat}) - c_1]$$

From Table 9.3-2 we have

$$k = 0.69 \left(\frac{D v^0}{z}\right)^{1/2}$$

We also know that the entering water is pure; that is, when

$$z = 0, \quad c_1 = 0$$

Combining these results and integrating, we find

$$\frac{c_1}{c_1(\text{sat})} = 1 - e^{-(Dz/l^2 v)^{1/2}}$$

Inserting the numerical values given,

$$z = \left(\frac{l^2 v}{D}\right)\left[\ln\left(1 - \frac{c_1}{c_1(\text{sat})}\right)\right]^2$$

$$= \left(\frac{(0.07 \text{ cm})^2 (3 \text{ cm/sec})}{1.8 \cdot 10^{-5} \text{ cm}^2/\text{sec}}\right) [\ln(1 - 0.1)]^2$$

$$= 9 \text{ cm}$$

This approximate calculation has been improved elaborately, even though its practical value is small (Sherwood et al., 1975).

Example 9.3-3: Measuring stomach flow

Imagine we want to estimate the average flow in the stomach by measuring the dissolution rate of a nonabsorbing solute present as a large spherical pill. From *in vitro* experiments we know that this pill's dissolution is accurately described with a mass transfer coefficient. How can we do this?

Solution. We first calculate the concentration c_1 of the dissolving solute in the stomach and then show how this is related to the flow. From a mass balance,

$$V \frac{dc_1}{dt} = \pi d^2 k[c_1(\text{sat}) - c_1]$$

where V is the stomach's volume, πd^2 is the pill's area, k is the mass transfer coefficient, and $c_1(\text{sat})$ is the solute's solubility. Because no solute is initially present,

$$c_1 = 0 \quad \text{when } t = 0$$

Integrating,

$$k = \frac{V}{\pi d^2 t} \ln \left(\frac{c_1(\text{sat})}{c_1(\text{sat}) - c_1} \right)$$

If we assume that stomach flow is essentially forced convection, we find, from Table 9.3-2,

$$\frac{kd}{D} = 2 + 0.6 \left(\frac{dv}{\nu} \right)^{1/2} \left(\frac{\nu}{D} \right)^{1/3}$$

where d is the pill diameter, D is the diffusion coefficient, and v is the unknown stomach flow. Combining and rearranging,

$$v = \frac{25}{9} \left(\frac{\nu^{1/3} D^{2/3}}{d} \right) \left[\frac{V}{\pi dt D} \ln \left(\frac{c_1(\text{sat})}{c_1(\text{sat}) - c_1} \right) - 2 \right]^2$$

which is the desired result. Note the assumptions in this problem: The flow is due to forced convection, the pill diameter is constant, and flow in and out of the stomach is negligible.

Example 9.3-4: Glucose uptake by red blood cells

The uptake of glucose across the red blood cell membrane has a reported maximum rate ranging from 0.1 to 5 μmole/cm^2-hr (Stein, 1967). Apparently these differences result from differences in experimental conditions. Using the correlation for liquid drops in Table 9.3-2, estimate the effect of mass transfer in the bulk to see when it could have affected these uptake rates. To make the estimation more quantitative, assume that a typical experiment is made in a beaker containing 100 cm^3 of red blood cells suspended in 1 liter of plasma. The beaker is stirred with a 1/50-hp motor. The cells originally contain little glucose. At time zero, radioactively tagged glucose is added and its uptake measured. The diffusion coefficient of glucose is about $6 \cdot 10^{-6}$ cm^2/sec, and the plasma viscosity is approximately that of water.

Solution. We are interested in the case in which glucose uptake is dominated by mass transfer. In this case, glucose will diffuse to the membrane and then almost instantaneously be taken into the cell. Thus,

$$N_1 = kc_1$$

where c_1 is the bulk concentration. If we can calculate k, then we can estimate N_1, the desired quantity. We see from Table 9.3-2 that for a suspension of liquid drops,

$$k = 0.13 \left(\frac{P}{V} \right)^{1/4} \rho^{-1/4} \nu^{-5/12} D^{2/3}$$

$$= 0.13 \left(\frac{1/50 \text{ hp}}{1,000 \text{ cm}^3} \frac{7.45 \cdot 10^9 \text{ g-cm}^2}{\text{hp-sec}^3} \right)^{1/4} \left(\frac{1 \text{ g}}{\text{cm}^3} \right)^{-1/4}$$

$$\cdot \left(\frac{0.01 \text{ cm}^2}{\text{sec}} \right)^{-5/12} \left(6 \cdot 10^{-6} \frac{\text{cm}^2}{\text{sec}} \right)^{2/3}$$

$$= 5.7 \cdot 10^{-3} \text{ cm/sec} = 21 \text{ cm/hr}$$

The flux is

$$N_1 = (21 \text{ cm/hr}) c_1$$

Whether or not diffusion outside the cells is significant depends on c_1, the amount of glucose used. If c_1 far exceeds 0.3 mmol/liter, then the flux due to diffusion will be much faster than that due to the cell membrane. The measurements will then truly represent membrane properties. However, if the glucose concentration is less than 0.3 mmol/liter, then the measurements will be functions both of the membrane and of mass transfer in plasma. Such restrictions can compromise measurements in biological systems.

Section 9.4. Dimensional analysis: the route to correlations

The correlations in the previous section provide a useful and compact way of presenting experimental information. Use of these correlations quickly gives reasonable estimates of mass transfer coefficients. However, when we find the correlations inadequate, we will be forced to make our own experiments and develop our own correlations. How can we do this?

The basic form of mass transfer correlations is easily developed using a method called dimensional analysis (Bridgeman, 1922; Becker, 1976). This method is easily learned via the two specific examples that follow. Before embarking on this description, I want to emphasize that most people go through three mental states concerning this method. At first they believe it is a route to all knowledge, a simple technique by which any set of experimental data can be greatly simplified. Next they become disillusioned when they have difficulties in the use of the technique. These difficulties commonly result from efforts to be too complete. Finally they learn to use the method with skill and caution, benefiting both from their past successes and from their frequent failures. I mention these three stages because I am afraid many may give up at the second stage and miss the real benefits involved. We now turn to the examples.

Aeration

Aeration is a common industrial process, and yet one in which there is often serious disagreement about correlations. This is especially true for deep-bed fermentors and for sewage treatment, where the rising bubbles can be the chief means of stirring. We want to study this process using the equipment shown schematically in Fig. 9.4-1. We plan to inject oxygen into a variety of aqueous solutions and measure the oxygen concentration in the

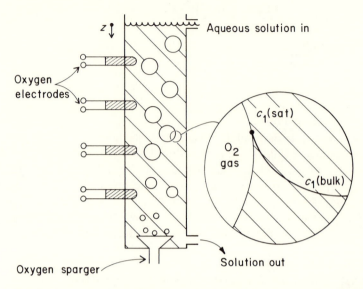

Fig. 9.4-1. An experimental apparatus for the study of aeration. Oxygen bubbles from the sparger at the bottom of the tower partially dissolve in the aqueous solution. The concentration in this solution is measured with electrodes that are specific for dissolved oxygen. The concentrations found in this way are interpreted in terms of mass transfer coefficients; this interpretation assumes that the solution is well mixed, except very near the bubble walls.

bulk with oxygen selective electrodes. We expect to vary the average bubble velocity v, the solution's density ρ and viscosity μ, the entering bubble diameter d, and the depth of the bed L.

We measure the steady-state oxygen concentration as a function of position in the bed. These data can be summarized as a mass transfer coefficient in the following way. From a mass balance, we see that

$$0 = -v \frac{dc_1}{dz} + ka[c_1(\text{sat}) - c_1] \tag{9.4-1}$$

where a is the total bubble area per column volume. This equation, a close parallel to the many mass balances in Section 9.1, is subject to the initial condition

$$z = 0, \quad c_1 = 0 \tag{9.4-2}$$

Thus,

$$ka = \frac{v}{z} \ln \left(\frac{c_1(\text{sat})}{c_1(\text{sat}) - c_1(z)} \right) \tag{9.4-3}$$

Ideally, we would like to measure k and a independently, separating the effects of mass transfer and geometry. This would be difficult here; so we report only the product ka.

Our experimental results now consist of the following:

$$ka = ka(v, \rho, \mu, d, z) \tag{9.4-4}$$

We assume that this function has the form

$$ka = [\text{constant}]v^\alpha \rho^\beta \mu^\gamma d^\delta z^\varepsilon \tag{9.4-5}$$

where both the constant in the square brackets and the exponents are dimensionless. Now the dimensions or units on the left-hand side of this equation must equal the dimensions or units on the right-hand side. We cannot have centimeters per second on the left-hand side equal to grams on the right. Because ka has dimensions of the reciprocal of time $(1/t)$, v has dimensions of length/time (L/t), ρ has dimensions of mass per length cubed (M/L^3), and so forth, we find

$$\frac{1}{t} \,[=]\, \left(\frac{L}{t}\right)^\alpha \left(\frac{M}{L^3}\right)^\beta \left(\frac{M}{Lt}\right)^\gamma (L)^\delta (L)^\varepsilon \tag{9.4-6}$$

The only way this equation can be dimensionally consistent is if the exponent on time on the left-hand side of the equation equals the sum of the exponents on time on the right-hand side:

$$-1 = -\alpha - \gamma \tag{9.4-7}$$

Similar equations hold for the mass:

$$0 = \beta + \gamma \tag{9.4-8}$$

and for the length:

$$0 = \alpha - 3\beta - \gamma + \delta + \varepsilon \tag{9.4-9}$$

Equations 9.4-7 to 9.4-9 give three equations for the five unknown exponents.

We can solve these equations in terms of the two key exponents and thus simplify Eq. 9.4-5. We choose the two key exponents arbitrarily. For example, if we choose the exponent on the viscosity γ and that on column height ε, we obtain

$$\alpha = 1 - \gamma \tag{9.4-10}$$
$$\beta = -\gamma \tag{9.4-11}$$
$$\gamma = \gamma \tag{9.4-12}$$
$$\delta = -\gamma - \varepsilon - 1 \tag{9.4-13}$$
$$\varepsilon = \varepsilon \tag{9.4-14}$$

Inserting these results into Eq. 9.4-5 and rearranging, we find

$$\left(\frac{kad}{v}\right) = [\text{constant}] \left(\frac{dv\rho}{\mu}\right)^{-\gamma} \left(\frac{z}{d}\right)^\varepsilon \tag{9.4-15}$$

The left-hand side of this equation is a type of Stanton number. The first term in parentheses on the right-hand side is the Reynolds number, and the second such term is a measure of the tank's depth.

This analysis suggests how we should plan our experiments. We expect to plot our measurements of Stanton number versus two independent variables: Reynolds number and z/d. We want to cover the widest possible range of independent variables. Our resulting correlation will be a convenient and compact way of presenting our results, and everyone will live happily ever after.

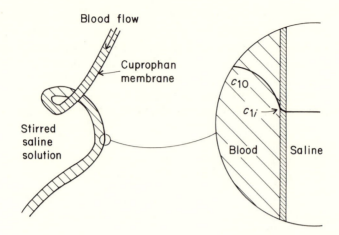

Fig. 9.4-2. Mass transfer in an artificial kidney. Arterial blood flows through a dialysis tube that is immersed in saline. Toxins in the blood diffuse across the tube wall and into the saline. If the saline is well stirred and if the tube wall is thin, then the rate of toxin removal depends on the concentration gradient in the blood. Experiments in this situation are easily correlated using dimensional analysis.

Unfortunately, it is not always that simple, for a variety of reasons (Aiba et al., 1973). First, we had to assume that the bulk liquid was well mixed, and it may not be. If it is not, we shall be averaging our values in some unknown fashion, and we may find that our correlation extrapolates unreliably. Second, we may find that our data do not fit an exponential form like Eq. 9.4-5. This can happen if the oxygen transferred is consumed in some sort of chemical reaction, which is true in aeration. Third, we do not know which independent variables are important. We might suspect that ka varies with tank diameter, or sparger shape, or surface tension, or the phases of the moon. Such variations can be included in our analysis, but they make it very complex.

Still, this strategy has produced a simple method of correlating our results. The foregoing objections are important only if they are shown to be so by experiment. Until then, we should use this easy strategy.

The artificial kidney

The second example to be discussed in this section is the mass transfer out of the tube shown schematically in Fig. 9.4-2. Such tubes are basic to the artificial kidney (Cooney, 1976). There, blood flowing in a loop of cellophane or another membrane is dialyzed against well-stirred saline solution. Toxins in the blood diffuse across the membrane into the saline, thus purifying the blood. This dialysis is often slow; it can take more than 40 hr per week. Increasing the mass transfer in this system would greatly improve its clinical value.

The first step in increasing this rate is to stir the surrounding saline rapidly. This mixing increases the rate of mass transfer on the saline side of the membrane, so that only a small part of the concentration difference is there, as shown in Fig. 9.4-2. In other words, we have decreased the resistance to mass transfer on the saline side. The second step in increasing the rate is to make the membrane as thin as possible. Although too thin a membrane would rupture, existing membranes are already so thin that the membrane thickness has only a minor effect. The result is that the concentration difference across the membrane is not the largest part of the overall concentration difference, again as shown in Fig. 9.4-2.

The rate of toxin removal now depends only on what happens in the blood. We want to correlate our measurements of toxin removal as a function of blood flow, tube size, and so forth. To do this, we find the flux for each case:

$$\text{flux } N_1 = \frac{\text{amount transferred}}{(\text{area})(\text{time})} \tag{9.4-16}$$

By definition,

$$N_1 = k(c_1 - c_{1i})$$
$$\doteq kc_1 \tag{9.4-17}$$

Because we know N_1 and c_1, we can find the mass transfer coefficient k.

As before, we recognize that the mass transfer coefficient of a particular toxin varies with the system's properties:

$$k = k(v, \rho, \mu, D, d) \tag{9.4-18}$$

where v, ρ, and μ are the velocity, density, and viscosity of the blood, D is the diffusion coefficient of the toxin in blood, and d is the diameter of the tube. We assume that this relation has the form

$$k = [\text{constant}]v^\alpha \rho^\beta \mu^\gamma D^\delta d^\varepsilon \tag{9.4-19}$$

where the constant is dimensionless. The dimensions or units on the left-hand side of this equation must equal the dimensions or units on the right-hand side; so

$$\frac{L}{t} [=] \left(\frac{L}{t}\right)^\alpha \left(\frac{M}{L^3}\right)^\beta \left(\frac{M}{Lt}\right)^\gamma \left(\frac{L^2}{t}\right)^\delta L^\varepsilon \tag{9.4-20}$$

This equation will be dimensionally consistent only if the exponent on the length on the left-hand side of the equation equals the sum of the exponents on the right-hand side:

$$1 = \alpha - 3\beta - \gamma + 2\delta + \varepsilon \tag{9.4-21}$$

Similar equations must hold for mass:

$$0 = \beta + \gamma \tag{9.4-22}$$

and for time:

$$-1 = -\alpha - \gamma - \delta \tag{9.4-23}$$

We solve these equations in terms of the exponents α and δ:

$$\alpha = \alpha \tag{9.4-24}$$
$$\beta = \alpha + \delta - 1 \tag{9.4-25}$$
$$\gamma = 1 - \alpha - \delta \tag{9.4-26}$$

$$\delta = \delta \tag{9.4-27}$$

$$\varepsilon = \alpha - 1 \tag{9.4-28}$$

We combine these results with Eq. 9.4-19 and collect terms:

$$\frac{kd\rho}{\mu} = [\text{constant}] \left(\frac{dv\rho}{\mu}\right)^{\alpha} \left(\frac{\mu}{\rho D}\right)^{-\delta} \tag{9.4-29}$$

By convention, we multiply both sides of this equation by the dimensionless quantity $\mu/\rho D$ to obtain

$$\frac{kd}{D} = [\text{constant}] \left(\frac{dv\rho}{\mu}\right)^{\alpha} \left(\frac{\mu}{\rho D}\right)^{1-\delta} \tag{9.4-30}$$

This equation is the desired correlation. The two remaining exponents, α and δ, are found experimentally. The result for other liquids and gases is (see Table 9.3-2)

$$\frac{kd}{D} = 1.86 \left(\frac{dv\rho}{\mu}\right)^{1/3} \left(\frac{\mu}{\rho D}\right)^{1/3} = 1.86 \left(\frac{dv}{D}\right)^{1/3} \tag{9.4-31}$$

The Sherwood number (kd/D) is proportional to the cube root of the Péclet number (dv/D); so the experiments show that viscosity is not important.

As in the first example, a key step in the analysis is the arbitrary choice of the two exponents α and δ. Any other pair of exponents could have been chosen, and would have given a completely equivalent correlation. However, the particular manipulations here are made so that the dimensionless groups found are consistent with traditional patterns. Such traditional patterns sometimes reflect experience and sometimes merely mirror convention. Multiplying both sides of Eq. 9.4-29 by $\mu/\rho D$ involves these factors.

Section 9.5. Mass transfer across interfaces

In the previous sections we used mass transfer coefficients as an easy way of describing diffusion occurring from an interface into a relatively homogeneous solution. These coefficients involved many approximations and sparked the explosion of definitions exemplified by Table 9.2-2. Still, they are a very easy way to correlate experimental results, or to make estimates using the published relations summarized in Table 9.3-2.

In this section we extend these definitions to transfer across an interface, from one well-mixed bulk phase into another different one. This case occurs much more frequently than does transfer from an interface into one bulk phase; indeed, I had trouble dreaming up examples earlier in this chapter. Transfer across an interface again sparks potentially major problems of unit conversion, but these problems are often simplified in special cases.

The basic flux equation

Presumably we can describe mass transfer across an interface in terms of the same type of flux equation as before:

$$N_1 = K\Delta c_1 \tag{9.5-1}$$

(a) Heat transfer

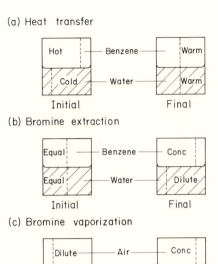

(b) Bromine extraction

(c) Bromine vaporization

Fig. 9.5-1. Driving forces across interfaces. In heat transfer, the amount of heat transferred depends on the temperature difference between the two liquids, as shown in (a). In mass transfer, the amount of solute that diffuses depends on the solute's "solubility" or, more exactly, on its chemical potential. Two cases are shown. In (b), bromine diffuses from water into benzene because it is much more soluble in benzene; in (c), bromine evaporates until its chemical potentials in the solutions are equal. This behavior complicates analysis of mass transfer.

where N_1 is the solute flux relative to the interface, K is called an "overall mass transfer coefficient," and Δc_1 is some appropriate concentration difference. But what is Δc_1?

Choosing an appropriate value of Δc_1 turns out to be difficult. To illustrate this, consider the three situations shown in Fig. 9.5-1. In the first, warm benzene is placed on top of cold water; after a while, the benzene is cooled and the water heated until they reach the same temperature. Equal temperature is the correct criterion for equilibrium, and the amount of energy transferred turns out to be proportional to the temperature difference between the liquids. Everything seems secure.

As a second example, shown in Fig. 9.5-1(b), imagine that a benzene solution of bromine is placed on top of water containing the same concentration of bromine. After a while, we find that the initially equal concentrations have changed, that the bromine concentration in the benzene is much higher than that in water. This is because the bromine is more soluble in benzene, so that its concentration in the final solution is higher.

This result suggests which concentration difference we can use in Eq. 9.5-1. We should not use the concentration in benzene minus the concentration in water; that is initially zero, and yet there is a flux. Instead, we can use the concentration actually in benzene minus the concentration that would be in benzene that was in equilibrium with the actual concentration in water. Symbolically,

$$N_1 = K[c_1(\text{in benzene}) - Hc_1(\text{in water})] \tag{9.5-2}$$

where H is a partition coefficient, the ratio at equilibrium of the concentration in benzene to that in water. Note that this does predict a zero flux at equilibrium.

A better understanding of this phenomenon may come from the third case, shown in Fig. 9.5-1(c). Here, bromine is vaporized from water into air. Initially, the bromine's concentration in water is higher than that in air; afterward, it is lower. Of course, this reversal of the concentration difference is an accident of the units used. For example, the concentration in the liquid might be expressed in moles per liter, and that in gas as a partial pressure in atmospheres; so it is not surprising that strange things happen.

As you think about this more carefully, you will realize that the units of pressure or concentration cloud a deeper truth: Mass transfer can be described in terms of more fundamental chemical potentials. If this were done, the peculiar concentration differences would disappear. However, chemical potentials turn out to be very difficult to use in practice, and so the concentration differences for mass transfer across interfaces will remain complicated by units.

The overall mass transfer coefficient

We want to include these qualitative observations in more exact equations. To do this, we consider the example of the gas–liquid interface in Fig. 9.5-2. In this case, gas on the left is being transferred into the liquid on the right. The flux in the gas is

$$N_1 = k_G(p_{10} - p_{1i}) \tag{9.5-3}$$

where k_G is the gas-phase mass transfer coefficient (typically in mol/cm²-sec-atm), p_{10} is the bulk pressure, and p_{1i} is the interfacial pressure. Because the interfacial region is thin, the flux across it will be in steady state, and the flux in the gas will equal that in the liquid. Thus,

$$N_1 = k_L(c_{1i} - c_{10}) \tag{9.5-4}$$

where the liquid-phase mass transfer coefficient k_L is typically in centimeters per second and c_{1i} and c_{10} are the interfacial and bulk concentrations, respectively.

We now need to eliminate the unknown interfacial concentrations from these equations. In almost all cases, equilibrium exists across the interface:

$$p_{1i} = Hc_{1i} \tag{9.5-5}$$

where H is a type of Henry's law or partition constant (here in cm³-atm/mol). Combining Eqs. 9.5-3 through 9.5-5, we can find the interfacial concentrations

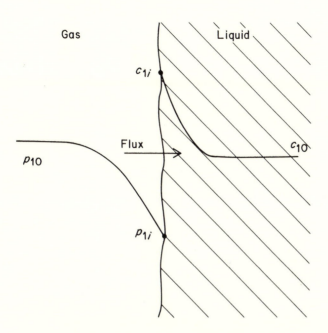

Fig. 9.5-2. Mass transfer across a gas–liquid interface. In this example, a solute vapor is diffusing from the gas on the left into the liquid on the right. Because the solute concentration changes both in the gas and in the liquid, the solute's flux must depend on a mass transfer coefficient in each phase. These coefficients are combined into an overall flux equation in the text.

$$c_{1i} = \frac{p_{1i}}{H} = \frac{k_G p_{10} + k_L c_{10}}{k_G H + k_L} \qquad (9.5\text{-}6)$$

and the flux

$$N_1 = \frac{1}{1/k_G + H/k_L} (p_{10} - H c_{10}) \qquad (9.5\text{-}7)$$

You should check the derivations of these results.

Before proceeding further, we make a quick analogy. This result is often compared to an electric circuit containing two resistances in series. The flux corresponds to the current, and the concentration difference $p_{10} - H c_{10}$ corresponds to the voltage. The resistance is then $1/k_G + H/k_L$, which is roughly a sum of two resistances in series. This is a good way of thinking about these effects. You must remember, however, that the resistances $1/k_G$ and $1/k_L$ are not directly added, but always weighted by partition coefficients like H.

We now want to write Eq. 9.5-7 in the form of Eq. 9.5-1. We can do this in two ways. First, we can write

$$N_1 = K_L(c_1^* - c_{10}) \qquad (9.5\text{-}8)$$

where

$$K_L = \frac{1}{1/k_L + 1/k_G H} \qquad (9.5\text{-}9)$$

and

$$c_1^* = \frac{p_{10}}{H} \qquad (9.5\text{-}10)$$

K_L is called an "overall liquid-side mass transfer coefficient," and c_1^* is the hypothetical liquid concentration that would be in equilibrium with the bulk gas. Alternatively,

$$N_1 = K_G(p_{10} - p_1^*) \qquad (9.5\text{-}11)$$

where

$$K_G = \frac{1}{1/k_G + H/k_L} \qquad (9.5\text{-}12)$$

and

$$p_1^* = Hc_1$$

K_G is an "overall gas-side mass transfer coefficient," and p_1^* is the hypothetical gas-phase concentration that would be in equilibrium with the bulk liquid.

We now turn to a variety of examples illustrating mass transfer across an interface. These examples have the annoying characteristic that they are initially difficult to do, but they are trivial after you understand them. Remember that most of the difficulty comes from that ancient but common curse: unit conversion.

Example 9.5-1: Oxygen mass transfer

Estimate the overall liquid-side mass transfer coefficient at 25°C for oxygen from water into air. In this estimate, assume that each individual mass transfer coefficient is

$$k = \frac{D}{0.01 \text{ cm}}$$

This relation is justified in Section 11.1.

Solution. For oxygen in air, the diffusion coefficient is 0.23 cm²/sec; for oxygen in water, the diffusion coefficient is $2.1 \cdot 10^{-5}$ cm²/sec. The Henry's law constant in this case is $4.4 \cdot 10^4$ atm. We need only calculate k_L and k_G and plug these values into Eq. 9.5-9. Finding k_L is easy:

$$k_L = \frac{D_L}{0.01 \text{ cm}} = \frac{2.1 \cdot 10^{-5} \text{ cm}^2/\text{sec}}{0.01 \text{ cm}}$$

$$= 2.1 \cdot 10^{-3} \text{ cm/sec}$$

Finding k_G and H is harder because of unit conversions. From Eq. 9.5-3 and Table 9.2-2,

$$k_G(p_{10} - p_{1i}) = k(c_{10} - c_{1i})$$

$$k_G = \frac{k}{RT} = \frac{D_G}{(0.01 \text{ cm})(RT)}$$

$$= \frac{0.23 \text{ cm}^2/\text{sec}}{(0.01 \text{ cm})(82 \text{ cm}^3\text{-atm/g-mol-}°K)(298°K)}$$

$$= 9.4 \cdot 10^{-4} \text{ g-mol/cm}^2\text{-sec-atm}$$

From the units of the Henry's law constant, we see that the value given implies

$$p_{1i} = H'x_{1i}$$

By comparison with Eq. 9.5-5,

$$p_{1i} = Hc_{1i} = (Hc)x_{1i}$$

Thus,

$$H = \left(\frac{H'}{c}\right) = \frac{4.4 \cdot 10^4 \text{ atm}}{1 \text{ g-mol/18 cm}^3} = 7.9 \cdot 10^5 \text{ atm-cm}^3/\text{g-mol}$$

Inserting these results into Eq. 9.5-9, we find

$$K_L = \frac{1}{1/k_L + 1/k_G H}$$

$$= \frac{1}{\dfrac{1}{2.1 \cdot 10^{-3} \text{ cm/sec}} + \dfrac{1}{(9.4 \cdot 10^{-4} \text{ g-mol/cm}^2\text{-sec-atm})(7.9 \cdot 10^5 \text{ cm}^3\text{-atm/g-mol})}}$$

$$= 2.1 \cdot 10^{-3} \text{ cm/sec}$$

The mass transfer is dominated by the liquid-side resistance. This would also be true if we calculated the overall gas-side resistance, a consequence of the slow diffusion in the liquid state. It is the usual case for problems of this sort.

Example 9.5-2: Perfume extraction

Jasmone ($C_{11}H_{16}O$) is a valuable material in the perfume industry, used in many soaps and cosmetics. Suppose we are recovering this material from a water suspension of jasmine flowers by an extraction with benzene. The aqueous phase is continuous; the mass transfer coefficient in the benzene drops is $3.0 \cdot 10^{-4}$ cm/sec; the mass transfer coefficient in the aqueous phase is $2.4 \cdot 10^{-3}$ cm/sec. Jasmone is about 170 times more soluble in benzene than in the suspension. What is the overall mass transfer coefficient?

Solution. For convenience, we designate all concentrations in the benzene phase with a prime, and all those in the water without a prime. The flux is

$$N_1 = k(c_{10} - c_{1i}) = k'(c'_{1i} - c'_{10})$$

The interfacial concentrations are in equilibrium:

$$c'_{1i} = Hc_{1i}$$

Eliminating these interfacial concentrations, we find

$$N_1 = \left[\frac{1}{1/k' + H/k} \right] (Hc_{10} - c_{10}')$$

The quantity in square brackets is the overall coefficient K' that we seek. This coefficient is based on a driving force in benzene. Inserting the values,

$$K' = \frac{1}{\dfrac{1}{3.0 \cdot 10^{-4} \text{ cm/sec}} + \dfrac{170}{2.4 \cdot 10^{-3} \text{ cm/sec}}}$$

$$= 1.3 \cdot 10^{-5} \text{ cm/sec}$$

Similar results for the overall coefficient based on a driving force in water are easily found.

Two points about this problem deserve mention. First, the result is a complete parallel to Eq. 9.5-12, but for a liquid–liquid interface instead of a gas–liquid interface. Second, mass transfer in the water dominates the process even though the mass transfer coefficient in water is larger, because jasmone is so much more soluble in benzene.

Example 9.5-3: Overall mass transfer coefficients in a packed tower

We are studying gas absorption into water at 2.2 atm total pressure in a packed tower containing Berl saddles. From earlier experiments with ammonia and methane, we believe that for both gases the mass transfer coefficient times the packing area per tower volume is 18 lb-mol/hr-ft^3 for the gas side and 530 lb-mol/hr-ft^3 for the liquid side. The values for these two gases may be similar, because methane and ammonia have similar molecular weights. However, their Henry's law constants are different: 9.6 atm for ammonia and 41,000 atm for methane. What is the overall gas-side mass transfer coefficient for each gas?

Solution. This is essentially a problem in unit conversion. Although you can extract the appropriate equations from the text, I always feel more confident if I repeat parts of the derivation.

The quantity we seek, the overall gas-side transfer coefficient K_G, is defined by

$$N_1 a = K_G a(y_{10} - y_1^*)$$
$$= k_G a(y_{10} - y_{1i})$$
$$= k_L a(x_{1i} - x_{10})$$

where y_1 and x_1 are the gas and liquid mole fractions. Note that the mass transfer coefficients used here are not those used in most of this text, but are consistent with the units given.

The interfacial concentrations are related by Henry's law:

$$p_{1i} = py_{1i} = Hx_{1i}$$

When these interfacial concentrations are eliminated, we find that

$$\frac{1}{K_G a} = \frac{1}{k_G a} + \frac{H/p}{k_L a}$$

In passing, we recognize that y_1^* must equal Hx_{10}/p.

We can now find the overall coefficient for each gas. For ammonia,

$$\frac{1}{K_G a} = \frac{1}{18 \text{ lb-mol/hr-ft}^3} + \frac{9.6 \text{ atm}/2.2 \text{ atm}}{530 \text{ lb-mol/hr-ft}^3}$$

$$K_G a = 16 \text{ lb-mol/hr-ft}^3$$

The gas-side resistance controls the rate. For methane,

$$\frac{1}{K_G a} = \frac{1}{18 \text{ lb-mol/hr-ft}^3} + \frac{41,000 \text{ atm}/2.2 \text{ atm}}{530 \text{ lb-mol/hr-ft}^3}$$

$$K_G a = 0.03 \text{ lb-mol/hr-ft}^3$$

The coefficient for methane is smaller and is dominated by the liquid-side mass transfer coefficient.

Section 9.6. Conclusions

This chapter presents an alternative model for diffusion, one using mass transfer coefficients rather than diffusion coefficients. The model is most useful for diffusion across phase boundaries. It assumes that large changes in the concentration occur only very near these boundaries and that the solutions far from the boundaries are well mixed. Such a description is called a lumped-parameter model.

In this chapter we have shown how experimental results can be converted into mass transfer coefficients. We have also shown how these coefficients can be efficiently organized as dimensionless correlations, and we have cataloged published correlations that are commonly useful. These correlations are compromised by problems with units that come out of a plethora of closely related definitions.

Mass transfer coefficients provide especially useful descriptions of diffusion in complex multiphase systems. They are basic to the analysis and design of industrial processes like absorption, extraction, and distillation. They should find major applications in the study of physiologic processes like membrane diffusion, blood perfusion, and digestion; physiologists and physicians do not often use these models, but would benefit from doing so.

Mass transfer coefficients are not useful in chemistry when the focus is on chemical kinetics or chemical change. They are not useful in studies of the solid state, where concentrations vary with both position and time, and lumped-parameter models do not help much. However, mass transfer coefficients are used in analyzing etching processes, like those used in making silicon chips.

All in all, the material in this chapter is a solid alternative for analyzing diffusion near interfaces. It is basic stuff for chemical engineers, but it is an unexplored method for many others. It repays careful study.

References

Aiba, S., Humphry, A. E., & Mills, N. F. (1973). *Biochemical Engineering,* 2nd ed. New York: Academic Press.

Becker, H. A. (1976). *Dimensionless Parameters – Theory and Methodology.* New York: Wiley.

Bennett, C. O., & Meyers, J. E. (1974). *Momentum, Heat and Mass Transfer,* 2nd ed. New York: McGraw-Hill.

Bird, R. B., Stewart, W. E., & Lightfoot, E. N. (1960). *Transport Phenomena.* New York: Wiley.

Bridgeman, P. W. (1922). *Dimensional Analysis.* New Haven: Yale University Press.

Calderbank, P. H. (1967). In: *Mixing,* ed. V. Uhl. New York: Academic Press.

Cooney, D. O. (1976). *Biomedical and Biological Transport Processes.* New York: Dekker.

McCabe, W. L., & Smith, J. C. (1975). *Unit Operations of Chemical Engineering,* 3rd ed. New York: McGraw-Hill.

Perry, R. H., & Chilton, C. H. (1973). *Chemical Engineers Handbook,* 5th ed. New York: McGraw-Hill.

Schlichting, H. (1979). *Boundary Layer Theory,* 7th ed. New York: McGraw-Hill.

Sherwood, T. K., Pigford, R. L., & Wilke, C. R. (1975). *Mass Transfer.* New York: McGraw-Hill.

Stein, W. D. (1967). *The Movement of Molecules Across Cell Membranes.* New York: Academic Press.

Treybal, R. E. (1980). *Mass Transfer Operations,* 3rd ed. New York: McGraw-Hill.

10 ABSORPTION AND DISTILLATION

The most common use of the mass transfer coefficients developed in Chapter 9 is for analytical description of large-scale separation processes like gas absorption and liquid–liquid extraction. These mass transfer coefficients describe the absorption of a solute vapor like SO_2 or NH_3 from air into water. They describe the extraction of waxes from lubricating oils, the leaching of copper from low-grade ores, and the efficiency of the distillation of xylene isomers.

Mass transfer coefficients are useful because they describe how fast these separations occur. They thus represent a step beyond thermodynamics, which establishes the maximum separations that are possible. They are a step short of analyses using diffusion coefficients, which have a more exact fundamental basis. Mass transfer coefficients are accurate enough to correlate experimental results from industrial separation equipment, and they provide the basis for designing new equipment.

The understanding of these industrial separation processes is clouded by complicated algebra. As a result, students often easily understand every step of an analysis, but have a poor perspective of the overall problem. Accordingly, we begin this chapter with simple cases of dilute solutions and then extend the analysis to more complicated cases common in industrial practice. This split is like that for diffusion, where the simple dilute limits of Chapter 2 gave way to the concentrated and general results in Chapter 3.

The two processes emphasized in this chapter are gas absorption and distillation. Gas absorption, an example of what is called "differential contacting," depends directly on mass transfer coefficients. Distillation, an example of what is called "stage-wise contacting," is less influenced by mass transfer, but is the most common separation process. More specifically, the absorption of a dilute vapor, the simplest case, is presented in Section 10.1. There, the description takes a very simple analytical form not unlike that in Section 9.1. Absorption of more concentrated vapors that produce nonideal solutions is covered in Section 10.2; this description uses graphic methods that are quantitatively archaic but that are useful for thinking. The equilibrium limits of distillation are outlined in Section 10.3, and the effects of mass transfer on stage efficiencies are given in Section 10.4. Other separation methods are handled with similar ideas, as sketched in Section 10.5. Although this chapter is synoptic, it supplies all the tools required for deeper study of special cases.

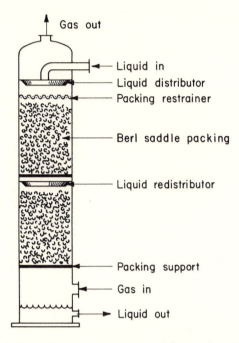

Fig. 10.1-1. A packed tower used for gas absorption. A gas mixture enters the bottom of the tower and flows out the top. Part of this mixture is absorbed by liquid flowing countercurrently, from top to bottom. When the gas mixture is dilute, the analysis is a simple example of the use of mass transfer coefficients.

Section 10.1. Absorption of a dilute vapor

The simplest case that uses mass transfer coefficients to analyze industrial separations is the absorption of a dilute vapor in a solvent gas using a nonvolatile liquid. For example, traces of ammonia in air can be removed by absorbing the ammonia with water. Small amounts of SO_2 in the stack gas of a power plant can be removed by absorption into solutions of lime.

Such processes are often called "scrubbing" or "gas washing." They commonly use large "packed towers," one of which is shown schematically in Fig. 10.1-1. A packed tower is essentially a piece of pipe set on its end and filled with inert material or "tower packing." Liquid poured into the top of the tower trickles down through the packing; gas pumped into the bottom of the tower flows counter-currently upward. The intimate contact between gas and liquid achieved in this way effects the gas absorption.

Designing a packed tower of this type involves three steps:

1   Overall mole balances. These specify how large a process is needed.

2   Tower cross-sectional area. Good contact between liquid and gas

occurs in a narrow range of gas and liquid flows. This range essentially determines the tower's diameter.

3   Tower height. Even with good flow, the solute must have enough time to diffuse from the gas into the liquid. This time depends on the height of the tower.

The first step depends on mass balances, and the second is controlled by fluid mechanics. The third, which is the most difficult, is that described by mass transfer coefficients. These steps will now be discussed sequentially.

Mole balances

The simplest aspects of absorption are the overall mole balances, which imply the overall size of the process to be designed. When the solute vapor being absorbed is dilute, the remaining gas is insoluble in the liquid. The liquid may also have a small vapor pressure. In this case, the gas flow and the liquid flow can be treated as constants throughout the column. The liquid and vapor concentrations are obviously related:

$$\begin{pmatrix} \text{solute entering} \\ \text{minus solute leaving} \\ \text{in gas stream} \end{pmatrix} = \begin{pmatrix} \text{solute leaving} \\ \text{minus solute entering} \\ \text{in liquid stream} \end{pmatrix} \qquad (10.1\text{-}1)$$

In symbolic terms, this can be written as

$$(GA)(y_{in} - y_{out}) = (LA)(x_{out} - x_{in}) \qquad (10.1\text{-}2)$$

where A is the tower's cross-sectional area; G and L are the molar fluxes of the gas and liquid, respectively; and y and x are the mole fractions of the absorbing solute in the gas and liquid, respectively. These mole fractions, different from the mass and mole concentrations used elsewhere in this book, are written without subscripts because only one solute is being transferred. Remember that this equation implies that G and L are constant, a result of the assumption that the solute is dilute.

Tower cross-sectional area

We can use this mass balance to establish the overall size of the absorption process. We next turn to the cross-sectional area of the tower. This area is governed by two factors: the type of packing in the tower and the nature of the fluid flow.

The chief purpose of the tower's packing is to provide a large contact area between the gas and the liquid. In addition, the packing must be inexpensive, must be chemically inert, and must offer minimum resistance to flow. Some common, commercially available packing materials are shown in Fig. 10.1-2 (Norton Co., 1980). Random packings like those shown are more frequently used in gas absorption; stacked packings of redwood slats are frequently used in cooling towers.

The flow through a specific packing is subject to three main problems. All three are characteristic of fluid flow and are almost always independent of

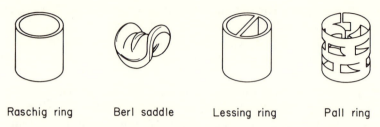

Raschig ring   Berl saddle   Lessing ring   Pall ring

Fig. 10.1-2. Common tower packings. These materials are used to fill the packed towers discussed in this section. Their chief purpose is to provide a very large surface area between the gas and liquid.

mass transfer itself. The first of these, called "channeling," occurs when the gas or liquid flow is much greater at some points than at others. Such channeling is undesirable, for it can substantially reduce mass transfer. It can be severe in stacked packing. It is usually minor in crushed solid packing and is minimal in commercially purchased random packing.

The other two flow problems affecting packed towers are called "loading" and "flooding." To see what these are, you should imagine a liquid flowing downward through a packed tower. Originally, there is no gas flow; when a small gas flow is introduced, the liquid flow is unaltered. However, when the gas flow is greatly increased, the liquid flow will be reduced; this condition is called "loading." At very high gas flow, the liquid stops flowing altogether and collects in the top of the column, this condition is called "flooding."

Flooding dramatically reduces mass transfer. It is avoided by reducing the liquid and gas fluxes. Of course, the total amounts of liquid and gas involved depend on how large a gas-absorption process is needed. The fluxes, that is, the flows per area, are reduced by increasing the cross-sectional area of the tower. This area is usually found using an empirical correlation like that in Fig. 10.1-3 (Eckert, 1975; Treybal, 1980). This correlation gives the liquid flux at flooding as a function of the ratio of liquid and gas flows used. Because this ratio is usually specified, the flux at flooding is easily calculated. One then designs the tower to operate at fluxes equal to half this flooding condition. This empirical choice of one-half flooding represents a balance of two factors: a lower flux implies an unnecessarily large tower, and a higher flux requires an unnecessarily large pump. Thus, the tower's cross-sectional area is chosen on the basis of fluid mechanics, and it includes the choice of appropriate fluxes.

Tower height

We now must estimate the height of the packed tower required for this absorption. This estimate includes the use of mass transfer coefficients, the raison d'être for this chapter. We begin by making a mole balance on the solute in the gas and the liquid over the differential volume of tower shown in Fig. 10.1-4:

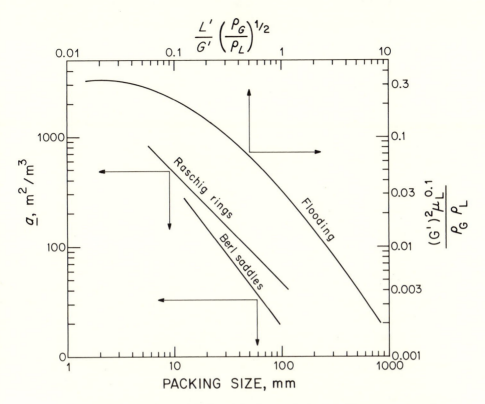

Fig. 10.1-3. Flooding and surface areas in a packed tower. The upper curve gives the flooding velocity in a packed tower as a function of the ratio of liquid and gas fluxes. Note that these fluxes L' and G' are written in mass units, not molar ones; this is because flooding is the result of fluid mechanics, not chemical factors. Note also that the ordinate in this figure is not dimensionless, but is written in SI units. The two lower curves give the surface area per packing volume for two common tower packings. [Adapted from Treybal (1980).]

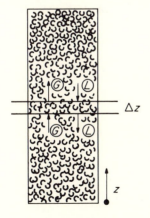

Fig. 10.1-4. Mole balances in a packed tower. The design of the height of a packed tower depends on mole balances made on a differential volume located at an arbitrary position z in the tower. In this simple analysis, the gas flux G and liquid flux L are both taken as constant, independent of z.

$$\begin{pmatrix} \text{accumulation} \\ \text{of solute} \end{pmatrix} = \begin{pmatrix} \text{flow of solute} \\ \text{in minus flow out} \end{pmatrix} \qquad (10.1\text{-}3)$$

$$0 = GA(y|_z - y|_{z+\Delta z}) + LA(x|_{z+\Delta z} - x|_z) \qquad (10.1\text{-}4)$$

As usual, we divide by the volume $A\Delta z$ and take the limit as this volume goes to zero:

$$0 = -G\frac{dy}{dz} + L\frac{dx}{dz} \qquad (10.1\text{-}5)$$

We find it convenient to rearrange this equation as

$$\frac{dx}{dy} = \frac{G}{L} \qquad (10.1\text{-}6)$$

Integrating, we find

$$x = x_0 + \frac{G}{L}(y - y_0) \qquad (10.1\text{-}7)$$

where the subscript 0 signals the mole fraction at the bottom of the tower (i.e., at $z = 0$). This mole balance on solute in the liquid and in the gas is the first of three key equations used in our analysis.

We next make a mole balance on the solute in the gas only:

$$\begin{pmatrix} \text{solute} \\ \text{accumulation} \end{pmatrix} = \begin{pmatrix} \text{solute flow} \\ \text{in minus flow out} \end{pmatrix} - \begin{pmatrix} \text{solute lost} \\ \text{by absorption} \end{pmatrix} \qquad (10.1\text{-}8)$$

In symbolic terms, this can be written as

$$0 = GA(y|_z - y|_{z+\Delta z}) - K_G a(A\Delta z)(c_1 - c_1^*) \qquad (10.1\text{-}9)$$

in which a represents the packing area per volume and K_G is the overall gas-phase mass transfer coefficient. The concentration c_1 is that in the bulk gas, and the concentration c_1^* is the concentration that the gas would have if it were in equilibrium with the liquid. Again, we divide this equation by the volume $A\Delta z$ and take the limit as this volume goes to zero; we also recognize that the total molar concentration c of the gas is constant. Thus, c_1 equals cy, c_1^* equals cy^*, and

$$0 = -G\frac{dy}{dz} - K_G ac(y - y^*) \qquad (10.1\text{-}10)$$

Integrating this equation gives the column height l:

$$l = \int_0^l dz = -\frac{G}{K_G ac}\int_{y_0}^{y_l} \frac{dy}{(y - y^*)} \qquad (10.1\text{-}11)$$

This equation, a mole balance on that part of the solute that is in the vapor, is the second key equation in our analysis.

We want to evaluate explicitly the integral in Eq. 10.1-11. To do so, we must know how y^* depends on y. This dependence, the third key relation of our analysis, is a thermodynamic equilibrium like

$$y^* = Hx \qquad (10.1\text{-}12)$$

where H is a Henry's law constant. The simplest case occurs when the solution is so dilute that H is a constant. Then we combine this equilibrium with Eq. 10.1-7:

$$y^* = H \left[x_0 + \frac{G}{L} (y - y_0) \right]$$ (10.1-13)

We substitute this equation into Eq. 10.1-11 and integrate:

$$l = \frac{G}{K_G a c} \int_{y_l}^{y_0} \frac{dy}{(1 - HG/L)y + H[(G/L)y_0 - x_0]}$$

$$= \frac{G}{K_G a c} \left(\frac{1}{1 - HG/L} \right) \ln \left(\frac{y_0 - Hx_0}{y_l - H[x_0 + (G/L)(y_l - y_0)]} \right)$$

$$= \frac{1}{K_G a c} \left(\frac{1}{1/G - H/L} \right) \ln \left(\frac{y_0 - Hx_0}{y_l - Hx_l} \right)$$ (10.1-14)

Solving for the tower height is as easy as plugging in the numbers.

This result merits reflection. First, although the analysis repeatedly exploits the assumption of dilute solution, the extension to concentrated solutions should be relatively straightforward. Second, we have implied mass transfer of a solute vapor from a gas into a liquid; such a process is called "gas scrubbing." We can repeat the identical analysis for mass transfer of a vapor from a liquid into a gas; such a reversed process is called "stripping." Third, we have written the preceding equations in terms of gas-phase mole fractions; we could write completely analogous equations for liquid-phase mole fractions.

These remarks may call attention to the overall mass transfer coefficient K_G used in these equations. This overall coefficient is slightly different than those used in Section 9.5. It may be defined as

$$\frac{1}{K_G} = \frac{1}{k_G} + \frac{H}{k_L} \frac{c(\text{gas})}{c(\text{liq})}$$ (10.1-15)

where H is given by Eq. 10.1-12 and k_G and k_L are defined by

$$N_1 = k_G[c_1(\text{gas}) - c_{1,i}(\text{gas})]$$

$$= k_L[c_{1,i}(\text{liq}) - c_1(\text{liq})]$$ (10.1-16)

Thus, k_G and k_L correspond to the definition for k in Table 9.2-2. Alternatively, K_G can be defined in terms of the k'' in Table 9.2-2:

$$\frac{1}{K_G c} = \frac{1}{k_G''} + \frac{H}{k_L''}$$ (10.1-17)

where c [$= c(\text{gas})$] is the molar concentration in the gas. In either case, K_G retains dimensions of velocity, and so for me it has a clearer physical significance. We shall use coefficients of this type throughout this chapter. We now turn to a simple illustration of these ideas.

Example 10.1-1: Carbon dioxide absorption

A short packed tower uses an organic amine at 14°C to absorb carbon dioxide. The entering gas, which contains 1.27% CO_2, would be in equilibrium with a solution of amine containing 7.3 mol% CO_2. The gas leaves containing 0.04% CO_2. The amine, flowing countercurrently, enters essentially pure. Gas flow is 2.31 g-mol/sec, liquid flow is 0.46 g-mol/sec, the tower's cross-sectional area is about 0.84 m², and the mass transfer coefficient times the surface area per volume is 0.022 sec^{-1}. How large a tower is needed to achieve this amount of CO_2 absorption?

Solution. This problem is a straightforward application of the equations developed earlier. We first make a mole balance on CO_2 (see Eq. 10.1-2):

$$AG(y_{in} - y_{out}) = AL(x_{out} - x_{in})$$
$$(2.31 \text{ mol/sec})(0.0127 - 0.0004) = (0.46 \text{ mol/sec})(x_{out} - 0)$$
$$x_{out} = 0.062$$

From the equilibrium condition at the inlet, we parallel Eq. 10.1-12:

$$y_{in} = Hx_{out}^*$$
$$0.0127 = H(0.073)$$
$$H = 0.174$$

Thus, from Eq. 10.1-14,

$$l = \left[\frac{m^3/10^3 \text{ liters}}{\dfrac{0.022}{\text{sec}} \left(\dfrac{\text{g-mol}}{(22.4 \text{ liters})(287/273)} \right)} \right] \left[\frac{1}{\dfrac{0.84 \text{ m}^2}{2.31 \text{ g-mol/sec}} + \dfrac{(0.84 \text{ m}^2)(0.174)}{0.46 \text{ g-mol/sec}}} \right]$$

$$\cdot \ln \left(\frac{0.0127 - (0.174)(0.062)}{0.0004 - 0} \right)$$

$$= 2.5 \text{ m}$$

The simplicity of this calculation is typical of those in dilute solutions.

Section 10.2. Absorption of a concentrated vapor

In this section we want to extend the preceding analysis to the case of a concentrated vapor. As before, we plan to accomplish this adsorption using a packed tower. As before, we must decide on an appropriate tower packing and on liquid and gas fluxes that will avoid flooding. As before, we depend on a variety of mole balances, though now for concentrated solutions.

Before we develop these new mass balances, we can benefit by looking at our analysis for a dilute vapor in a somewhat different way. This analysis depended on three key equations. A first key equation came from a balance on both liquid and gas (see Eq. 10.1-6):

$$\frac{dy}{dx} = \frac{L}{G} \tag{10.2-1}$$

We integrated this easily (see Eq. 10.1-7):

$$y = y_0 + \frac{L}{G}(x - x_0) \tag{10.2-2}$$

This mole balance, shown in Fig. 10.2-1, is called the "operating line." A second key equation is the equilibrium condition (see Eq. 10.1-12):

$$y^* = Hx \tag{10.2-3}$$

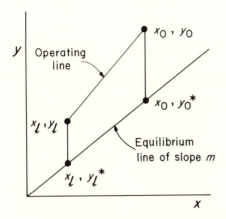

Fig. 10.2-1. Designing an absorption tower for a dilute vapor. The height of the tower is closely related to the area of the trapezoid shown. However, for a dilute vapor, this area is easily calculated analytically using the equations of Section 10.1. The equilibrium line shown here is based on thermodynamics, and the operating line reflects mole or mass balances.

This thermodynamic relation, shown in Fig. 10.2-1, is called the "equilibrium line." Finally, from a mass balance on the gas alone, we found (see Eq. 10.1-11)

$$l = \frac{G}{K_G a c} \int_{y_l}^{y_0} \frac{dy}{y - y^*}$$

(10.2-4)

We combined Eqs. 10.2-2 through 10.2-4 and integrated to find the tower's height l.

Instead, we could have made use of the graph in Fig. 10.2-1. We begin on the operating line at the point (x_l, y_l) and move vertically until we hit the equilibrium line at the point (x_l, y_l^*). From these points, we know $y_l - y_l^*$. We repeat this procedure for a lot of values of y. We then use these results to integrate Eq. 10.2-4 by either graphical or numerical means. We thus find the tower height l.

At first glance, this method of calculation using operating and equilibrium lines may seem archaic, an anachronism from the days of slide rules. Certainly, this method was developed to circumvent the elaborate integrals that are often encountered in the analysis of large-scale mass transfer processes (American Institute of Chemical Engineers, 1981). These integrals can now be routinely handled with computers.

Still, operating and equilibrium lines remain the focus of everyone's thinking. This is not just the result of mental inertia or academic sloth. Instead, this focus has remained valuable because the operating line summarizes a mass balance, and the equilibrium line is a statement based on the second law of thermodynamics. Such a split can make thinking about separation processes easier and more rational.

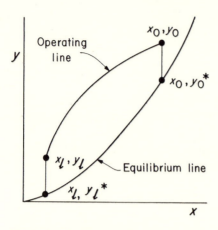

Fig. 10.2-2. Designing an absorption tower for a concentrated vapor. The height of the tower is again related to the area of the figure shown. This height is often more easily found graphically or numerically than analytically. The curvatures of the equilibrium and operating lines reflect the fact that both gas and liquid are concentrated solutions.

We want to extend this analysis to absorption in concentrated solutions. We begin with a mole balance on both gas and liquid on the control volume in Fig. 10.1-3. The result is a parallel to Eq. 10.1-5:

$$0 = -\frac{d}{dz}(Gy) + \frac{d}{dz}(Lx) \tag{10.2-5}$$

Before, the flux of gas G and that of liquid L were nearly constant because the absorbing species was always dilute. Now, however, we expect that

$$G = G_0 \left(\frac{1}{1-y}\right) \tag{10.2-6}$$

where G_0 is the flux of the nonabsorbing gas. For example, if we are using water to absorb SO_2 out of air, G_0 is the flux of air. Similarly,

$$L = L_0 \left(\frac{1}{1-x}\right) \tag{10.2-7}$$

where L_0 is the flux of the nonvolatile liquid. When we combine these equations and integrate, we find

$$y = \frac{\left(\dfrac{y_0}{1-y_0}\right) + \dfrac{L_0}{G_0}\left(\dfrac{x}{1-x} - \dfrac{x_0}{1-x_0}\right)}{1 + \left(\dfrac{y_0}{1-y_0}\right) + \dfrac{L_0}{G_0}\left(\dfrac{x}{1-x} - \dfrac{x_0}{1-x_0}\right)} \tag{10.2-8}$$

This mole balance is the operating line for a concentrated vapor, the analogue of Eq. 10.1-7 or 10.2-2. It reduces to these equations as the concentrations become small. However, in general, its shape is more like that in Fig. 10.2-2.

The next step is the specification of a new equilibrium relation analogous to Eq. 10.1-12 or 10.2-3:

$$y^* = y^*(x) \tag{10.2-9}$$

In general, this relation is not written in an analytical form, but simply presented as a table or graph of experimental results. The important point is that y^* and x are no longer directly proportional, related by a single, constant Henry's law coefficient. Instead, they vary nonlinearly, as exemplified by the equilibrium line in Fig. 10.2-2.

The final step is a mole balance on the gas in a differential tower volume:

$$0 = -\frac{d}{dz}(Gy) - K_G ac(y - y^*) \tag{10.2-10}$$

We combine this result with Eq. 10.2-6 to find

$$0 = -\frac{G_0}{(1-y)^2}\frac{dy}{dz} - K_G ac(y - y^*) \tag{10.2-11}$$

or

$$l = \int_0^l dz = \frac{G_0}{K_G ac} \int_{y_l}^{y_0} \frac{dy}{(1-y)^2(y - y^*)} \tag{10.2-12}$$

This result for concentrated solutions reduces to Eq. 10.1-11 or 10.2-4 for dilute solutions, where $1 - y$ is about unity.

The tower height l can be found by integrating Eq. 10.2-12, using values of y and y^* read from Fig. 10.2-2. Because this equation is complicated, it is often broken into two parts that are evaluated separately:

$$l = (HTU)(N) \tag{10.2-13}$$

where HTU is the "height of a transfer unit," given by

$$HTU = \frac{G_0}{K_G ac} \tag{10.2-14}$$

and N is the number of transfer units, equal to

$$N = \int_{y_l}^{y_0} \frac{dy}{(1-y)^2(y - y^*)} \tag{10.2-15}$$

This two-part equation is commonly used, especially in the older literature (Sherwood, 1937; Perry & Chilton, 1973). I find it somewhat confusing, for it implies thinking of nonequilibrium gas absorption in terms appropriate to near-equilibrium distillation. Still, I find myself using these forms when I solve practical problems, even though I distrust myself for doing so. I am reassured when I think of N as a measure of the difficulty of the separation, and HTU as a measure of the efficiency of the equipment.

The integration in Eq. 10.2-12 or Eq. 10.2-15 is straightforward, but is limited by two key assumptions made during the analysis. These assumptions can have subtle implications, and so merit review. First, we are assuming absorption of a single vapor from an inert gas into a nonvolatile liquid. The gas is inert in the sense that only negligible amounts dissolve in the liquid; the liquid is nonvolatile in the sense that only negligible amounts evaporate into the gas. These approximations underlie Eq. 10.2-6 and lead to the factors involving $(1 - y)^{-1}$ and $(1 - y)^{-2}$ in the analysis. In passing, we should mention that some textbooks use a slightly different set of assumptions, and so use slightly different equations (King, 1971; Sherwood et al., 1975; McCabe & Smith, 1975; Treybal, 1980). The differences are caused by these factors and are not often important.

The second key assumption in this analysis is that a mass transfer coefficient can adequately express the mass transfer in a concentrated solution. In other words, it implies that mass transfer coefficients are independent of concentration differences, though they certainly do depend on variables like the Reynolds and Schmidt numbers. This turns out to be only a first approximation. In general,

$$N_1 = K_G \Delta c_1 + K'_G (\Delta c_1)^2 + \cdots \qquad (10.2\text{-}16)$$

where K'_G is a new correction factor for concentrated solution.

The reason why mass transfer coefficients may not be accurate in concentrated solution is because mass transfer itself creates convection. Such convection is like that caused by diffusion, as explained in detail in Section 3.1. Extensions of mass transfer coefficients to more concentrated solutions or, more strictly, to situations of fast mass transfer are outlined in Section 11.5. However, these extensions are not often used in practice, even when they are known in principle. Instead, practicing engineers tend to use empirical correlations of actual experimental data. These are usually reliable.

Example 10.2-1: Ammonia scrubbing

A gas mixture at 0°C and 1 atm, flowing at 1.20 m³/sec, and containing 37% NH_3, 16% N_2, and 47% H_2 is to be scrubbed with water containing a little sulfuric acid and at 0°C. The exit gas should contain 1% NH_3 and the exit liquid 23 mol% NH_3.

Design a packed tower to carry out this task. The tower should use $\frac{1}{2}$-inch Berl saddles, which have a surface area per volume of 460 m²/m³. It should operate at 50% of flooding. Pilot-plant data suggest that the overall gas-side mass transfer coefficient in this tower will be 0.032 m/sec; this value is larger than normal because of the chemical reaction of ammonia with water.

In this design, answer the following specific questions: (a) What is the flow of pure water into the top of the tower? (b) What tower diameter should be used? (c) How tall should the tower be?

Solution. (a) We first find the total flow AG_0 of the nonabsorbed gases (i.e., of N_2 and H_2):

$$AG_0 = 0.63 \left(\frac{1.20 \text{ m}^3/\text{sec}}{22.4 \text{ m}^3/\text{kg-mol}} \right)$$

$$= 0.0338 \text{ kg-mol/sec}$$

We then find the ammonia transferred:

$$\left(\begin{array}{c}NH_3\\ absorbed\end{array}\right) = 0.37 \left(\frac{1.20 \text{ m}^3/\text{sec}}{22.4 \text{ m}^3/\text{kg-mol}}\right) - \left(\frac{0.01(0.0338 \text{ kg-mol/sec})}{0.99}\right)$$

$$= 0.0195 \text{ kg-mol/sec}$$

From this, we find the desired water flow AL_0:

$$AL_0 = \left(\frac{0.77}{0.23}\right)(0.0195 \text{ kg-mol/sec})$$

$$= 0.0652 \text{ kg-mol/sec}$$

(b) The risk of flooding is greatest at the bottom of the tower where the flows are greatest. Moreover, because flooding is determined by fluid mechanics, it depends on mass flows, not molar flows. To make this conversion, we first find that the average molecular weight of the gas is 11.7. Then we see that

$$\left(\begin{array}{c}\text{total flow}\\ \text{of gas}\end{array}\right) = \frac{11.7 \text{ kg}}{\text{kg-mol}}\left(\frac{0.0338 \text{ kg-mol/sec}}{0.63}\right)$$

$$= 0.628 \text{ kg/sec}$$

The average molecular weight of the liquid stream (neglecting any H_2SO_4) is 17.8, so

$$\left(\begin{array}{c}\text{total flow}\\ \text{of liquid}\end{array}\right) = \frac{17.8 \text{ kg}}{\text{kg-mol}}\left(\frac{0.0652 \text{ kg-mol/sec}}{0.77}\right)$$

$$= 1.51 \text{ kg/sec}$$

Thus,

$$\left(\begin{array}{c}\text{liquid flow}\\ \text{gas flow}\end{array}\right)\sqrt{\rho_y/\rho_x} = \frac{1.51 \text{ kg/sec}}{0.628 \text{ kg/sec}}\sqrt{0.522 \text{ kg/m}^3/10^3 \text{ kg/m}^3}$$

$$= 0.055$$

Remembering that the values found in Fig. 10.1-2 are not dimensionless, but imply SI units, we find that

$$\left(\begin{array}{c}\text{gas flux}\\ \text{at flooding}\end{array}\right)^2 = \frac{0.27 \ \rho_x\rho_y}{\mu^{0.1}}$$

$$= \frac{0.27(10^3 \text{ kg/m}^3)(0.522 \text{ kg/m}^3)}{(0.001 \text{ kg/m-sec})^{0.1}}$$

or

$$\left(\begin{array}{c}\text{gas flux}\\ \text{at flooding}\end{array}\right) = 16.7 \text{ kg/m}^2\text{-sec}$$

Because we want to operate at 50% flooding, our flux should be half this value, or about 8.38 kg/m²-sec. We now can find the tower's diameter:

$$\frac{\pi}{4}d^2 = \frac{0.628 \text{ kg/sec}}{8.38 \text{ kg/m}^2\text{-sec}}$$

$$d = 0.309 \text{ m} \equiv 30.9 \text{ cm}$$

The tower's diameter is about 1 ft.

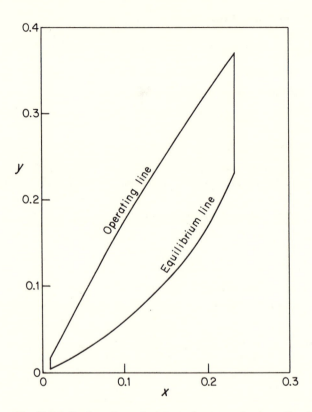

Fig. 10.2-3. Design of a packed tower for ammonia absorption. The tower's height can be calculated from Eq. 10.1-12 using values of $y - y^*$ found as the distances between the operating and equilibrium lines. This calculation involves graphical or numerical integration. The tower's diameter is estimated from fluid mechanics, as explained in the text.

(c) The calculation of the tower's height can begin with Eqs. 10.2-14 and 10.2-15. From the former,

$$HTU = \frac{G_0}{K_G ac}$$

$$= \frac{0.0338 \text{ kg-mol/sec/}[(\pi/4)(0.309 \text{ m})^2]}{(0.032 \text{ m/sec})(460 \text{ m}^2/\text{m}^3)(1 \text{ kg-mol/22.4 m}^3)}$$

$$= 0.686 \text{ m}$$

To find the number of units N, we first plot values of y versus x using Eq. 10.2-8, shown as the operating line in Fig. 10.2-3. We also plot y^* versus x, shown as the equilibrium line in this figure. We then read off values of y^* versus y and integrate Eq. 10.2-15 from $y_0 = 0.37$ to $y_l = 0.01$. The result is

$$N = 13$$

From Eq. 10.2-13,

$$l = (HTU)(N)$$
$$= (0.686 \text{ m})(13) = 9 \text{ m}$$

Problems of stripping gases are very similar except that the operating line falls below the equilibrium line.

Section 10.3. Fundamentals of distillation

Distillation is the process of heating a liquid solution to drive off a vapor, and then collecting and condensing this vapor. It is the most common method for chemical separation, the workhorse of the chemical process industries. Distillation columns are ubiquitous; they are the towers that rise brightly lighted from chemical plants, and they are the stills used by moonshiners.

Diffusion influences distillation, but it does not dominate it. In distillation, the liquid and vapor are usually contacted in individual stages. In these stages, the liquid and vapor are intimately mixed, so that mass transfer proceeds quickly. As a result, liquid and vapor in a single stage are close to being in equilibrium. After we design a distillation column on this basis, we may drag in an efficiency, a fudge factor reflecting departures from equilibrium caused by mass transfer. However, it is the equilibrium limit that dominates our thinking about distillation. This is the antithesis of the analysis of gas absorption; there, mass transfer was central to the entire analysis.

The different intellectual approach for analysis of distillation gives me a problem in writing this part of the book. I cannot neglect distillation, because it is too important; but diffusion does not greatly influence this analysis. As a result, I am faced with the pedagogical equivalent of l'Hospital's rule, where I must somehow take the product of an important process and a minor influence. I have done this by writing two sections. This section is a very brief synopsis of distillation theory, which is detailed elsewhere (Smith, 1963; King, 1979; McCabe & Smith, 1975; Perry & Weissberger, 1976). The next is a discussion of the effect on distillation of mass transfer.

The basic problem

The most common problem in distillation is designing a distillation column. We are usually given the flow and the concentration of the entering stream, and we are frequently told what flow and concentration of product is desired. We can easily calculate other flows and concentrations from mass balances. We want to find how large a distillation column is needed.

The calculation of the column's size reflects the basic equipment used for industrial distillation. A typical column is shown schematically in Fig. 10.3-1. The material entering the column is called the feed; that leaving the top of the column is called the distillate; that coming out the bottom is referred to as "bottoms." That part of the column above the feed is called the rectifying

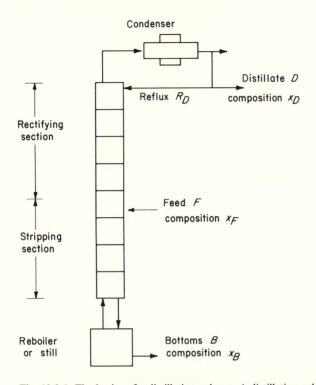

Fig. 10.3-1. The basics of a distillation column. A distillation column separates a liquid mixture by vaporizing the more volatile component. Usually, the column is fed in the center, the bottoms are reboiled and the vapor recycled, and the distillate is condensed and partially recycled.

section; that below is called the stripping section. The feed to the column is often preheated, so that it enters as a hot liquid, sometimes containing vapor. The distillate vapor is condensed, and a fraction is recycled as liquid into the top of the tower. The bottoms are partially boiled in the still, creating a vapor flow upward through the tower.

This superficial description makes distillation sound much like gas absorption. Both have a flow of vapor upward and a flow of liquid downward. The differences between these processes come not from these flows but from the columns themselves.

An industrial distillation column usually contains a series of "trays" or "plates." Most modern columns contain sieve trays that permit large amounts of vapor to bubble up from the trays below and shunt liquid in a flow down through the column. The close contact between liquid and vapor achieved in these frothy trays is why liquid and vapor are so nearly in equilibrium during distillation. The most common alternative to the sieve tray is the bubble cap tray, which effects the same close contact between liquid and vapor. A few columns, especially smaller ones in laboratories and pilot plants, are filled with packing like that used in absorption. However, the idealization of distillation columns almost always involves a cascade of discrete stages, even though these may not physically exist. In sieve tray and bubble cap columns, they do; in packed distillation columns, they do not; in both cases, we shall pretend that they do.

The mole balances

We begin our analysis by making a variety of mole balances on the distillation column. We choose mole rather than mass balances to anticipate the fact that vapor–liquid equilibrium data are almost always given in molar terms. For simplicity, we restrict our analysis to binary distillation. We then write overall and species mole balances on the entire column using the notation in Fig. 10.3-1:

$$F = D + B \tag{10.3-1}$$

$$Fx_F = Dx_D + Bx_B \tag{10.3-2}$$

Arbitrarily, we assume that these mole fractions refer to the more volatile component. These overall balances must hold no matter what happens inside of the column.

We next develop mole balances over parts of the column itself using the notation suggested by Fig. 10.3-2. We initially consider only what is happening between the top of the tower and the nth plate, which is above the feed plate. Over this region, we assume that the vapor flow G and the liquid flow L are constant between the top of the tower and this nth plate. We define x_n and y_n as the liquid and vapor mole fractions of the more volatile component on the nth stage. Note that these definitions are different than those used elsewhere in this book, where subscripts like these refer to component n. Because the more volatile component is moving up the column, x_n is greater than x_{n+1}, and y_n is greater than y_{n+1}. Less logically, we respect convention and call the composition of the vapor leaving the first stage y_a ($= y_1$) and the composition of the liquid entering the first stage x_a ($= x_0$).

We now write mole balances between the nth stage and the top of the column (see the control volume in Fig. 10.3-2). We assume that the column is operating in steady state, so that

$$L + G = L + G \tag{10.3-3}$$

This is merely a restatement of our assumption that both liquid and vapor flows are constant throughout the column. The mass balance on the volatile component is more interesting:

$$Lx_a + Gy_{n+1} = Lx_n + Gy_a \tag{10.3-4}$$

or

$$y_{n+1} = \left[y_a - \frac{L}{G} x_a \right] + \left(\frac{L}{G} \right) x_n \tag{10.3-5}$$

This important result relates the vapor composition on the $(n+1)$th plate y_{n+1} to the liquid composition on the nth plate x_n. It says that y_{n+1} varies linearly

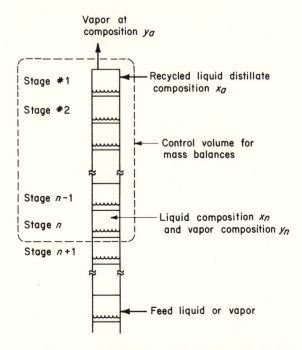

Fig. 10.3-2. The distillation column itself. The analysis depends on mole balances between the top of the column and the *n*th stage (i.e., over that part of the system surrounded by the dotted line). The analysis given is simplified by the assumption that the gas flux *G* and the liquid flux *L* are constants.

with x_n; the slope of this line is L/G, and the intercept is the quantity in square brackets.

These mole balances on the rectifying section have parallels in the lower stripping section. Differences exist because the feed results in different flow rates of vapor G' and liquid L'. Parallel to the foregoing, we assume that G' and L' are constant through the stripping section. The mole balance that results is

$$y_{n+1} = \left[y_b - \frac{L'}{G'} x_b \right] + \left(\frac{L'}{G'} \right) x_n \qquad (10.3\text{-}6)$$

where N is the total number of stages in the column, $y_b \,(= y_{N+1})$ is the vapor composition entering the bottom of the column, and $x_b \,(= x_N)$ is the liquid composition leaving this stage.

The column's inlet and outlets

The final group of mole balances are those at the ends of the column and that at the feed plate. First, we consider the behavior at the top of the column, including the condenser. This condenser can be operated in different ways. In one limit, all of the vapor removed from the top of the column can be condensed. Part of the condensate is the product distillate; the re-

(a) Total condenser

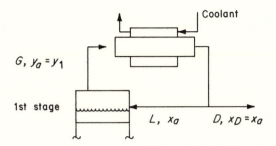

(b) Reboiler

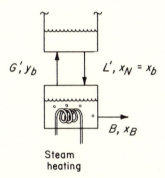

Steam
heating

Fig. 10.3-3. The ends of the distillation column. The vapor leaving the top of the column is condensed, as suggested by (a). Part of this condensate is recycled to the tower as liquid. The liquid leaving the bottom of the tower is partially vaporized in the reboiler, shown in (b). This reboiler effectively serves as an extra stage in the column.

mainder, the reflux, is returned to the top of the tower. In this limit, the vapor composition y_a leaving the top plate will equal the liquid composition entering the top plate x_a. This is shown schematically in Fig. 10.3-3(a).

Alternatively, the condenser at the top of the tower may first condense some of the vapor and reflux this easily condensed fraction. A condenser operating in this way has a delightful name: a "dephlegmator." The remaining vapor is completely condensed and becomes the distillate. In most elementary texts, the liquid and vapor leaving the dephlegmator are assumed in equilibrium. This is equivalent to assuming that the partial condenser is an additional stage, and it is a significant approximation. Moreover, the analysis of partial condensers like this is complex (see Section 17.4). As a result, we refer those interested in the effects of partial condensers to the more specialized books on distillation (Holland, 1963; Treybal, 1980). We shall restrict our discussion to total condensers.

For the case of a total condenser, we can rewrite Eq. 10.3-5 in a form that will be useful later. From Fig. 10.3-3(a), we see that $x_D = x_a = y_a$ and that

$$G = D + L \tag{10.3-7}$$

When we combine this with Eq. 10.3-5, we obtain

$$y_{n+1} = \left(\frac{1}{1 + R_D}\right) x_D + \frac{R_D}{1 + R_D} x_n \tag{10.3-8}$$

where R_D $(= L/D)$ is called the reflux ratio. This important result is called "the operating line of the rectifying section." Essentially a mole balance, it is more useful than Eq. 10.3-5 because it is written in terms of experimentally accessible quantities. The concentration x_D is usually given as part of the problem, and the reflux ratio R_D can be changed with the twist of a valve. In addition, because R_D is positive, Eq. 10.3-8 says that a plot of y_{n+1} versus x_n has a positive intercept and a slope less than unity. We shall come back to this equation later.

We next consider what happens in the reboiler at the bottom of the column. There, the liquid leaving the column is boiled and partially evaporated. The vapor is returned to the column, and the liquid is removed as product. To a reasonable approximation, vapor and liquid are in equilibrium; so the reboiler simply acts as an additional stage. The notation used in this region is shown in Fig. 10.3-3(b).

We find it useful to rewrite Eq. 10.3-6 in terms of the events around the reboiler. Mole balances give

$$L' = G' + B \tag{10.3-9}$$
$$L'x_b = G'y_b + Bx_B \tag{10.3-10}$$

Combining with Eq. 10.3-4, we obtain

$$y_{n+1} = -\left(\frac{B}{G'}\right) x_b + \left(1 + \frac{B}{G'}\right) x_n \tag{10.3-11}$$

This result, called the "operating line of the stripping section," is also important and will be used later. Note that in contrast to the rectifying section, a plot of y_{n+1} versus x_n in the stripping section has a negative intercept and a slope greater than unity.

Finally, we consider the feed plate in more detail. The nature of the feed strongly influences what happens in the column. If the feed is entirely saturated liquid, it simply joins the liquid flowing down the column; if it is entirely saturated vapor, it will join the vapor stream and flow upward. If the feed is a liquid cooled below saturation, then it will both join the liquid stream and condense some vapor to produce additional liquid. Analyzing these changes requires both mass and energy balances around the feed plate.

In this book, we assume that the feed is always saturated liquid. As a result, we find

$$\begin{aligned} G' &= G \\ &= L + D \\ &= D(R_D + 1) \end{aligned} \tag{10.3-12}$$

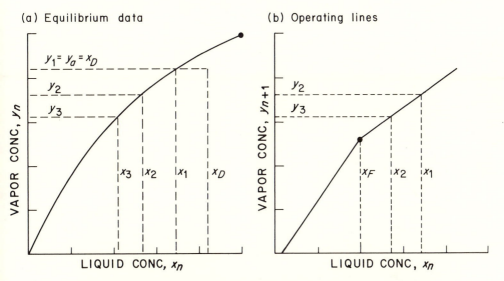

Fig. 10.3-4. Sizing the distillation column. This calculation depends on difference equations rather than differential equations, and so is more easily presented in geometrical terms. It involves combining the thermodynamic information shown in (a) with the mole balances represented in (b). This calculation is usually made using a combined graph like that in fig. 10.3-5, but this combination is more difficult to follow initially.

We can use this result to rewrite the operating line for the stripping section:

$$y_{n+1} = -\left(\frac{B}{D(R_D + 1)}\right) x_B + \left(1 + \frac{B}{D(R_D + 1)}\right) x_n \qquad (10.3\text{-}13)$$

We have all the pieces for the puzzle of finding the number of plates in the column.

Sizing the column

We now want to combine these operating lines with vapor–liquid equilibrium data, and so calculate the number of stages required in our separation. This combination is conveniently presented in geometric terms rather than in the differential equations that are the basis of the rest of this book.

To begin, we graph the vapor–liquid equilibrium data, as shown in Fig. 10.3-4(a). These data relate the vapor composition on a particular stage y_n to the liquid composition on that same stage x_n. In addition, for a particular recycle ratio R_D, we graph the two operating lines as shown in Fig. 10.3-4(b). We can easily show that these lines intersect at x_F, the composition of the feed stream. These lines relate the vapor composition y_{n+1} on one stage to the liquid composition x_n on an adjacent stage.

We recognize that $x_D = y_1$, and we use Fig. 10.3-4(a) to find x_1, the liquid

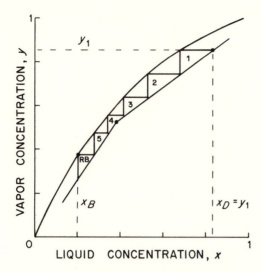

Fig. 10.3-5. Sizing the distillation column. The calculation in the preceding figure can be made in a much easier way by graphing both equilibrium and operating lines on the same graph. The result is the familiar staircase shown in all discussions of distillation. Novices sometimes forget that the "risers" on this stairc se reflect equilibrium, and the "treads" represent mole balances.

composition on the first stage. Using this value, we can find y_2 from Fig. 10.3-4(b). With y_2, we can find x_2 from Fig. 10.3-4(a). Then we use x_2 to find y_3 from Fig. 10.3-4(b). We continue this calculation, jumping back and forth until we reach x_B. The number of jumps corresponds to the number of stages. This calculation is straightforward but clumsy.

We can make this calculation in a more elegant way by plotting equilibrium and operating lines on the same graph, as shown in Fig. 10.3-5. This graph, simply the superposition of the two parts of Fig. 10.3-4, demonstrates the most common method of sizing a distillation column. The integers correspond to the stages numbered from the top of the column. For the separation shown, the column has five stages plus the reboiler (RB). I should caution you that this type of "staircase" graph is easier to make than to understand. The "risers" of the staircase reflect vapor–liquid equilibrium, and the "treads" represent mole balances (or operating lines).

Several additional aspects of this calculation deserve mention. First, decreasing the reflux ratio R_D will decrease the slope of the operating line in the rectifying section and increase that in the stripping section. Eventually, the intersection of these lines will collide with the equilibrium line. The slopes of the operating lines at this point give the minimum reflux at which a separation is possible. This reflux is never used, for it requires an infinite number of stages. Columns are often operated at reflux ratios about 60% above this minimum; so the idea is useful.

The antithesis of this case is that of the minimum number of stages. This

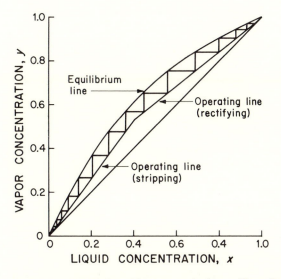

Fig. 10.3-6. Distillation of benzene and toluene. The column is calculated to require 13 stages plus the reboiler. Obviously, more complete separations or more accurate calculations may require expanding parts of this graph. The calculation made here assumes that vapor and liquid are in equilibrium on each stage, an assumption explored in detail in Section 10.4.

occurs when the reflux is very large, much greater than the distillate or bottoms. In this case, the operating lines of both sections are the same and equal to the 45-degree diagonal. Large refluxes require a minimum in equipment but a very large amount of reboiling. Capital costs are low, but energy costs are high. As a result, large refluxes are used only in small laboratory columns where products of extremely high purity are desired.

Finally, we should repeat that this analysis is based on four chief assumptions:

1   The molar liquid and vapor flow rates are constant.
2   The distillate is completely condensed.
3   The feed is a saturated liquid.
4   The liquid and vapor in each stage are at equilibrium.

The approximations introduced by the first three assumptions are discussed in texts on distillation, and ways to avoid them are detailed. The fourth assumption is of greatest interest here, for it implies infinitely rapid diffusion. Its accuracy is the subject of the next section of this book.

Example 10.3-1: Distillation of benzene and toluene

We wish to distill 3,500 moles per hour containing 40 mol% benzene into streams containing 97 mol% benzene and 98 mol% toluene. The column uses a total condenser and reflux ratio of 3.5. How many stages will be required?

Solution. We begin by making mole balances on the entire column. From Eqs. 10.3-1 and 10.3-2,

$$3,500 \text{ mol/hr} = D + B$$

$$0.4(3,500 \text{ mol/hr}) = 0.97D + 0.02B$$

Thus, D is 1,400 mol/hr, and B is 2,100 mol/hr. The operating lines, found from Eqs. 10.3-8 and 10.3-13, are plotted as shown in Fig. 10.3-6. The equilibrium line is added, and the difference calculation is made as shown. The result is that the column should have 14 stages, including the reboiler. The feed should be added to the eighth stage.

Section 10.4. Stage efficiencies

The analysis of mass transfer by stage-wise contacting always depends on one huge assumption: that the two streams leave the stage in equilibrium. For example, the analysis of distillation sketched in the previous section assumes that the vapor composition y_n and the liquid composition x_n leave the nth stage in equilibrium with each other. This assumption is the basis of the entire analysis, including the graphical methods exemplified by Figs. 10.3-5 and 10.3-6.

This assumption can be easily checked experimentally. To do so, we recognize that the vapor compositions y_{n+1} and y_n can be measured directly. The liquid composition x_n can also be measured experimentally, and the vapor composition y_n^* can be found from this value. We can use these measured values to define an efficiency η (Murphee, 1932):

$$\eta = \frac{y_n - y_{n+1}}{y_n^* - y_{n+1}} \tag{10.4-1}$$

If liquid and vapor are truly in equilibrium, then y_n equals y_n^*, and the efficiency is unity, or 100%. If liquid and vapor do not have time to reach equilibrium, then the efficiency is less than 100%.

This definition, called a "Murphee efficiency for the gas phase," is but one way in which we can describe how closely one stage reaches equilibrium. We might use other definitions, including a Murphee efficiency for the liquid phase. However, that used here is the most common, and it serves to focus this discussion.

The Murphee efficiencies have two characteristics that we want to discuss. First, they are obviously related in some way to the speed of diffusion, expressed as mass transfer coefficients. Second, they must somehow affect the design of distillation columns sketched in the previous section. We discuss these points sequentially.

Stage efficiencies and mass transfer

Intuitively, we expect that the Murphee efficiencies should be a function of mass transfer and of flow. If mass transfer between liquid and vapor is fast, these phases should almost be in equilibrium, and the effi-

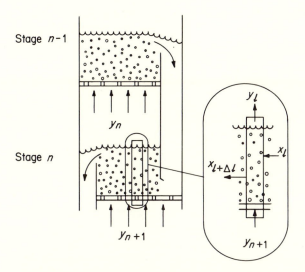

Fig. 10.4-1. Mass transfer on a single stage. Vapor and liquid on a particular stage of a distillation column approach equilibrium at a rate determined by mass transfer. They will essentially reach equilibrium only when the mass transfer is very rapid or the time spent on the stage is long. Putting these generalizations on a quantitative basis begins with mole balances on the small volume shown in the inset.

ciency should approach unity. If the liquid and vapor flow slowly past each other, then these phases again should almost be in equilibrium, and again the efficiency should be about 100%.

Putting these intuitions into a more quantitative form requires a more definite physical model like that in Fig. 10.4-1. In this model, we assume that the gas phase entering stage n is well mixed. We also assume that the liquid is well mixed vertically, although its composition may vary in the l direction, across the stage. As a result, the vapor composition leaving at some point y_l may vary across the plate; the average value of y_l is y_n, the composition going up to the next plate.

On this basis, we write mole balances on the volatile species in the vapor in the control volume $A\Delta h$:

$$(\text{accumulation}) = \begin{pmatrix} \text{volatile solute in} \\ \text{minus solute out} \\ \text{by convection} \end{pmatrix} + \begin{pmatrix} \text{solute gained by} \\ \text{mass transfer} \\ \text{from liquid} \end{pmatrix} \qquad (10.4\text{-}2)$$

or

$$0 = \left(AGy \Big|_h - AGy \Big|_{h+\Delta h} \right) + [K_G c(aA\Delta h)(y_i^* - y)] \qquad (10.4\text{-}3)$$

where A and h are the control volume's cross-sectional area and height, respectively, G is the molar gas flux, K_G is the overall gas-side mass transfer coefficient, c is the molar concentration in the gas, a is the surface area per volume, and y_i^* is the gas composition in equilibrium with the local liquid composition x_l. Dividing by $A\Delta h$ and taking the limit as this volume goes to zero,

$$0 = -G\frac{dy}{dh} + K_Gca(y_l^* - y) \tag{10.4-4}$$

This equation is subject to the constraints that y_l^* is constant at a given value of l and that

$$h = 0, \quad y = y_{n+1} \tag{10.4-5}$$
$$h = h, \quad y = y_l \tag{10.4-6}$$

Integrating,

$$\frac{y_l^* - y_l}{y_l^* - y_{n+1}} = e^{-(K_Gcah/G)} \tag{10.4-7}$$

This result gives the variation of the exiting vapor as a function of the mass transfer and the flow, just as we expected.

The mass transfer coefficient so slickly introduced in the analysis merits reflection. It is described as the overall gas-side mass transfer coefficient, defined as (see Eq. 10.1-15)

$$\frac{1}{K_G} = \frac{1}{k_G} + \frac{H}{k_L}\frac{c(\text{gas})}{c(\text{liq})} \tag{10.4-8}$$

where k_G and k_L are the individual mass transfer coefficients for gas and liquid, respectively, $c(\text{gas})$ and $c(\text{liq})$ are the total molar concentrations of these phases, and H is a Henry's law constant (see Eq. 10.1-12):

$$y^* = Hx \tag{10.4-9}$$

Other definitions can also be used; that introduced here means that the mass transfer coefficient still has dimensions of velocity.

No matter what the definition, K_G is a function of H. But in almost every distillation, the vapor and liquid compositions at equilibrium are not linearly related, because the equilibrium line is curved. As a result, the Henry's law constant H is not a constant at all, but varies with composition. We often are forced to use a different value of H. We shall return to this point in the numerical examples later.

We now want to relate this result to the Murphee efficiency. Such a relation depends on the flow on the plate. Two limiting cases are of interest: The liquid can flow so tumultuously that it is well mixed, or the liquid can flow so regularly that it mixes only vertically. These two limits give different results.

If the liquid is well mixed, then it is of constant composition, and x_l equals x_n. As a result, y_l^* equals y_n^*, and y_l equals y_n. We then combine Eq. 10.4-1 with Eq. 10.4-7 to find

$$\eta = 1 - e^{-(K_Gcah/G)} \tag{10.4-10}$$

The efficiency increases if the mass transfer coefficient or the bubble area per volume increases. It decreases as the gas flow increases. It is independent of the liquid flow, unless this flow increases the depth of fluid h.

If the liquid is in plug flow, then its composition varies across the plate. By

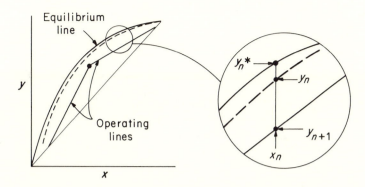

Fig. 10.4-2. Using Murphee efficiencies. Murphee efficiencies effectively lower the solid equilibrium line to the dashed line. They do so because the vapor composition y_n leaving a particular stage is less than the equilibrium value y_n^*. Remember that these efficiencies rarely have the same value in each stage.

assuming that each differential volume of liquid remains on the tray for the same length of time, we can show that (Lewis, 1936)

$$\eta = \left(\frac{L}{G}\right)\left\{\exp\left[\frac{HG}{L}\left(1 - e^{-(K_G cah/G)}\right)\right] - 1\right\} \qquad (10.4\text{-}11)$$

The efficiencies in the plug-flow limit are higher than those in the well-mixed limit, just as conversions in a plug-flow reactor are higher than those in a continuously stirred tank reactor. Other cases falling between the well-mixed and plug-flow limits have also been considered; these involve use of a dispersion coefficient to try to account for partial mixing (American Institute of Chemical Engineers, 1958).

Stage efficiencies and column design

In most of our efforts to design distillation towers, we usually know the gas and liquid fluxes G and L. We have available methods of estimating k_G and k_L, and we have some ideas of the flow on the plate. These estimates permit calculation of the Murphee efficiency η and can be used to improve the design of distillation towers.

This improved calculation has three steps. First, values of the Henry's law constant H are found as a function of concentration using the definition in Eq. 10.4-9. Second, these values are used to calculate K_G and η from Eqs. 10.4-8, 10.4-10, and 10.4-11. This provides us with estimates of η as a function of concentration x.

Finally, we graph the equilibrium and operating lines as usual. We remember that for a given value of x_n, the vertical distance between the equilibrium and operating lines is $y_n^* - y_{n+1}$, as shown in the inset of Fig. 10.4-2. But from Eq. 10.4-1, this distance times the Murphee efficiency is $y_n - y_{n+1}$. Thus, we can plot a new, nonequilibrium line, shown as the dashed curve in Fig. 10.4-2. When we design a distillation column, we can use this dashed curve and the operating lines to calculate the number of stages required. This calculation is illustrated by one of the examples that follow.

Calculations based on Murphee efficiencies are about as far as mass transfer models can be pushed. These calculations may not always be reliable, even though they are based on a huge number of experimental results. The reason is that a single overall mass transfer coefficient may be inadequate to describe all aspects of the flow and diffusion occurring in a single stage. Still, the value of any scientific effort is the product of the importance of the problem and the quality of the solution. Distillation is very important; although concepts of efficiency are certainly imperfect, they seem to me to remain valuable.

Example 10.4-1: Finding mass transfer coefficients from stage efficiencies

On one tray of an acetone–water distillation we find that y_n equals 0.84, x_n equals 0.70, and y_{n+1} equals 0.76. The stage is at 59°C and 1 atm; it has a vapor flux of 0.146 kg-mol/m²-sec and a liquid flux of 0.33 kg-mol/m²-sec; it has a liquid depth of 1.27 cm and is believed to be well mixed. What are the Murphee efficiency and the mass transfer coefficient on this tray?

Solution. From vapor–liquid equilibrium data for acetone–water, we find that when x_n equals 0.70, y_n^* equals 0.874. The Murphee efficiency is then found from Eq. 10.4-1:

$$\eta = \frac{y_n - y_{n+1}}{y_n^* - y_{n+1}}$$
$$= \frac{0.84 - 0.76}{0.874 - 0.76}$$
$$= 0.70$$

If the tray is well mixed, then the relation between the efficiency and the mass transfer coefficient is that in Eq. 10.4-10, which is easily rearranged:

$$K_G a = -\left(\frac{G}{ch}\right) \ln(1 - \eta)$$
$$= -\frac{0.146 \text{ kg-mol/m}^2\text{-sec}}{\left(\frac{\text{kg-mol}}{22.4 \text{ m}^3}\right)\left(\frac{273}{332}\right)(0.0127 \text{ m})} \ln(1 - 0.70)$$
$$= 377/\text{sec}$$

We cannot calculate the mass transfer coefficient itself without knowing the surface area per volume a.

Example 10.4-2: Distillation design using Murphee efficiencies

A solution containing 47 mol% carbon disulfide in carbon tetrachloride is to be separated in a distillation tower operated at very high reflux. The distillate and bottoms should contain 97% CS_2 and 5% CS_2, respec-

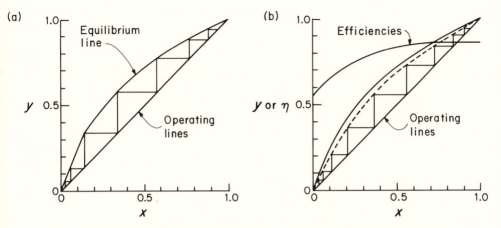

Fig. 10.4-3. Distillation of carbon disulfide and carbon tetrachloride. If liquid and vapor reached equilibrium, this separation would require seven stages, including the reboiler. When the stage efficiencies are considered, the separation requires nine stages. Note how the stage efficiency drops at the lower end of the column as the result of higher values of the Henry's law parameter H.

tively. The average column temperature is 60°C, and the average liquid density is 1.5 g/cm^3. The trays in the tower, which are of a proprietary design, have a gas flux of 0.059 kg-mol/m^2-sec and a liquid depth of 1.0 cm. The mass transfer in these systems is not infinitely fast, but is characterized by a $k_G a$ of 440/sec and a $k_L a$ of 1.7/sec. Find the number of stages required in this column when (a) equilibrium is in fact reached on each stage and (b) the column is affected by mass transfer.

Solution. (a) Because the system is operating at a very high reflux, both operating lines become equal to the diagonal. We then can plot these lines and the equilibrium line as shown in Fig. 10.4-3(a). By making the usual graphic calculation, we find that the system should contain six stages plus the reboiler.

(b) When the mass transfer is considered, we must correct the equilibrium line for the inefficiencies of the distillation. To do this, we must first find the overall mass transfer coefficients from Eq. 10.4-8:

$$\frac{1}{K_G a} = \frac{1}{k_G a} + \frac{H}{k_L a}\left(\frac{c(\text{gas})}{c(\text{liq})}\right)$$

$$= \frac{\text{sec}}{440} + \frac{H}{1.7 \text{ sec}} \; \frac{\left(\dfrac{\text{g-mol}}{22.4 \cdot 10^3 \text{ cm}^3}\right)\left(\dfrac{273}{273 + 60}\right)}{\left(\dfrac{1.5 \text{ g/cm}^3}{[0.47(76.14) + 0.53(158.67)]\text{g/g-mol}}\right)}$$

$$= 0.00227 \text{ sec} + (0.00172 \text{ sec})H$$

We can find the values of H for different values of x using the equilibrium curves in Fig. 10.4-3. For example, for $x = 0.2$, $H = 1.41$, and

$$K_G a = 213/\text{sec}$$

This and other values are then used to find the efficiency η from Eq. 10.4-7. For example, for $x = 0.2$,

$$\eta = 1 - \exp\left(-\frac{K_G ach}{G}\right)$$

$$= 1 - \exp\left[\frac{-\left(\dfrac{213}{\text{sec}}\right)\left(\dfrac{\text{kg-mol}}{22.4 \text{ m}^3}\right)\left(\dfrac{273}{273 + 60}\right)(0.01 \text{ m})}{0.059 \text{ kg-mol/m}^2\text{-sec}}\right]$$

$$= 0.73$$

Other values are shown in Fig. 10.4-3(b). Using these values, we calculate the new y–x curve in this same figure. Using this, we find that the number of stages required is now eight stages plus the reboiler.

Section 10.5. Conclusions

This chapter explains how mass transfer coefficients are used to design industrial equipment. Such designs idealize the equipment in two very different ways, sometimes called differential contactors and stage-wise contactors. The former is exemplified by gas absorption and the latter by distillation. How mass transfer coefficients are used depends on the method of contacting chosen.

Mass transfer coefficients are central to the design of differential contactors like gas absorbers. In this case, a gas mixture flows into the bottom of a packed tower; a liquid enters at the top of the tower and absorbs one component of the gas. At no point are gas and liquid in equilibrium. The analysis depends on a mole balance written as a differential equation and on gas–liquid equilibrium data.

When these relations are summarized graphically, the mole balance is called the "operating line," and the equilibrium data are termed the "equilibrium line." For dilute solutions, these lines are straight, and the size of the column can be found by simple integration. For more concentrated solutions, the equilibrium and operating lines are curved, and the size of the column must be found by graphical or numerical integration.

Mass transfer coefficients are second-order effects in the design of stage-wise contactors like distillation towers. Vapor and liquid again flow counter-currently, but through a series of discrete stages or trays. On each tray, vapor and liquid are close to being in equilibrium. The usual analysis is based on the approximation that they are actually in equilibrium. As before, this analysis depends on a mole balance and vapor–liquid equilibrium data; as before, these are usually represented graphically; as before, they are called the "operating line" and the "equilibrium line." For dilute solutions, these

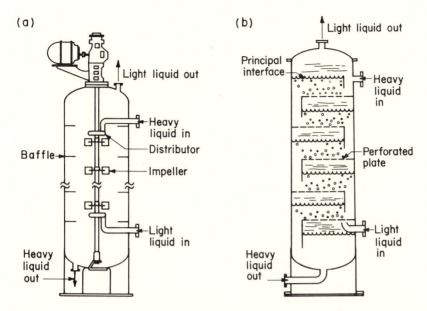

Fig. 10.5-1. Liquid–liquid extraction. This separation can be effected either with a differential contactor like that in (a) or with staged equipment like that in (b). If differential contacting is used, the analysis follows that for gas absorption. If staged contacting is the choice, the analysis parallels that for distillation. In both cases, the calculations are phrased in terms of equilibrium and operating lines. [Adapted from Treybal (1980).]

lines are straight; for concentrated solutions, they are curved. In either case, the size of the column is found by a difference calculation, which may be either graphical or numerical. Corrections for departures from equilibrium are made using "Murphee efficiencies," which depend on mass transfer coefficients.

Other industrial mass transfer operations are analyzed using the same two types of ideas. An example is liquid–liquid extraction. This method can involve either differential contacting, shown in Fig. 10.5-1(a), or stage-wise contacting, shown in Fig. 10.5-1(b). If the equipment provides differential contacting, the analysis is like that of gas absorption. Additional complexities come from overall mass transfer coefficients that are affected by flow on both sides of the liquid–liquid interface. If the equipment provides stage-wise contacting, then the analysis is like that for distillation. Additional complexity comes from the fact that these extractions always involve three components.

But there are no new ideas. Sure, liquid–liquid extraction is more complex than simple gas absorption or distillation, but designing the equipment still hinges on operating and equilibrium lines. Mass transfer coefficients are still central to differential contactors and a second-order effect for stage-wise contractors. Calculations are easy for dilute solutions and harder for concentrated solutions. The basic strategy you use remains the same, not only for liquid extraction but also for other industrial mass transfer operations.

Indeed, if you read about any industrial mass transfer processes, you will find the discussion couched in the terms used in this chapter. I find these advanced discussions tedious to read, for they always refer back to distillation and absorption for key equations. If you need to understand any industrial processes, first master the material in this chapter, and then refer to more specialized monographs.

References

American Institute of Chemical Engineers (1958). *Bubble Tray Design Manual*. New York: AIChE.

American Institute of Chemical Engineers (1981). *Computer Program Manual*. New York: AIChE.

Eckert, J. S. (1975). *Chemical Engineering (New York)*, **82,** 70.

Holland, C. D. (1963). *Multicomponent Distillation*. Englewood Cliffs, N.J.: Prentice-Hall.

King, C. J. (1971). *Separation Processes*. New York: McGraw-Hill.

Lewis, W. K., Jr. (1936). *Industrial and Engineering Chemistry*, **28,** 399.

McCabe, W., & Smith, J. C. (1975). *Unit Operations of Chemical Engineering*, 3rd ed. New York: McGraw-Hill.

Murphee, E. V. (1932). *Industrial and Engineering Chemistry*, **24,** 519.

Norton Company, Chemical Processing Products Division (1980).

Perry, E. S., & Weissberger, A. (1976). *Techniques of Organic Chemistry. Vol. 4 – Distillation*. New York: Wiley Interscience.

Perry, R. H., & Chilton, C. H. (1973). *Chemical Engineers Handbook*, 5th ed. New York: McGraw-Hill.

Sherwood, T. K. (1937). *Absorption and Extraction*. New York: McGraw-Hill.

Sherwood, T. K., Pigford, R. L., & Wilke, C. R. (1975). *Mass Transfer*. New York: McGraw-Hill.

Smith, B. D. (1963). *Design of Equilibrium Stage Processes*. New York: McGraw-Hill.

Treybal, R. E. (1980). *Mass Transfer Operations*, 3rd ed. New York: McGraw-Hill.

11 FORCED CONVECTION

In this chapter we want to connect mass transfer coefficients, diffusion coefficients, and fluid flow. In seeking these connections, we are combining this third section of the book, which deals with mass transfer, with the first two sections, which dealt with diffusion.

To find these connections, we need to specify the type of fluid flow involved. In this chapter we assume that this flow is largely determined by factors other than diffusion, factors like pressure gradients and wetted area. Such flows, called forced convection, exist whether or not diffusion occurs. Flows engendered by density differences caused by diffusion itself are called free convection and are treated in Chapter 12.

This chapter uses two different kinds of tools: simple physical models and elaborate analytical mathematics. The simple physical models pretend that the maelstrom of fluid motions is in fact diffusion across a thin film or diffusion into a semi-infinite slab. Under these conditions, the effect on diffusion of the convective flow is easily calculated. These simple calculations are found by experiment to bracket the observed behavior; so they supply physical insight into mass transfer.

The second type of tool, the elaborate mathematics, results from nearly 100 years of research on heat transfer. Because heat and mass transfer are described with similar equations, many have based their careers on analogies between the two. Unfortunately, these analogies are less useful than is routinely implied. For example, one important heat transfer problem is the heating of a fluid flowing in a shell-tube heat exchanger. The mathematically parallel mass transfer problem involves a fluid dissolving the tube's walls, an uncommon situation. As a counterexample, an important mass transfer process is liquid–liquid extraction. The parallel heat transfer problem of hot liquid drops suspended in a cold liquid solvent is not important. Thus, analogies between heat and mass transfer often provide more historical perspective than new information.

In this chapter we first discuss the simple models and then move on to the mathematically more sophisticated problems. Mass transfer across a thin film and into a semi-infinite fluid is discussed in Sections 11.1 and 11.2, respectively. Mass transfer in the boundary layer near a flat plate, outlined in Section 11.3, is the best of the heat transfer parallels. Mass transfer into a fluid flowing in a pipe is described in Section 11.4. Fast mass transfer that

281

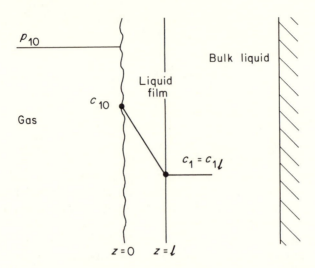

Fig. 11.1-1. The film theory for mass transfer. In this model, the interfacial region is idealized as a hypothetical film or "unstirred layer." Mass transfer involves diffusion across this thin film. Note that the constant value p_{10} implies no resistance to mass transfer in the gas.

generates additional convection leads to the more complicated results reported in Section 11.5.

Section 11.1. The film theory

The simplest theory for interfacial mass transfer, shown schematically in Fig. 11.1-1, assumes that a stagnant film exists near every interface. This film, also called an "unstirred layer," is almost always hypothetical, for fluid motions commonly occur right up to even a solid interface (Campbell & Hanratty, 1982). Nonetheless, such a hypothetical film, suggested first in 1904 (Nernst, 1904), gives the simplest model of the interfacial region.

We now imagine that a solute present at high dilution is slowly diffusing across this film. The restriction to high dilution allows us to neglect the diffusion-induced convection perpendicular to the interface; so we can use the simple results in Chapter 2, rather than the more complex ones in Chapter 3. The steady-state flux across this thin film can be written in terms of the mass transfer coefficient:

$$N_1 \doteq k(c_{1i} - c_1) \qquad (11.1-1)$$

in which N_1 is the flux relative to the interface, k is the mass transfer coefficient, and c_{1i} and c_1 are the interfacial and bulk concentrations in the fluid to the right of the interface. The flux across this film can also be calculated in terms of the diffusion coefficient (see Section 2.2):

$$N_1 \doteq n_1|_{z=0} \doteq j_1|_{z=0} = \frac{D}{l}(c_{1i} - c_1) \qquad (11.1-2)$$

The approximation that the total flux n_1 equals the diffusion flux j_1 reflects the assumption that the solution is dilute. If we compare Eq. 11.1-1 and Eq. 11.1-2, we see that

$$k = \frac{D}{l} \tag{11.1-3}$$

This result can be dignified as

$$\left(\begin{array}{c}\text{Sherwood}\\\text{number}\end{array}\right) = \frac{kl}{D} = 1 \tag{11.1-4}$$

Such dignity seems silly now but will be useful later.

This simplest theory says that the mass transfer coefficient is directly proportional to the diffusion coefficient. Doubling diffusion will double mass transfer. However, the mass transfer coefficient still varies in some unknown fashion with variables like fluid viscosity and stirring, because these variations are lumped into the unknown film thickness l. This thickness is almost never known a priori, but must be found from measurements of k and D. But if we cannot predict k from the film theory, what value has this theory?

The film theory is valuable for two reasons. First, it provides simple physical insight into mass transfer, for it shows in very simple terms how resistance to mass transfer might occur near an interface. Second, it often accurately predicts changes in mass transfer caused by other factors like chemical reaction or concentrated solution. As a result, the film theory is the picture around which most people build their ideas. In fact, we have already implicitly used it in the correlations of mass transfer coefficients in Chapter 9. These correlations are almost always written in terms of the Sherwood number:

$$\left(\begin{array}{c}\text{Sherwood}\\\text{number}\end{array}\right) = \frac{\left(\begin{array}{c}\text{mass transfer}\\\text{coefficient}\end{array}\right)\left(\begin{array}{c}\text{a character-}\\\text{istic length}\end{array}\right)}{\left(\begin{array}{c}\text{diffusion}\\\text{coefficient}\end{array}\right)} = F\left(\begin{array}{c}\text{other system}\\\text{variables}\end{array}\right) \tag{11.1-5}$$

By using a characteristic length, we imply a form equivalent to Eq. 11.1-3:

$$\left(\begin{array}{c}\text{mass transfer}\\\text{coefficient}\end{array}\right) = \frac{(\text{diffusion coefficient})}{(\text{some characteristic length})}\left(\begin{array}{c}\text{a correction}\\\text{factor } F\end{array}\right) \tag{11.1-6}$$

In many theories, we predict a mass transfer coefficient divided by the fluid velocity, so that a Stanton number seems to be the natural variable. Still, we religiously rewrite our results in terms of a Sherwood number, thus genuflecting toward the film theory.

Example 11.1-1: Finding the film thickness

Carbon dioxide is being scrubbed out of a gas using water flowing through a packed bed of 1-cm Berl saddles. The carbon dioxide is absorbed at a rate of $2.3 \cdot 10^{-6}$ mol/cm²-sec. The carbon dioxide is present at a partial pressure of 10 atm, the Henry's law coefficient H is 600 atm, and the diffusion coefficient of carbon dioxide in water is $1.9 \cdot 10^{-5}$ cm²/sec. Find the film thickness.

Solution. We first find the interfacial concentration of carbon dioxide:

$$p_1 = Hx_1 = H\left(\frac{c_{1i}}{c}\right)$$

$$10 \text{ atm} = 600 \text{ atm} \left(\frac{c_{1i}}{(\text{g-mol})/(18 \text{ cm}^3)}\right)$$

Thus, c_{1i} equals $9.3 \cdot 10^{-4}$ mol/cm³. We use Eq. 11.1-1 to find the mass transfer coefficient:

$$N_1 = k(c_{1i} - c_1)$$
$$2.3 \cdot 10^{-6} \text{ mol/cm}^2\text{-sec} = k \, [(9.3 \cdot 10^{-4} \text{ mol/cm}^3) - 0]$$

As a result, k is $2.5 \cdot 10^{-3}$ cm/sec. Finally, we use Eq. 11.1-3 to find the film thickness:

$$l = \frac{D}{k} = \frac{1.9 \cdot 10^{-5} \text{ cm}^2/\text{sec}}{2.5 \cdot 10^{-3} \text{ cm/sec}}$$
$$= 0.0076 \text{ cm}$$

Values around 10^{-2} cm are typical of many mass transfer processes.

Section 11.2. Penetration and surface-renewal theories

These theories provide a better physical picture of mass transfer than the film theory in return for a modest increase in mathematics. The net gain in understanding is often worth the price. Moreover, although the physical picture is still limited, similar equations can be derived from other, more realistic physical pictures.

Penetration theory

The model basic to this theory, suggested in 1935 (Higbie, 1935), is that of the film shown schematically in Fig. 11.2-1. As before, we define the mass transfer coefficient into this film as

$$N_1 = k(c_{1i} - c_1) \tag{11.2-1}$$

where N_1 is the flux across the interface, c_{1i} is the solute concentration in the liquid in equilibrium with the well-stirred gas, and c_1 is the constant solute concentration far into the liquid.

The diffusion flux shown in Fig. 11.2-1 can be calculated using the arguments in Section 2.5. The key assumption in this section is that the falling film is very thick. Other important assumptions are that in the z direction, diffusion is much more important than convection, and that in the x direction, diffusion is much less important than convection.

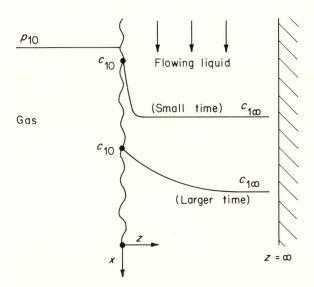

Fig. 11.2-1. The penetration theory for mass transfer. Here, the interfacial region is imagined to be a very thick film continuously generated by flow. Mass transfer now involves diffusion into this film. In this and other theories, the interfacial concentration in the liquid is assumed to be in equilibrium with that in the gas.

These assumptions lead to an equation for the interfacial flux:

$$N_1 = n_1|_{z=0} \doteq j_1|_{z=0} = \sqrt{Dv_{\max}/\pi x}\,(c_{1i} - c_1) \qquad (11.2\text{-}2)$$

One should remember that here N_1 is the flux at the interface and that the flux will have smaller values within the fluid. Moreover, it is a point value for some specific x. The interfacial flux averaged over x is given by

$$N_1 = \frac{1}{WL} \int_0^L \int_0^W n_1|_{z=0}\, dy\, dx \qquad (11.2\text{-}3)$$

Combining this with the foregoing, we see that the average flux is

$$N_1 = 2\sqrt{Dv_{\max}/\pi L}\,(c_{1i} - c_1) \qquad (11.2\text{-}4)$$

Comparing this with Eq. 11.2-1, we see that the mass transfer coefficient is

$$k = 2\sqrt{Dv_{\max}/\pi L} \qquad (11.2\text{-}5)$$

The quantity $L/v_{\max}$, sometimes called the "contact time," may not be exactly known a priori in complicated situations, just as the film thickness l was unknown in film theory. In addition, doubling the diffusion coefficient here increases the mass transfer coefficient by a factor of $\sqrt{2}$. In contrast, according to the film theory, doubling D doubles k. These two predictions tend to bracket the experimentally observed results.

Equation 11.2-5 can be written in terms of dimensionless groups using the fact that the average velocity in the film v^0 is two-thirds the maximum velocity $v_{\max}$. The result is

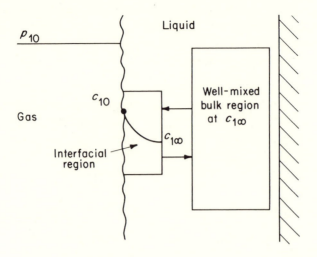

Fig. 11.2-2. The surface-renewal theory for mass transfer. This approach tries to apply the mathematics of the penetration theory to a more plausible physical picture. The liquid is pictured as two regions, a large well-mixed bulk and a surface region that is renewed so fast that it behaves as a thick film. The surface renewal is caused by liquid flow.

$$\frac{kL}{D} = \left(\frac{6}{\pi}\right)^{1/2} \left(\frac{Lv^0}{D}\right)^{1/2} = \left(\frac{6}{\pi}\right)^{1/2} \left(\frac{Lv^0}{\nu}\right)^{1/2} \left(\frac{\nu}{D}\right)^{1/2} \tag{11.2-6}$$

or

$$\binom{\text{Sherwood}}{\text{number}} = \left(\frac{6}{\pi}\right)^{1/2} \binom{\text{Péclet}}{\text{number}}^{1/2} = \left(\frac{6}{\pi}\right)^{1/2} \binom{\text{Reynolds}}{\text{number}}^{1/2} \binom{\text{Schmidt}}{\text{number}}^{1/2} \tag{11.2-7}$$

The use of the Sherwood number suggests a film theory, even though nothing like a film exists here.

Surface-renewal theory

The penetration theory is often found to be in reasonable agreement with many experimental results. This agreement is somewhat embarrassing, for the physical model and the assumptions of the theory are restrictive. We have apparently derived a good answer, but quite possibly from the wrong basis.

The embarrassment over the success of the penetration theory can be allayed by deriving similar results from other physical models. One popular alternative is the surface-renewal theory, of which a synopsis is given next (Dankwerts, 1951). A summary of other similar theories is given elsewhere (Sherwood et al., 1975).

The model used for the surface-renewal theory is shown schematically in Fig. 11.2-2. The specific geometry used in the film and penetration theories is replaced with the vaguer picture of two regions. In one "interfacial" region, mass transfer occurs by means of the penetration theory. However, small volumes or elements of this interfacial region are not static, but are constantly exchanged with new elements from a second "bulk" region. This idea of replacement or "surface renewal" makes the penetration theory a part of the overall process.

The mathematical description of this surface renewal depends on the length of time that small fluid elements spend in the interfacial region. The concept suggests the definition

$$E(t)\ dt = \left(\begin{array}{c} \text{the probability that a given surface} \\ \text{element will be at the surface for} \\ \text{time } t \end{array} \right) \qquad (11.2\text{-}8)$$

The quantity $E(t)$ is the residence time distribution used so often in the description of the chemical kinetics of stirred reactors (Levenspiel, 1972; Carberry, 1976). Obviously, the sum of these probabilities is unity:

$$\int_0^\infty E(t)\ dt = 1 \qquad (11.2\text{-}9)$$

We now assume that the transfer of different interfacial elements into the bulk region is random. Stated another way, we assume that the interfacial region is uniformly accessible, so that any surface element is equally likely to be withdrawn. In this case, the fraction of surface elements θ remaining at time t must be

$$\theta = e^{-t/\tau} \qquad (11.2\text{-}10)$$

in which τ is a characteristic constant. This fraction θ must also be the sum of the probabilities from time t to infinity

$$\theta = \int_t^\infty E(t)\ dt \qquad (11.2\text{-}11)$$

Thus, the residence time distribution of surface elements is

$$E(t) = \frac{e^{-t/\tau}}{\tau} \qquad (11.2\text{-}12)$$

The physical significance of τ in these equations is an average residence time for an element in the interfacial region.

Armed with these probabilities, we can average the mass transfer coefficient over time. In the interfacial region, the flux is that for diffusion into an infinite slab:

$$n_1|_{z=0} \doteq \sqrt{D/\pi t}\ (c_{1i} - c_1) \qquad (11.2\text{-}13)$$

Of course, the interfacial region is anything but infinite; but when the surface is quickly renewed and τ is small, it momentarily behaves as if it were. The average flux is then

$$N_1 = \int_0^\infty E(t) n_1|_{z=0}\ dt = \sqrt{D/\tau}\ (c_{1i} - c_1) \qquad (11.2\text{-}14)$$

By comparison with Eq. 11.2-1, we see that

$$k = \sqrt{D/\tau} \qquad (11.2\text{-}15)$$

As in the penetration theory, doubling the diffusion coefficient increases the mass transfer coefficient by $\sqrt{2}$.

At this point, some conclude that the surface-renewal theory is much ado about nothing. It predicts the same variation with diffusion as the penetration theory. Moreover, the new residence time τ is as unknown as the film thickness l introduced in the film theory. This new theory may at first seem of small value.

The value of the surface-renewal theory is that the simple math basic to the penetration theory is extended to a more realistic physical situation. Although the result is less exact than one might wish, the surface-renewal theory does suggest reasonable ways to think about mass transfer in complex situations. Such thoughts can lead to more effective correlations and to better models.

Indeed, it is this guide for thinking that is the point of all three simple theories of mass transfer. No theory is completely satisfactory, for each introduces an unknown parameter. However, each provides a simple way in which to think about mass transfer. This guide for simple thought is why these ideas have had an intellectual impact and have been subjected to so many serious "extensions" and "improvements."

Example 11.2-1: Finding the adjustable parameters of the penetration and surface-renewal theories

What are the contact time L/v_{max} and the surface residence time τ for the carbon dioxide scrubber described in Example 11.1-1?

Solution. In this earlier example, we were given that D was $1.9 \cdot 10^{-5}$ cm²/sec, and we calculated that k was $2.5 \cdot 10^{-3}$ cm/sec. Thus, from Eq. 11.2-5,

$$2.5 \cdot 10^{-3} \text{ cm/sec} = 2 \sqrt{[(1.9 \cdot 10^{-5} \text{ cm}^2/\text{sec})/\pi](v_{max}/L)}$$

$$\frac{L}{v_{max}} = 3.9 \text{ sec}$$

Similarly, from Eq. 11.2-15,

$$2.5 \cdot 10^{-3} \text{ cm/sec} = \sqrt{(1.9 \cdot 10^{-5} \text{ cm}^2/\text{sec})/\tau}$$

$$\tau = 3.0 \text{ sec}$$

Deciding whether these values are physically realistic requires additional information.

Section 11.3. Boundary layer theory

We now turn to more complete descriptions of mass transfer, descriptions which use exact physical pictures of specific situations, rather than models approximately applicable to many situations. These more complete descriptions require much more mathematical work than the simple models, but they give superior results for these situations (Levich, 1962; Skelland, 1974; Schlichting, 1979).

The majority of the more complete descriptions of mass transfer are based on parallels with earlier studies of fluid mechanics and heat transfer. Those who did much of the early work on mass transfer did so in the shadow of the intellectual cathedral of ideas erected in these other fields. These early mass transfer workers were intimidated, cowed especially by the insights of boundary layer theory. Even now, no self-respecting graduate course in mass transfer can ignore boundary layer theory; some such courses arrogantly contain little else. Such bias owes less to utility than to historical intimidation.

A qualitative overview

In this section we discuss the common example of a laminar boundary layer formed near a sharp-edged plate. This plate is made of a sparingly soluble solute immersed in a rapidly flowing solvent. We want to find the rate at which solute dissolves. In more scientific terms, we want to calculate how the Sherwood number varies with the Reynolds and Schmidt numbers.

This physical situation is shown in the oft-quoted schematic in Fig. 11.3-1. The plate is immersed in a smoothly flowing fluid. The flow is disrupted by the drag caused by the plate; this region of disruption, called a "boundary layer," becomes larger as the flow proceeds down the plate. The boundary layer is usually defined as the locus of distances over which 99% of the disruptive effect occurs. Such specificity is obviously arbitrary.

While this flow pattern develops, the sparingly soluble solute dissolves off the plate. This dissolution produces the concentration profiles shown in Fig. 11.3-1(b). The distance that the solute penetrates produces a new boundary layer, but this layer is not the same as that observed for flow. In general, these two layers will influence each other, and the problem will be very complex indeed. However, when the dissolving solute is only sparingly soluble, the boundary layer caused by the flow is unaffected.

The basic scheme by which the mass transfer coefficient is calculated is shown in Fig. 11.3-2. Most of the work involves finding the boundary layer caused by flow. The key step is assuming that the flow varies as a power series in the boundary layer thickness, and then using the boundary conditions to find the constants in this series. By roughly parallel arguments, we can find the concentration boundary layer δ_c as a function of δ. From this point it is only a short step to the mass transfer coefficient.

This calculation is not difficult, but it is long. I tend to get lost in the individual steps and forget the overall scheme in Fig. 11.3-2. I shall try to keep you from getting lost by frequent reference to this figure.

The boundary layer for flow

Our first step, step A in Fig. 11.3-2, is to assume a profile for the fluid flowing parallel to the flat plate:

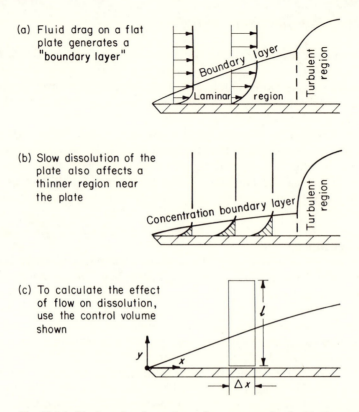

(a) Fluid drag on a flat plate generates a "boundary layer"

(b) Slow dissolution of the plate also affects a thinner region near the plate

(c) To calculate the effect of flow on dissolution, use the control volume shown

Fig. 11.3-1. The boundary layer theory for mass transfer. In this theory, both the flow and the diffusion are analyzed for specific geometries like that of a flat plate. The results are accurate for the specific case, but are purchased with considerable mathematical effort.

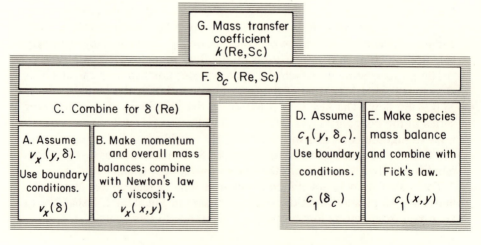

G. Mass transfer coefficient k (Re, Sc)

F. δ_c (Re, Sc)

C. Combine for δ (Re)

A. Assume $v_x(y, \delta)$. Use boundary conditions.

$v_x(\delta)$

B. Make momentum and overall mass balances; combine with Newton's law of viscosity.

$v_x(x, y)$

D. Assume $c_1(y, \delta_c)$. Use boundary conditions.

$c_1(\delta_c)$

E. Make species mass balance and combine with Fick's law.

$c_1(x, y)$

Fig. 11.3-2. A schematic outline of the boundary layer calculations. Individual steps in these calculations are straightforward, but the collection and combination of these steps can be difficult. This illustration, a map of the ways these steps are assembled, should provide a guide through this intellectual swamp.

$$v_x = a_0 + a_1 y + a_2 y^2 + a_3 y^3 + \cdots \tag{11.3-1}$$

where y is the distance normal to the plate and the a_i are constants. These constants are independent of y, but may vary with x. The velocity described by this equation is subject to four boundary conditions:

$$y = 0, \quad v_x = 0 \tag{11.3-2}$$

$$\frac{\partial^2 v_x}{\partial y^2} = 0 \tag{11.3-3}$$

$$y = \infty, \quad v_x = v^0 \tag{11.3-4}$$

$$\frac{\partial v_x}{\partial y} = 0 \tag{11.3-5}$$

The first of these conditions says that the fluid sticks to the plate, and the second says that because the plate is solid, the stress on it is a constant. The other two conditions say that far from the plate, the plate has no effect.

As an approximation, we can replace Eqs. 11.3-4 and 11.3-5 with

$$y = \delta, \quad v_x = v^0 \tag{11.3-6}$$

$$\frac{\partial v_x}{\partial y} = 0 \tag{11.3-7}$$

Now we can find the constants in Eq. 11.3-1: a_0 is zero by Eq. 11.3-2; a_2 is zero by Eq. 11.3-3; and so

$$\frac{v_x}{v^0} = \frac{3}{2}\left(\frac{y}{\delta}\right) - \frac{1}{2}\left(\frac{y}{\delta}\right)^3 \tag{11.3-8}$$

At this point I always feel that I have gotten something for nothing, until I remember that δ is an unknown function of x. Moreover, we should remember that Eq. 11.3-1 was arbitrarily truncated and that more accurate calculations might require more terms.

We now turn to step B, the overall mass and momentum balances on the control volume shown in Fig. 11.3-1(c). In so doing, we shun the creative slickness of the preceding for more recognizable physical arguments. A mass balance on the control volume is

$$(\text{accumulation}) = \begin{pmatrix}\text{flow in minus flow out in} \\ \text{the } x \text{ direction}\end{pmatrix}$$

$$+ \begin{pmatrix}\text{dissolution} \\ \text{in from plate}\end{pmatrix} - \begin{pmatrix}\text{flow out} \\ \text{top of plate}\end{pmatrix} \tag{11.3-9}$$

Because the solute is sparingly soluble, the dissolution has a negligible effect on the overall mass balance. Because the process is in steady state, the accumulation is zero. Thus,

$$0 = \left(\int_0^l W\rho v_x \, dy\right)_x - \left(\int_0^l W\rho v_x \, dy\right)_{x+\Delta x} - \rho v_y W\Delta x \tag{11.3-10}$$

in which W is the width of the plate. Dividing by $W\Delta x$, taking the limit as this area goes to zero, and assuming ρ is constant gives

$$v_y = -\frac{d}{dx}\int_0^l v_x \, dy \tag{11.3-11}$$

This overall mass balance can be supplemented by an x-momentum balance:

$$(\text{accumulation}) = \begin{pmatrix} x\text{-momentum flow in minus flow out} \\ \text{in the } x \text{ direction} \end{pmatrix}$$

$$+ \begin{pmatrix} x\text{-momentum flow in minus flow out} \\ \text{in the } y \text{ direction} \end{pmatrix}$$

$$+ \begin{pmatrix} \text{forces acting in} \\ \text{the } x \text{ direction} \end{pmatrix} \tag{11.3-12}$$

which in symbolic terms becomes

$$0 = \left[\left(\int_0^l W v_x v_x \, dy \right)_x - \left(\int_0^l W v_x v_x \, dy \right)_{x+\Delta x} \right]$$
$$+ (0 - \rho v^0 v_y W \Delta x) + (\tau_0 W \Delta x) \tag{11.3-13}$$

Again dividing by $W\Delta x$ and taking the limit as this area goes to zero,

$$-\tau_0 = -\frac{d}{dx} \int_0^l \rho v_x^2 \, dy - \rho v^0 v_y \tag{11.3-14}$$

In this result, τ_0 represents the shear stress generated by the fluid on the plate. We now can use Eq. 11.3-11 to eliminate v_y from Eq. 11.3-14:

$$-\tau_0 = \frac{d}{dx} \int_0^l (v^0 - v_x)\rho v_x \, dy \tag{11.3-15}$$

When y is greater than the boundary layer δ, v^0 equals v_x. Moreover, the stress τ_0 is given by Newton's law of viscosity. Thus,

$$\mu \frac{\partial v_x}{\partial y}\bigg|_{y=0} = \frac{d}{dx} \int_0^\delta (v^0 - v_x)\rho v_x \, dy \tag{11.3-16}$$

The left-hand side of this equation comes from the shear force; the right-hand side represents momentum convection, rewritten with the help of the overall mass balance.

We now combine Eqs. 11.3-8 and 11.3-16 to find an equation for the boundary layer:

$$\delta \frac{d\delta}{dx} = \frac{140}{13} \left(\frac{\mu}{\rho v^0} \right) \tag{11.3-17}$$

This combination, which corresponds to step C in Fig. 11.3-2, is subject to the boundary condition

$$x = 0, \quad \delta = 0 \tag{11.3-18}$$

Integration is straightforward:

$$\frac{\delta}{x} = \left(\frac{280}{13} \right)^{1/2} \left(\frac{x v^0 \rho}{\mu} \right)^{-1/2} \tag{11.3-19}$$

Because we know δ, we now know v_x, and so we have solved the fluid mechanics of the boundary layer formation.

At this point we make one quick digression. The foregoing calculation is approximate, subject to a variety of assumptions of uncertain quality. We should check these assumptions by experiment. One way to do so is to rewrite our equations in terms of a friction factor f, defined as

$$\tau_0 = -\mu \left. \frac{\partial v_x}{\partial y} \right|_{y=0} = f[\tfrac{1}{2}\rho(v^0)^2] \tag{11.3-20}$$

When we insert Eqs. 11.3-8 and 11.3-19 into this definition, we find

$$f = \frac{3\mu}{\delta v^0 \rho}$$

$$= 0.646 \left(\frac{xv^0}{\mu}\right)^{-1/2} \tag{11.3-21}$$

This prediction is about 3% below that observed experimentally, justifying our assumptions (Blasius, 1908; Nikuradse, 1942).

The boundary layer for concentration

We now want to calculate the shape of the concentration profile shown in Fig. 11.3-1(b) in terms of the concentration boundary layer. In so doing, we assume that the boundary layer for flow is unchanged by the dissolving plate, thereby implying that the dissolving solute is only sparingly soluble.

We first find the concentration profile as a function of this new boundary layer, as suggested by step D in Fig. 11.3-2. We assume

$$c_1 = a_0 + a_1 y + a_2 y^2 + a_3 y^3 + \cdots \tag{11.3-22}$$

We know this function must be subject to the boundary conditions

$$y = 0, \quad c_1 = c_{1i} \tag{11.3-23}$$

$$\frac{\partial^2 c_1}{\partial y^2} = 0 \tag{11.3-24}$$

$$y = \infty, \quad c_1 = 0 \tag{11.3-25}$$

$$\frac{\partial c_1}{\partial y} = 0 \tag{11.3-26}$$

The first two conditions indicate that both concentration and flux are constant at the plate's surface. The last two conditions apply deep into the fluid, where the plate has no effect; they can be replaced by

$$y = \delta_c, \quad c_1 = 0 \tag{11.3-27}$$

$$\frac{\partial c_1}{\partial y} = 0 \tag{11.3-28}$$

Using these conditions, we can quickly discover

$$\frac{c_1}{c_{1i}} = 1 - \frac{3}{2}\left(\frac{y}{\delta_c}\right) + \frac{1}{2}\left(\frac{y}{\delta_c}\right)^3 \tag{11.3-29}$$

This type of manipulation should seem familiar, for it is completely parallel to that used to find the flow boundary layer.

The next step is to find the solute's mass balance to complete step E in Fig. 11.3-2. Again, we develop this mass balance on the control volume $lW\Delta x$ shown in Fig. 11.3-1(c):

$$\left(\begin{matrix} \text{solute} \\ \text{accumulation} \end{matrix}\right) = \left(\begin{matrix} \text{solute convection in minus} \\ \text{that out in the } x \text{ direction} \end{matrix}\right)$$

$$+ \left(\begin{matrix} \text{solute diffusion in minus} \\ \text{that out in the } y \text{ direction} \end{matrix}\right) \qquad (11.3\text{-}30)$$

Notice that we yet again are making our common assumption that mass transfer in the x direction is dominated by forced convection, but that mass transfer in the y direction takes place only by diffusion. On this basis, we can write Eq. 11.3-30 in symbolic terms as

$$0 = \left[\left(\int_0^l Wc_1v_x \, dy\right)_x - \left(\int_0^l Wc_1v_x \, dy\right)_{x+\Delta x}\right] + (N_1W\Delta x - 0) \qquad (11.3\text{-}31)$$

Dividing by the area $W\Delta x$ and taking the limit as this area becomes small, we find

$$-N_1 = n_1|_{y=0} = -\frac{d}{dx}\int_0^l c_1v_x \, dy \qquad (11.3\text{-}32)$$

This equation is simplified in two ways: by replacing n_1 with Fick's law and by remembering that c_1 is zero from δ_c to l. The result is

$$D\frac{\partial c_1}{\partial y}\bigg|_{y=0} = -\frac{d}{dx}\int_0^{\delta_c} c_1v_x \, dy \qquad (11.3\text{-}33)$$

At this point, many readers may have forgotten our purpose, which is to calculate the mass transfer coefficient a priori for a flat plate slowly dissolving in laminar flow. We have completed steps A to E in Fig. 11.3-2; we have only two steps to go.

We combine Eqs. 11.3-8, 11.3-19, 11.3-29, and 11.3-33 to find a differential equation for the concentration boundary layer. In this combination, which is step F in Fig. 11.3-2, we assume that δ_c is much smaller than δ, an approximation whose chief justification must be experimental. The result can be written as

$$\frac{4}{3}x\frac{d}{dx}\left(\frac{\delta_c}{\delta}\right)^3 + \left(\frac{\delta_c}{\delta}\right)^3 = \left(\frac{D\rho}{\mu}\right) \qquad (11.3\text{-}34)$$

If δ_c is smaller, it must develop slower; so Eq. 11.3-34 must be subject to the condition

$$x = 0, \quad \frac{\delta_c}{\delta} = 0 \qquad (11.3\text{-}35)$$

Integration is easy:

$$\left(\frac{\delta_c}{\delta}\right)^3 = \left(\frac{D\rho}{\mu}\right) + bx^{-3/4} \qquad (11.3\text{-}36)$$

The term containing x in this equation is negligible in most situations. Thus, combining Eq. 11.3-19 and Eq. 11.3-36,

$$\left(\frac{\delta_c}{x}\right) = 4.64\left(\frac{\mu}{xv^0\rho}\right)^{1/2}\left(\frac{D\rho}{\mu}\right)^{1/3} \qquad (11.3\text{-}37)$$

The novelty of this result is the appearance of the Schmidt number $\mu/\rho D$.

The mass transfer coefficient

The final manipulation in this analysis (step G in Fig. 11.3-2) is the reformulation of the boundary layer as the mass transfer coefficient. Such a manipulation is straightforward. By definition,

$$N_1 = k(c_{1i} - 0) \qquad (11.3\text{-}38)$$

and by Fick's law,

$$N_1 = n_1 \Big|_{y=0} \doteq -D \frac{\partial c_1}{\partial y} \Big|_{y=0} \qquad (11.3\text{-}39)$$

We have the concentration profile in terms of δ_c in Eq. 11.3-29; so

$$N_1 = \frac{3Dc_{1i}}{2\delta_c} \qquad (11.3\text{-}40)$$

Note again how this result parallels the film theory, with $2\delta_c/3$ equivalent to the film thickness. Combining this with Eq. 11.3-37, we find

$$N_1 = 0.323 \frac{Dc_{1i}}{x} \left(\frac{xv^0\rho}{\mu}\right)^{1/2} \left(\frac{\mu}{\rho D}\right)^{1/3} \qquad (11.3\text{-}41)$$

Comparing this with Eq. 11.3-38,

$$\frac{kx}{D} = 0.323 \left(\frac{xv^0\rho}{\mu}\right)^{1/2} \left(\frac{\mu}{\rho D}\right)^{1/3} \qquad (11.3\text{-}42)$$

This is the desired result for the boundary layer theory near a flat plate. For the coefficient averaged over the length L, the result is similar:

$$\frac{\bar{k}L}{D} = 0.626 \left(\frac{Lv^0\rho}{\mu}\right)^{1/2} \left(\frac{\mu}{\rho D}\right)^{1/3} \qquad (11.3\text{-}43)$$

The prediction agrees with experiments for a flat plate when the boundary layer is laminar, which occurs when the local Reynolds number ($xv^0\rho/\mu$) is less than 300,000 (Sherwood et al., 1975). At higher Reynolds numbers, theories of turbulent boundary layers must be used. In the laminar region, note that k and $\bar{k}$ vary with the two-thirds power of the diffusion coefficient, which is a value midway between the linear variation of the film theory and the square-root variation of the penetration and surface-renewal theories.

Example 11.3-1: Calculation of mass transfer coefficients from boundary layer theory

Water flows at 10 cm/sec over a sharp-edged plate of benzoic acid. The dissolution of benzoic acid is diffusion-controlled, with a diffusion coefficient of $1.00 \cdot 10^{-5}$ cm²/sec. Find (a) the distance at which the laminar boundary layer ends, (b) the thickness of the flow and concentration boundary layers at that point, and (c) the local mass transfer coefficients at the edge and at the position of transition, as well as the average mass transfer coefficient over this length.

Solution. (a) The length before turbulence begins can be found from

$$\frac{xv^0\rho}{\mu} = \frac{x(10 \text{ cm/sec})(1 \text{ g/cm}^3)}{0.01 \text{ g/cm-sec}} = 300,000$$

Thus, the transition occurs when x is 300 cm.

(b) The boundary layer for flow can be found from Eq. 11.3-19:

$$\delta = \left(\frac{280}{13}\right)^{1/2} x \left(\frac{\mu}{xv^0\rho}\right)^{1/2}$$

$$= \left(\frac{280}{13}\right)^{1/2} (300 \text{ cm}) \left(\frac{0.01 \text{ g/cm-sec}}{(300 \text{ cm})(10 \text{ cm/sec})(1 \text{ g/cm}^3)}\right)^{1/2}$$

$$= 2.5 \text{ cm}$$

The boundary layer for concentration is given by Eq. 11.3-36:

$$\delta_c = \left(\frac{D\rho}{\mu}\right)^{1/3} \delta$$

$$= \left(\frac{(1.00 \cdot 10^{-5} \text{ cm}^2/\text{sec})(1 \text{ g/cm}^3)}{0.01 \text{ g/cm-sec}}\right)^{1/3} (2.5 \text{ cm})$$

$$= 0.25 \text{ cm}$$

The concentration boundary layer is thinner, just as was assumed in the derivation given earlier.

(c) The local and average mass transfer coefficients can be found from Eqs. 11.3-42 and 11.3-43. At the sharp edge of the plate, the local mass transfer coefficient is infinity. Where the transition to turbulent flow occurs, the value is given by

$$k = 0.323 \frac{D}{x} \left(\frac{xv^0\rho}{\mu}\right)^{1/2} \left(\frac{\mu}{\rho D}\right)^{1/3}$$

$$= 0.323 \left(\frac{1.00 \cdot 10^{-5} \text{ cm}^2/\text{sec}}{300 \text{ cm}}\right)$$

$$\cdot (300,000)^{1/2} \left(\frac{0.01 \text{ g/cm-sec}}{(1 \text{ g/cm}^3)(1.00 \cdot 10^{-5} \text{ cm}^2/\text{sec})}\right)^{1/3}$$

$$= 5.9 \cdot 10^{-5} \text{ cm/sec}$$

The calculations are completely straightforward, but the values found are instructive. The average value is twice this, about $11.8 \cdot 10^{-5}$ cm/sec. Remember that these calculations are restricted to a flat plate and that the entire derivation must be repeated for each new geometry.

Section 11.4. The Graetz–Nusselt problem

Thus far, we have modeled mass transfer as diffusion across thin films, into thick films, or away from dissolving plates. We now turn to mass transfer across the walls of a pipe containing a fluid in laminar flow, as shown in Fig. 11.4-1. We want to know how the mass transfer coefficient varies with the fluid's flow and the solute's diffusion. In other words, we want to find the

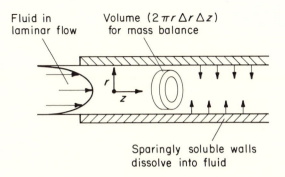

Fig. 11.4-1. The Graetz–Nusselt problem. In this case, a pure solvent flowing laminarly in a cylindrical tube suddenly enters a section where the tube's walls are dissolving. The problem is to calculate the wall's dissolution rate, and hence the mass transfer coefficient. The problem's solutions, based on analogies with heat transfer, are elegant but rarely useful.

dissolution rate as a function of quantities like Reynolds and Schmidt numbers.

Like boundary layer theory, this problem was originally solved for heat transfer. It is often referred to as the Graetz–Nusselt problem (Graetz, 1880; Nusselt, 1909, 1910). The problem again assumes a sparingly soluble solute, so that the velocity profile is parabolic, as expected for laminar flow. The detailed solution depends on the exact boundary conditions involved. The results for several common boundary conditions are tabulated in Table 11.4-1 (Lévêque, 1928; Jakob, 1957).

In this section we detail the derivation only for the case of fixed solute concentration at the wall of a short tube. The restriction to short tubes simplifies the mathematics. The choice of a constant wall concentration implies that the sparingly soluble solute dissolves very quickly, so that its concentration at the wall is essentially at saturation. This situation is more common than that of constant wall flux.

The mass balance and its solution

The steady-state continuity equation for the solute in a constant-density fluid is

$$\begin{pmatrix} \text{solute} \\ \text{accumulation} \end{pmatrix} = \begin{pmatrix} \text{solute in minus solute out} \\ \text{by diffusion} \end{pmatrix}$$
$$+ \begin{pmatrix} \text{solute in minus solute out} \\ \text{by convection} \end{pmatrix} \tag{11.4-1}$$

or, in symbolic terms,

$$0 = D \left(\frac{1}{r} \frac{\partial}{\partial r} r \frac{\partial c_1}{\partial r} + \frac{\partial^2 c_1}{\partial z^2} \right) - v_z \frac{\partial c_1}{\partial z} \tag{11.4-2}$$

Table 11.4-1. *The Graetz–Nusselt problem and its parallels*

	Short tubes	Long tubes
Fixed wall concentration	$\dfrac{c_1}{c_{10}} = \dfrac{\Gamma\left(\frac{2}{3}, \xi^3\right)}{\Gamma\left(\frac{2}{3}\right)}, \quad \xi = (R-r)\left(\dfrac{4v_0}{9DRz}\right)^{1/3}$	$\dfrac{c_1}{c_{10}} = 1 - 1.477 j_0 \left(\dfrac{r}{R}\right) e^{-3.658 Dz/v_0 R^2} + 0.810 j_1 \left(\dfrac{r}{R}\right) e^{-22.178 Dz/v_0 R^2}$ $- \cdots \quad (j_i \text{ tabulated in literature; Jakob, 1957})$
Fixed dissolution rate at the wall	$\dfrac{c_1}{N_1 R/D} = \left(\dfrac{9zD}{4v_0 R^2}\right)^{1/3}\left[\dfrac{e^{-\xi^3}}{\Gamma\left(\frac{4}{3}\right)} - \xi\left(1 - \dfrac{\Gamma\left(\frac{2}{3}, \xi^3\right)}{\Gamma\left(\frac{2}{3}\right)}\right)\right]$	$\dfrac{c_1}{N_1 R/D} = \dfrac{2Dz}{R^2 v_0} + \left(\dfrac{r}{R}\right)^2 - \dfrac{1}{4}\left(\dfrac{r}{R}\right)^4 - \dfrac{7}{24}$

This equation can be found either by a mass balance on the washer-shaped region in Fig. 11.4-1 or by the appropriate simplification of the general mass balances given in Table 3.4-2.

For short tubes, solute diffusion occurs mainly near the wall, and the bulk of the fluid near the tube's axis is pure solvent. As a result, axial diffusion is small, and $\partial^2 c_1/\partial z^2$ can be neglected:

$$0 = \frac{D}{r}\frac{\partial}{\partial r} r \frac{\partial c_1}{\partial r} - 2v^0 \left[1 - \left(\frac{r}{R}\right)^2\right]\frac{\partial c_1}{\partial z} \tag{11.4-3}$$

We define a new coordinate:

$$s = R - r \tag{11.4-4}$$

Rewriting the mass balance in terms of this variable,

$$0 = D\frac{\partial^2 c_1}{\partial s^2} - \frac{4v_0 s}{R}\frac{\partial c_1}{\partial z} \tag{11.4-5}$$

In this, we have used the fact that s/R is much less than unity near the wall; so the wall curvature and the nonlinear velocity can be ignored.

The boundary conditions for this differential equation are the following:

$$z = 0, \quad \text{all } s, \quad c_1 = 0 \tag{11.4-6}$$
$$z > 0, \quad s = 0, \quad c_1 = c_{1i} \tag{11.4-7}$$
$$z > 0, \quad s = \infty, \quad c_1 = 0 \tag{11.4-8}$$

The intriguing condition is, of course, the third one, because this suggests that the tube is short.

To solve this problem, we define the new dimensionless variable

$$\xi = s\left(\frac{4v^0}{9DRz}\right)^{1/3} \tag{11.4-9}$$

The differential equation now becomes

$$0 = \frac{d^2 c_1}{d\xi^2} + 3\xi^2 \frac{dc_1}{d\xi} \tag{11.4-10}$$

subject to

$$\xi = 0, \quad c_1 = c_{1i} \tag{11.4-11}$$
$$\xi = \infty, \quad c_1 = 0 \tag{11.4-12}$$

One integration gives

$$\frac{dc_1}{d\xi} = (\text{constant})e^{-\xi^3} \tag{11.4-13}$$

A second integration and use of the boundary conditions results in

$$c_1 = c_{1i}\frac{\displaystyle\int_\xi^\infty e^{-\xi^3}\, d\xi}{\Gamma\left(\frac{4}{3}\right)} \tag{11.4-14}$$

The numerator in this expression is called the incomplete gamma function. This completes the first part of the problem, the calculation of the concentration profile.

The mass transfer coefficient

To find the mass transfer coefficient k, we again compare the definition

$$N_1 = k(c_{1i} - 0) \tag{11.4-15}$$

with the value found from Fick's law:

$$N_1 = n_1 \Big|_{r=R} \doteq -D \frac{\partial c_1}{\partial r} \Big|_{r=R} = D \frac{\partial c_1}{\partial s} \Big|_{s=0}$$

$$= D \left(\frac{4v^0}{9DRz}\right)^{1/3} \frac{\partial c_1}{\partial \xi} \Big|_{\xi=0} = D \frac{\left(\dfrac{4v^0}{9DRz}\right)^{1/3}}{\Gamma\left(\frac{4}{3}\right)} (c_{1i} - 0) \tag{11.4-16}$$

Comparison gives

$$k = \frac{D}{\Gamma\left(\frac{4}{3}\right)} \left(\frac{4v^0}{9DRz}\right)^{1/3} \tag{11.4-17}$$

As in the penetration theory, this k is a local value located at fixed z. If we average this coefficient over a pipe length L, we find, after rearrangement,

$$\frac{\bar{k}L}{D} = \left(\frac{2}{3\Gamma\left(\frac{4}{3}\right)}\right)\left(\frac{v^0 L^2}{DR}\right)^{1/3} = \left(\frac{2}{3\Gamma\left(\frac{4}{3}\right)}\right)\left(\frac{Lv^0\rho}{\mu}\right)^{1/3}\left(\frac{\mu}{\rho D}\right)^{1/3}\left(\frac{L}{R}\right)^{1/3} \tag{11.4-18}$$

This is

$$\left(\begin{array}{c}\text{Sherwood}\\\text{number}\end{array}\right) = \left(\frac{2}{3\Gamma\left(\frac{4}{3}\right)}\right)\left(\begin{array}{c}\text{Reynolds}\\\text{number}\end{array}\right)^{1/3}\left(\begin{array}{c}\text{Schmidt}\\\text{number}\end{array}\right)^{1/3}\left(\frac{\text{length}}{\text{radius}}\right)^{1/3} \tag{11.4-19}$$

The mass transfer coefficient here varies with the diffusion coefficient, as in the boundary layer theory, but it varies differently with the fluid flow.

Example 11.4-1: Mass transfer of benzoic acid

Water is flowing at 6.1 cm/sec through a pipe 2.3 cm in diameter. The walls of a 14-cm section of this pipe are made of benzoic acid, whose diffusion coefficient in water is $1.00 \cdot 10^{-5}$ cm²/sec. Find the average mass transfer coefficient over this section.

Solution. Instead of thinking this problem out sensibly, let us dash through it in typical student fashion, and then see if the answer makes sense. The answer is probably in one of the most recent equations, like Eq. 11.4-18. This gives

$$\bar{k} = \left(\frac{2}{3\Gamma\left(\frac{4}{3}\right)}\right)\frac{D}{L}\left(\frac{v^0 L^2}{DR}\right)^{1/3}$$

$$= (0.746)\frac{1.00 \cdot 10^{-5} \text{ cm}^2/\text{sec}}{14 \text{ cm}}\left(\frac{(6.1 \text{ cm/sec})(14 \text{ cm})^2}{(1.00 \cdot 10^{-5} \text{ cm}^2/\text{sec})(1.15 \text{ cm})}\right)^{1/3}$$

$$= 2.5 \cdot 10^{-4} \text{ cm/sec}$$

That calculation is fast enough to be finished before beer on a Sunday night. But is it right?

Well, the answer lies midway between the film and boundary layer theories, so that much is reassuring. The basic equation used (as well as those that preceded it) assumed that the flow was laminar. Let us see. The Reynolds number is

$$\frac{dv^0\rho}{\mu} = \frac{(2.3 \text{ cm})(6.1 \text{ cm/sec})(1 \text{ g/cm}^3)}{0.01 \text{ g/cm-sec}} = 1,400$$

That is less than 2,100, the transition to turbulent flow. So far, so good. However, the foregoing equations also assumed that the pipe was very short. Is this pipe short? Well, 14 cm *seems* short, but who knows? Maybe we should go back and use the more general equation for long pipes, given in Table 11.4-1. That may take longer, but it will be safer. But when we look, we see that this equation is very complex, and parts of it are not given here, but in other books. Looking in other books is a pain. There must be a better way.

Is the pipe short? Well, in the derivation, we see that it is short if

$$s \ll R$$

(i.e., diffusion never penetrates very far from the wall). By dimensional arguments or by analogies with Chapter 2, we guess a characteristic value of s:

$$s = \sqrt{4DL/v^0} \ll R$$

$$\sqrt{4(1.00 \cdot 10^{-5} \text{ cm}^2/\text{sec})(14 \text{ cm})/(6.1 \text{ cm/sec})} \ll 1.15 \text{ cm}$$

$$\begin{pmatrix} 0.01 \text{ cm} \\ \text{diffusion has} \\ \text{penetrated} \end{pmatrix} \ll \begin{pmatrix} 1.15\text{-cm} \\ \text{radius of} \\ \text{pipe} \end{pmatrix}$$

Thus, the benzoic acid dissolution does take place in a short pipe, where short is defined in terms of diffusion. The example is simple enough to allow the Sunday beer. More important, the moral to be drawn is that mathematics is just a tool limited by assumptions, assumptions that often need quantitative checks.

Section 11.5. Theories for concentrated solutions

By this time, we should recognize one omnipresent assumption in all of the foregoing theories: the assumption of a dilute solution. Restricting our arguments to dilute solution allows a focus on diffusion and a neglect of the convection that diffusion itself can generate. In terms of this book, the restriction to dilute solution uses the simple ideas in Chapter 2, not the more complex concepts in Chapter 3.

The intellectual basis in dilute solution is that underlying correlations of mass transfer coefficients like those given in Chapter 9. These correlations are routinely applied to mass transfer in concentrated solutions. The correlations are often successful, especially when the mass transfer is slow. This is because the volume average velocity in the fluid remains small even though the solution is concentrated. Again, dilute-solution results are more applicable than we might at first expect.

In a few cases, however, these simple ideas of mass transfer fail. This failure is most commonly noticed as a mass transfer coefficient k that depends on the driving force. In other words, we define as before

$$N_1 = k\Delta c_1 \tag{11.5-1}$$

As expected, we find that k is a function of Reynolds and Schmidt numbers. However, when mass transfer is fast, k may also be a function of Δc_1. Stated another way, if we double Δc_1, we do not double N_1 even when the Reynolds and Schmidt numbers are constant.

These shortcomings lead to alternative definitions of mass transfer coefficients that include the effects of diffusion-induced convection. One such definition is (Bird et al., 1960)

$$\begin{aligned} N_1 &= k(c_{1i} - c_1) - c_{1i}v^0 \\ &= k(c_{1i} - c_1) + c_{1i}(\bar{V}_1 N_1 + \bar{V}_2 N_2) \end{aligned} \tag{11.5-2}$$

in which k is now the mass transfer coefficient for rapid mass transfer and v^0 is the velocity at the interface. Note that the convective term is defined in terms of c_{1i}, the interfacial concentration, not in terms of some average value.

We want to calculate this new coefficient, just as we calculated the dilute coefficient in earlier sections of this chapter. In general, we might expect to repeat the whole chapter, producing an entirely new series of equations for the film, penetration, surface-renewal, and boundary layer theories. However, these calculations not only would be difficult but also would retain the unknown parameters like film thickness and contact time.

To avoid these inaccuracies, we adopt a new strategy, one that we shall use more later, especially in the study of coupled mass transfer and chemical reaction. We assume that from experiments we know the mass transfer coefficient in dilute solution. We then calculate the corrections caused by fast mass transfer in concentrated solution, as predicted by the film theory (Lewis & Chang, 1928). In other words, we find the ratio of the mass transfer coefficient in concentrated solution to that in dilute solution. The ratio found from the film theory turns out to be close to that found from other theories. Thus, the film theory gives a reasonable estimate of the changes engendered by high solute concentration.

To make this calculation, imagine a thin film like that shown schematically in Fig. 11.5-1. A mass balance on a thin shell Δz thick shows that the total flux is a constant:

$$0 = -\frac{dn_1}{dz} \tag{11.5-3}$$

Integrating and combining with Fick's law gives

$$n_1 = -D\frac{dc_1}{dz} + c_1v^0 = -D\frac{dc_1}{dz} + c_1(\bar{V}_1 n_1 + \bar{V}_2 n_2) \tag{11.5-4}$$

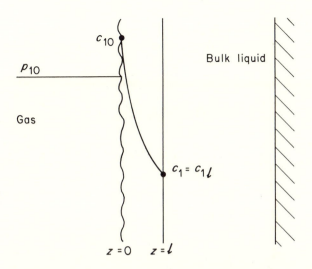

Fig. 11.5-1. The film theory for fast mass transfer. When mass transfer is rapid, the formulations of mass transfer given in earlier parts of this chapter can break down. This is because the diffusion process itself can generate convection normal to the interface. As a result, the simple concentration profile shown in Fig. 11.1-1 for the film theory becomes more complicated. Still, correction factors for fast mass transfer based on this simple theory turn out to be reasonably accurate.

The first term on the right-hand side is the flux due to diffusion, and the second term is the flux due to diffusion-induced convection. This equation is subject to the boundary conditions

$$z = 0, \quad c_1 = c_{1i} \tag{11.5-5}$$

$$z = l, \quad c_1 = c_1 \tag{11.5-6}$$

In dilute solution, c_1 is small, and the diffusion-induced convection is negligible; so Eq. 11.5-4 is easily integrated to give Eq. 11.1-2. As a result, the mass transfer coefficient in dilute solution k^0 is D/l.

In concentrated solutions, c_1 is large, and no easy simplifications are possible. However, because n_1 and n_2 are constants, v^0 is as well, and Eq. 11.5-4 can be integrated to give

$$\frac{c_1 - n_1/v^0}{c_{1i} - n_1/v^0} = e^{v^0 l/D} \tag{11.5-7}$$

This equation can be rearranged:

$$N_1 = n_1 \bigg|_{z=0} = \left(\frac{v^0}{e^{v^0 l/D} - 1} \right) (c_{1i} - c_1) + c_{1i} v^0 \tag{11.5-8}$$

If we compare this result with Eq. 11.5-2, we find

$$k = \frac{v^0}{e^{v^0 l/D} - 1} \tag{11.5-9}$$

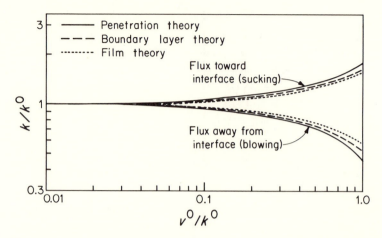

Fig. 11.5-2. Correction factors for rapid mass transfer. This figure gives the mass transfer coefficient k as a function of the interfacial convection v^0. In dilute solution, v^0 is small, and k approaches the slow mass transfer limit k^0. In concentrated solution, k may reach a new value, although estimates of this value from different theories are about the same (the boundary layer theory shown is for a Schmidt number of 1,000).

We then can eliminate the unknown l by using the dilute-solution coefficient k^0 given by Eq. 11.1-2:

$$\frac{k}{k^0} = \frac{v^0/k^0}{e^{v^0/k^0} - 1} \tag{11.5-10}$$

or, in terms of a power series,

$$k = k^0 \left(1 - \frac{v^0}{2k^0} + \frac{(v^0)^2}{12(k^0)^2} - \cdots \right) \tag{11.5-11}$$

Note that the mass transfer coefficient k can be either increased or decreased in concentrated solution. If a large convective flow blows from the interface into the bulk, then v^0 is positive and k is less than k^0. If a large flow sucks into the interface from the bulk, then v^0 is negative and k is greater than k^0. These changes are made clearer by the following example.

Before proceeding to this example, we compare these film results with those found from other mass transfer theories (Bird et al., 1960) in Fig. 11.5-2. All theories give similar results. To be sure, more complex theories provide greater detail in the form of a Schmidt-number dependence, but this is rarely a major factor. The corrections given by the film theory are sufficient in many cases.

Example 11.5-1: Fast benzene evaporation

Benzene is evaporating from a flat plate into pure flowing air. Using the film theory, estimate how much a concentrated solution increases the mass transfer rate beyond that expected for a simple theory. Then calculate

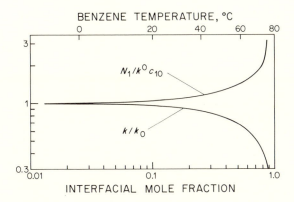

Fig. 11.5-3. Fast benzene evaporation. This figure gives the flux N_1 and mass transfer coefficient k for fast benzene evaporation through stagnant air relative to values expected for slow mass transfer. These estimates are based on the film theory, although other estimates would be similar.

the resulting change in the mass transfer coefficient defined by Eq. 11.5-2. In other words, find N_1/k^0c_{1i} and k/k^0 as a function of the concentration of benzene at the surface of the plate.

Solution. The benzene evaporates off the plate into air flowing normal to the plate. Thus, n_2 is zero. Moreover, if air and benzene behave as ideal gases, $\bar{V}_1 = \bar{V}_2 = c^{-1}$, and $v^0 = n_1/c = N_1/c$. In addition, $c_1 = 0$. Thus, from Eq. 11.5-4,

$$N_1 = n_1 \Big|_{z=0} = \frac{Dc}{l} \ln \left(\frac{c}{c - c_{1i}} \right)$$

In dilute solution, the flux will be Dc_{1i}/l, and k^0 will be D/l; thus,

$$\frac{N_1}{k^0 c_{1i}} = -\frac{1}{x_{1i}} \ln(1 - x_{1i})$$

Values for this flux are given in Fig. 11.5-3. The values of mass transfer coefficient can be found by combining these results with Eq. 11.5-10:

$$\frac{k}{k^0} = \frac{N_{1/ck^0}}{e^{N_1/ck^0} - 1} = -\left(\frac{1 - x_{10}}{x_{10}} \right) \ln(1 - x_{10})$$

Note that k/k^0 is less than unity. Values of this ratio are also given in Fig. 11.5-3.

Section 11.6. Conclusions

By this time, the catalogue of theories given in this chapter may cloud our perspective. We may understand each step and each equation but still be confused about the arguments used. As a result, we can gain insight by stepping back and seeing what we have accomplished.

Table 11.6-1. *Theories for mass transfer coefficients*

Method	Basic form	f(flow)	f(D)	Advantages	Disadvantages
Film theory	$k = \dfrac{D}{l}$	—	1.0	Simple; often good base for ideas	Film thickness l is unknown
Penetration theory	$k = 2\sqrt{Dv^0/\pi L}$	$\frac{1}{2}$	$\frac{1}{2}$	Simplest including flow	Contact time (L/v^0) often unknown
Surface-renewal theory	$k = \sqrt{D/\tau}$	—	$\frac{1}{2}$	Similar math to penetration theory, but better physical picture	Surface-renewal rate (τ) is unknown
Boundary layer theory	$k = 0.626\dfrac{D}{L}\left(\dfrac{Lv^0}{\nu}\right)^{1/2}\left(\dfrac{\nu}{D}\right)^{1/3}$	$\frac{1}{2}$	$\frac{2}{3}$	Much better physical picture	Laminar flow past flat plate of length L only; math hard.
Graetz–Nusselt theory	$k = \left(\dfrac{2}{3\Gamma\left(\frac{4}{3}\right)}\right)\dfrac{D}{L}\left(\dfrac{L^2 v^0}{DR}\right)^{1/3}$	$\frac{1}{3}$	$\frac{2}{3}$	Exact result for short tubes	Valid only in laminar flow

We began this chapter by trying to predict mass transfer coefficients from our knowledge of diffusion. Such an effort can improve our understanding of interfacial mass transfer and give a more complete picture of concentration versus position than is commonly available. However, trying to analyze every different situation is very difficult indeed; so we must make major approximations in the geometry by trying to approximate many physical situations with simple models. Our success depends on our skill in picking models. Those that we have detailed in this chapter are summarized in Table 11.6-1.

Which of the models we used is correct? The answer depends on whether we are pessimistic or optimistic. If we are pessimistic, we conclude that none of the models is right. None gives a simple result that contains no empirical parameters and that is applicable to all physical situations. All are restricted to dilute solution. If we are optimistic, we can conclude that all the models work well. Of course, they work well only when the physical situation is close to that which is assumed, which frankly is what we should expect. The restriction to dilute solution can be removed if we deem this necessary.

However, there is a more general and more powerful conclusion implied by the summary in Table 11.6-1. All the predictions cluster around experimentally observed values. Most predict a Sherwood number that varies with the Reynolds number to about the one-half power and with the Schmidt number to the one-third power. Deviation from this range is not that frequent. Thus, the physical chemistry of interfacial mass transfer seems to forgive our approximations, for all answers are similar.

This is an important result. It does not say that our calculations can replace careful experimental studies of interfacial mass transfer. It does say that we can use simple models to provide insight into what happens in this process. In general, the simplest models, like the film and penetration theories, are the most useful. These successfully predict the changes in mass transfer caused by convection or by chemical reaction. In general, more elaborate approaches, like the boundary layer theory, are difficult to apply and produce minor gains.

References

Bird, R. B., Stewart, W. E., & Lightfoot, E. N. (1960). *Transport Phenomena*. New York: Wiley.

Blasius, H. (1908). *Zeitschrift für Mathematik und Physik,* **56,** 1.

Campbell, J. A., & Hanratty, T. J. (1982). *American Institute of Chemical Engineers Journal,* **28,** 988.

Carberry, J. J. (1976). *Chemistry and Catalytic Reaction Engineering*. New York: McGraw-Hill.

Dankwerts, P. V. (1951). *Industrial and Engineering Chemistry,* **43,** 1460.

Graetz, L. (1880). *Zeitschrift für Mathematik und Physik,* **25,** 316, 375.

Higbie, R. (1935). *Transactions of the American Institute of Chemical Engineers,* **31,** 365.

Jakob, M. (1957). *Heat Transfer,* 3rd ed. New York: Wiley.

Levenspiel, O. (1972). *Chemical Reaction Engineering,* 2nd ed. New York: Wiley.

Lévêque, M. A. (1928). *Annales des Mines,* **13,** 201, 305, 381.

Levich, V. (1962). *Physiochemical Hydrodynamics*. Englewood Cliffs, N.J.: Prentice-Hall.

Lewis, W. K., & Chang, K. C. (1928). *Transactions of the American Institute of Chemical Engineers,* **21,** 127.

Nernst, W. (1904). *Zeitschrift für Physikalische Chemie,* **47,** 52.

Nikuradse, J. (1942). *Laminare Reibungsschichten an der langsangestromten Platte*. Berlin: Zentrale für wiss. Berichtswesen.

Nusselt, W. (1909). *Zeitschrift des Vereines der Deutscher Ingenieure,* **53,** 1750; (1910) **54,** 1154.

Schlichting, H. (1979). *Boundary Layer Theory,* 7th ed. New York: McGraw-Hill.

Sherwood, T. K., Pigford, R. L., & Wilke, C. O. (1975). *Mass Transfer*. New York: McGraw-Hill.

Skelland, A. H. (1974). *Diffusional Mass Transfer*. New York: Wiley Interscience.

12 FREE CONVECTION

Mass transfer can produce density gradients that in turn cause flow. Such a flow is called "free convection," to distinguish it from "forced convection" caused by stirrers, pumps, and the like.

As an example of free convection, imagine that a crystal of salt is held just under the surface of initially stagnant water. The salt dissolves and diffuses away from the crystal, producing a solution near the crystal that is denser than the water itself. This solution flows downward, and the resulting flow increases the rate of mass transfer. Thus, mass transfer and flow couple to accelerate dissolution. Other examples where this coupling is important include such disparate processes as cloud formation, dendrite production, and cooling by sweating.

This short chapter analyzes a few of the interactions between mass transfer and free convection. One important question turns out to be whether free convection occurs at all. This is equivalent to asking if the system is stable. If it is, then free convection does not occur, and the description of mass transfer depends on a diffusion analysis like that in Chapter 2. Such a description can use experiments or mathematical calculations. If the system is unstable, then free convection does occur, and it is more easily described with mass transfer coefficients like those in Chapter 9. This description is usually based on experiment, although it may occasionally use complex numerical calculations.

The first two sections of this chapter are concerned with the question of stability. In Section 12.1 we calculate when free convection will occur in a vertical tube in which a dense solution is diffusing downward. This problem, one of the simplest for which a complete analysis is available, has been used as a method of measuring diffusion coefficients. In Section 12.2 we summarize the cases in which the system's stability regarding free convection has been carefully studied, and we use examples to illustrate the effects involved.

The second important question involving free convection concerns the nature of plumes, like those from a smokestack. We often want to know the size and concentration of such plumes as part of studies of pollution. A simplified analysis of plumes in stagnant surroundings is given in Section 12.3. In this analysis, we emphasize the simplest cases that have been studied, attempting to supply physical insight instead of murky detail.

Section 12.1. Free convection in a vertical tube

One simple example of free convection engendered by diffusion occurs in the apparatus shown schematically in Fig. 12.1-1(a). The apparatus consists of two well-stirred reservoirs. The upper reservoir contains a dense solution, but the lower one is filled with less dense solvent. Because solution and solvent are miscible, solute diffuses from the upper reservoir into the lower one.

We want to know if the difference in densities between solution and solvent causes flow. From our experience, we expect that flow will occur if the tube diameter is very large. After all, gin tends to rise to the surface of a summer's gin-and-tonic without completely mixing, and vinegar falls below oil in salad dressing. Intuitively, we expect that such flows will cease if the tube diameter becomes very small. More speculatively, we might guess that whether or not flow occurs depends inversely on viscosity; high viscosity means less chance of flow.

To analyze this problem more completely, we write a mass balance on the solute, an overall mass balance on all species present, and a momentum balance to describe the flow. We then imagine small perturbations in the concentration or in the flow. If our balances indicate that these small perturbations get smaller with time, then the system is stable. If these perturbations grow with time, then the system is unstable, and free convection will occur.

We first write these balances for the unperturbed system in which no free convection exists (Wooding, 1959). These are

$$0 = D\nabla^2 \bar{c}_1 - \bar{\mathbf{v}} \cdot \nabla \bar{c}_1 \tag{12.1-1}$$

$$0 = -\nabla \cdot \bar{\mathbf{v}} \tag{12.1-2}$$

$$0 = -\nabla \bar{p} + \bar{\rho}\mathbf{g} \tag{12.1-3}$$

where the overbars refer to the unperturbed system. The solution of these equations for the situation shown in Fig. 12.1-1 is that expected:

$$\frac{\bar{c}_1 - c_{10}}{c_{1l} - c_{10}} = \frac{\bar{\rho} - \rho_0}{\rho_l - \rho_0} = \frac{z}{l} \tag{12.1-4}$$

$$\bar{\mathbf{v}} = 0 \tag{12.1-5}$$

$$\bar{p} = p_0 + \int_0^z \bar{\rho}g \, dz \tag{12.1-6}$$

Although we do not need the details of these solutions in the following, I find them reassuring.

The corresponding equations for an incompressible but perturbed system are

$$\frac{\partial c_1}{\partial t} = D\nabla^2 c_1 - \mathbf{v} \cdot \nabla c_1 \tag{12.1-7}$$

$$0 = -\nabla \cdot \mathbf{v} \tag{12.1-8}$$

$$\rho \frac{\partial \mathbf{v}}{\partial t} = \mu\nabla^2\mathbf{v} - \nabla p + \rho\mathbf{g} \tag{12.1-9}$$

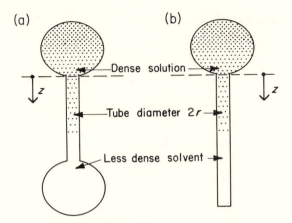

Fig. 12.1-1. Free convection in a vertical tube. In both cases shown, a dense solution is diffusing downward into a less dense solute. If the density difference is large, the fluids will flow together quickly. The problem is to calculate the maximum density difference that can be tolerated without convective flow. Case (a) is that analyzed in the text; case (b) has been used as a way of determining diffusion coefficients.

We now rewrite these relations in terms of the perturbations themselves. For example, for the mass balance, we define

$$c_1 = \bar{c}_1 + c_1' \qquad (12.1\text{-}10)$$

$$\mathbf{v} = \mathbf{v}' \qquad (12.1\text{-}11)$$

where the primes signify perturbations from stable values of Eqs. 12.1-4 through 12.1-6. Remember that the stable value of the velocity is zero. We then insert these definitions in Eq. 12.1-7, subtract Eq. 12.1-1, and neglect terms involving the squares of perturbations:

$$\frac{\partial c_1'}{\partial t} = D\nabla^2 c_1' - v_z' \frac{d\bar{c}_1}{dz} \qquad (12.1\text{-}12)$$

This procedure is like that in Section 4.3. Equation 12.1-12 is subject to the boundary condition that the tube walls are solid:

$$r = R_0, \quad \frac{\partial c_1'}{\partial r} = 0 \qquad (12.1\text{-}13)$$

Similar arguments lead to the modified momentum balance:

$$\left(\mu\nabla^2 - \frac{\partial}{\bar{\rho}\partial t}\right)\mathbf{v}' = -\nabla p' + \rho'\mathbf{g} = -\nabla p' + \mathbf{g}\beta c_1' \qquad (12.1\text{-}14)$$

in which the primed quantities are again perturbations and β $(= \partial\rho/\partial c_1)$ is the density increase caused by the solute. Equation 12.1-14 is subject to the condition

$$r = R_0, \quad v_z' = 0 \qquad (12.1\text{-}5)$$

This says that there is no vertical flow at the wall.

Equations 12.1-12 and 12.1-14 must now be solved simultaneously. A simple solution requires two chief assumptions. The first is that the time derivatives in these equations can be neglected; this is equivalent to the assertion that marginal stability can exist (Wooding, 1959). The second assumption is that the perturbations have their largest effects normal to the z direction; this implies that any convection cells that occur will be long (Chandrasekhar, 1961). I find these assumptions reasonable, but hardly obvious. Because they are justified by experiment, they are tributes to the genius of G. I. Taylor (1954), who had the gall to present the answers to this problem without derivation.

The velocity v'_z calculated from these assumptions is

$$v'_z = \sum_{n=0}^{\infty} \left[A_n J_n\left(\text{Ra}^{1/4}\,\frac{r}{R_0}\right) + B_n I_n \left(\text{Ra}^{1/4}\,\frac{r}{R_0}\right) \right] \cos n\theta \qquad (12.1\text{-}16)$$

where A_n and B_n are constants, characteristics of the nature of any perturbations, J_n and I_n are nth-order Bessel functions, and Ra is a dimensionless ratio called the Rayleigh number:

$$\begin{aligned} Ra &= \left(\frac{gR_0^4}{\mu D}\right)\left(\beta \, \frac{dc_1}{dz}\right) \\ &= \left(\frac{gR_0^4}{\mu D}\right) \frac{d\rho}{dz} \end{aligned} \qquad (12.1\text{-}17)$$

Taylor (1954) found that the first nonzero roots for A and B that are consistent with the boundary conditions in Eqs. 12.1-13 and 12.1-15 occur when

$$\text{Ra} = 67.94 \qquad (12.1\text{-}18)$$

This value corresponds to the case in which solution is falling down one half of the tube and solvent is rising through the other half.

This critical value of the Rayleigh number provides the limit for the stability in the vertical tube. When the density difference is so small that the Rayleigh number is less than 67.94, free convection will not occur. When the density difference is so large that the Rayleigh number exceeds 67.94, then free convection does occur. This result also supports our intuitive speculations at the beginning of this section. The chances for free convection decrease sharply as the tube diameter decreases. They also decrease as the viscosity or the diffusion coefficient increases. In every case, the change of a given variable required to spark free convection can be predicted from this critical Rayleigh number.

Section 12.2. The necessary conditions for free convection

Other cases in which free convection may occur are shown in Table 12.2-1. These results, which include those developed in the previous section, have been verified by experiment. In some cases, like the dissolving sphere, free convection will always occur; in others, like the horizontal slit with the denser solution below, free convection never occurs.

Table 12.2-1. *Free convection under a variety of situations*

Physical situation	Critical Rayleigh number, Ra	Variables	Remarks
Horizontal plates with dense solution below	$\dfrac{l^4 g}{\mu D}\dfrac{d\rho}{dz} = \infty$	l = gap between plates	Free convection never occurs
Horizontal plates with dense solution above	$\dfrac{l^4 g}{\mu D}\dfrac{d\rho}{dz} = 1{,}708$	l = gap between plates	Free convection occurs only when Ra exceeds the value given (Rayleigh, 1916; Chandrasekhar, 1961)
Vertical plates with dense solution on one side	$\dfrac{l^4 g}{\mu D}\dfrac{d\rho}{dz} = 0$	l = gap between plates	Free convection always occurs
Gas absorbing into a horizontal layer of solvent, and producing a dense solution	$\dfrac{l^4 g}{\mu D}\dfrac{d\rho}{dz} = 110$	l = thickness of liquid layer	Problems like this are important in oceanography and meteorology (Chandrasekhar, 1961)
Vertical tube with dense solution below	$\dfrac{R_0^4 g}{\mu D}\dfrac{d\rho}{dz} = \infty$	R_0 = tube radius	Free convection never occurs
Vertical tube with dense solution above	$\dfrac{R_0^4 g}{\mu D}\dfrac{d\rho}{dz} = 67.94$	R_0 = tube radius	Free convection occurs when Ra exceeds the value given (Taylor, 1954)
Vertical tube filled with a porous medium and with a dense solution above	$\dfrac{R_0^2 P g}{\mu D \varepsilon}\dfrac{d\rho}{dz} = 3.39$	R_0 = tube radius P = permeability, defined by $v = (P/\mu)(\nabla p)$ ε = void fraction	This result finds application in geological problems (Wooding, 1959)
Dissolving sphere	$\dfrac{R_0^4 g}{\mu D}\dfrac{d\rho}{dr} = 0$	R_0 = sphere radius	Free convection always occurs

The cases of greatest interest are those in which free convection is possible but not inevitable. Whether or not convection occurs depends on the Rayleigh number Ra, defined as

$$\text{Ra} = \left(\frac{l^4 g}{\mu D}\right)\frac{d\rho}{dz} \tag{12.2-1}$$

where l is some characteristic dimension in the system. Other dimensionless groups are found by experiment to influence the stability, but the Rayleigh number remains by far the most important (Turner, 1973). In physical terms, it can be regarded as the ratio of buoyancy forces tending to cause flow to other processes tending to resist flow. The buoyancy force is related to $l^4 g$ $d\rho/dz$; the other processes are momentum and mass transfer and are signaled by μ and D. Alternatively, the Rayleigh number can be regarded as the ratio of a buoyancy velocity $(l^3 g/\mu)(d\rho/dz)$ to a diffusion velocity D/l (i.e., as a form of Péclet number).

Whatever physical significance is chosen, the Rayleigh number provides the criterion for determining if free convection occurs. We turn to examples to illustrate this.

Example 12.2-1: Using instabilities to measure diffusion coefficients

The apparatus in Fig. 12.1-1(b) and the appropriate equation in Table 12.2-1 can be used in a simple method of measuring diffusion coefficients. To make such a measurement, we set up the apparatus so that free convection will occur. When the convection stops, the Rayleigh number has a value of 67.94 or less everywhere in the tube. All parameters in the equation are easily measured experimentally, except for the diffusion coefficient; thus, this coefficient can be calculated.

Imagine we make such an experiment using a 0.056-cm tube filled with a colored aqueous solution of 0.010-g/cm^3 910-Å polystyrene latex spheres. After one day at 25°C, we find that free convection has stirred the tube's contents to a depth of 5.4 cm. The value of β (= $d\rho/dc_1$) in this system is 0.032; the other solution properties are essentially those of water.

Use these data to calculate the diffusion coefficient of the latex, and compare the result with estimates based on the Stokes–Einstein equation (Eq. 5.2-1).

Solution. From Table 12.2-1 or Eq. 12.1-18, we see that when convection occurs,

$$\frac{gR_0^4}{\mu D}\left(\beta \frac{dc_1}{dz}\right) = 67.94$$

Integrating this relation, we find that

$$D = \frac{gR_0^4 \beta}{67.94\mu}\left(\frac{c_{10}}{L}\right)$$

where c_{10} is the concentration in the reservoir and L is the depth to which free convection has stirred the solution. Inserting the values given,

$$D = \frac{(980 \text{ cm/sec}^2)(0.028 \text{ cm})^4}{67.94(0.0089 \text{ g/cm-sec})} \frac{(0.032 \text{ g/g})(0.010 \text{ g/cm}^3)}{5.4 \text{ cm}}$$

$$= 5.9 \cdot 10^{-8} \text{ cm}^2/\text{sec}$$

From the Stokes–Einstein equation for a solute of radius R_0,

$$D = \frac{k_B T}{6\pi\mu R_0}$$

$$= \frac{(1.38 \cdot 10^{-16} \text{ g-cm}^2/\text{sec}^2\text{-°K})(298°\text{K})}{6\pi(0.0089 \text{ g/cm-sec})(455 \cdot 10^{-8} \text{ cm})}$$

$$= 5.4 \cdot 10^{-8} \text{ cm}^2/\text{sec}$$

In this case, the free-convection experiment is within 10% of the expected value.

However, the method does have a potentially major error: The free convection may overshoot the critical Rayleigh number, mixing the system more than the analysis predicts. To my surprise, the available experiments suggest that this may not happen (Lowell & Anderson, 1982).

Example 12.2-2: Free convection in a diaphragm cell

The diaphragm cell, a popular method of measuring diffusion, consists of two stirred compartments separated by a porous diaphragm. Initially, one compartment is filled with solution and the other with solvent. Diffusion then occurs through the diaphragm. After a known time, the concentration in each compartment is measured and used to calculate the diffusion coefficient. Details of the equipment and the analysis are given in Example 2.2-4.

Over the years, there has been much discussion over the best configuration of these cells. The best experiments most frequently use a horizontal diaphragm with the denser solution in the top compartment (e.g., Robinson & Stokes, 1960). Many good experiments use a horizontal diaphragm with the dense solution in the bottom compartment. Experiments using a vertical diaphragm can given anomalous results.

Discuss these observations in terms of free convection. Make the discussion specific by assuming that the solution is 1-M KCl, the solvent is water, the pores in the diaphragm are $10 \cdot 10^{-4}$ cm in diameter and 0.3 cm long, and the cell has a diameter of 2 cm and a length of 8 cm.

Solution. The three situations involved are shown schematically in Fig. 12.2-1. In case (a), where the dense solution is on top, there is the potential for free convection. This convection depends on the Rayleigh number. In the pores, this is

$$\text{Ra} = \left(\frac{R_0^4 g}{\mu D}\right) \frac{d\rho}{dz}$$

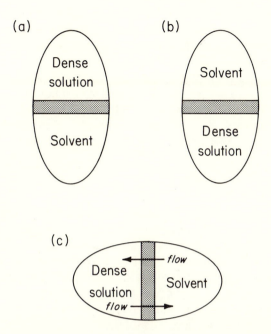

Fig. 12.2-1. Free convection in a diaphragm cell. This simple method of measuring diffusion involves two well-stirred solutions separated by a thin porous diaphragm. When this diaphragm is horizontal, as in (a) or (b), the measurements are much more reliable than when it is vertical, as in (c). Interestingly, measurements with the denser solution on top tend to be best. These generalizations result from free convection, as explained in the text.

$$= \left(\frac{(5 \cdot 10^{-4} \text{ cm})^4(980 \text{ cm/sec}^2)}{10^{-2} \text{ g/cm-sec})(10^{-5} \text{ cm}^2/\text{sec})} \right) \frac{0.046 \text{ g/cm}^3}{0.3 \text{ cm}}$$

$$\doteq 10^{-4}$$

Because this is much less than the critical value in Table 12.2-1, free convection will not occur. In the cell itself,

$$\text{Ra} = \left(\frac{(1 \text{ cm})^4(980 \text{ cm/sec}^2)}{(10^{-2} \text{ g/cm-sec})(10^{-5} \text{ cm}^2/\text{sec})} \right) \frac{0.046 \text{ g/cm}^3}{4 \text{ cm}}$$

$$\doteq 10^{+8}$$

This is much greater than the critical value; so free convection will occur. Thus, free convection augments stirring in the cell's compartments, but does not affect the diffusion in the pores. This is why this cell configuration is most effective.

In case (b), free convection will never occur, because the denser solution is always lower. This is fine in the pores, but it may inhibit stirring in the compartments. This is apparently why data in case (b) are slightly less reliable than in case (a).

In case (c), convection may occur by flow, as shown in the diagram: Fluid flows from right to left through the top of the diaphragm and from left to right through the bottom. If you have trouble visualizing this flow, imagine that the heavier solution has the density of mercury, that the lighter solution is water, and that the diaphragm's pores are large indeed. Obviously the mercury will try to slither under the water. This slithering is less dependent on diffusion than on the viscosity in the pores. Note that such a flow will *always* occur if the solutions have different densities and if the diaphragm is vertical. Unfortunately, the results of many published reports of carefully measured "diffusion coefficients" are compromised by such flows (Toor, 1967).

Section 12.3. Plumes

In this section we discuss the chief characteristics of plumes – fluids discharged at different concentrations and temperatures than the surroundings. One familiar form of a plume is that made by a smokestack. To reduce ground-level pollution at the base of the stack, we routinely discharge hot gases, which rise because they are relatively light. The effective height of our smokestack is thus its actual height plus the distance that the buoyant gases rise. The higher they rise, the more effectively these gases can be dispersed. This increased dispersion is quantitatively described using the effective stack height along with the dispersion coefficients given in Chapter 4.

In this section we estimate how the plume's size, concentration, and velocity change as the plume rises (Morton et al., 1956; Morton, 1971; Csanady, 1973). These estimates include entrainment coefficients, quantities whose function is like that of the mass transfer coefficients in Chapters 9 and 11. Whereas the analysis is certainly approximate, the results allow estimation of the concentration changes produced by changing quantities like the flow rate up the stack.

We first consider a plume like that shown in Fig. 12.3-1. (Although this figure implies that the plume rises, like that from a smokestack, the following arguments apply equally to the falling plumes that can occur in lakes and oceans.) From our experience, we know that the plume gets wider as it rises. Moreover, this rise is turbulent, flapping back and forth. This turbulence is not shown in the figure, and we shall ignore it to discuss only the average concentration.

We consider here only one limiting case, that of a cylindrically symmetric plume rising in steady state without any wind. In this case, the force per time lifting the plume ($\pi \rho_0 Q$) is

$$\pi \rho_0 Q = \int_0^{2\pi} \int_0^\infty v_z (\rho_0 - \rho) g r \, dr \, d\theta \qquad (12.3\text{-}1)$$

where v_z is the plume's velocity, ρ_0 and ρ are the densities of the ambient air and of the plume, respectively, g is the acceleration due to gravity, and r is the radial distance from the plume's center. Because the plume is in steady

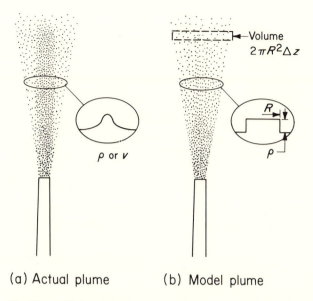

(a) Actual plume   (b) Model plume

Fig. 12.3-1. Plumes. Plumes occur frequently, pouring out of smokestacks or into lakes or oceans. An actual plume has roughly Gaussian concentration and velocity profiles, as suggested by (a). In our approximate model, we assume that these profiles are rectangular pulses, as shown in (b).

state, Q is a constant. The concentration of any solute vapor in the plume is related to the density difference $\rho_0 - \rho$; we expect that this difference gets smaller as the plume rises and that the density or concentration profile has a roughly Gaussian shape. These expectations are borne out by experiment. We also expect that the plume will slow down as it rises, that v_z will vary inversely with height. Because Q is constant, this means that the plume must get broader, which again is consistent with experience.

We want to develop a simple model predicting the plume's size, velocity, and concentration as a function of height. To do so, we assume that the plume consists of a homogeneous core rising through an infinite sink of well-mixed air of constant physical properties. As a result, the equation for Q is simplified:

$$\pi \rho_0 Q = [(\pi R_0^2 v)(\rho_0 - \rho)]g \qquad (12.3\text{-}2)$$

where R_0, v, and ρ are the radius, velocity, and density of the plume's core, functions of height but not of time. The differences between this model and the actual situation can be seen by comparing parts (a) and (b) in Fig. 12.3-1. In the actual situation, the velocity varies with both r and z. In the model, the velocity is a constant within the plume but zero outside of it, implying that turbulent fluctuations are much more important than viscous effects. In the actual situation, the concentration profile is Gaussian. In the model, it is a rectangular pulse that is quaintly called a "top-hat approximation" (Turner, 1973).

We focus on a differential volume of this well-mixed model plume located at z. If we make a volume balance on this volume, we find

$$\begin{pmatrix} \text{accumulation} \\ \text{of volume} \end{pmatrix} = \begin{pmatrix} \text{volume} \\ \text{flow in} \end{pmatrix} - \begin{pmatrix} \text{volume} \\ \text{flow out} \end{pmatrix} + \begin{pmatrix} \text{volume} \\ \text{entrained} \\ \text{by plume} \end{pmatrix} \qquad (12.3\text{-}3)$$

or, in symbolic terms,

$$0 = (\pi R_0^2 v)_z - (\pi R_0^2 v)_{z+\Delta z} + (2\pi R_0 \Delta z)k \qquad (12.3\text{-}4)$$

where k is the mass transfer coefficient characterizing entrainment by the plume of ambient air. Dividing by Δz and taking the limit as Δz goes to zero, we find

$$\frac{d}{dz}(R_0^2 v) = 2R_0 k \qquad (12.3\text{-}5)$$

In a similar way, we can make a momentum balance on this differential volume:

$$\begin{pmatrix} \text{accumulation} \\ \text{of momentum} \end{pmatrix} = \begin{pmatrix} \text{momentum} \\ \text{flow in} \end{pmatrix} - \begin{pmatrix} \text{momentum} \\ \text{flow out} \end{pmatrix} + \begin{pmatrix} \text{buoyancy} \\ \text{forces} \end{pmatrix} \qquad (12.3\text{-}6)$$

or

$$0 = (\pi R_0^2 \rho v^2)_z - (\pi R_0^2 \rho v^2)_{z+\Delta z} + (\pi R_0^2 \Delta z)(\rho_0 - \rho)g \qquad (12.3\text{-}7)$$

Again, we divide by Δz and take the limit as Δz approaches zero:

$$\frac{d}{dz}(\rho R_0^2 v^2) = R_0^2 g(\rho_0 - \rho) \qquad (12.3\text{-}8)$$

Because we expect plume velocity and plume width to be more important than density changes, we approximate this as

$$\rho_0 \frac{d}{dz}(R_0^2 v^2) = R_0^2 g(\rho_0 - \rho) \qquad (12.3\text{-}9)$$

This says that changes in the plume's momentum result only from buoyancy forces.

We now want to calculate the plume's size, velocity, and concentration from Eqs. 12.3-2, 12.3-5, and 12.3-9. To do so, we need two boundary conditions (Seinfeld, 1975). The first is that the plume comes from a small source:

$$z = 0, \quad R_0 = 0 \qquad (12.3\text{-}10)$$

The second condition is that the plume initially has no momentum:

$$z = 0 \quad \rho_0 R_0^2 v^2 = 0 \qquad (12.3\text{-}11)$$

We must also know how the mass transfer coefficient varies as a function of velocity. We expect that more entrainment comes from a higher velocity, and so blithely assume

$$k = \alpha v \qquad (12.3\text{-}12)$$

where the dimensionless quantity α is called an "entrainment coefficient." These equations can now be combined and integrated to find

$$R_0 = \left(\frac{6}{5}\alpha\right) z \tag{12.3-13}$$

$$v = \frac{5}{6\alpha}\left(\frac{9\alpha Q}{10}\right)^{1/3} z^{-1/3} \tag{12.3-14}$$

$$\frac{\rho_0 - \rho}{\rho_0} = \frac{1}{g}\left(\frac{5Q}{6\alpha}\right)\left(\frac{9\alpha Q}{10}\right)^{1/3} z^{-5/3} \tag{12.3-15}$$

The same variations of R_0, v, and ρ with z can be found from dimensional arguments (Batchelor, 1954).

This analysis is obviously approximate, limited to one special case. It can be extended to include the effect of wind perpendicular to the plume or the effect of an inversion ceiling caused by the decrease in the air's density with height. Still, the simple results obtained here are in reasonable agreement with actual observations (Csanady, 1973). As a result, they give a useful way of making quick estimates of a plume's key characteristics.

From our perspective, the interesting aspect of the analysis is the use of a mass transfer coefficient. We have already discussed mass transfer coefficients in free-convection problems in Section 9.3, but those coefficients were for interfacial transport. Here, there is no actual interface; there is only the imaginary boundary between the plume's core and its surroundings. We have already discussed the interaction of the mass transfer coefficient and the velocity in Chapter 11; but there the velocity was a consequence of forced convection, imposed independent of the mass transfer. Here, the velocity and the mass transfer affect each other.

The mass transfer coefficient used here is especially interesting because it is assumed to be independent of the diffusion coefficient and directly proportional to velocity. This independence of diffusion illustrates how a diffusion-based idea can be valuable elsewhere, just as the Gaussian decay of a pulse described in Section 2.4 was successfully applied to dispersion in Section 4.1. The proportionality between mass transfer and velocity reiterates the Reynolds analogy (see Section 17.2), developed 75 years before this theory of plumes. Indeed, the entrainment factor α is a form of Stanton number, and so parallels many theories of simultaneous heat and mass transfer. To check the physical consequences of α, we turn to an example.

Example 12.3-1: Reducing pollutant concentrations

Imagine that the concentration in a plume has dropped 10 times when it is 12 m in diameter and 100 m above a smokestack. How much will it have dropped 400 m above the stack? How much will it be changed at 400 m by adding warm inert gases to the stream so that the flow rate in the stack is doubled but the initial density is unchanged? What will the plume's radius be at 400 m in each case?

Solution. These questions are easily answered using Eqs. 12.3-13 through 12.3-15. From the last of these,

$$(\rho_0 - \rho) \propto z^{-5/3}$$

Thus, the density difference will be reduced by $4^{-5/3}$ times, to 10% of its initial value. If the concentration is reduced in proportion to the density, its value will also be 10 times lower.

If inert gases are added to the plume in a manner described, there will be two effects; the solute concentrations will be cut in half, and the buoyance (represented by Q) will be doubled. From Eq. 12.3-15, we see that

$$(\rho_0 - \rho) \propto Q^{4/3}$$

so $\rho_0 - \rho$ will be 2.5 times greater than with the smaller flow. The net effect is an increase in solute concentration of a factor of 2.5/2, an increase of about 25%. Thus, the plume travels higher before entrainment reduces the density difference. This leads to a larger value of z_0 in expressions like those in Table 4.1-1, and hence a more effective dispersion.

Finally, we turn to the plume's radius. From Eq. 12.3-13, we see that

$$R_0 = (\tfrac{6}{5} \alpha) z$$

$$6 \text{ m} = (\tfrac{6}{5} \alpha) (100 \text{ m})$$

Thus, α is 0.05. As a result, the plume has a diameter of about 40 m at a height of 400 m. Because R_0 is independent of Q, feeding inerts into the stack does not affect this diameter.

Section 12.4. Conclusions

This short chapter discusses two aspects of free convection. The first is whether free convection will in fact occur; answering this question leads to a linear stability theory. Here, diffusion can help to damp out free convection by reducing any density differences involved. The second aspect is the nature of simple plumes; describing these plumes involves use of a mass transfer coefficient across an imaginary interface. This aspect is independent of the diffusion coefficient, but it uses arguments like those for other models of mass transfer.

By making this chapter short, I do not mean to imply that free convection is an uninteresting or underdeveloped subject. It is not. I find it especially fascinating when applied to meteorology, oceanography, and pollution. However, it involves considerably more fluid mechanics than diffusion, and so it is at the edge of the scope of this book. I intend this chapter as an intellectual sorbet in what is already a rich feast of ideas.

References

Batchelor, G. K. (1954). *Quarterly Journal of the Royal Meteorological Society,* **80,** 339.
Chandrasekhar, S. (1961). *Hydrodynamic and Hydromagnetic Stability.* Oxford: Clarendon Press.
Csanady, G. T. (1973). *Turbulent Diffusion in the Environment.* Boston: Kluwer Boston.
Lowell, M. E., & Anderson, J. L. (1982). *Chemical Engineering Communications,* **18,** 93.
Morton, B. R. (1971). *Journal of Geophysical Research,* **76,** 7409.

Morton, B. R., Taylor, G. I., & Turner, J. S. (1956). *Proceedings of the Royal Society of London, Series A,* **234,** 1.

Lord Rayleigh (1916). *Philosophical Magazine,* **32,** 529.

Robinson, R. A., & Stokes, R. L. (1960). *Electrolyte Solutions.* London: Butterworth.

Seinfeld, J. H. (1975). *Air Pollution.* New York: McGraw-Hill.

Taylor, G. I. (1954). *Proceedings of the Royal Society of London, Series B,* **67,** 857.

Toor, H. L. (1967). *Industrial and Engineering Chemistry Fundamentals,* **6,** 454.

Turner, J. S. (1973). *Buoyancy Effects in Fluids.* Cambridge University Press.

Wooding, R. A. (1959). *Proceedings of the Royal Society of London, Series A,* **252,** 120.

PART IV

Diffusion coupled with other processes

13 GENERAL QUESTIONS AND HETEROGENEOUS CHEMICAL REACTIONS

In the previous chapters we have discussed how diffusion involves physical factors. We calculated the gas diffusion through a polymer film, or sized a packed absorption column, or found how diffusion coefficients were related to mass transfer coefficients. In every case, we were concerned with physical factors like the film's thickness, the area per volume of the column's packing, or the fluid flow in the mass transfer. We were rarely concerned with chemical change, except when this change reached equilibrium, as in the dyeing of wool.

In this chapter we begin to focus on chemical changes and their interaction with diffusion. We are particularly interested in cases in which diffusion and chemical reaction occur at roughly the same speed. When diffusion is much faster than chemical reaction, then only chemical factors influence the reaction rate; these cases are detailed in books on chemical kinetics. When diffusion is not much faster than reaction, then diffusion and kinetics interact to produce very different effects.

The interaction between diffusion and reaction can be a large, dramatic effect. It is the reason for stratified charge in automobile engines, where imperfect mixing in the combustion chamber can reduce pollution. It is the reason for the size of a human spermatozoon. It can reduce the size needed for an absorption tower by 100 times. The interaction between diffusion and reaction can even produce diffusion across membranes from a region of low concentration into a region of high concentration.

In this and the two following chapters, we explore interactions between diffusion and chemical reaction. For heterogeneous reactions, we shall find that diffusion and reaction occur by steps in series, steps that can produce results much like mass transfer across an interface. For homogeneous reactions, we shall find that diffusion and reaction occur by steps partially in parallel, steps that are different than processes considered before. In both cases, we shall find that non-first-order stoichiometries lead to unusual results.

We begin with two surprisingly subtle questions. First, in Section 13.1 we discuss whether a chemical reaction is heterogeneous or homogeneous. The answer turns out to be a question of judgment; we must decide which aspects of the chemistry to ignore and choose the type of description leading to the simplest result. This choice leads in Section 13.2 to the concept of a "diffusion-controlled reaction,'' an idea with as many manifestations as Vishnu. After these general questions, we turn to a more explicit discussion of heterogeneous reactions. In Section 13.3 we calculate simplest results for first-order heterogeneous reactions. We extend these results to reactions producing ash in Section 13.4 and to different stoichiometries in Section 13.5. The results supply a synopsis of the heterogeneous results.

Section 13.1. Is the reaction heterogeneous or homogeneous?

We first want to discuss the difference between heterogeneous and homogeneous reactions. On first inspection, this difference seems obvious. Heterogeneous reactions must involve two different phases, with the chemical reactions occurring at the interface. Homogeneous reactions take place in a single phase, and the reaction occurs throughout.

In practice, this distinction is less obvious. As an example, imagine a spherical coal particle burning in a fluidized bed. All the reaction initially takes place on the sphere's surface. This initial reaction is best modeled as heterogeneous. If it were first-order in, for example, oxygen, then the rate equation might be

$$\begin{pmatrix} \text{combustion rate} \\ \text{per particle} \\ \text{area} \end{pmatrix} = \kappa_1 \begin{pmatrix} \text{oxygen concentration} \\ \text{at the surface} \end{pmatrix} \qquad (13.1\text{-}1)$$

in which κ_1 is the heterogeneous rate constant. If the oxygen concentration is in moles per volume, the constant κ_1 has dimensions of length per time, the same dimensions as the mass transfer coefficient.

However, as the combustion proceeds, the particle may become porous, and the chemical reaction may occur not only at the surface but on all pore walls throughout the particle. In some cases the pore area may far exceed the particle's superficial surface area. The combustion is now occurring throughout the particle as if the reaction were homogeneous; its rate is best modeled as

$$\begin{pmatrix} \text{combustion rate} \\ \text{per particle} \\ \text{volume} \end{pmatrix} = \kappa_1 \begin{pmatrix} \text{oxygen concentration} \\ \text{per volume} \end{pmatrix} \qquad (13.1\text{-}2)$$

The oxygen concentration can be defined either per pore volume or per particle volume. In either case, the homogeneous rate constant has units of reciprocal time. Such constants are fixtures of elementary chemistry textbooks (e.g., Laidler, 1963; Hammes, 1978).

Thus, combustion of the coal particle can be modeled as heterogeneous or homogeneous, depending on how the coal is burning. The choice of a model for the reaction is usually subjective, but rarely explicitly stated in the research literature. Instead, it must be inferred. The key for inference is the continuity equation (Aris, 1975; Butt, 1980). If this equation has the form

$$\frac{\partial c_1}{\partial t} = D\nabla^2 c_1 - \nabla \cdot c_1 \mathbf{v}^0 \tag{13.1-3}$$

then any reactions appear in the boundary conditions. Such reactions are being modeled as heterogeneous. If the continuity equation is

$$\frac{\partial c_1}{\partial t} = D\nabla^2 c_1 - \nabla \cdot c_1 \mathbf{v}^0 + r_1 \tag{13.1-4}$$

the reactions r_1 can occur in every differential volume in this system. Such a reaction is being modeled as homogeneous.

In your own work, you must use your judgment about which model is most appropriate. This judgment can be tested with the following spectrum of examples:

1. *Particles of low-grade lead sulfide are roasted to produce a porous lead oxide ash.* At the center of the ore particles is a core of unreacted sulfide. The oxygen permeability in this core is much less than in the ash. Thus, this reaction is best modeled as heterogeneous, occurring at the interface between ore and ash.

2. *Traces of ammonia are scrubbed out of air with water.* The reaction to produce ammonium hydroxide will take place within the liquid phase, and the reaction is best modeled as homogeneous.

3. *Aspirin dissolution.* The aspirin hydrates at the solid surface and then diffuses away into the solution. This case is best treated as a heterogeneous hydration followed by diffusion.

4. *Ethane is dehydrogenated on a single platinum crystal.* The reaction takes place largely on the crystal's surface and is best modeled as heterogeneous.

5. *Ethane is dehydrogenated on a porous platinum catalyst.* Here, the ethane can diffuse through the catalyst pores at a rate similar to diffusion in the surrounding gas. The reaction takes place on pore walls through the catalyst pellet. This reaction is best treated as homogeneous, even though the chemistry is very similar to the previous example.

6. *Sulfur oxides are scrubbed out of stack gas with an aqueous lime slurry.* Here, sulfur oxides and dissolved lime both will have about the same diffusion coefficients in the water. Lime is not very soluble; so its concentration is low. You need to know more chemistry before you can guess where the reaction takes place, and whether it is best modeled as heterogeneous or homogeneous.

We discuss models of heterogeneous reactions in this chapter and models of homogeneous reactions in the next chapter. This artificial division finesses the embarrassing question of which model we should use. This is usually not a major problem. We shall find that the choice between heterogeneous and homogeneous models is often obvious, one that we shall make almost automatically.

Section 13.2. What is a "diffusion-controlled reaction"?

Throughout science and engineering we find references to diffusion-controlled mass transfer and diffusion-controlled chemical reaction. Those using these terms often have very specific cases in mind or are not aware of how broadly and loosely these terms are used. In this short section we want to describe the cases to which these terms most commonly refer. Each of these cases will be analyzed in detail later.

A diffusion-controlled process always involves various sequential steps. For example, the dehydrogenation of ethane on a single platinum crystal involves the diffusion of the ethane to the solid followed by the reaction on the solid surface. The reaction between protons and hydroxyl ions in water first has these species diffusing together and then reacting.

Reactions like these are said to be diffusion-controlled when the diffusion steps take much longer than the reaction steps. Four cases in which this is true are shown in Fig. 13.2-1. For the heterogeneous reaction in Fig. 13.2-1(a), reagent diffuses to the surface; the reagent quickly reacts to form product, and the product diffuses away (Hougen & Watson, 1947; Aris, 1975). The overall rate is determined by the diffusion steps weighted by the equilibrium constant of the surface reaction. Cases like these, exemplified by ethane dehydrogenation on a single catalyst crystal, are the subject of the later sections of this chapter.

A second, very different type of diffusion-controlled reaction occurs in the case of a porous catalyst, shown schematically in Fig. 13.2-1(b) (Carberry, 1976; Satterfield, 1980). Here, the reagent must diffuse into the pores to reach catalytically active sites, where it reacts quickly. The overall rate of reaction depends on this diffusion. This case, central not only to catalysis but also to many scrubbing and extraction systems, is a subject of the next chapter.

Both these processes can be "diffusion-controlled," but the ways in which this control is exerted are very different. One way to see this difference is to examine electrical analogues of the two processes. For the heterogeneous reaction, this analogue is just the three resistors in series shown in Fig. 13.2-2(a). For the porous catalyst, the analogue is the much more elaborate arrangement shown in Fig. 13.2-2(b). Unlike case (a), this combination is not simply a sum of diffusion and reaction steps in series. It has some characteristics of diffusion and reaction in parallel. In my own lectures I have sometimes urged students to think in these helpful but inexact terms.

In each of these cases, the reactions are said to be diffusion-controlled if the resistance to diffusion is much greater than the resistance to reaction. However, these cases will clearly lead to very different combinations of these resistances.

In addition to these two cases, other very different situations can also be diffusion-controlled. Two more examples are the case of facilitated diffusion across membranes and that of reactions controlled by Brownian motion. In the facilitated diffusion case, shown schematically in Fig. 13.2-1(c), one

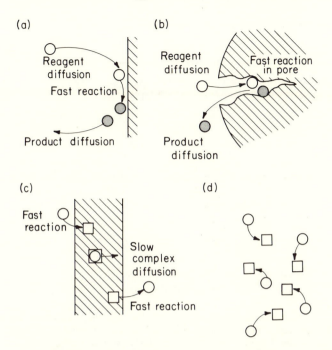

Fig. 13.2-1. Four types of diffusion-controlled reactions. In (a), a reagent slowly diffuses to a solid surface and quickly reacts there; this case occurs frequently in electrochemistry. In (b), a reagent slowly diffuses into the pores of a catalyst pellet, quickly reacting all along the way; this reaction is modeled as if it were homogeneous. In (c), a circular solute quickly reacts with a mobile carrier, thus facilitating the solute's diffusion across the membrane. In (d), two solutes rapidly react by diffusing together after a perturbation caused by fast reaction methods.

solute quickly reacts with a second carrier solute to form a complex; this complex then diffuses across the membrane (Lonsdale, 1982). The overall transport rate is governed by complex diffusion weighted by the equilibrium constant for complex formation. This case is discussed in Section 15.2.

Still another diffusion-controlled process is called in chemistry a "diffusion-controlled reaction," although it is very different from the other reactions described earlier. In this process, shown schematically in Fig. 13.2-1(d), the system is an initially homogeneous mixture of two types of molecules (Bradley, 1975; Pilling, 1975). These species react instantaneously whenever they collide, so that their reaction rate is controlled by their molecular motion, that is, by their diffusion. This process is described in detail in Section 14.4; a similar dispersion-controlled process is described in Section 14.5.

By this point, we should try to find the common thread through this tweed of "diffusion control." The key feature in all these cases is the coupling

Heterogeneous reaction on a flat catalyst surface

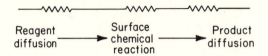

Reagent
diffusion

Surface
chemical
reaction

Product
diffusion

"Homogeneous" reaction within a porous catalyst

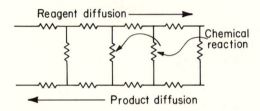

Reagent diffusion

Chemical
reaction

Product diffusion

Fig. 13.2-2. Electrical analogues of two reactions affected by diffusion. These two cases correspond to those shown in Fig. 13.2-1(a) and 13.2-1(b), respectively. If the reactions are diffusion-controlled, then the resistances to chemical reaction will be relatively small.

between chemical kinetics and diffusion. In every case, the overall rate is a function of the diffusion coefficient. Sometimes this rate depends on little else; more frequently, it also includes aspects of chemical dynamics. In either case, the idea of "diffusion control" is obviously indefinite without reference to a more specific situation. Make sure you know which definition is being implied before beginning.

Section 13.3. Diffusion and first-order heterogeneous reactions

After these general concerns, we turn to the analysis of diffusion and heterogeneous chemical reaction. The simplest case is the first-order mechanism shown schematically in Fig. 13.3-1. The reaction mechanism, the standard against which other ideas are measured, depends on three sequential steps. First, reagent diffuses to the surface; second, it reacts reversibly at the surface; finally, product diffuses away from the surface. The first and third steps depend on physical factors like reagent flow and fluid viscosity. The second step depends largely on chemistry, including adsorption and electron transfer.

This particular case occurs with what for me is surprising frequency. The most common practical example is an electrochemical reaction (Bard & Faulkner, 1980). For example, anions diffuse to the anode; these anions react there, and any products diffuse away. Because this kind of reaction

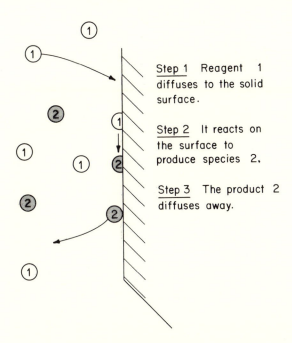

Step 1 Reagent 1 diffuses to the solid surface.

Step 2 It reacts on the surface to produce species 2.

Step 3 The product 2 diffuses away.

Fig. 13.3-1. Diffusion and heterogeneous chemical reaction. The reaction involved is first-order and reversible. The overall reaction rate depends on a sum of resistances, not unlike those involved in interfacial mass transfer.

often takes place in aqueous solution and at moderately high voltage, its rate is often governed by the diffusion steps.

We want to calculate the overall rate of the reaction shown in Fig. 13.3-1. To do so, we must calculate the rate of each of the three steps shown (Aris, 1975). More specifically, at steady state the overall per-area reaction rate r_2 equals the diffusion fluxes:

$$r_2 = n_1 = k_1(c_1 - c_{1i})$$
$$= -n_2 = k_3(c_{2i} - c_2) \tag{13.3-1}$$

in which c_1 and c_2 are the bulk concentrations, c_{1i} and c_{2i} are the concentrations at the surface, and k_1 and k_3 are the mass transfer coefficients of steps 1 and 3 in Fig. 13.3-1. In passing, note that the concentrations c_{1i} and c_{2i} have the same units as c_1 and c_2. For example, they might be moles per cubic decimeter.

The surface reaction, step 2 in Fig. 13.3-1, is first-order:

$$\text{species } 1 \underset{\kappa_{-2}}{\overset{\kappa_2}{\rightleftarrows}} \text{species } 2 \tag{13.3-2}$$

The rate constants κ_2 and κ_{-2} refer to the forward and reverse reactions, respectively. Such a reaction is described by the rate equation

$$r_2 = \kappa_2 c_{1i} - \kappa_{-2} c_{2i} \tag{13.3-3}$$

Because this reaction rate has units of moles per area per time and the concentrations have units of moles per volume, the rate constants both have units of length per time.

Two interesting points in these equations merit emphasis. First, the units used in these equations bother some readers, who feel that a surface reaction should be written in terms of surface concentrations. Such concentrations would have units of moles per area. Some make a fuss over this; they rewrite everything in terms of these surface concentrations, or claim that the surface concentrations are in equilibrium with c_{1i} and c_{2i} in the bulk. Because none of this affects the form of the final result, the argument is tangential.

The second interesting idea comes from comparing Eq. 13.3-3 with Eq. 13.3-1. In Eq. 13.3-3, the forward and reverse reaction rate constants are different. These differences lead to an equilibrium constant for the chemical reaction:

$$K_2 = \frac{\kappa_2}{\kappa_{-2}} \qquad\qquad (13.3\text{-}4)$$

In contrast, in Eq. 13.3-1, the rate constants of the forward and reverse steps are the same. As a result, the equilibrium constant for the diffusion step 1 is

$$K_1 = \frac{k_1}{k_1} = 1 \qquad\qquad (13.3\text{-}5)$$

In other words, diffusion is like a reversible first-order reaction with an equilibrium constant of unity.

We now return to our objective, finding the overall reaction rate. We can measure the bulk concentrations c_1 and c_2, but not the surface concentrations c_{1i} and c_{2i}. Accordingly, we combine Eqs. 13.3-1 and 13.3-3 to eliminate these unknowns. This combination is a complete parallel to that in Section 9.5, simple but algebraically elaborate. The result is

$$r_2 = n_1 = [K]\left(c_1 - \frac{c_2}{K_2}\right) \qquad\qquad (13.3\text{-}6)$$

in which the overall rate constant K is given by

$$K = \frac{1}{1/k_1 + 1/\kappa_2 + 1/k_3 K_2} \qquad\qquad (13.3\text{-}7)$$

This is resistances in series again. It looks just like mass transfer across an interface; the heterogeneous reaction is just another step in the series.

Some parallels between heterogeneous reaction and interfacial mass transfer are obvious. Steps 1, 2, and 3 do occur in series, and resistances like $1/k_1$ and $1/\kappa_2$ do add to the total resistance $1/K$. Moreover, the quantity c_2/K_2 is chemically equivalent to the concentration of species 1 that would be in equilibrium with the existing bulk concentration of species 2. I find this chemical equivalence easier to grasp than the physical one of c_1^* and p_1^*, those elusive pseudo-concentrations that characterize interfacial transport.

Another parallel between heterogeneous reaction and interfacial transport is more subtle. The reaction equilibrium constant K_2 is roughly parallel to the

Henry's law coefficients H that characterize phase changes. Both K_2 and H vary widely, easily covering a range of 10^6 or more for different systems. Thus, it is the equilibrium constraint K_2 that determines the relative impact of k_1 and k_3. The implications of this are best seen in the examples that follow.

Example 13.3-1: Limits of a first-order heterogeneous reaction

What is the overall rate for a first-order heterogeneous reaction under each of the conditions (a) fast stirring, (b) high temperature, and (c) an irreversible reaction? Express this rate as r_2.

Solution. Each of these cases is a limit of Eqs. 13.3-6 and 13.3-7. For rapid stirring, k_1 and k_3 become very large. Thus,

$$r_2 = [\kappa_2] \left(c_1 - \frac{c_2}{K_2} \right)$$

In this case, physics is unimportant and chemistry is omnipotent. For the case of high temperature, κ_2 and κ_{-2} become much larger than k_1 and k_3. Thus,

$$r_2 = \left[\frac{1}{1/k_1 + 1/k_3 K_2} \right] \left(c_1 - \frac{c_2}{K_2} \right)$$

The effect of the reaction is still very much there, but as the equilibrium constant. Finally, for an irreversible reaction, K_2 becomes infinite, and

$$r_2 = \left[\frac{1}{1/k_1 + 1/\kappa_2} \right] c_1$$

Only in this case are the resistances so simply additive. In other cases, these resistances are weighted with equilibrium constants.

Example 13.3-2: The rate of ferrocyanide oxidation

We are studying electrochemical kinetics using a flat platinum electrode 0.3 cm long immersed in a flowing aqueous solution. In one series of experiments, the solution is 1-M KCl containing traces of potassium ferrocyanide. The ferrocyanide is reduced by means of the reaction

$$Fe(CN)_6^{4-} \rightarrow Fe(CN)_6^{3-} + e^-$$

When the solution is flowing at 70 cm/sec and the potential is at some fixed value, the overall mass transfer coefficient is about 0.0087 cm/sec. Estimate the rate constant of this reaction, assuming that the solution has the properties of water.

Solution. The overall mass transfer coefficient K found experimentally is

$$K = \frac{1}{1/k_1 + 1/\kappa_2}$$

We want to find κ_2; so we must calculate k_1. We can do this from Eq. 11.3-43 and the diffusion coefficient in Table 6.1-1:

$$k_1 = 0.626 \left(\frac{D}{L}\right)\left(\frac{Lv}{\nu}\right)^{1/2}\left(\frac{\nu}{D}\right)^{1/3}$$

$$= 0.626 \left(\frac{1.0 \cdot 10^{-5} \text{ cm}^2/\text{sec}}{0.3 \text{ cm}}\right)\left(\frac{(0.3 \text{ cm})(70 \text{ cm}/\text{sec})}{10^{-2} \text{ cm}^2/\text{sec}}\right)^{1/2}\left(\frac{10^{-2} \text{ cm}^2/\text{sec}}{10^{-5} \text{ cm}^2/\text{sec}}\right)^{1/3}$$

$$= 0.96 \cdot 10^{-2} \text{ cm/sec}$$

Inserting this value and that for K into the preceding equation, we find

$$0.0087 \text{ cm/sec} = \frac{1}{\dfrac{\text{sec}}{0.0096 \text{ cm}} + \dfrac{1}{\kappa_2}}$$

$$\kappa_2 = 0.10 \text{ cm/sec}$$

Obviously, other values will be found at other potentials.

This type of experiment can be used to give reliable values of the mass transfer coefficient, but not the rate constant. The reason is that the electrode surface is usually contaminated in some fashion. Instead, electrochemists commonly use the more reliable method of cyclic voltometry, in which the potential is not held constant, but cycled sinusoidally. The results are complex, but they give considerable qualitative information about the chemistry involved (Bard & Faulkner, 1980).

Example 13.3-3: Cholesterol solubilization in bile

Bile is the body's detergent, responsible for solubilization of water-insoluble materials. It is the key to fat digestion and the principal route of cholesterol excretion. Indeed, the failure of bile to effect excretion of available cholesterol is responsible for the formation of cholesterol gallstones.

Pharmacological experiments have shown that gallstones can be dissolved without surgery by feeding patients specific components of bile. These experiments have sparked the study of the dissolution rates of these gallstones. These rates are conveniently studied with a spinning disc of radioactively tagged cholesterol, like that described in Section 3.4. In one experiment with such a disc, the cholesterol dissolution rate was found to be $5.37 \cdot 10^{-9}$ g/cm^2-sec in a solution containing 5 wt% sodium taurodeoxycholate, a 4:1 molar ratio of this bile salt to lecithin, and 0.15-M NaCl. The solubility of cholesterol in this solution is $1.48 \cdot 10^{-3}$ g/cm^3, and the diffusion coefficient is about $2 \cdot 10^{-6}$ cm^2/sec. The disc was 1.59 cm in diameter, spinning rapidly with a Reynolds number of 11,200 and a Schmidt number of 4,250. The kinematic viscosity of this bile is about 0.036 cm^2/sec, and the density of a cholesterol-saturated solution is $1.0 \cdot 10^{-5}$ g/cm^3 greater than bile containing no cholesterol.

Find the rate of the surface reaction, assuming that this reaction is irreversible. Then find the dissolution rate for a 1-cm gallstone in unstirred bile (Tao et al., 1974).

Solution. As in the previous example, we first need to find the mass transfer coefficient in this solution. From Table 9.3-2,

$$k_1 = 0.62 \frac{D}{d} \left(\frac{d^2\omega}{\nu}\right)^{1/2} \left(\frac{\nu}{D}\right)^{1/3}$$

$$= 0.62 \frac{2 \cdot 10^{-6} \text{ cm}^2/\text{sec}}{1.59 \text{ cm}} (11,200)^{1/2}(4,250)^{1/3}$$

$$= 1.34 \cdot 10^{-3} \text{ cm/sec}$$

The overall rate constant is the sum of this resistance and that of the surface reaction:

$$K = \frac{1}{1/k_1 + 1/\kappa_2}$$

$$\frac{5.37 \cdot 10^{-9} \text{ g/cm}^2\text{-sec}}{1.48 \cdot 10^{-3} \text{ g/cm}^3} = \frac{1}{\dfrac{\text{sec}}{1.34 \cdot 10^{-3} \text{ cm}} + \dfrac{1}{\kappa_2}}$$

Thus,

$$\kappa_2 = 3.6 \cdot 10^{-6} \text{ cm/sec}$$

Note that the experimentally measured rate is dominated by the surface reaction.

We now want to use this surface rate constant to find the dissolution rate if this model bile is flowing past a spherical stone in unstirred bile. This bile is probably affected by free convection. Thus, from Table 9.3-2,

$$\frac{k_1 d}{D} = 2 + 0.6 \left(\frac{d^3 \Delta\rho g}{\rho\nu^2}\right)^{1/4} \left(\frac{\nu}{D}\right)^{1/3}$$

$$\frac{k_1(1 \text{ cm})}{2 \cdot 10^{-6} \text{ cm}^2/\text{sec}}$$

$$= 2 + 0.6 \left(\frac{(1 \text{ cm})^3(1.0 \cdot 10^{-5} \text{ g/cm}^3)(980 \text{ cm/sec}^2)}{(1 \text{ g/cm}^3)(0.036 \text{ cm}^2/\text{sec})^2}\right)^{1/4} (4,250)^{1/3}$$

$$k_1 = 3.6 \cdot 10^{-5} \text{ cm/sec}$$

We then can find the overall rate as before:

$$K = \frac{1}{\dfrac{\text{sec}}{3.6 \cdot 10^{-5} \text{ cm}} + \dfrac{\text{sec}}{3.6 \cdot 10^{-6} \text{ cm}}}$$

$$= 3.3 \cdot 10^{-6} \text{ cm/sec}$$

In unstirred bile, the rate is also controlled by reaction.

Section 13.4. Finding the mechanism of irreversible heterogeneous reactions

The simple ideas of the previous section are the benchmark of this entire chapter, the standard against which more complex concepts are compared. Our basic strategy was to assume a simple mechanism and calculate the

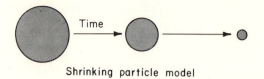

Shrinking particle model

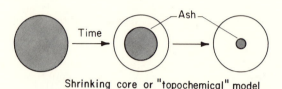

Shrinking core or "topochemical" model

Fig. 13.4-1. Two basic models for heterogeneous reaction. Many solid–gas non-catalytic reactions follow one of these two limiting models. Note that "ash" can be any solid product. Some characteristics of these models are given in Table 13.4-1.

overall reaction rate for this situation. The result was a close parallel to the results for interfacial mass transfer.

However, in some ways, the strategy of the previous section is misleading, for it implies that the mechanism is known. This is often not the case. In many situations, we already have experimental results, and we want to find which mechanisms are consistent with these results. In other words, the arguments of the previous section are backward.

In this section we want to explore how the mechanism of an irreversible reaction can be found from the overall reaction rate, instead of the other way around. This exploration can be very complicated, hampered by elaborate algebra. As a result, we consider only special cases of the two types of heterogeneous reactions shown in Fig. 13.4-1. These types differ in the products produced by the reaction. In some cases, these products are fluid, and hence diffuse away. More commonly, the products form a layer of ash around an unreacted core; this second case is sometimes called a "topochemical model."

To reduce algebraic complexity, we consider only limits in which the overall rate is controlled by a single diffusion or reaction step (Levenspiel, 1972; Glassman, 1977). The five limits we consider are tabulated in Table 13.4-1. In cases A and C, we assume that the surface reaction

$$\begin{pmatrix} \text{gaseous} \\ \text{species 1} \end{pmatrix} + \begin{pmatrix} \text{solid} \\ \text{species 2} \end{pmatrix} \rightarrow \begin{pmatrix} \text{various} \\ \text{products} \end{pmatrix} \qquad (13.4\text{-}1)$$

is described by the rate equation

$$r_2 = -\kappa_2 c_2 c_1 \qquad (13.4\text{-}2)$$

Note that the concentration of the solid, c_2, is essentially constant, so that the reaction behaves as if it were first-order. In such an equation, the rate

Table 13.4-1. *Common types of heterogeneous reactions*

Physical situation	Rate-controlling step	Size = f(time, reagent)	Size = f(temperature)	Size = f(flow)	Remarks
A Shrinking particle	Reaction	$r \propto c_1 t$	Strong temperature variation	Independent of flow	Other reaction stoichiometries can be found easily
B Shrinking particle	External diffusion	$r \propto (c_1 t)^{1/2}$ (small particles) $r \propto (c_1 t)^{3/4}$ (larger particles)	Weak	Independent for small particles only	The exact variation with flow depends on the mass transfer coefficient
C Shrinking core[a]	Reaction	$r \propto c_1 t$	Strong temperature variation	Independent of flow	This is the same as case A, except for ash formation
D Shrinking core[a]	External diffusion	$r \propto c_1 t$	Weak	Usually about square root of flow	This case is uncommon
E Shrinking core[a]	Ash diffusion	$r \propto (c_1 t)^{1/2}$	Weak	Independent of flow	This case is common, an interesting contrast with the previous one

[a] This is often called the "topochemical model."

constant κ_2 has dimensions of $L^4/mol\text{-}t$. We now can write a mass balance on one particle of radius r:

$$\frac{d}{dt}\left(\frac{4}{3}\pi r^3 c_2\right) = 4\pi r^2 r_2 \tag{13.4-3}$$

Combining the previous two equations,

$$\frac{d}{dt} r = -\kappa_2 c_1 \tag{13.4-4}$$

This equation, subject to the conditions that c_1 is constant and r is initially R_0, is easily integrated:

$$r = R_0 - \kappa_2 c_1 t \tag{13.4-5}$$

The particle size is proportional to time, whether the reaction produces gaseous or solid products.

The other three cases are more interesting. When diffusion outside of a shrinking particle is rate-controlling (case B), the key variable becomes the mass transfer coefficient. This coefficient is a function of particle size; for example, for a single particle, it is often assumed to be (see Table 9.3-2)

$$\frac{kd}{D} = 2.0 + 0.6 \left(\frac{dv}{\nu}\right)^{1/2}\left(\frac{\nu}{D}\right)^{1/3} \tag{13.4-6}$$

where d is the particle diameter, D is the diffusion coefficient of the reacting gas, v is the fluid's velocity, and ν is the fluid's kinematic viscosity. For very small particles, this implies

$$k = \frac{D}{r} \tag{13.4-7}$$

a relation derived in Section 2.4. The mass balance on a single particle is now

$$\frac{d}{dt}\left(\frac{4}{3}\pi r^3 c_2\right) = -(4\pi r^2)\frac{Dc_1}{r} \tag{13.4-8}$$

Again, this can be integrated, with the condition that r is initially R_0, to give

$$r^2 = R_0^2 - \left(\frac{2Dc_1}{c_2}\right)t \tag{13.4-9}$$

For larger particles, the mass transfer coefficient is

$$k = \left(\frac{0.42v^{1/2}D^{2/3}}{\nu^{1/6}}\right)\frac{1}{r^{1/2}} \tag{13.4-10}$$

and the resulting variation of r is

$$r^{3/2} = R_0^{3/2} - \left(\frac{0.64v^{1/2}D^{2/3}}{\nu^{1/6}}\right)\left(\frac{c_1}{c_2}\right)t \tag{13.4-11}$$

Equations 13.4-9 and 13.4-11 are both for a diffusion-controlled reaction on the surface of a shrinking particle.

The results are different for a particle of constant size with a shrinking core of unreacted material. In such a topochemical model, there are two diffusional resistances in series. First, material must diffuse from the bulk to the particle's surface; second, it must diffuse from the surface through ash to the unreacted core.

When diffusion in the bulk controls (case D), the reaction rate is again determined by the mass transfer coefficient around the particle. Because the particle size is constant, this mass transfer coefficient is also constant. The mass balance is still

$$\frac{d}{dt}\left(\frac{4}{3}\pi r^3 c_2\right) = -(4\pi r^2)kc_1 \tag{13.4-12}$$

This can be easily integrated using the same initial size of particle R_0:

$$r = R_0 - \left(k\frac{c_1}{c_2}\right)t \tag{13.4-13}$$

Although this result has the same variation with time as do the cases where surface reaction controls, it shows a dependence on flow. It also shows a much smaller variation with temperature, for mass transfer coefficients vary much less with temperature than reaction-rate constants.

When ash diffusion controls (case E), the mass transfer coefficient depends on the thickness of the ash layer. The usual assumption is that this coefficient is

$$k = \frac{D}{R_0 - r} \tag{13.4-14}$$

in which D is now an effective value for diffusion through the ash. This assumption is tricky, for it implies that diffusion across the ash layer is a steady-state process. At the same time, we are assuming that the particle size varies with time. These assumptions imply that the diffusion through the ash is much faster than the combustion of the entire particle.

The mass balance on the particle now becomes

$$\frac{d}{dt}\left(\frac{4}{3}\pi r^3 c_2\right) = -(4\pi r^2)\frac{Dc_1}{R_0 - r} \tag{13.4-15}$$

Integrating this result yields, after some rearrangement,

$$r = R_0 - \left(\frac{2Dc_1}{c_2}t\right)^{1/2} \tag{13.4-16}$$

The particle size now varies with the square root of time. The differences between these cases and others in this section are considered in the example that follows.

Example 13.4-1: Mechanisms of coal gasification with steam

The gasification of coal particles is being studied in a batch fluidized-bed reactor. In this reactor, the steam concentration is held constant, and the average particle size is monitored versus time. A plot of the logarithm of this size versus time has a slope of 0.79; the slope does show some variation with the gas flow used to maintain fluidization. What is the rate-controlling step for this gasification?

Solution. The variation of reaction rate with flow indicates that the process is controlled by diffusion outside of the particle. By inference, the resistance of any ash formed is apparently negligible. At the same time, the variation of particle radius with time is in the range suggested by case B in Table 13.4-1; checking this point further requires using a mass transfer correlation in fluid beds parallel to Eq. 13.4-6. Although no dependence on temperature is mentioned, we would expect this dependence to be small.

Section 13.5. Heterogeneous reactions of unusual stoichiometries

In the previous sections we discussed how the overall rate of reaction was affected by the rates of diffusion and reaction. The rate of diffusion could be altered by changes in factors like fluid flow or diffusion coefficient or ash thickness. The overall rate of reaction was always assumed to be first-order, always doubling when the reagent concentration was doubled.

In this section we want to consider two examples of other chemistry that can alter the simple combinations of diffusion and reaction found before. The first example is an irreversible second-order reaction. The second involves fast reactions of concentrated reagents and products.

A second-order heterogeneous reaction

The case considered here, shown schematically in Fig. 13.5-1, involves two sequential steps. The first of these steps is simple mass transfer, but the second step is a second-order irreversible chemical reaction. As before, we want to calculate the overall rate r_1 of this heterogeneous reaction. In mathematical terms, this rate is

$$r_1 = n_1 = k_1(c_1 - c_{1i}) \tag{13.5-1}$$

where c_1 and c_{1i} are again the bulk and interfacial concentrations, respectively. This rate is also

$$r_1 = \kappa_2 c_{1i}^2 \tag{13.5-2}$$

We can combine these two equations to find the unknown interfacial concentration:

$$c_{1i} = \frac{k_1}{2\kappa_2}(\sqrt{1 + 4\kappa_2 c_1/k_1} - 1) \tag{13.5-3}$$

This can, in turn, be used to find the overall reaction rate:

$$r_1 = k_1 c_1 \left[1 - \frac{k_1}{2\kappa_2 c_1}(\sqrt{1 + 4\kappa_2 c_1/k_1} - 1)\right] \tag{13.5-4}$$

This expression is obviously very different from the corresponding result for a first-order reaction, given by Eq. 13.3-6.

The conclusion drawn from this result is that resistances in series are no longer additive. Indeed, this is true whenever any of the resistances are not first-order. In electrical or thermal systems, resistances are almost always first-order. However, in chemical systems, resistances will not be first-order when there are non-first-order chemical reactions. This occurs frequently.

Heterogeneous reactions in concentrated solutions

In the simple cases in Sections 13.3 and 13.4, we assumed that the mass transfer coefficient k_1 was independent of the reaction rate. This is

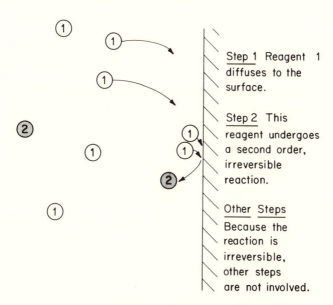

Fig. 13.5-1. A second-order heterogeneous reaction. When the simple stoichiometry used earlier in this chapter is not followed, the overall rate no longer depends on a simple sum of the resistances of the various steps.

actually an implicit approximation, valid only in dilute solution or for reagents producing a mole of product for every mole of reagent. To see where this approximation might be inaccurate, consider the two gas–solid reactions shown in Fig. 13.5-2. In the first, a reagent is split into many smaller parts:

$$\text{species } 1 \xrightarrow{\kappa_2} m(\text{species 2}) \tag{13.5-5}$$

In the second, the converse occurs:

$$m(\text{species 1}) \xrightarrow{\kappa_2} \text{species 2} \tag{13.5-6}$$

The first of these is an idealization of a cracking reaction, and the second of a reforming reaction.

The difficulty in both cases is that the mass transfer step must include both diffusion and convection. In the case of the cracking reaction, this convection is away from the surface, so that the reacting species must diffuse against the current, swimming upstream to reach the reactive surface. For the reforming reaction, the opposite is true; the reacting species are buoyed along, swept toward the surface by the reaction.

To estimate the size of this effect, we idealize the region near the reactive surface as a thin stagnant gas film of thickness l, as in Section 11.5. At the outside of this film, located at $z = 0$, the reagent concentration is the bulk value, and the product concentration is zero. At the solid surface, at $z = l$, the reaction occurs. The overall rate of reaction r_1 in this film is

$$r_1 = n_1 = -\frac{n_2}{m} = \kappa_2 c_{1i} \tag{13.5-7}$$

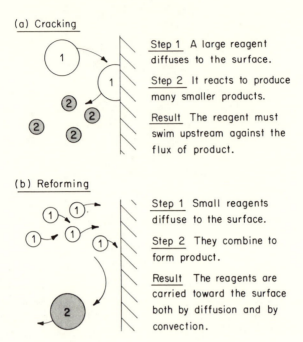

(a) Cracking

Step 1   A large reagent diffuses to the surface.

Step 2   It reacts to produce many smaller products.

Result   The reagent must swim upstream against the flux of product.

(b) Reforming

Step 1   Small reagents diffuse to the surface.

Step 2   They combine to form product.

Result   The reagents are carried toward the surface both by diffusion and by convection.

Fig. 13.5-2. Diffusion-induced convection can alter heterogeneous reaction rates. In concentrated solutions, unusual stoichiometry can alter the rate of mass transfer and hence change the overall rate of reaction.

If the stoichiometric coefficient m is greater than unity, we have the cracking reaction; if m is less than unity, we have the reforming reaction; if m equals unity, we have the simple case discussed in Section 13.3.

We now must calculate the flux n_1. In dilute solutions, this hinged on Eq. 13.3-1; but in these concentrated solutions, we must return to the continuity equation

$$0 = -\frac{dn_1}{dz} \tag{13.5-8}$$

Integrating and combining with Fick's law,

$$n_1 = -D\frac{dc_1}{dz} + c_1 v^0 \tag{13.5-9}$$

For a gas,

$$cv^0 = n_1 + n_2 = (1 - m)n_1 \tag{13.5-10}$$

where c is the total concentration. Inserting this result into the previous equation and integrating to find n_1, we have

$$r_1 = n_1 = -\frac{k_1 c}{(m - 1)} \ln\left(\frac{1 + (m - 1)n_1/\kappa_2 c}{1 + (m - 1)c_{10}/c}\right) \tag{13.5-11}$$

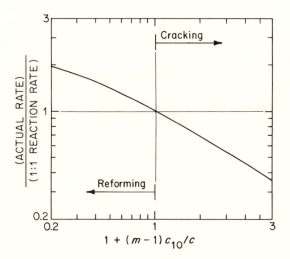

Fig. 13.5-3. Effect of diffusion-induced convection. This graph gives the change in reaction rate for a diffusion-controlled reaction producing m moles of product per mole of reagent. The product concentration in the bulk is zero; that of the reagent is c_{10}.

in which k_1 ($= D/l$) is the mass transfer coefficient. The rate relative to that in dilute solution is shown for a variety of stoichiometries in Fig. 13.5-3 for the limit of diffusion control (i.e., $\kappa_2 c_{10}/n_1$ is large).

Example 13.5-1: Limiting behavior of a second-order heterogeneous reaction

Describe what happens to Eq. 13.5-4 if the dimensionless group $\kappa_2 c_1/k_1$ is either very large or very small.

Solution. When $\kappa_2 c_1/k_1$ is very large, the reaction will become diffusion-controlled. Under these conditions,

$$\lim_{\kappa_2 c_1/k_1 \to \infty} r_1 = k_1 c_1 \left\{ 1 - \frac{k_1}{2\kappa_2 c_1} \sqrt{4\kappa_2 c_1/k_1} \right\} = k_1 c_1$$

On the other hand, when $\kappa_2 c_1/k_1$ is small, the reaction will be unaffected by diffusion:

$$\lim_{\kappa_2 c_1/k_1 \to 0} r_1 = k_1 c_1 \left\{ 1 - \frac{k_1}{2\kappa_2 c_1} \left[1 + \frac{1}{2}\left(\frac{4\kappa_2 c_1}{k_1}\right) - \frac{1}{8}\left(\frac{4\kappa_2 c_1}{k_1}\right)^2 \right.\right.$$
$$\left.\left. + \frac{1}{16}\left(\frac{4\kappa_2 c_1}{k_1}\right)^3 - \cdots 1 \right] \right\}$$
$$= \kappa_2 c_1^2 \left(1 - \frac{2\kappa_2 c_1}{k_1} + \cdots \right)$$

In this case, the chemical reaction controls the overall rate.

Example 13.5-2: Thermal cracking of gas oil

Thermal cracking of a gas oil is being studied on a hot plate immersed in a rapidly flowing gas stream. The plate is so hot that the reaction is essentially diffusion-controlled. The molecular weight of the product is only 23% of that of the reagent. By how much will convection introduced by cracking change the reaction rate?

Solution. In this case, the change in molecular weight implies that

1 molecule of gas oil $\rightarrow$ 1/0.23 molecules of smaller size

Moreover, because the gas oil is undiluted, c_{10}/c is unity. Thus, from Fig. 13.5-3, the actual rate will be about 45% of that expected from reactions or mass transfer coefficients measured in dilute solution.

Section 13.6. Conclusions

This chapter has two principal parts. In the first part (Sections 13.1 and 13.2), the focus is on the modeling of systems containing reaction and diffusion. One question concerns whether a reaction is heterogeneous or homogeneous; the answer depends more on the physical geometry involved and less on the chemistry at a molecular level. A second question concerns what diffusion-controlled reactions are; the answer is that they are reactions in which the time for diffusion is much more than that for chemical change. However, how these times are combined depends on the specific situation involved.

The second part of the chapter (Sections 13.3 through 13.5) is concerned with heterogeneous reactions. The key point is that the overall rate frequently varies with a sum of resistances in series. The results are similar to those involved in interfacial mass transport, but with chemical equilibrium constants replacing the Henry's law constants. Although this simplicity can be compromised by funny stoichiometry or by concentrated solutions, the analogy with interfacial mass transfer is useful and worth remembering.

References

Aris, R. (1975). *The Mathematical Theory of Diffusion and Reaction in Permeable Catalysts*. Oxford: Clarendon Press.

Bard, A. J., & Faulkner, L. R. (1980). *Electrochemical Methods*. New York: Wiley.

Bradley, J. N. (1975). *Fast Reactions*. Oxford: Clarendon Press.

Butt, J. B. (1980). *Reaction Kinetics and Reactor Design*. Englewood Cliffs, N.J.: Prentice-Hall.

Carberry, J. J. (1976). *Chemical and Catalytic Reaction Engineering*. New York: McGraw-Hill.

Glassman, I. (1977). *Combustion*. London: Academic Press.

Hammes, G. G. (1978). *Chemical Kinetics*. New York: Academic Press.

Hougen, O. A., & Watson, K. M. (1947). *Kinetics and Catalysis*. New York: Wiley.

Laidler, K. J. (1963). *Reaction Kinetics*. London: Pergamon Press.

Levenspiel, O. (1972). *Chemical Reaction Engineering,* 2nd ed. New York: Wiley.

Lonsdale, H. K. (1982). *Journal of Membrane Science,* **10,** 81.

Pilling, M. J. (1975). *Reaction Kinetics*. Oxford: Clarendon Press.

Satterfield, C. N. (1980). *Heterogeneous Catalysis in Practice*. New York: McGraw-Hill.

Tao, J. C., Cussler, E. L., & Evans, D. F. (1974). *Proceedings of the National Academy of Sciences of the United States of America,* **71,** 3917.

14 HOMOGENEOUS CHEMICAL REACTIONS

Diffusion rates can be tremendously altered by chemical reactions. Indeed, these alterations are among the largest effects discussed in this book, routinely changing the mass fluxes by orders of magnitude. The effects of a chemical reaction depend on whether the reaction is homogeneous or heterogeneous. This question can be difficult to answer. In well-mixed systems, the reaction is heterogeneous if it takes place at an interface and homogeneous if it takes place in solution. In systems that are not well mixed, diffusion clouds this simple distinction, as detailed in Section 13.1.

The overall rate of a homogeneous reaction is determined by a nonlinear combination of effects of diffusion and chemical reaction. The effects of such a reaction on the rates of mass transfer are analyzed in the first two sections of this chapter. In Section 14.1 we describe the simplest case, that of a first-order irreversible chemical reaction. We also summarize extensions of this case, extensions that produce significant gains only at the cost of major effort. In Section 14.2 we describe some results for second-order reactions.

In Section 14.3 we turn to the effect of diffusion on the rate of a chemical reaction in a porous catalyst. This effect occurs when reagents diffuse into the catalyst interior more slowly than they can react. As a result, the rate of chemical reaction is less than it would be if all of the catalyst's surface were exposed. Such decreases in reaction rate often compromise the performance of industrial reactors, and so they are of commercial importance. Note that these heterogeneous reactions are treated mathematically as if they were homogeneous.

The last two sections in this chapter are concerned with reactions commonly described as "fast" or "diffusion-controlled." In Section 14.4 we discuss chemical reactions whose rates are controlled not by chemical kinetics but by Brownian motion of the reagents. These reactions are studied by suddenly changing the temperature or pressure and measuring the decay of the resulting perturbation. In Section 14.5 we investigate the speed of second-order reactions in turbulent flow. If these reactions are fast, their reaction rates are determined not by chemical kinetics but by the turbulent dispersion summarized in Chapter 4. These reactions can be described with mathematics like those for diffusion, and so are best treated here.

346

Section 14.1. Mass transfer with first-order chemical reactions

Chemical reaction increases the rate of interfacial mass transfer. The reaction reduces the reagent's local concentration, thus increasing its concentration gradient and its flux. Moreover, because chemical reaction rates can be very fast, the increase in mass transfer can be large.

In this section we want to calculate the increased mass transfer caused by a first-order, irreversible chemical reaction. This special case is the limit with which more elaborate calculations are compared. As a result, we shall go over the calculation in considerable detail so that its nuances will be explicitly stated.

One might wonder why we make such a fuss over first-order reactions. After all, these reactions are uncommon. Real chemical reactions involve two reagents, like sodium hydroxide plus hydrochloric acid, or methane plus oxygen. This focus on first-order reactions may seem a scientific ploy, emphasizing problems we can solve rather than problems that are important.

This skepticism has some justification, for there certainly are important reactions that are not first-order. However, in many cases, all but one of the reagents will be present in excess; in stoichiometric terms, only one of the reagents is limiting. In this case, we can accurately approximate the reactions as first-order. For the examples given earlier, we might have

$$\begin{pmatrix} \text{reaction} \\ \text{rate} \end{pmatrix} = [\kappa_1 c_{HCl}]c_{NaOH} \qquad (14.1\text{-}1)$$

for excess hydrogen chloride, or

$$\begin{pmatrix} \text{reaction} \\ \text{rate} \end{pmatrix} = [\kappa_1 c_{NaOH}]c_{HCl} \qquad (14.1\text{-}2)$$

for excess sodium hydroxide, or

$$\begin{pmatrix} \text{reaction} \\ \text{rate} \end{pmatrix} = [\kappa_1 c_{O_2}]c_{CH_4} \qquad (14.1\text{-}3)$$

for excess oxygen. In each case, the quantity in square brackets is a pseudo-first-order reaction-rate constant, with, of course, a different numerical value in each case. Under the circumstances given, each of these reactions can be treated as if it were first-order.

We are interested in how any first-order chemical reaction alters the mass transfer in industrial equipment. For example, imagine that we are scrubbing ammonia out of air with water, using equipment like that shown in Fig. 10.1-1. To increase our equipment's capacity, we are considering adding small amounts of hydrogen chloride to the water. We want to predict the effect of this acid. However, the a priori prediction of mass transfer in a scrubber is a tremendously difficult problem, requiring a rediculously expensive numerical calculation.

As a result, we are much better off to use existing experimental correlations for mass transfer without reaction and to calculate a correction factor for the chemical reaction. Calculating this correction turns out to be easy for

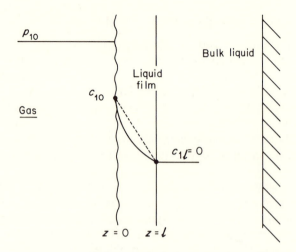

Fig. 14.1-1. Mass transfer with first-order chemical reaction. Species 1 is being absorbed from a gas into a liquid. If this species reacts in the liquid, its concentration profile changes from the dashed line to the solid line. Such a reaction increases the rate of mass transfer. The picture shown here implies the film theory.

a first-order system. Moreover, we make good use of the 50 years of empirical correlations carefully obtained for industrial equipment.

Irreversible reactions

To calculate the correction to mass transfer due to reaction, we again adopt the simple film model shown in Fig. 14.1-1 (Hatta, 1928, 1932). In this model, a liquid is in contact with a well-mixed gas containing the material to be absorbed. The liquid is not well mixed. Near its surface, there is a thin film across which the absorbing species 1 is diffusing steadily. At the gas–liquid interface, this solute species is in equilibrium with the gas; at the other side of the film, its concentration is zero.

We can easily write a mass balance on this film. If there is no chemical reaction, this is

$$0 = -\frac{d}{dz} n_1 \doteq -\frac{dj_1}{dz} = D \frac{d^2c_1}{dz^2} \tag{14.1-4}$$

This is subject to the boundary conditions

$$z = 0, \quad c_1 = c_{1i} \tag{14.1-5}$$
$$z = l, \quad c_1 = 0 \tag{14.1-6}$$

Integration and evaluation of the diffusion flux is easy, just as it was in Sections 2.2 and 11.1:

$$c_1 = c_{1i}\left(1 - \frac{z}{l}\right) \tag{14.1-7}$$

as shown by the dotted line in Fig. 14.1-1. The flux is

$$j_1 = \frac{D}{l}(c_{1i} - 0) \tag{14.1-8}$$

The mass transfer coefficient is the same old friendly value:

$$k^0 = \frac{D}{l} \tag{14.1-9}$$

where the superscript 0 indicates no chemical reaction. Note that we have again implicitly made the familiar assumption of dilute solution.

However, if there is a first-order chemical reaction, the mass balance becomes

$$0 = D\frac{d^2c_1}{dz^2} - \kappa_1 c_1 \tag{14.1-10}$$

Integration of this equation gives

$$c_1 = ae^{\sqrt{\kappa_1/D}\,z} + be^{-\sqrt{\kappa_1/D}\,z} \tag{14.1-11}$$

Evaluation of the integration constants a and b using the boundary conditions in Eqs. 14.1-5 and 14.1-6 gives

$$\frac{c_1}{c_{1i}} = \frac{\sinh[\sqrt{\kappa_1/D}\,(l - z)]}{\sinh[\sqrt{\kappa_1/D}\,l]} \tag{14.1-12}$$

This concentration profile is curved like the solid line in Fig. 14.1-1. Note that

$$\lim_{\kappa_1 \to 0} \frac{c_1}{c_{1i}} = \frac{\sqrt{\kappa_1/D}\,(l - z) + \cdots}{\sqrt{\kappa_1/D}\,l + \cdots}$$

$$= 1 - \frac{z}{l} \tag{14.1-13}$$

As the reaction gets slow, the concentration profile approaches the usual film result in Eq. 14.1-7.

The flux in the presence of reaction is found by combining this concentration profile with Fick's law:

$$j_1 = -D\frac{dc_1}{dz}$$

$$= \sqrt{D\kappa_1}\,c_{1i}\left(\frac{\cosh[\sqrt{\kappa_1/D}\,(l - z)]}{\sinh[\sqrt{\kappa_1/D}\,l]}\right) \tag{14.1-14}$$

At the interface, where $z = 0$, this is

$$j_1 = [\sqrt{D\kappa_1}\,\coth(\sqrt{\kappa_1/D}\,l)]c_{1i} \tag{14.1-15}$$

Thus, in the case of chemical reaction, the mass transfer coefficient is

$$k = \sqrt{D\kappa_1}\,\coth(\sqrt{\kappa_1/D}\,l) \tag{14.1-16}$$

This result reduces to Eq. 14.1-9 as κ_1 becomes small.

The shortcoming of this result is that the film thickness l is unknown. However, we can dodge this if we remember that we did not want to find the mass transfer coefficient itself, but only the correction to mass transfer caused by chemical reaction. To do this, we combine Eq. 14.1-9 and Eq. 14.1-16 to find

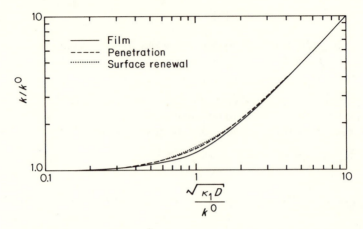

Fig. 14.1-2. Mass transfer corrected for first-order reaction. For slow reaction, the mass transfer coefficient is unchanged; for fast reaction, it equals $\sqrt{\kappa_1 D}$. Note that all theories give very similar results. [From Sherwood et al. (1975), with permission.]

$$\frac{k}{k^0} = \sqrt{D\kappa_1/(k^0)^2} \ \coth \ \sqrt{D\kappa_1/(k^0)^2} \tag{14.1-17}$$

This is the desired result. Two limits are instructive. First, when the reaction is slow, κ_1 is small, and

$$\frac{k}{k^0} = 1 + \frac{D\kappa_1}{(k^0)^2} + \cdots \tag{14.1-18}$$

Second, when the chemical reaction is fast, κ_1 is large, and

$$k = \sqrt{D\kappa_1} \tag{14.1-19}$$

The mass transfer coefficient now has nothing to do with k^0, but is simply the square root of the diffusion coefficient times the rate constant. I find this one of the most charming results in engineering.

Of course, these results did assume the film model. Because this model is an unsatisfying method for calculating mass transfer coefficients, we might expect it to be inaccurate. Instead, we can, with considerable effort, show that corrections for chemical reaction are all nearly the same, independent of the specific model chosen. Some of these are shown in Fig. 14.1-2 (Dankwerts, 1970; Sherwood et al., 1975). Differences are minor, and the use of the film theory is justified.

This coupling between diffusion and reaction means that the mass transfer coefficient in a rapidly reacting system will vary sharply with temperature. If the reaction rate doubles every 10°C, then the mass transfer will double every 20°C. In contrast, doubling the mass transfer coefficient when no reaction is present usually takes about 50°C.

Reversible reactions

Interestingly, the correspondence among these various models, which is so close for irreversible reactions, is completely destroyed for reversible reactions. The calculation of the general case of reversible reactions is tedious and best left in more specialized references (Olander, 1960; Huang & Kuo, 1965; Sherwood et al., 1975). One simple and interesting limit occurs when the reaction is so fast that it reaches equilibrium. For a thin film of thickness l, the mass transfer coefficient in the presence of such a reaction is

$$k = \frac{D}{l} \tag{14.1-20}$$

as shown in Example 2.2-3. In other words, a fast reversible chemical reaction has no effect in a thin film. For a thick slab exposed for a time t, the mass transfer coefficient is

$$k = \sqrt{4D(1 + K)/\pi t} \tag{14.1-21}$$

found by averaging over time the results of Example 2.3-3. Thus, in the case of the thick slab, the mass transfer is increased in proportion to the square root of the reaction's equilibrium constant.

To illustrate the size of the effects involved in both irreversible and reversible reactions, we now turn to specific examples.

Example 14.1-1: Mass transfer required for kinetic studies

We are planning a series of experiments of the reactions of methyl iodide with pyridine and similar compounds:

$$CH_3I + \underset{N}{\bigcirc} \rightarrow \underset{N^+}{\bigcirc} + I^-$$

We are going to contact these reagents in a small laboratory reactor by bubbling methyl iodide vapor diluted with nitrogen through benzene solutions containing around 0.1 mole per liter pyridine. We expect the rate constant of this and similar reactions to be about $1.46 \cdot 10^{-4}$ liter/mol-sec at 60°C (Grimm et al., 1931). How large must the mass transfer coefficient be to make sure we are studying the chemical kinetics?

Solution. If the methyl iodide vapor is present at much lower concentrations than the pyridine, we can approximate this as a first-order reaction whose rate r_1 is

$$r_1 = -[\kappa' c_{\text{pyridine}}] c_{\text{CH}_3\text{I}}$$
$$= -[(1.46 \cdot 10^{-4} \text{ liter/mol-sec})(0.1 \text{ mol/liter})] c_{\text{CH}_3\text{I}}$$
$$= -\frac{1.46 \cdot 10^{-5}}{\text{sec}} c_{\text{CH}_3\text{I}}$$

From Eq. 14.1-17 or Fig. 14.1-2, we see that the transition from usual mass transfer to the reaction-limited case occurs when

$$\frac{D\kappa_1}{(k^0)^2} = 1$$

The diffusion coefficient of methyl iodide in benzene is about $2.0 \cdot 10^{-5}$ cm^2/sec. Thus,

$$\frac{(1.5 \cdot 10^{-5}/\text{sec})(2.0 \cdot 10^{-5} \text{ cm}^2/\text{sec})}{(k^0)^2} = 1$$

$$k^0 = 1.7 \cdot 10^{-5} \text{ cm/sec}$$

Because mass transfer coefficients are usually around 10^{-3} cm/sec, we may have trouble studying the kinetics in this way.

Example 14.1-2: Finding the reaction-rate constant from mass transfer data

In studies with a wetted-wall absorption column, we find that the mass transfer coefficient for chlorine into water is $16 \cdot 10^{-3}$ cm/sec. The chlorine presumably is irreversibly reacting with the water:

$$\text{Cl}_2 + \text{H}_2\text{O} \rightarrow \text{Cl}^- + \text{H}^+ + \text{HOCl}$$

From similar experiments with nonreacting systems, we expect that the mass transfer coefficient without reaction is around $1 \cdot 10^{-3}$ cm/sec. What is the rate constant for this reaction?

Solution. To solve this problem, we must make two assumptions. First, we assume that the reaction kinetics are first-order and irreversible. In other words, we linearize the reaction:

$$r_{\text{Cl}} = -[\kappa' c_{\text{H}_2\text{O}}]c_{\text{Cl}_2} = -[\kappa_1]c_{\text{Cl}_2}$$

We identify the quantity in brackets with a first-order rate constant κ_1, thus assuming that the water concentration changes very little. Second, we assume that because the coefficient with reaction is higher than expected, mass transfer is influenced by reaction. Because the diffusion coefficient of Cl_2 in water is $1.25 \cdot 10^{-5}$ cm^2/sec, we find

$$k = \sqrt{\kappa_1 D}$$

$$16 \cdot 10^{-3} \text{ cm/sec} = \sqrt{\kappa_1(1.25 \cdot 10^{-5} \text{ cm}^2/\text{sec})}$$

$$\kappa_1 = 20 \text{ sec}^{-1}$$

The value obtained from a more complete study of mass transfer is 14 sec^{-1}; that found from fast-reaction techniques is 25 sec^{-1} (Brian et al., 1962).

Example 14.1-3: Mass transfer of corrosion inhibitors into marble

In an effort to inhibit the decay of marble, we have been studying the uptake of various fluorides. We find for one of these fluorides that the mass transfer coefficient across a 0.02-cm sheet of porous marble is $1.5 \cdot 10^{-7}$ cm/

sec. We find that the mass transfer coefficient found over 1 month into a thick slab of the same marble is also $1.5 \cdot 10^{-7}$ cm/sec. Is the fluoride reacting with the marble? How large is any reaction?

Solution. If there is no chemical reaction, we should be able to calculate one mass transfer coefficient from the other. Specifically, from Eq. 14.1-17, we see that

$$D = kl = (1.5 \cdot 10^{-7} \text{ cm/sec})(0.02 \text{ cm})$$
$$= 3 \cdot 10^{-9} \text{ cm}^2/\text{sec}$$

Because this experiment takes so long, we expect that any ionic reactions will be reversible and will tend to equilibrium. Thus, we insert this value for D into Eq. 14.1-21 and find for no reaction ($K = 0$) that

$$k = \sqrt{4D/\pi t} = \left(\frac{4(3 \cdot 10^{-9} \text{ cm}^2/\text{sec})}{\pi(2.6 \cdot 10^6 \text{ sec})} \right)^{1/2}$$
$$= 0.4 \cdot 10^{-7} \text{ cm/sec}$$

This is lower by a factor of 4 than that observed. The difference could be explained if there were a rapid reversible adsorption of the fluoride by the marble. If such a reaction had an equilibrium constant of about 16, the two measurements would be in agreement.

Example 14.1-4: Variation of mass transfer with fluid flow

Imagine a spinning disc of reagent 1 immersed in a dilute solution containing reagent 2. We plan to measure reagent 1 lost from the disc as a function of the rotation speed of the disc. How will this rate vary if the reagent dissolves and then irreversibly reacts, that is, if the reaction is homogeneous? How will it vary if the reaction is heterogeneous?

Solution. The answer in this case depends on how the mass transfer coefficient varies with fluid flow, or, in more general terms, on how the Sherwood number varies with the Reynolds number. This variation depends on the specific experimental situation. For the spinning disc described in Section 3.4,

$$\frac{k^0 d}{D} = a \left(\frac{dv}{\nu} \right)^{1/2}$$

The quantity a includes variables like the Schmidt number.

For a first-order irreversible homogeneous reaction, this flow dependence can be combined with Eq. 14.1-17 to give

$$k = \sqrt{D\kappa_1} \coth \left[\frac{1}{a} \left(\frac{\kappa_1 dv}{Dv} \right)^{1/2} \right]$$

At low flow, the hyperbolic cotangent is a constant, and thus k is also constant. At high flow, the hyperbolic cotangent approaches the reciprocal of its argument, and k varies with the square root of flow. This behavior is shown in Fig. 14.1-3.

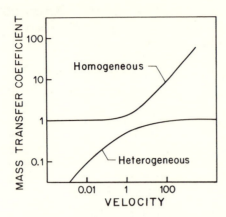

Fig. 14.1-3. Mass transfer for different types of reactions. At low flow, the mass transfer coefficient is dominated by chemical kinetics if the reaction is homogeneous, but it is independent of kinetics if the reaction is heterogeneous. At high flow, the reverse is true. The units of both axes are arbitrary.

The behavior for a heterogeneous reaction is completely different, a special case of Eq. 13.3-7:

$$\frac{1}{k} = \frac{1}{\kappa_2} + \frac{1}{k_3}$$

$$= \frac{1}{\kappa_2} + \frac{1}{a}\left(\frac{dv}{D^2v}\right)^{1/2}$$

Note that the same symbol κ_i is used here to represent a heterogeneous rate constant. The results in this case are also given in Fig. 14.1-3.

The completely different variation with flow that results provides an easy way to distinguish between heterogeneous and homogeneous reactions. I have found it especially useful in biochemical systems, where ambiguity between the two types of reactions is frequent.

Section 14.2. Mass transfer with second-order chemical reactions

Like first-order reactions, second-order reactions can enhance interfacial mass transfer. Unlike the situation with first-order reactions, this enhancement cannot be easily calculated. Because second-order reactions are common and important, we resort to a variety of limiting cases to predict mass transfer coefficients in these situations.

The reason that predictions are difficult for second-order reactions is again best illustrated by the film theory, as shown in Fig. 14.2-1(a). The mass balances in this film are

$$0 = D_1 \frac{d^2c_1}{dz^2} - \kappa_1 c_1 c_2 \tag{14.2-1}$$

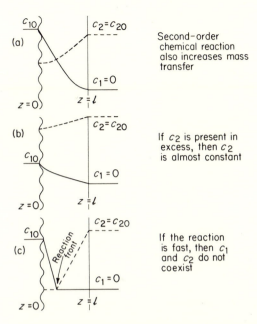

Second-order chemical reaction also increases mass transfer

If c_2 is present in excess, then c_2 is almost constant

If the reaction is fast, then c_1 and c_2 do not coexist

Fig. 14.2-1. Mass transfer with second-order chemical reaction. The mass transfer coefficients for the general case (a) cannot be easily calculated. They can be found for the special cases (b) and (c).

and

$$0 = D_2 \frac{d^2 c_2}{dz^2} - \kappa_1 c_1 c_2 \tag{14.2-2}$$

The boundary conditions are typically

$$z = 0, \quad c_1 = c_{1i}, \quad \frac{dc_2}{dz} = 0 \tag{14.2-3}$$

$$z = l, \quad c_1 = 0, \quad c_2 = c_{2l} \tag{14.2-4}$$

Unfortunately, solving these equations is difficult because of the nonlinear reaction term. Various numerical solutions are available (Secor & Beutler, 1967; Aris, 1975), but I never really understand what they are saying.

A more satisfying strategy is to consider three limiting cases. The most obvious limit, shown in Fig. 14.2-1(b), occurs when reagent 2 is present in excess, so that the second-order reaction is equivalent to a first-order reaction. This limit was discussed in the previous section.

A second, more interesting limit occurs when the reaction is very fast and irreversible. Here, finite concentrations of the two reagents cannot coexist, but simultaneously disappear at the reaction front, shown schematically in Fig. 14.2-1(c). The result is like two film theories, slapped one on top of the other.

Finding the mass transfer in this case is easy. For example, for the reaction

$$\begin{pmatrix} \text{species} \\ 1 \end{pmatrix} + \nu \begin{pmatrix} \text{species} \\ 2 \end{pmatrix} \rightarrow \text{(products)} \tag{14.2-5}$$

we have

$$n_1 = \frac{D_1}{z_c}(c_{1i}) \tag{14.2-6}$$

$$n_2 = -\frac{D_2}{l - z_c}(c_{2l}) \tag{14.2-7}$$

and

$$\nu n_1 + n_2 = 0 \tag{14.2-8}$$

The distance z_c is the location of the reaction front. Combining these results to eliminate z_c, we find

$$n_1 = \left[\frac{D_1}{l}\left(1 + \frac{D_2 c_{2l}}{\nu D_1 c_{1i}}\right)\right]c_{1i} \tag{14.2-9}$$

The quantity in the square brackets corresponds to the mass transfer coefficient with chemical reaction. Remembering that the mass transfer coefficient without reaction k^0 equals D_1/l, we have

$$\frac{k}{k^0} = 1 + \frac{D_2 c_{2l}}{\nu D_1 c_{1i}} \tag{14.2-10}$$

the desired result. Again, we can extend this result to, for example, penetration and surface-renewal theories of mass transfer (Sherwood et al., 1975). I believe that these extensions rarely produce significant improvements.

The third limit occurs when the second-order reaction is very fast and reversible, so that it essentially reaches equilibrium. The exact form of the result depends on the stoichiometry. As an example, consider the reaction

$$\text{(species 1)} + \text{(species 2)} \rightleftarrows \text{(species 3)} \tag{14.2-11}$$

so

$$c_3 = K c_1 c_2 \tag{14.2-12}$$

where K is the equilibrium constant of the reaction. If species 2 and 3 are nonvolatile, their fluxes are zero at the gas–liquid interface. The calculation of the mass transfer in this case is similar to that for facilitated diffusion, given in detail in Section 15.2. Accordingly, only the result is given here (Olander, 1960):

$$\frac{k}{k^0} = 1 + \frac{D_3}{D_1}\left(\frac{K c_{2l}}{1 + K(D_3/D_2)c_{1i}}\right) \tag{14.2-13}$$

where k/k^0 represents the correction to the mass transfer coefficient caused by this kind of chemical reaction. Note that this is the same as Eq. 14.2-10 for an irreversible reaction producing 1 mole of product (i.e., for K infinite and ν unity).

Example 14.2-1: Oxygen uptake by a synthetic blood

Oxygen uptake by blood is faster than oxygen uptake by water because of the reaction of oxygen and hemoglobin. Many chemists have dreamed of inventing a new compound capable of fast, selective reaction with oxygen. Aqueous solutions of this compound, of molecular weight around 500, could then be used as the basis of a process for oxygen separation from air. Such a compound would complex with oxygen at low temperatures, but would give up the oxygen at high temperatures.

How concentrated would this solution have to be to increase the oxygen concentration in water 50 times? How much faster would oxygen mass transfer into this solution be? The diffusion coefficient of oxygen in water is $2.10 \cdot 10^{-5}$ cm²/sec; that of the new compound would be about $5 \cdot 10^{-6}$ cm²/sec.

Solution. The solubility of oxygen in water is $3 \cdot 10^{-7}$ mol/cm³; so we want a solubility of $1.5 \cdot 10^{-5}$ mol/cm³. This implies approximately a 0.7 wt% solution of our new compound. The rate of mass transfer can now be estimated from Eq. 14.2-10:

$$\frac{k}{k^0} = 1 + \frac{(5 \cdot 10^{-6} \text{ cm}^2/\text{sec})(1.5 \cdot 10^{-5} \text{ mol/cm}^3)}{(2.1 \cdot 10^{-5} \text{ cm}^2/\text{sec})(3 \cdot 10^{-7} \text{ mol/cm}^3)}$$

$$= 13$$

There is about a 13-fold increase in rate.

Example 14.2-2: Sulfur dioxide absorption in a packed tower

We are using an absorption tower 12 m high and 2 m in diameter to remove sulfur dioxide from a process gas. From previous experiments on nonreacting systems, we know that when the tower uses 20°C water at the desired rate, the gas-side mass transfer is characterized by a $k_G^0 a$ of 1.7 sec⁻¹, and the liquid-side mass transfer by a $k_L^0 a$ of $3.8 \cdot 10^{-3}$ sec⁻¹. We also know that under the current process conditions, sulfur dioxide is present at a partial pressure of around 10 mm Hg, producing a solution that contains 0.1 wt%.

We want to know how much the mass transfer in this column will be improved if we replace water with dilute solutions of sodium hydroxide. Estimate the size of these improvements as a function of NaOH concentration.

Solution. Because acid–base reactions like this are essentially instantaneous, we can estimate the improved mass transfer in the liquid from Eq. 14.2-10. Because c_{1i} in this equation is the interfacial value, we are forced to parallel the derivation in Section 9.5. The flux across the interface must be

$$j_1 = k_G(p_1 - p_{1i})$$

$$= k_L(c_{1i} - 0)$$

$$= k_L^0 \left(c_{1i} + \frac{D_2 c_{2l}}{\nu D_1}\right)$$

where the subscripts 1 and 2 refer to SO_2 and NaOH, respectively. We know that the gas and liquid concentrations at the interface are in equilibrium:

$$p_{1i} = Hc_{1i}$$

We now can find the interfacial concentration:

$$c_{1i} = \frac{k_G^0 p_1 - k_L^0 (D_2 c_{2l}/\nu D_1)}{k_G^0 H + k_L^0}$$

Note that this interfacial concentration is zero when

$$c_{2l} \geq \left(\frac{k_G \nu D_1}{k_L^0 D_2}\right) p_1$$

When it is greater than zero, the flux is

$$j_1 = K_L(p_1/H)$$

where

$$K_L a = \frac{1 + D_2 c_{2l} H/\nu D_1 p_1}{1/k_G^0 a H + 1/k_L^0 a}$$

From the problem statement, we know that $k_L^0 a$ is $3.8 \cdot 10^{-3}$ sec^{-1}. We can find $k_G a$ by converting units:

$$k_G a = \frac{1.7/\text{sec}}{(82 \text{ atm-cm}^3/\text{mol-°K})(293°K)}$$

$$= 7.1 \cdot 10^{-5} \text{ mol/cm}^3\text{-sec-atm}$$

The Henry's law constant involves other unit conversions:

$$p_1 = Hc_1$$

$$(10 \text{ mm Hg}) \left(\frac{\text{atm}}{760 \text{ mm Hg}}\right) = H \left(\frac{0.001 \text{ g}}{\text{cm}^3}\right)\left(\frac{\text{mol}}{64 \text{ g}}\right)$$

$$H = 840 \text{ cm}^3\text{-atm/mol}$$

The diffusion coefficient for NaOH found from Table 6.6-1 is $2.1 \cdot 10^{-5}$ cm^2/sec; that for SO_2 is about $1.9 \cdot 10^{-5}$ cm^2/sec. To produce sulfite, $\nu = 2$.

Using these values, we find the c_{1i} equals zero when c_2 is 0.44 mol/liter. The change in the overall mass transfer coefficient is

$$\frac{K_L a}{K_L^0 a} = 1 + \frac{D_2 c_{2l} H}{\nu D_1 p_1}$$

$$= 1 + \frac{(2.1 \cdot 10^{-5} \text{ cm}^2/\text{sec})c_2(840 \text{ cm}^3\text{-atm/mol})(\text{liter}/1,000 \text{ cm}^3)}{2(1.9 \cdot 10^{-5} \text{ cm}^2/\text{sec})(10/760 \text{ atm})}$$

$$= 1 + (35 \text{ liter/mol})c_2$$

The results are shown in Table 14.2-1. At very low hydroxide concentrations, the coefficient approaches the limit of no reaction; after an increase of a factor of 16, it becomes limited by mass transfer in the gas phase.

Table 14.2-1. *Increases of sulfur dioxide (1) mass transfer effected with sodium hydroxide (2)*

c_2 (mol/liter)	$K_1 a/K_1^0 a$
0.0	1
0.01	1.4
0.1	4.5
0.5	16
2.0	16

Example 14.2-3: Carbon dioxide absorption with amines

In many industrial processes, carbon dioxide must be removed from a gas mixture. This removal is frequently accomplished by scrubbing with aqueous solutions of pH 8 to 10 containing compounds like monoethanol amine (Astarita, 1966; Astarita et al., 1983):

$$NH_2CH_2CH_2OH$$

Carbon dioxide can react with hydroxyl and the amine groups. However, the pK_a of the hydroxyl group is about 11, so that this reaction will be important only at a pH above 11. Such extremely basic conditions occur infrequently.

Reaction with the amine group involves a plethora of possibilities:

$$CO_2 + H_2O \rightleftarrows H_2CO_3 \qquad\qquad K = 10^3$$
$$H_2CO_3 \rightleftarrows H^+ + HCO_3^- \qquad\qquad K = 4 \cdot 10^{-7} \text{ mol/liter}$$
$$HCO_3^- \rightleftarrows H^+ + CO_3^{2-} \qquad\qquad K = 4 \cdot 10^{-11} \text{ mol/liter}$$
$$RNH_2 + H^+ \rightleftarrows RNH_3^+ \qquad\qquad K = 3 \cdot 10^9 \text{ mol/liter}$$
$$RNH_2 + HCO_3^- \rightleftarrows RNH_3^+ + CO_3^{2-} \qquad K = 8 \cdot 10^{-2} \text{ mol/liter}$$
$$RNH_2 + HCO_3^- \rightleftarrows RNHCOO^- + H_2O, \quad K = 50 \text{ mol/liter}$$

In these equations, RNH_2 represents the monoethanol amine, and the K's are the equilibrium constants of the various reactions. For example, the last one is

$$[RNHCOO^-] = 50[RNH_2][HCO_3^-]$$

where the square brackets signify concentrations. Note that water does not appear in these equilibria because it is assumed to be present in excess.

How will the rate of mass transfer in these systems vary with the total amine concentration?

Solution. The key aspect in this problem is making some sense out of all the possible chemical alternatives. First, we need to know in which forms the carbon dioxide occurs. For example,

$$[H^+][HCO_3^-] = 4 \cdot 10^{-7}[H_2CO_3]$$

Over the pH range to be used, $[H^+]$ will be 10^{-8} to 10^{-10} mol/liter; thus, $[HCO_3^-]$ will far exceed $[H_2CO_3]$. Similar reasoning shows that $[HCO_3^-]$ will exceed $[CO_3^{2-}]$. Turning to the reactions with amine, we note that we expect $[RNH_2]$ to be relatively large, certainly greater than 10^{-2} mol/liter. Thus, $[RNH_3^+]$ and $[RNCOO^-]$ are the chief reaction products, and the overall reaction involved here is

$$CO_2 + 2RNH_2 \rightleftarrows RNH_3^+ + RNHCOO^-$$

This reaction is restricted to the pH range studied.

We next need to approximate the kinetics involved in this reaction. Because it is an acid–base reaction, we expect the reaction to be instantaneous, described by Eq. 14.2-10:

$$\frac{k}{k^0} = 1 + \frac{D_2 c_2}{\nu D_1 c_{1i}}$$

where 1 and 2 refer to carbon dioxide and amine at the boundaries of the interfacial reaction. In a given experiment, we expect c_{1i} to be fixed and

$$c_2 = \bar{c}_2(1 - 2\theta)$$

where $\bar{c}_2$ is the total amine concentration and θ is the fraction of the amine already combined with carbon dioxide. The factor of 2 is stoichiometric, reflecting the overall reaction in which one carbon dioxide molecule combines with two amines. Thus, we expect

$$\frac{k}{k^0} = 1 + \frac{D_2 \bar{c}_2}{2D_1 c_{1i}} (1 - 2\theta)$$

This prediction is verified for industrial absorption towers (Astarita et al., 1983), for which

$$\frac{k}{k^0} = 1 + \frac{5.56 \text{ liters}}{\text{mol}} \bar{c}_2(1 - 2\theta)$$

This holds over the pH range of 8 to 10; outside of this range, other chemical reactions are more important. Note that the difficulty in this problem, and many like it, is not in the mathematics but in the chemistry.

Section 14.3 Effectiveness factors

Many industrial reactions are carried out by contacting reagent gases with pellets of porous solid catalysts. Such catalysts are frequently of silica, impregnated with small quantities of noble metals like platinum. The chemical reactions in these catalysts are heterogeneous, occurring mainly at clusters of dispersed metal atoms. Such catalysts are a bulwark of the petrochemical industry.

The reaction rates in these catalysts are expected to be proportional to the total surface area available for reaction. If we double the surface area per mass of catalyst, we expect to double the reaction rate. If we keep the surface area constant, we expect to keep the reaction rate constant. This simple behavior is often observed.

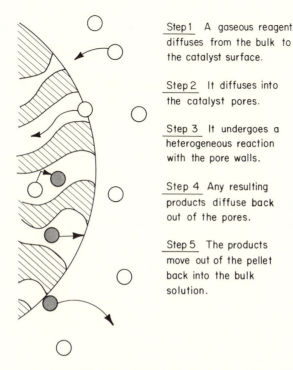

Step 1 A gaseous reagent diffuses from the bulk to the catalyst surface.

Step 2 It diffuses into the catalyst pores.

Step 3 It undergoes a heterogeneous reaction with the pore walls.

Step 4 Any resulting products diffuse back out of the pores.

Step 5 The products move out of the pellet back into the bulk solution.

Fig. 14.3-1. Diffusion and reaction in a porous catalyst. Although the reaction is actually heterogeneous, it is analyzed as if it is homogeneous. The result is a very different rate equation than that in Chapter 13.

However, in some cases the rate constant will drop as the size of the catalyst particle increases, even when the total catalyst area increases. These decreases in reaction rate are the result of slow diffusion into the pores of the catalyst pellet. This diffusion is part of the overall reaction sequence shown schematically in Fig. 14.3-1. This sequence has five steps. The first and fifth steps represent mass transfer from the external phase; they are described using the mass transfer correlations given in Chapter 9. The third step describes the actual chemical change; this step includes adsorption and surface reaction.

The second and fourth steps are those of interest here. These steps involve diffusion of reagent through the catalyst's pores to the active sites, and diffusion of products away from these sites. For large pores, the diffusion coefficient in these steps will equal that in the surroundings. For small pores, the diffusion may be altered by interactions with the boundary (see Chapter 7).

Before we mathematically analyze these steps, we review how the reaction rate varies under different process conditions. Such a "field guide to reaction rates" is shown in Table 14.3-1 for the special case of a packed bed.

Table 14.3-1. *A field guide to the rate-controlling step in porous catalysts*

Rate-controlling step[a]	With flow v[b]	With pellet size[b]	With temperature	With number of active sites S	Remarks
1	$\propto v^{1/2}$	$\propto d^{-1/2}$	Very small	—	If the reaction is irreversible and flow-dependent, this step controls
2	—	$\propto d^{-1}$	Small	—	If the reaction is irreversible, flow-independent, but pellet-size-dependent, this step controls
3	—	—	Large	$\propto S$	If the reaction varies sharply with temperature and is independent of pellet size, this step controls
4	—	$\propto d^{-1}$	Small	—	This case is influenced by the reaction equilibrium constant
5	$\propto v^{1/2}$	$\propto d^{-1/2}$	Very small	—	If the reaction is reversible and flow-dependent, this step is important

[a] See Fig. 14.3-1 for illustrations of these steps.
[b] The variations given are of the reaction rate per catalyst area. These variations are typical, but they may not hold quantitatively for some geometries.

The reaction rate will depend on flow through the bed only if step 1 or step 5 controls the rate. The rate will be independent of this flow but will vary with pellet size if step 2 or step 4 is slowest. It will vary sharply with temperature only if step 3 is rate-controlling. For more elaborate analyses for rate-controlling steps, the reader is referred to books on catalysis (Levenspiel, 1972; Aris, 1975; Carberry, 1976).

We want to calculate the variation of reaction rate with the size of the catalyst pellet. This effect is most easily found as a relative rate, expressed as an effectiveness factor η (Damköhler, 1937; Thiele, 1939):

$$\eta = \frac{\left(\begin{array}{c}\text{actual reaction rate}\\ \text{including diffusion}\end{array}\right)}{\left(\begin{array}{c}\text{reaction rate if diffusion}\\ \text{were instantaneous}\end{array}\right)} \qquad (14.3\text{-}1)$$

As before, we shall calculate this rate for the simplest case of a first-order irreversible reaction occurring on a spherical catalyst pellet. Then we shall quickly review other cases.

We begin by making a steady-state mass balance on the thin spherical shell shown schematically in Fig. 14.3-2:

$$0 = (\text{diffusion in minus that out}) + (\text{reaction})$$

$$0 = \left(4\pi r^2 j_1 \bigg|_r - 4\pi r^2 j_1 \bigg|_{r+\Delta r}\right) - (4\pi r^2 \Delta r \kappa_1 c_1) \qquad (14.3\text{-}2)$$

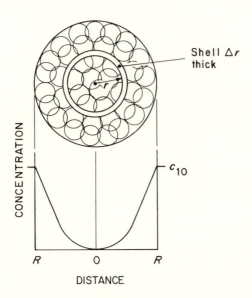

Fig. 14.3-2. A model for chemical reaction in a porous catalyst. The catalyst pellet is analyzed as if it is a homogeneous sphere. The key to the analysis is a mass balance on the spherical shell shown.

in which κ_1 is a first-order reaction-rate constant to be described in more detail later. If we divide by Δr, take the limit as Δr goes to zero, and combine the result with Fick's law, we find

$$0 = \frac{D}{r^2}\frac{d}{dr}r^2\frac{dc_1}{dr} - \kappa_1 c_1 \tag{14.3-3}$$

Note that we have implicitly made the assumption of dilute solution, or, more exactly, of a zero volume average velocity. We have also assumed that the diffusion coefficient is a constant effective value, dependent on pellet properties like tortuosity, but independent of reagent concentration.

The rate constant κ_1 deserves reflection. This rate constant is for an apparently homogeneous reaction, and it has units of reciprocal time. Reactions in porous catalysts are heterogeneous and should be described by a heterogeneous rate constant whose units are like velocity. However, because we are concerned with the variation of reagent concentration through the pellet, we are modeling this heterogeneous reaction per catalyst area as a homogeneous reaction per catalyst volume. The relation between these rate constants is

$$\kappa_1(\text{homogeneous}) = a\left[\frac{1}{1/k_1 + 1/\kappa_2(\text{heterogeneous})}\right] \tag{14.3-4}$$

in which a is the catalyst area per volume, k_1 is the mass transfer coefficient from the bulk of the pore to the pore wall, and the quantity in brackets is the overall heterogeneous rate constant given in Example 13.3-1. In the rest of this section, κ_1 will stand for the homogeneous value.

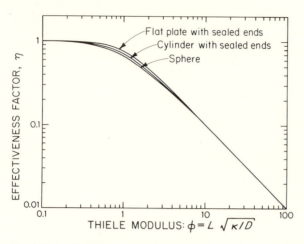

Fig. 14.3-3. The effectiveness factor for a first-order reversible reaction. This factor, defined by Eq. 14.3-1, gives the changes in reaction caused by diffusion in a catalyst pellet. In this figure, L is the volume of the pellet divided by its exterior surface area. With this definition, the factor does not vary much with pellet shape. [From Levenspiel (1972), with permission.]

The foregoing mass balance is subject to the boundary conditions

$$r = 0, \quad \frac{dc_1}{dr} = 0 \tag{14.3-5}$$

$$r = R_0, \quad c_1 = c_{10} \tag{14.3-6}$$

Integration of Eq. 14.3-3 and combination with these conditions closely parallels that in Section 14.1.

The concentration profile in the catalyst pellet is found to be

$$\frac{c_1}{c_{10}} = \frac{R_0}{r} \frac{\sinh(\phi \, r/R_0)}{\sinh \phi} \tag{14.3-7}$$

in which ϕ is the Thiele modulus, defined as

$$\phi = \sqrt{\kappa_1 R_0^2/D} \tag{14.3-8}$$

To find the actual reaction rate, we use this profile to find the flux at the surface:

$$\begin{pmatrix} \text{actual} \\ \text{reaction} \\ \text{rate} \end{pmatrix} = -4\pi R_0^2 D \left. \frac{dc_1}{dr} \right|_{r=R_0}$$

$$= -4\pi R_0 D c_{10}(\phi \coth \phi - 1) \tag{14.3-9}$$

The reaction rate without diffusion effects is $-[(\frac{4}{3})\pi R_0^3]\kappa_1 c_{10}$. Dividing this rate into the preceding equation gives the desired effectiveness factor:

$$\eta = \frac{3}{\phi^2} (\phi \coth \phi - 1) \tag{14.3-10}$$

A plot of this factor is shown in Fig. 14.3-3 as a function of Thiele modulus.

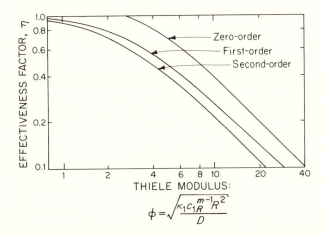

Fig. 14.3-4. Effectiveness factor for different reaction orders. The Thiele modulus in this case varies with the surface concentration c_{1R}. Note that even a zero-order reaction is affected by these diffusion effects. [From Aris (1975), with permission.]

The results in Eq. 14.3-10 and Fig. 14.3-3 are useful and commonly quoted. They allow calculation of the reduction of chemical reaction caused by mass transfer in the pores. They are restricted to first-order irreversible reactions in spherical catalyst pellets. This restriction has provided a focus for scores of underemployed professors, who have happily extended the analysis to other geometries and other types of reactions. These extensions are very similar to the simplest case (Weisz, 1973). Changes in geometry make little difference, as shown in Fig. 14.3-3; extensions to other types of reactions are shown in Fig. 14.3-4 (Satterfield & Sherwood, 1963).

Calculations for nonisothermal reactions produce the most interesting effects, as shown in Fig. 14.3-5. For exothermic reactions, effectiveness factors can exceed unity; that is, the reaction rate in a larger particle with diffusion effects is greater than the rate in a smaller particle without diffusion effects. This happens because the center of the larger particle gets very hot. The increased temperature increases the reaction rate by more than the diffusion decreases it. For endothermic reactions, the results are less interesting, for the reaction rate is decreased by the cool catalyst pellet as well as by the diffusion (Carberry, 1976).

Example 14.3-1: The reaction rate in a large catalyst pellet

In a packed-bed laboratory reactor using 0.2-cm catalyst pellets, we have found that a particular reaction behaves as if it is first order, with an apparent rate constant of 8.6 sec^{-1}. Diffusion coefficients at the reactor temperature and pressures are around 0.027 cm^2/sec. We want to scale up this reactor. To keep the flow patterns constant, we plan to use 1-cm pellets to get the same particle Reynolds number. What is the true rate constant of this reaction? What will the rate constant of the new reaction be?

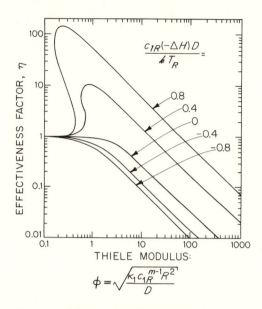

Fig. 14.3-5. Effectiveness factors for nonisothermal reactions. For nonisothermal reactions, two new variables are important. One, E/RT, is the Arrhenius factor giving the change of reaction rate with temperature. In this figure, this factor equals 20. The second variable shown in the figure includes the heat of reaction ΔH and the particle's thermal conductivity $\textit{k}$. [From Weisz & Hicks (1962), with permission.]

Solution. The units of this rate constant indicate that it is defined per unit volume of reactor, not per mass of catalyst. To answer this question, we first assume that the larger reactor will have the same catalyst area per reactor volume and that the reaction is approximately isothermal. Then the reaction rate should be the same as before, unless it is affected by diffusion. To check for this, we calculate the Thiele modulus:

$$\phi = \sqrt{\kappa_1 R_0^2/D}$$
$$= \sqrt{(8.6 \text{ sec}^{-1})(0.1 \text{ cm})^2/(0.027 \text{ cm}^2/\text{sec})} = 1.8$$

Because the Thiele modulus is already greater than unity, diffusion is already important.

To find the actual rate constant, we refer to Eq. 14.3-1 or Eq. 14.3-10 to see

$$\frac{\kappa_1(\text{apparent})}{\kappa_1(\text{actual})} = \frac{3}{\phi^2} [\phi \coth \phi - 1]$$

Remember that ϕ must be based on the actual value. Thus, we find

$$\kappa_1(\text{actual}) = 11.5 \text{ sec}^{-1}$$

Using this value, we can easily find for the 1-cm spheres:

$$\kappa_1(\text{apparent}) = 3.0 \text{ sec}^{-1}$$

Thus, we need to use smaller pellets or a larger reactor.

(a) Apparatus

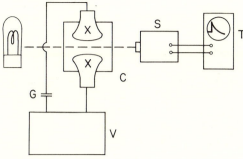

(b) Typical output

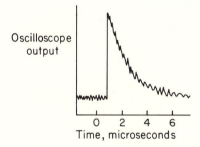

Fig. 14.4-1. The temperature-jump apparatus. Two large electrodes (X) are placed in a small cell (C) containing the solution of interest. The electrodes are charged with a power supply (V) until a spark flashes across the gap (G). The sudden current rapidly increases the cell's temperature, perturbing any reaction equilibria. These perturbations are recorded with the spectrophotometer (S) and displayed on an oscilloscope as a trace (T). A typical trace is also shown.

Section 14.4. Diffusion-controlled fast reactions

In this section we turn to chemical reactions that take place faster than a few milliseconds, often faster than microseconds. The most common example is the reaction of acid and base; other good examples are the combustion of methane and the action of the enzyme urease. These reactions occur so fast that in industrial situations they are always diffusion-controlled. However, they are of major scientific interest, providing an extremely active area for research.

To see how these fast reactions can be studied, we turn to the specific example of the temperature-jump apparatus shown in Fig. 14.4-1 (Hague, 1971; Bradley, 1975). In this apparatus, a cell containing perhaps 0.3 cm^3 of conducting solution is suddenly heated by discharging a capacitor through the solution. This heating, typically about 10°C, shifts any reactions in the cell away from equilibrium. These reactions then move to a new equilibrium at the new, higher temperature. The speed with which they reach this new state is measured with a spectrophotometer attached to an oscilloscope.

One reaction studied with this temperature-jump apparatus is

$$H^+ + OH^- \rightleftharpoons H_2O \qquad (14.4\text{-}1)$$

To study this reaction, the cell might be filled with an aqueous solution of KCl and a pH indicator. The KCl makes the solution conducting, so that capacitors can discharge; the indicator must be chosen so that its color change is still more rapid than the acid–base reaction. When the temperature is suddenly increased, the ionization of water is slightly increased, so that the indicator's color will slightly change. This change, monitored by the oscilloscope, can be used to find the rate constant κ_1.

In the remainder of this section we shall describe how the oscilloscope trace can be related to the rate constant and how the rate constant varies with the diffusion coefficient.

Finding the rate constant

The experimental signal found on the oscilloscope is rarely a smooth, exactly defined curve. Far more frequently it contains considerable noise, more than that shown in Fig. 14.4-1(b). One method of reducing this noise is to repeat the experiment and to average electronically the various signals obtained. Even so, the signal remains inexactly known.

As a result, the signals obtained in this sort of experiment are almost always analyzed as if they are first-order exponential decay. In other words, the logarithm of the signal's intensity is plotted versus time; the slope of this plot is the measured parameter. This slope, which has the units of reciprocal time, is a pseudo-first-order rate constant. For convenience, these slopes are rarely reported; instead, their reciprocals, called "relaxation times τ," are the data given.

We want to relate these relaxation times to kinetic rate constants. Such relations depend on the particular stoichiometry involved. As an illustration, we consider that in Eq. 14.4-1,

$$(\text{species 1}) + (\text{species 2}) \underset{\kappa_{-1}}{\overset{\kappa_1}{\rightleftharpoons}} (\text{species 3}) \qquad (14.4\text{-}2)$$

Details of other cases are given elsewhere (Bradley, 1975; Hammes, 1978). For this case, a mass balance on the cell gives

$$\frac{dc_1}{dt} = -\kappa_1 c_1 c_2 + \kappa_{-1} c_3 \qquad (14.4\text{-}3)$$

We now let c_1' be the small perturbation in concentration caused by the change in temperature:

$$c_1 = \bar{c}_1 + c_1' \qquad (14.4\text{-}4)$$
$$c_2 = \bar{c}_2 + c_1' \qquad (14.4\text{-}5)$$
$$c_3 = \bar{c}_3 - c_1' \qquad (14.4\text{-}6)$$

where the $\bar{c}_i$ are the equilibrium concentrations at the new, higher temperature. Combining these relations,

$$\frac{dc_1'}{dt} = -\kappa_1(\bar{c}_1 + c_1')(\bar{c}_2 + c_1') + \kappa_{-1}(\bar{c}_3 - c_1') \qquad (14.4\text{-}7)$$

However, at equilibrium,

$$0 = \kappa_1 \bar{c}_1 \bar{c}_2 - \kappa_{-1} \bar{c}_3 \qquad (14.4\text{-}8)$$

We now subtract these equations and, recognizing that c_1' is small, neglect any terms in $(c_1')^2$:

$$\frac{dc_1'}{dt} = -[\kappa_1(\bar{c}_1 + \bar{c}_2) + \kappa_{-1}]c_1' \qquad (14.4\text{-}9)$$

Integrating, we see that

$$\ln c_1' = (\text{constant}) - [\kappa_1(\bar{c}_1 + \bar{c}_2) + \kappa_{-1}]t \qquad (14.4\text{-}10)$$

Thus, the relaxation time τ is

$$\frac{1}{\tau} = \kappa_1(\bar{c}_1 + \bar{c}_2) + \kappa_{-1} \qquad (14.4\text{-}11)$$

In cases like this, we also know the equilibrium constant $K (= \kappa_1/\kappa_{-1})$. Thus, from measurements of τ, we can find both the forward rate constant κ_1 and the reverse rate constant κ_{-1}. Note also that for this stoichiometry, τ^{-1} varies linearly with $\bar{c}_1 + \bar{c}_2$, permitting another experimental check of this argument (Hammes, 1978).

Relating the relaxation time and the diffusion coefficient

At this point, we have described the temperature-jump method, a fast-reaction technique, and we have found relations between the relaxation times found from this method and the rate constants implied by the reaction's stoichiometry. In general, these rate constants will depend on factors like electronic structure. However, if the two reagents are very reactive, then they will react whenever they collide. In this case, their reaction rate depends not on electronic structure but on how often they collide (i.e., on their diffusion).

To explore this dependence of reaction rate on diffusion, we consider the forward reaction in Eq. 14.4-2. Imagine that species 2 is dilute and stationary. We then choose as a system a volume $1/c_2\tilde{N}$ having a molecule of species 2 at its center. We assume that whenever a molecule of species 1 reaches this center, it reacts. We then write a mass balance on species 1 in this system (Pilling, 1975):

$$\left(\frac{\text{reaction}}{\text{volume}}\right)\left(\frac{\text{system}}{\text{volume}}\right) = \left(\begin{array}{c}\text{diffusion}\\\text{flux into}\\\text{system}\end{array}\right)$$

$$(r_1)\left(\frac{1}{c_2\tilde{N}}\right) = 4\pi r^2 j_1 \qquad (14.4\text{-}12)$$

where r is the approximate radius of the system. The flux of species 1 for such a system has already been calculated (see Eq. 2.4-24):

$$j_1 = \left(\frac{D_1\sigma_{12}}{r^2}\right)c_1 \qquad (14.4\text{-}13)$$

in which σ_{12} is the collision distance between the two species and c_1 is the concentration in the bulk, far away from species 2. Thus,

$$r_1 = 4\pi D_1 \sigma_{12} \tilde{N} c_1 c_2 \tag{14.4-14}$$

But species 2 is not stationary; if it too is allowed to move, this relation must be modified:

$$r_1 = 4\pi (D_1 + D_2) \sigma_{12} \tilde{N} c_1 c_2 \tag{14.4-15}$$

Now the forward reaction rate is

$$r_1 = \kappa_1 c_1 c_2 \tag{14.4-16}$$

Thus, the forward rate constant for this type of diffusion-controlled reaction is

$$\kappa_1 = 4\pi (D_1 + D_2) \sigma_{12} \tilde{N} \tag{14.4-17}$$

This is the desired result.

The foregoing derivation, due to Smoluchowski (1917) (see also Debye, 1942), is approximate. Most obviously, it neglects any potential surrounding the two species, a potential that might greatly accelerate or inhibit their interaction. It also ignores molecular shape. Because different-shaped reagents might need to rotate before they react with each other, the diffusion coefficients in Eq. 14.4-17 might include rotational contributions. Still, this derivation provides a first approximation for diffusion-controlled fast reactions.

Example 14.4-1: Diffusion-controlled reactions
Which of the following reactions is diffusion-controlled?
(a) Reaction of a proton and a hydroxyl:
$$H^+ + OH^- \rightarrow H_2O, \quad \kappa_1 = 1.4 \cdot 10^{11} \text{ liter/mol-sec}$$
(b) Reaction of ethylenediaminetetraacetic acid:
$$OH^- + EDTA^{3-} \rightarrow EDTA^{4-}, \quad \kappa_1 = 3.8 \cdot 10^7 \text{ liter/mol-sec}$$

Solution. As a rule of thumb, reactions are diffusion-controlled if their rate constants are faster than 10^9 liter/mol-sec. To make this somewhat more quantitative, we can insert into Eq. 14.4-17 estimates of diffusion coefficients from Chapters 5 and 6 and sizes from solid-state radii, and thus estimate the rate constants if the reactions are diffusion-controlled. For reaction (a),

$$\kappa_1 = 4\pi (D_1 + D_2) \sigma_{12} \tilde{N}$$

$$= 4\pi (9.3 + 5.3) \left(10^{-5} \frac{cm^2}{sec} \right) (2.8 \cdot 10^{-8} \text{ cm}) \left(\frac{liter}{10^3 \text{ cm}^3} \right) \left(\frac{6.02 \cdot 10^{23}}{mol} \right)$$

$$= 3 \cdot 10^{10} \text{ liter/mol-sec}$$

The experimental value is still more rapid, apparently because of the electrostatic interaction (Eigen, 1964); thus, reaction (a) is diffusion-controlled. For reaction (b),

$$\kappa_1 = 4\pi (5.3 + 0.8) \left(10^{-5} \frac{cm^2}{sec} \right) (5 \cdot 10^{-8} \text{ cm}) \left(\frac{liter}{10^3 \text{ cm}^3} \right) \left(\frac{6.02 \cdot 10^{23}}{mol} \right)$$

$$= 2 \cdot 10^{10} \text{ liter/mol-sec}$$

This diffusion-based estimate is much faster than the experimentally observed rate constant. Thus, the kinetics of reaction (b) involve more than diffusion.

Section 14.5. Dispersion-controlled fast reactions

As the final topic in this chapter, we consider the rates of chemical reaction in turbulent flow. Such flow produces rapid mixing, so that the fluid appears homogeneous. Such mixing turns out to be only macroscopic. In other words, if we take 10 samples, each of 1 cm^3, we find that the average concentrations of the samples differ by only a few tenths of a percent. However, if we take 10 samples of 10^{-12} cm^3, we find that their concentrations vary widely. For example, if we are mixing acid and base, we might find that some samples contain 10^{-2}-mol/liter H$^+$, and others have 10^{-12}-mol/liter H$^+$.

Such microscopic heterogeneity can sharply reduce the rate of a chemical reaction. Such reductions are not automatically bad. For example, in an automobile engine, we may wish to slow the combustion in order to reduce the maximum temperature, and hence the formation of nitrogen oxide pollutants. It is important to know how this altered reaction rate can be estimated. Such estimates are the subject of this section.

In this analysis, we consider only reactions whose overall rates are controlled by turbulent dispersion. In other words, if the reagents could be very rapidly and completely mixed, their reaction rates would occur in microseconds. Such dispersion-controlled reactions are sometimes also called "diffusion-controlled," even though their rates are independent both of diffusion coefficients and of molecular motion. This misnomer is used only because the mathematical descriptions of diffusion and dispersion are similar, as discussed in Chapter 4.

The analysis of reaction rates in turbulent flow is a difficult problem. Instead of trying to solve this entire problem, we calculate the relations between turbulent reactions and turbulent mixing. In many ways, this strategy is similar to that used in the first section of this chapter. There, we calculated the change in mass transfer coefficient caused by chemical reaction and then combined this with mass transfer correlations to find the desired result. Here, we find the change in reaction rate caused by turbulence and then combine this with information about turbulent mixing.

The analysis has three different steps. First, we review the concentration fluctuations caused by turbulent mixing. Second, we examine fluctuations in a reacting system. Finally, we use these fluctuations to find the conversion in a chemical reactor.

Concentration fluctuations without reaction

We first consider a turbulent mixer like that shown schematically in Fig. 14.5-1. The mixer is basically an annular jet. Species 1 flows at a con-

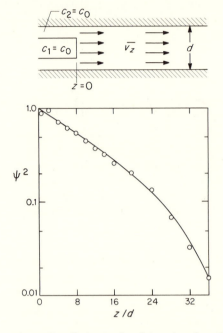

Fig. 14.5-1. Concentration fluctuations in turbulent mixing. In the jet shown at the top, two solutions containing the same concentration of different solutes are combined in turbulent flow. The solutes do not completely mix right away; instead, their concentrations show significant fluctuations characterized by ψ (see Eq. 14.5-6). [Data from Brodkey (1974).]

centration c_0 out of the core of the jet; species 2 at this same concentration flows out of the annular gap. The total flow in the jet is the same as that in the gap.

The concentrations of the two species after mixing begins are given by the mass balances

$$\frac{\partial c_i}{\partial t} = D\nabla^2 c_i - \mathbf{v} \cdot \nabla c_i \quad (i = 1, 2) \tag{14.5-1}$$

These equations assume that the two species have the same diffusion coefficient. Note also that they are not time-averaged, and so they include fluctuations in the concentrations. The equations for the two species are subject to the same boundary conditions, except at $z = 0$. There, the conditions differ, because c_1 is in the jet and c_2 is in the gap. To remove this difference, we define

$$\theta = \frac{c_0 - c_1}{c_0} = \frac{c_2}{c_0} \tag{14.5-2}$$

Each mass balance is now

$$\frac{\partial \theta}{\partial t} = D\nabla^2 \theta - \mathbf{v} \cdot \nabla \theta \tag{14.5-3}$$

For either solute, θ equals zero in the jet and unity in the gap; so both mass balances are subject to the same boundary conditions.

Because the equations for both species are the same, they must have the same solutions:

$$\frac{c_0 - c_1}{c_0} = \frac{c_2}{c_0} = f(z, t, D, \ldots) \tag{14.5-4}$$

For example, they must have the same time averages:

$$\frac{c_0 - \bar{c}_1}{c_0} = \frac{\bar{c}_2}{c_0} \tag{14.5-5}$$

The overbar indicates a time-averaged quantity, just as it did in Section 4.3. Subtracting these last two equations, squaring the result, and time-averaging, we see that

$$\overline{\left(\frac{c_1'}{c_0}\right)^2} = \overline{\left(\frac{c_2'}{c_0}\right)^2} = \Psi^2 \tag{14.5-6}$$

in which c_i' $(= c_i - \bar{c}_i)$ is the fluctuation of the concentration of i. This shows that the fluctuations of the concentrations of these two species are the same. The values of Ψ, the average percent of concentration fluctuation, can be measured experimentally; typical values are shown in Fig. 14.5-1 (Brodkey, 1974).

Concentration fluctuations with reaction

We next examine how these conclusions are altered when a second-order chemical reaction is present. The mass balances are now

$$\frac{\partial c_1}{\partial t} = D\nabla^2 c_1 - \mathbf{v} \cdot \nabla c_1 - \kappa_1 c_1 c_2 \tag{14.5-7}$$

$$\frac{\partial c_2}{\partial t} = D\nabla^2 c_2 - \mathbf{v} \cdot \nabla c_2 - \kappa_1 c_1 c_2 \tag{14.5-8}$$

Again, we should stress that these equations have not been time-averaged. If we subtract one from the other, we find

$$\frac{\partial}{\partial t}(c_2 - c_1) = D\nabla^2(c_2 - c_1) - \mathbf{v} \cdot \nabla(c_2 - c_1) \tag{14.5-9}$$

Further, if we define

$$\theta = \frac{c_0 + c_2 - c_1}{2c_0} \tag{14.5-10}$$

and insert this into Eq. 14.5-9, we get Eq. 14.5-3. Again, θ is zero in the jet and unity in the gap; all other boundary conditions on θ are just like those in the nonreactive problem. Again, because the differential equation and boundary conditions are the same, the solution in the reactive problem must be that in the nonreactive problem:

$$\left(\frac{c_0 + c_2 - c_1}{2c_0}\right)_{\text{rxn}} = \left(\frac{c_0 - c_1}{c_0}\right)_{\text{rxn}}^{\text{no}} \tag{14.5-11}$$

where we have added the subscript rxn to emphasize the different situations to which these equations refer. If we time-average this result, we find

$$\left(\frac{c_0 + \bar{c}_2 - \bar{c}_1}{2c_0}\right)_{\text{rxn}} = \left(\frac{c_0 - \bar{c}_1}{c_0}\right)_{\text{no} \atop \text{rxn}} \tag{14.5-12}$$

Subtracting this from the previous equation, we find that

$$\left(\frac{c_2' - c_1'}{2c_0}\right)_{\text{rxn}} = -\left(\frac{c_1'}{c_0}\right)_{\text{no} \atop \text{rxn}} \tag{14.5-13}$$

Now, if the chemical reactions are very fast, the local concentrations c_1 and c_2 cannot coexist; one of these concentrations will always be zero. As a result, the overall reaction rate cannot be influenced by chemical kinetics, only by turbulent flow. Accordingly (Toor, 1969),

$$\left(\frac{c_1'}{c_0}\right)_{\text{rxn}} = \left(\frac{c_1'}{c_0}\right)_{\text{no} \atop \text{rxn}} \tag{14.5-14}$$

When we multiply Eq. 14.5-13 by this identity and time-average, we find that

$$\overline{\left(\frac{c_2' c_1'}{c_0^2}\right)_{\text{rxn}}} = -\overline{\left(\frac{c_1'}{c_0}\right)_{\text{no} \atop \text{rxn}}^2} = -\Psi^2 \tag{14.5-15}$$

where Ψ is the function defined in Eq. 14.5-6. Thus, we know how fluctuations couple in this system.

Finding the reaction rate

We now use this knowledge of fluctuations to find the rate of a dispersion-controlled chemical reaction. As before, we begin with the mass balance for this system, given by Eqs. 14.5-7 and 14.5-8. We assume that the reactor is operating in a steady state and then time-average these equations:

$$0 = D\nabla^2 \bar{c}_1 - \bar{\mathbf{v}} \cdot \nabla \bar{c}_1 - \nabla \cdot \overline{\mathbf{v}' c_1'} - \kappa_1(\bar{c}_1 \bar{c}_2 + \overline{c_1' c_2'}) \tag{14.5-16}$$

A similar equation for c_2 is easily developed. The last term on the right-hand side represents the chemical reaction modified to include the effect of turbulent flow. We expect diffusion to be swamped by flow; so $D\nabla^2 \bar{c}_1$ will be negligible. We expect one-dimensional flow; so $\bar{\mathbf{v}}$ has only a z component. We do not expect dispersion to be significant; so we neglect $\nabla \cdot \overline{\mathbf{v}' c_1'}$. Thus, we have left

$$0 = -\bar{v}\frac{d\bar{c}_1}{dz} - \kappa_1(\bar{c}_1 \bar{c}_2 + \overline{c_1' c_2'}) \tag{14.5-17}$$

In words, this simple mass balance restates that the reagent flowing in minus that flowing out equals the amount reacted.

We next recognize that because the initial concentrations of the two reagents are the same, $\bar{c}_1$ equals $\bar{c}_2$. We also know $\overline{c_1' c_2'}$ from Eq. 14.5-15. Thus,

$$\frac{d}{dz}\bar{c}_1 = -\left[\frac{\kappa_1 c_0^2}{\bar{v}}\right]\left[\left(\frac{\bar{c}_1}{c_0}\right)^2 - \Psi^2\right] \tag{14.5-18}$$

Because Ψ is known from the nonreactive mixing experiments, we can integrate this equation to find the composition versus position in this reactor. The integration is often numerical, but always straightforward.

Typical results are given in Fig. 14.5-2. Three points about these results should be mentioned. First, the foregoing derivation is for stoichiometric

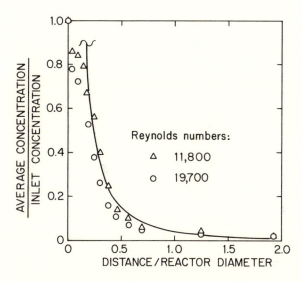

Fig. 14.5-2. Conversion in a dispersion-controlled reactor. The reactor in this case consisted of 94 feed tubes of 0.1-mol/liter NaOH and 94 feed tubes 0.1-mol/liter HCl, all entering a tubular reactor 3.2 cm in diameter. The mixing function found at large distances is $\psi^2 = 2.6 \cdot 10^{-3}(d/z)^{3.2}$, in which d is the reactor diameter and z is the distance downstream. [Data from Mao & Toor (1971).]

feeds, so at the reactor entrance, $\bar{c}_1 = c_0$ and $\bar{c}_2 = c_0$. Results for nonstoichiometric feeds are also known (Toor, 1969). If one reagent is present in excess, then the reaction will essentially be first-order. First-order reactions are unaffected by turbulent fluctuations (see Section 4.3).

A second point is that the reactions described here are so fast that the reaction rate is governed by the turbulence, not by the chemistry. Such a limit requires defining what is meant by a fast reaction. To discover this definition, we can write Eq. 14.5-18 in dimensionless terms:

$$\frac{d(\bar{c}_1/c_0)}{d(z/l)} = - \left[\frac{\kappa_1 c_0 l}{\bar{v}}\right]\left[\left(\frac{\bar{c}_1}{c_0}\right)^2 - \Psi^2\right] \qquad (14.5-19)$$

where l is the length that characterizes the turbulent mixing. The dimensionless quantity in the first set of square brackets, called the first Damköhler number, is the key parameter. When this number is small, the contributions of Ψ are negligible, and the reaction rate is controlled by chemical kinetics. When the first Damköhler number is large, the Ψ is the same magnitude as $\bar{c}_1/c_0$, and the reaction is controlled by dispersion.

Finally, we should mention some of the characteristics of Ψ. This function is essentially independent of the particular reagents involved, and so it need not be remeasured for every new reaction (Toor, 1969; Brodkey, 1974). However, it does depend on the reactor geometry, and hence on the size and shape of any mixing jets. To measure Ψ, we need not carry out a reaction. Instead, we can simply mix hot and cold water, use measurements of temperature fluctuations to define Ψ, and then integrate Eq. 14.5-18 to find the conversion. This is possible because turbulent mixing of energy and turbulent mixing of mass are very similar, a point explored further in Chapter 17.

Example 14.5-1: Conversion of a dispersion-controlled reaction

We are using the research reactor described in Fig. 14.5-2 to study the reaction.

$$\text{maleic acid} + OH^- \rightarrow \text{maleate}^- + H_2O$$

The rate constant of this reaction is about $3 \cdot 10^8$ liters/mol-sec, and the feed concentrations are stoichiometric, equal to 0.01 mol/liter. The Reynolds number is 12,000. Estimate the conversion in this reactor at a position 1 cm from the reactor inlet.

Solution. To see if this reaction is dispersion-controlled, we must calculate the first Damköhler number in Eq. 14.5-19. At a Reynolds number of 12,000, the fluid velocity in this reactor of 3.2-cm diameter must be 38 cm/sec. From the data in Fig. 14.5-2, we see that these reactions are about half finished at distances of 1 cm. Thus, the first Damköhler number is

$$\frac{\kappa_1 c_0 l}{\bar{v}} = \frac{(3 \cdot 10^8 \text{ liters/mol-sec})(0.01 \text{ mol/liter})(1 \text{ cm})}{38 \text{ cm/sec}}$$

$$= 7 \cdot 10^4$$

Because this is much greater than unity, the reaction is dispersion-controlled.

We now look at the data in Fig. 14.5-2. From these, we see that for the reaction

$$H^+ + OH^- \rightarrow H_2O$$

conversion at this distance is 64%. Because the maleic acid reaction is also dispersion-controlled, its conversion will also be 64%. Thus, the key factor in problems like this is not the chemistry of the reactants but the function Ψ produced by mixing in the reactor.

Section 14.6. Conclusions

The material in this chapter describes how homogeneous chemical kinetics can alter rates of mass transfer and how mass transfer can alter rates of chemical reaction. For example, the mass transfer of a very reactive solute can be increased to a limit of $\sqrt{\kappa_1 D}$, where κ_1 is the reaction-rate constant and D is the diffusion coefficient. In contrast, the rate of reaction in a catalyst pellet of radius R_0 can be reduced if the Thiele modulus $\sqrt{\kappa_1 R_0^2/D}$ is large (i.e., if D is small). Reaction-rate constants in water rarely exceed 10^9 liters/mol-sec, for in that range the reaction is not affected by chemistry, but only by Brownian motion of the reagents.

These effects are large, producing changes of rates of several orders of magnitude. Their analysis is often complex, easy only for the simplest reaction mechanisms. Still, there are enough of these simple mechanisms that the expected behavior can be bracketed by several limits. Solving more elaborate equations for complex mechanisms does not seem that important. Instead, when you work on problems like this, concentrate on how the chemistry can be approximated, not on how the mathematics can be made more complex.

References

Aris, R. (1975). *The Mathematical Theory of Diffusion and Reaction in Permeable Catalysts*. Oxford: Clarendon Press.

Astarita, G. (1966). *Mass Transfer and Chemical Reaction*. Amsterdam: Elsevier.

Astarita, G., Savage, D. W., & Bisio, A. (1983). *Gas Treating with Chemical Solvents*. New York: Wiley.

Bradley, J. N. (1975). *Fast Reactions*. Oxford: Clarendon Press.

Brian, P. L. T., Vivian, J. E., & Habib, A. G. (1962). *American Institute of Chemical Engineers Journal*, **8**, 205.

Brodkey, R. S. (1974). *Turbulence in Mixing Operations*. New York: Academic Press.

Carberry, J. J. (1976). *Chemical and Catalytic Reaction Engineering*. New York: McGraw-Hill.

Damköhler, G. (1937). *Deutsch Chemie Ingenieur*, **3**, 430.

Dankwerts, P. V. (1970). *Gas-Liquid Reactions*. New York: McGraw-Hill.

Debye, P. (1942). *Transactions of the Electrochemical Society*, **82**, 265.

Eigen, M. (1964). *Angewandte Chemie*, **3**, 1.

Grimm, H. G., Reif, H., & Wolff, H. (1931). *Zeitschrift für Physikalishe Chemie*, **B13**, 301.

Hague, D. N. (1971). *Fast Reactions*. London: Wiley Interscience.

Hammes, G. G. (1978). *Chemical Kinetics*. New York: Academic Press.

Hatta, S. (1928). *Tohoku Imperial University Technical Reports*, **8**, 1; (1932) **10**, 119.

Huang, C. J., & Kuo, C. H. (1965). *American Institute of Chemical Engineers Journal*, **11**, 901.

Levenspiel, O. (1972). *Chemical Reaction Engineering*. New York: Wiley.

Mao, K.-W., & Toor, H. L. (1971). *Industrial Engineering and Chemistry Fundamentals*, **10**, 192.

Olander, D. R. (1960). *American Institute of Chemical Engineers Journal*, **6**, 233.

Pilling, M. J. (1975). *Reaction Kinetics*. Oxford: Clarendon Press.

Satterfield, C. N., & Sherwood, T. K. (1963). *The Role of Diffusion in Catalysis*. Reading, Mass.: Addison-Wesley.

Secor, R. M., & Beutler, J. A. (1967). *American Institute of Chemical Engineers Journal*, **13**, 365.

Sherwood, T. K., Pigford, R. L., & Wilke, C. R. (1975). *Mass Transfer*. New York: McGraw-Hill.

Smoluchowski, M. V. (1917). *Zeitschrift für Physikalische Chemie*, **92**, 129.

Thiele, E. W. (1939). *Industrial and Engineering Chemistry*, **31**, 916.

Toor, H. L. (1969). *Industrial and Engineering Chemistry*, **8**, 655.

Weisz, P. B. (1973). *Science*, **179**, 433.

Weisz, P. B., and Hicks, J. S. (1962). *Chemical Engineering Science*, **17**, 265.

15 MEMBRANES AND OTHER DIFFUSION BARRIERS

Thin membranes can sharply alter rates of mass transfer. These altered rates occur because of physical and chemical interactions between the membrane and the diffusing components. The physical interactions include those encountered in filtration, where different components pass through the membrane at different rates determined largely by the components' sizes. The most common example is reverse osmosis. For example, in the purification of sea water, water molecules pass through a cellulose acetate membrane, but salt ions cannot. These physical interactions are usually described either with phenomenological equations or with mechanical models.

The chemical interactions in membranes produce a variety of unusually dramatic effects. These effects often result from a coupling of diffusion and chemical reaction within the membrane, which is the reason that this chapter is included here in Part IV. These reactions generate fluxes that are much greater than expected, that are not proportional to the concentration gradient, and that can occur backward, from a region of low solute concentration into a region of high solute concentration.

Membranes with chemical interactions can lead to a variety of interesting separations (Lonsdale, 1982). They can separate methane from hydrogen, oxygen from air, or ethylene from ethane. They can selectively separate metal ions like copper and chromium. Moreover, these membranes behave much like those in living systems. They supply an explanation for sugar diffusion in the intestine and for oxygen diffusion in blood.

These membrane effects are the subject of this chapter. The physical effects of pressure and concentration differences are given in Section 15.1; those of concentration and electrical-potential differences are given in Section 15.2. We have emphasized the simplest cases, because this area is otherwise burdened with elaborate algebra.

The chemical effects across membranes involve only concentration differences, but these are coupled with nonlinear chemical reactions. In Section 15.3 we discuss facilitated diffusion, which is the most important example of the chemical interactions. More complicated chemical interactions are listed in Section 15.4, including those responsible for membranes that can concentrate specific solutes. Finally, in Section 15.5 we describe other diffusion barriers where interesting effects occur, including tooth decay.

Membrane diffusion effects

Driving force / Flux	Pressure drop	Concentration difference	Electrical potential
Fluid flow	Permeability (Darcy's law permeability)	Osmosis	Electro-osmosis
Solute flux	Ultra-filtration	Diffusion (diffusive permeability)	Electro-phoresis
Total current	Streaming current	Diffusion current	Electrical current

Fig. 15.1-1. Membrane diffusion effects. The fluxes and forces that commonly occur across membranes are shown here. The four in the upper left-hand corner are discussed in this section; those in the lower right-hand corner are the subject of the next section. This figure does not include the effects of chemical reactions described in Sections 15.3 and 15.4.

Section 15.1. Physical factors in membranes

Mass transfer across membranes can occur because of gradients in pressure, concentration, or electrical potential. These gradients produce not only diffusion but also the other effects named in Fig. 15.1-1. In this section we discuss the four of these effects in the crosshatched region; in the next section we discuss the four in the stippled region.

Three basic effects

Experiments with membranes depend on phenomena that are unimportant in most other systems. Theories for membrane behavior automatically include these phenomena, thus implicitly assuming that everyone knows what they are. When I began research in this area, I did not understand these phenomena; because I suspect others may be in my original position, I am going to describe them here.

The first of these phenomena is osmotic pressure. This pressure occurs whenever a membrane separates a solvent from a solution, as illustrated schematically in Fig. 15.1-2. For simplicity, we regard the membrane as permeable to solvent, but completely impermeable to solute. In this case, the pure solvent has a higher free energy than that in the solution, and so it will tend to flow from left to right. It will continue to flow until the pressure in the solution rises enough to hold back this flow. When the flow ceases, the system is in equilibrium, with the osmotic pressure difference giving a measure of the concentration.

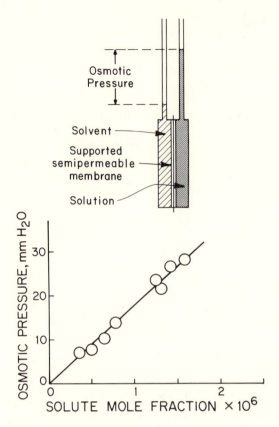

Fig. 15.1-2. Osmotic pressure. At equilibrium, the pressure on the solution must be higher than that on the solvent. This increased pressure, a measure of solution free energy, is often simply related to solute concentration. The specific data shown are for a polyethylene glycol (mol. wt. ca. 20,000) in water.

We want to know the relation between osmotic pressure and concentration more explicitly. To find this, we recognize that at equilibrium, the solvent's chemical potential must be constant:

$$\mu_2(T, p) = \mu_2(T, p + \Delta\Pi) \tag{15.1-1}$$

where $\Delta\Pi$ is the osmotic pressure. The chemical potential of the pure solvent is, of course, already the standard state, but that of the solution must be corrected for both solute concentration and pressure:

$$\mu_2^0(T, p) = \mu_2^0(T, p + \Delta\Pi) + RT \ln x_2$$
$$= \mu_2^0(T, p) + \bar{V}_2(\Delta\Pi) + RT \ln(1 - x_1) \tag{15.1-2}$$

where the superscript 0 signifies the standard state, $\bar{V}_2$ is the partial molar volume of solvent, and x_2 and x_1 are the mole fractions of solvent and solute, respectively. Note that Eq. 15.1-2 implies an ideal solution. Thus, osmotic pressure and solute concentration are related by

$$\bar{V}_2(\Delta\Pi) = -RT \ln(1 - x_1) \doteq RTx_1 + \cdots \tag{15.1-3}$$

or, for a dilute solution where $\bar{V}_2 \doteq c^{-1}$,

$$\Delta\Pi = RTC_1 + \cdots \tag{15.1-4}$$

This relation, restricted to dilute ideal solutions, is called van't Hoff's law. When it was first proposed, it excited enormous interest, for it looked so similar to the ideal-gas law (van't Hoff, 1887). As its restrictions to dilute ideal solutions were realized, it has been deemphasized. For our purposes, it remains a central precept.

The second key experimental observation is for diffusion across a thin membrane at constant temperature and pressure. The flux j_1 due to this diffusion is like that calculated in Section 2.2:

$$j_1 = \frac{DH}{l}(C_{10} - 0) \tag{15.1-5}$$

in which D is the diffusion coefficient, H is the partition coefficient between membrane and adjacent solution, and l is the membrane thickness. This thin-film result has been used again and again throughout this book. For membranes, it is convenient to write this in a different form:

$$j_1 = \omega\Delta\Pi \tag{15.1-6}$$

where the osmotic pressure difference now equals $RT(C_{10} - 0)$, and the solute permeability ω is given by

$$\omega = \frac{DH}{lRT} \tag{15.1-7}$$

This solute permeability typically has dimensions of moles per area-time-atmosphere, so it is another form of mass transfer coefficient like those in Chapter 9. However, ω is the currency in biophysics, and mass transfer coefficients are the coin of engineering, and these currencies are almost never exchanged (Kedem & Katchalsky, 1958).

The third experimental result comes not from diffusion but from fluid mechanics. If pressure is applied to a solvent on one side of a porous membrane, the solvent will flow across the membrane, as shown in Fig. 15.1-3. For slow flow, the flow is proportional to the pressure difference:

$$j_v = L_p \Delta p \tag{15.1-8}$$

This result is called Darcy's law, and L_p is called a hydraulic permeability or a Darcy's law permeability. Just as ω is constant only in dilute solutions, so L_p is constant only for slow flows (Batchelor, 1967). The total flow j_v typically has dimensions of volume per area-time, so L_p has units of length per time-atmosphere.

Flux equations

The flux equations for membrane transport can now be developed by combining the observations about osmotic pressure, membrane diffusion, and fluid flow. To do this, we consider the situation shown schematically in Fig. 15.1-4. In this situation, a concentrated solution at high pressure is

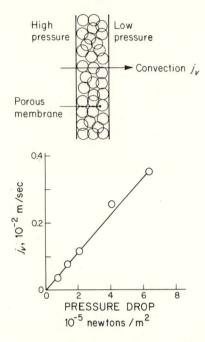

Fig. 15.1-3. Darcy's law. The convection of a pure fluid across a porous membrane is linearly proportional to the pressure drop if the pores are small. The data given are for water flowing in a packed bed that contains 0.5-μm spheres and is 0.014 cm long.

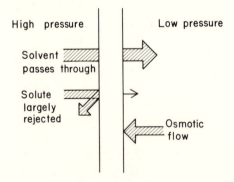

Fig. 15.1-4. Transport across a semipermeable membrane. Here, the solvent flux can be reduced by osmotic pressure, and the solute flux can be altered by convection. Describing this transport process requires at least three independent coefficients.

being forced across a membrane into a dilute solution at lower pressure. The membrane is more permeable to solvent than to solute, and so a concentration difference develops. This concentration difference in turn produces an osmotic pressure opposing the flow.

We want flux equations describing the total flux across the membrane and the flux of solute alone. The total flux j_v must be that caused by the applied pressure minus that caused in the opposite direction by the osmotic pressure:

$$j_v = L_p(\Delta p - \sigma \Delta \Pi) \tag{15.1-9}$$

where L_p is the permeability based on Darcy's law, defined so that j_v has dimensions of length/time, and σ is a new "reflection coefficient" characteristic of the membrane. If the membrane is permeable to solvent but completely impermeable to solute, σ equals unity. If the membrane is equally permeable to both solute and solvent, σ equals zero. The solute flux j_1 is defined relative to the membrane and is again viewed as that due to diffusion plus that caused by convection:

$$j_1 = \omega \Delta \Pi + (1 - \sigma')\bar{C}_1 j_v \tag{15.1-10}$$

in which $\bar{C}_1$ is the average solute concentration $(C_{10} + C_{11})/2$ and σ' is a new transport coefficient to be discussed in detail later. The flux j_1 has the expected dimensions of moles per area-time.

These two new flux equations form a curious contrast with the common result for binary diffusion:

$$\mathbf{n}_1 = -D\nabla c_1 + c_1 \mathbf{v}^0 \tag{15.1-11}$$

In some ways, Eq. 15.1-10 is similar to Eq. 15.1-11, for $\mathbf{n}_1$ and j_1 both describe the solute flux, and ∇c_1 and $\Delta \Pi_1$ both reflect changes in solute concentration. The quantities $\mathbf{v}^0$ and j_v represent the amount of convection.

At the same time, there are profound differences between Eqs. 15.1-9 and 15.1-10 for membranes and Eq. 15.1-11 for binary diffusion. These differences are reflected by the number of transport coefficients involved. For binary diffusion, there is one: D. For membrane transport, there are four: L_p, σ, ω, and $1 - \sigma'$. This is because there is only one force, the concentration gradient, responsible for binary diffusion. In contrast, there are two independent forces, the concentration difference and the pressure difference, involved in membrane transport. Thus, membrane transport is somewhat like ternary diffusion, where the three components are the solute, the solvent, and the membrane.

There is an alternative way of formulating these flux equations that is parallel to ternary diffusion and is based on irreversible thermodynamics (deGroot & Mazur, 1962; Katchalsky & Curran, 1967). The basic strategy is identical with that given in Section 8.2; so details are not given here. The key results are

$$j_v = (L_p)\Delta p + (-L_p \sigma)\Delta \Pi \tag{15.1-12}$$

and

$$j_D = \frac{j_1}{\bar{C}_1} - j_v$$

$$= (-L_p \sigma')\Delta p + \left(\frac{\omega}{\bar{C}_1} + L_p \sigma \sigma'\right)\Delta \Pi \tag{15.1-13}$$

Although these equations are inconvenient in practice, they have interesting theoretical implications. These center on the relation between the coefficients σ and σ'. If the Onsager reciprocal relations are valid across membranes, then these two are equal. In this case, the number of coefficients necessary to describe membrane diffusion will drop from four to three. This potential simplification, abstrusely justified by irreversible thermodynamics, has made Eqs. 15.1-12 and 15.1-13 popular.

However, a variety of studies show that the Onsager reciprocal relations are not always valid across membranes (Mason et al., 1972; Cussler, 1976). Some of these studies use phenomenological arguments that are difficult for me to follow; others are based on particular chemical mechanisms. All show instances in which σ and σ' are not equal. This is significant, because irreversible thermodynamics claims to be a general theory, valid without the restrictions of a particular model. As a result, I avoid Eqs. 15.1-12 and 15.1-13 and use Eqs. 15.1-9 and 15.1-10. I assume that σ and σ' are not equal, unless there is accurate experimental evidence that they are.

Example 15.1-1: Fruit juice concentration

Eastern European farmers produce a variety of fruit juices, which they wish to dehydrate to prolong shelf life and facilitate transportation. One dehydration method is to put the juice in a plastic bag and drop the bag into brine at 10°C. If the bag is permeable to water, but not to salt or juice components, then osmotic flow will concentrate the juice.

Is the osmotic pressure generated in this way significant? Assume that the juice contains solids equivalent to 1 wt% sucrose and that the brine contains 35 g sodium chloride per 100 g water.

Solution. For simplicity, assume that the juice and brine are ideal solutions. Then, from Eq. 15.1-3, the osmotic pressure difference is

$$\Delta\Pi = \frac{RT}{\bar{V}_2} \ln \left(\frac{(1 - x_1)_{\text{juice}}}{(1 - x_1)_{\text{brine}}} \right)$$

$$= \frac{(0.082 \text{ liter-atm/mol-}°\text{K})(283°\text{K})}{0.018 \text{ liter/mol}}$$

$$\cdot \ln \left(\frac{1 - \dfrac{(0.01 \text{ g})/(342 \text{ g/mol})}{(0.01 \text{ g})/(342 \text{ g/mol}) + (0.99)/(18 \text{ g/mol})}}{1 - \dfrac{2[(35 \text{ g})/(58.5 \text{ g/mol})]}{2[(35 \text{ g})/(58.5 \text{ g/mol})] + (100 \text{ g})/(18 \text{ g/mol})}} \right)$$

$$= 250 \text{ atm}$$

This large pressure can cause rapid dehydration.

Example 15.1-2: Finding membrane coefficients for ultrafiltration

To study the transport properties of glucose and water across an ultrafiltration membrane, we clamp a piece of the membrane across one end of a tube 0.86 cm in diameter and immerse the tube 2.59 cm into a large beaker of buffered water. We then fill the tube with a 0.03-mol/liter glucose solution. The result is somewhat like the diaphragm-cell apparatus described in Sections 2.2 and 8.3.

We make two experiments, both at 25°C. In the first, we adjust the tube so that the solution and solvent levels are initially the same, and we leave both solution and solvent open to the atmosphere. We find that the solute concentration drops 0.4% in 1.62 hr, and the total volume of solution increases 0.35%. In the second experiment, we initially adjust the solution level to be the same as the solvent level and leave the solution open to the atmosphere, but pull a vacuum of 733 mm Hg on the solvent. In this case, after 0.49 hr, the solute concentration decreases 0.125%, and the solution volume also decreases 0.05%. Find the transport coefficients across this membrane.

Solution. These experiments have been made over such short times that neither volume nor concentration has changed much. As a result, we can use the flux equations directly, without a messy integration. We first consider the total flow. In the first experiment, there is no applied pressure difference; so, from Eqs. 15.1-4 and 15.1-9,

$$j_v = -L_p\sigma(C_1 RT)$$

$$= -(0.0035)\left(\frac{\pi(0.43 \text{ cm})^2(2.59 \text{ cm})}{\pi(0.43 \text{ cm})^2(1.62 \text{ hr})(3600 \text{ sec/hr})}\right)$$

$$= -L_p\sigma[(0.03 \text{ mol/liter})(0.082 \text{ liter-atm/mol-°K})(298°K)]$$

$$L_p\sigma = 2.1 \cdot 10^{-6} \text{ cm/sec-atm}$$

In the second experiment, there are both pressure and osmotic differences; so

$$j_v = L_p\Delta p_1 - L_p\sigma\Delta\Pi_1$$

$$(0.0005)\left(\frac{2.59 \text{ cm}}{(0.490 \text{ hr})(3,600 \text{ sec/hr})}\right)$$

$$= L_p\left(\frac{733 \text{ mm Hg}}{760 \text{ mm Hg/atm}}\right)$$

$$- \left(\frac{2.1 \cdot 10^{-6} \text{ cm}}{\text{sec-atm}}\right)\left[\left(\frac{0.03 \text{ mol}}{\text{liter}}\right)\left(\frac{0.082 \text{ liter-atm}}{\text{mol-°K}}\right)(298°K)\right]$$

Thus,

$$L_p = 2.4 \cdot 10^{-6} \text{ cm/atm-sec}$$

and

$$\sigma = 0.90$$

About 90% of the glucose is rejected by this membrane.

Next, we turn to the glucose flux. From Eq. 15.1-10, we have

$$j_1 = \omega\Delta\Pi_1 + (1 - \sigma')\bar{C}_1 j_v$$

Using the values of j_v from the foregoing, we obtain

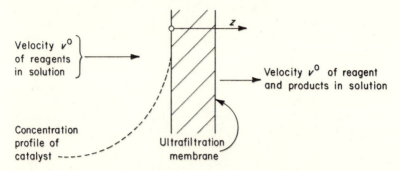

Fig. 15.1-5. Concentration polarization in ultrafiltration. The membrane is permeable to the reagents, but not to the catalyst. The resulting catalyst concentration profile causes an osmotic gradient inhibiting reagent flow.

$$0.004\left(\frac{(2.59 \text{ cm})(3 \cdot 10^{-5} \text{ mol/cm}^3)}{(1.62 \text{ hr})(3,600 \text{ sec/hr})}\right)$$

$$= \omega[(0.03 \text{ mol/liter})(0.082 \text{ liter-atm/mol-°K})(298°)]$$

$$+ (1 - \sigma')(0.015 \text{ mol/1,000 cm}^3)(-1.55 \cdot 10^{-6} \text{ cm/sec})$$

for the first experiment, and

$$0.00125\left(\frac{(2.59 \text{ cm})(3 \cdot 10^{-5} \text{ mol/cm}^3)}{(0.490 \text{ hr})(3,600 \text{ sec/hr})}\right)$$

$$= (0.03 \text{ mol/liter})(0.083 \text{ liter-atm/mol-°K})(298°\text{K})$$

$$+ (1 - \sigma')(0.015 \text{ mol/1,000 cm}^3)(0.73 \cdot 10^{-6} \text{ cm/sec})$$

for the second. Solving these,

$$\omega = 7.4 \cdot 10^{-11} \text{ mol/cm}^2\text{-sec-atm}$$

$$\sigma' = 0.95$$

Whether or not σ' actually equals σ in this case depends on the experimental errors involved.

Example 15.1-3: Concentration polarization in a membrane reactor

Homogeneous catalysis has not been widely used in industry, not because the catalysts are ineffective but because they are expensive and difficult to recover after the reaction. One way around these problems is the membrane reactor shown in Fig. 15.1-5. The ultrafiltration membrane used will pass all components except the catalyst, which is added only at the start of the experiment. However, the flux of solvent, reactants, and products is reduced by the concentration profile of the catalyst near the membrane. This reduction is called concentration polarization.

Find the steady-state concentration profile of the catalyst in terms of the total catalyst per membrane area M. Then calculate the flux through the membrane, including the amount it is reduced by this concentration polarization.

Solution. At steady state, a mass balance on the catalyst in solution gives

$$0 = -\frac{dn_1}{dz}$$

Integration shows that the flux n_1 is a constant. At the membrane surface, this flux is zero; so

$$n_1 = 0 = -D\frac{dc_1}{dz} + v^0 c_1$$

Integrating this, we find the concentration profile

$$c_1 = (\text{constant})e^{v^0 z/D}$$

Remember that according to the coordinates in Fig. 15.1-5, z is always negative. From the problem statement, we know that

$$\int_{-\infty}^{0} c_1 \, dz = M$$

so the desired concentration profile is

$$c_1 = \frac{Mv^0}{D} e^{v^0 z/D}$$

The flux through this membrane is reduced because of osmotic effects:

$$j_v = L_p(\Delta p - \sigma \Delta \Pi)$$

The total flux j_v is just v^0, and because the catalyst cannot get through the membrane at all, σ is unity. Moreover, $\Delta \Pi$ can be found from Eq. 15.1-4. Using the maximum value of c_1 at $z = 0$,

$$j_v = L_p(\Delta p - c_1 RT)$$
$$\doteq L_p\left(\Delta p - \frac{RTMv^0}{D}\right)$$

The first term in the parentheses represents the driving force without catalyst; the second is the reduction in the flux caused by the concentration polarization.

Section 15.2. Electrolyte diffusion across membranes

We next turn to the diffusion of electrolytes across membranes, a complex topic. The complexity has several sources: the electrolyte's ionization, the applied electric field, and electrostatic coupling in multicomponent mixtures. These effects are superimposed on top of the osmotic and pressure differences described in the earlier section.

To illustrate these processes, we consider first the diffusion of a single electrolyte without an applied electric field. We then consider the effects of applying such a field and of forming ion pairs. Finally, we mention results for polyelectrolytes, where complications like Donnan equilibria are important. In each case, we assume that the solutions are isobaric and dilute, so that solvent transport is small, thus finessing questions about osmotic pressure.

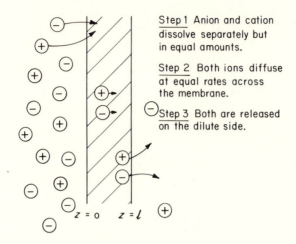

Step 1 Anion and cation dissolve separately but in equal amounts.

Step 2 Both ions diffuse at equal rates across the membrane.

Step 3 Both are released on the dilute side.

Fig. 15.2-1. 1-1 Electrolyte diffusion across an uncharged membrane. When no external potential is applied, there is no current, and cation and anion diffuse at the same speed. The diffusion coefficient is a harmonic average of those of the ions, but the partition coefficient is a geometric average.

Such simplifications mean missing effects like electro-osmosis, but they seem the only route to a clear description.

1-1 Electrolyte without an electric field

The results for this simplest case, shown schematically in Fig. 15.2-1, are largely implicit in the material of Section 6.2. The diffusion of each ion is given by

$$-j_i = D_i \frac{dc_i}{dz} + z_i c_i \frac{\mathcal{F}}{RT} \frac{d\psi}{dz} \tag{15.2-1}$$

For a 1-1 electrolyte of charge $|z|$ without an applied electric field, the equations for the two ions are easily combined and integrated to give the diffusion potential:

$$\psi_0 - \psi_l = \frac{RT}{|z|\mathcal{F}} \left(\frac{D_2 - D_1}{D_2 + D_1} \right) \ln\left(\frac{C_{10}}{C_{1l}} \right) \tag{15.2-2}$$

where the subscripts 1 and 2 refer to the cation and anion, respectively, and the subscripts 0 and l refer to the different sides of the membrane interfaces. This, in turn, can be used to simplify the flux equation:

$$-j_1 = D \frac{dc_1}{dz} = \left(\frac{2}{1/D_1 + 1/D_2} \right) \frac{dc_1}{dz} \tag{15.2-3}$$

This equation is easily integrated across the thin film to give

$$j_1 = \frac{DH}{l} (C_{10} - C_{1l}) \tag{15.2-4}$$

where $C_{10} - C_{1l}$ is the concentration difference across the membrane and H is the partition coefficient at the membrane interfaces. Just as D is a diffusion coefficient averaged between the ions, so H must be a partition coefficient averaged between the ions.

What sort of an average partition coefficient is H? To answer this question, we must turn to the definition of equilibrium across the interface, which for the cation 1 is

$$\mu_1 = \mu_1'$$

$$\mu_1^0 + RT \ln C_1 + z_1 \mathcal{F} \psi = (\mu_1^0)' + RT \ln c_1 + z_1 \mathcal{F} \psi' \tag{15.2-5}$$

where the primed and unprimed variables refer to the membrane and to the adjacent solution, respectively. Note that this equation implicitly assumes ideal solutions. A similar equality exists for the anion 2:

$$\mu_2^0 + RT \ln C_2 + z_2 \mathcal{F} \psi = (\mu_2^0)' + RT \ln c_2 + z_2 \mathcal{F} \psi' \tag{15.2-6}$$

We can eliminate the potentials ψ and ψ' by adding and rearranging these equations to get

$$c_1 c_2 = H_1 H_2 C_1 C_2 \tag{15.2-7}$$

where

$$H_i = e^{[(\mu_i^0)' - \mu_i^0]/RT} \tag{15.2-8}$$

Because we have only one 1-1 electrolyte, cation and anion concentrations are equal, and

$$c_1 = HC_1 = \sqrt{H_1 H_2}\, C_1 \tag{15.2-9}$$

Thus, the partition coefficient is a *geometric* average of the properties of the two ions. In contrast, the diffusion coefficient in Eq. 15.2-4 is a *harmonic* average of the properties of the two ions. Except for this point, the membrane diffusion of a 1-1 electrolyte proceeds as expected.

1-1 Electrolyte with an external electric field

When an external electric potential is applied across the membrane, the process is more complicated (Robinson & Stokes, 1960; Schultz, 1980). To examine this potential more fully, consider a membrane separating two well-stirred aqueous solutions like those in Fig. 15.2-2. Even without an externally applied electric field, the different electrolyte concentrations still produce a potential that can be calculated from Eq. 15.2-5 or 15.2-6:

$$\mu_i^0 + RT \ln C_{i0} + z_i \mathcal{F} \psi_0 = \mu_i^0 + RT \ln C_{il} + z_i \mathcal{F} \psi_l \tag{15.2-10}$$

or

$$(\psi_0 - \psi_l) = \left(\frac{RT}{z_i \mathcal{F}}\right) \ln\left(\frac{C_{1l}}{C_{10}}\right) \tag{15.2-11}$$

This is called the Nernst potential, and it is caused by a difference in electrolyte concentrations. Note that this difference is an equilibrium property, unlike the dynamic diffusion potential given in Eq. 15.2-2. When an external electric field is applied, the total applied potential is the sum of the electrode and Nernst potentials:

$$(\psi_0 - \psi_l) = (\psi_0 - \psi_l)_{\text{electrodes}} + \frac{RT}{z_i \mathcal{F}} \ln\left(\frac{C_{il}}{C_{i0}}\right) \tag{15.2-12}$$

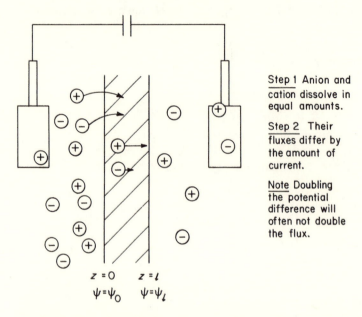

Step 1 Anion and cation dissolve in equal amounts.

Step 2 Their fluxes differ by the amount of current.

Note Doubling the potential difference will often not double the flux.

Fig. 15.2-2. Electrolyte diffusion across a membrane and in an applied field. When an electric field is applied, cation and anion diffusion will occur at different rates. This difference is closely related to the current. Note that the situation shown requires adding and removing ions at the electrodes.

An applied field also generates a current:

$$i = \mathcal{F}(z_1 j_1 + z_2 j_2) \tag{15.2-13}$$

Equations for non-1-1 electrolytes are fancier algebraically, but the same conceptually.

We now want to combine Eq. 15.2-1 and Eq. 15.2-12 and integrate across the thin film. This integration is a formidable task leading to a clumsy transcendental solution (MacInnes, 1961; Lakshminarayanaiah, 1969; Sten-Knudsen, 1978). Instead, many successfully use an approximation, due to Goldman (1943), that the total potential difference is a constant, that $d\psi/dz$ equals $(\psi_l - \psi_0)/l$. On this basis, the result is

$$j_i = \frac{D_i H_i z_i \mathcal{F}(\psi_0 - \psi_l)}{lRT} \left(\frac{C_{i0} - C_{il} e^{z_i \mathcal{F}(\psi_l - \psi_0)/RT}}{1 - e^{z_i \mathcal{F}(\psi_l - \psi_0)/RT}} \right) \tag{15.2-14}$$

where H_i is the single-ion partition coefficient given by Eq. 15.2-8. This equation is approximate, most likely to be accurate for thin membranes with little fixed charge (Schultz, 1980). Even so, the flux is not linear in either the concentration or the potential difference. This is explored further in the examples that follow.

An associated electrolyte without an electric field

So far, we have assumed that any ions in water will produce similar ions within the membrane. This may not be true. For example, imagine that

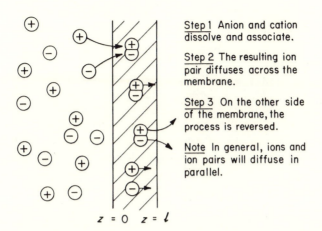

$$z = 0 \quad z = l$$

Step 1 Anion and cation dissolve and associate.

Step 2 The resulting ion pair diffuses across the membrane.

Step 3 On the other side of the membrane, the process is reversed.

Note In general, ions and ion pairs will diffuse in parallel.

Fig. 15.2-3. Electrolyte diffusion across a membrane of low dielectric constant. In such a membrane, much of the ionic solute may be present not as ions with positive or negative charge but as ion pairs of no net charge. The flux then can vary with the square of the concentration difference across the membrane.

a single 1-1 electrolyte is diffusing from one aqueous solution across a lipophilic membrane into another aqueous solution. If the dielectric constant of the membrane is low, any diffusing ions will tend to associate, forming ion pairs of no net charge (Sollner & Shean, 1964). Thus, we have simultaneous diffusion of ions and ion pairs, as shown in Fig. 15.2-3.

To calculate the steady-state flux in this case, we write continuity equations for cations 1 and ion pairs 12 (Duffey et al., 1978):

$$0 = D \frac{d^2c_1}{dz^2} - r_{12} \tag{15.2-15}$$

$$0 = D_{12} \frac{d^2c_{12}}{dz^2} + r_{12} \tag{15.2-16}$$

These equations can be added and integrated twice to give

$$j_1 + j_{12} = \frac{D}{l}(c_{10} - c_{1l}) + \frac{D_{12}}{l}(c_{12,0} - c_{12,l}) \tag{15.2-17}$$

where the integration constant $j_1 + j_{12}$ represents the total flux of ions and ion pairs, and the concentrations given are those within the membrane but at the interfaces. Because ion pairs form electrostatically, their rate of formation is so fast that it approaches equilibrium, and

$$c_{12} = Kc_1c_2 = Kc_1^2 \tag{15.2-18}$$

where K is the association constant. Thus,

$$j_1 + j_{12} = \frac{DH}{l}(C_{10} - C_{1l}) + \frac{D_{12}H^2K}{l}(C_{10}^2 - C_{1l}^2) \tag{15.2-19}$$

where H is defined by Eq. 15.2-9. When C_{1l} is zero and C_{10} is small, the diffusing solute will be ionic, and the flux will be proportional to C_{10}. When

C_{1l} is zero and C_{10} is large, diffusion of ion pairs will be paramount, and the flux will vary with C_{10}^2.

Polyelectrolytes

Finally, we want to mention the effects of proteins and other poly-electrolytes, either when these are in solution on one side of the membrane or when these are components of the membrane itself. At equilibrium, equal concentrations of any small electrolytes will not exist throughout these systems. For example, if a membrane containing fixed anionic charges is placed in a sodium chloride solution, the sodium chloride concentration in the membrane will be different than if the membrane had no charge, even if all solutions are behaving ideally. These new "Donnan equilibria" are the natural result of chemical potentials described by Eq. 15.2-5. They produce a variety of potentials and fluxes known as Donnan effects. For details, you should consult more specialized texts (Lakshminarayanaiah, 1969; Sten-Knudsen, 1978).

Example 15.2-1: Salt diffusion across an uncharged membrane

Potassium chloride is diffusing at 25°C from a 0.1-M aqueous solution through a water-insoluble membrane 0.0022 cm thick into a 0.01-M aqueous solution. The partition coefficients between water and the membrane are $5.2 \cdot 10^{-4}$ for the potassium ion and $9.4 \cdot 10^{-4}$ for the chloride ion. The ionic diffusion coefficients are $2.1 \cdot 10^{-5}$ cm²/sec and $1.6 \cdot 10^{-5}$ cm²/sec for K^+ and Cl^-, respectively. (a) What is the flux across the membrane with no applied potential? (b) What is the Nernst potential between the two solutions, as measured with Ag–AgCl electrodes? What is the diffusion potential? (c) If an electrode potential $\psi_0 - \psi_l$ of 100 mV is applied, what are the cation and anion fluxes? What is the current?

Solution. (a) The ionic diffusion coefficients can be combined to give

$$D = \frac{2}{1/D_1 + 1/D_2} = \frac{2}{(1/2.1 + 1/1.6) \cdot 10^5 \text{ sec/cm}^2}$$
$$= 1.8 \cdot 10^{-5} \text{ cm}^2/\text{sec}$$

The permeability is

$$H = \sqrt{H_1 H_2}$$
$$= \sqrt{(5.2 \cdot 10^{-4})(9.4 \cdot 10^{-4})} = 7.0 \cdot 10^{-4}$$

The flux can now be found from Eq. 15.2-5:

$$j_1 = j_2 = \left[\frac{DH}{l}\right](C_{10} - C_{1l})$$

$$= \left[\frac{(1.8 \cdot 10^{-5} \text{ cm}^2/\text{sec})(7.0 \cdot 10^{-4})}{0.0022 \text{ cm}}\right][(10^{-4} \text{ mol/cm}^3) - (10^{-5} \text{ mol/cm}^3)]$$

$$= 5.2 \cdot 10^{-10} \text{ mol/cm}^2\text{-sec}$$

(b) The Nernst potential for the cation is found from Eq. 15.2-11:

$$\psi_0 - \psi_l = \frac{RT}{z_i \mathscr{F}} \ln\left(\frac{C_{1l}}{C_{10}}\right)$$

$$= \frac{(8.31 \text{ joules/mol-}^\circ\text{K})(298^\circ\text{K})(\ln 0.01/0.10)}{(+1)\ 96{,}500\ \dfrac{\text{coulombs}}{\text{mol}} \left[\dfrac{1 \text{ watt}}{\text{volt (coulombs/sec)}}\right] \dfrac{\text{joule/sec}}{\text{watt}}}$$

$$= -0.059 \text{ volt} \equiv -59 \text{ mV}$$

The result for the anion is +59 mV. This factor is electrochemically ubiquitous, the same for each decade of concentration in an ideal solution. It occurs especially frequently in studies with specific ion electrodes.

The diffusion potential can be calculated from Eq. 15.2-2:

$$\psi_0 - \psi_l = \frac{RT}{|z_i|\mathscr{F}} \left(\frac{D_2 - D_1}{D_1 + D_2}\right) \ln\left(\frac{C_{10}}{C_{1l}}\right)$$

$$= \frac{(8.31 \text{ joules/mol-}^\circ\text{K})(298^\circ\text{K})}{96{,}500 \text{ joules/mol-volt}} \left(\frac{9.4 - 5.2}{5.2 + 9.4}\right) \ln\left(\frac{0.10}{0.01}\right)$$

$$= 0.017 \text{ volt} = +17 \text{ mV}$$

Note that the diffusion potential depends on the diffusion coefficients.

(c) The total applied potential can be calculated from Eq. 15.2-12:

$$\psi_0 - \psi_l = 100 \text{ mV} - \frac{59 \text{ mV}}{z_i}$$

For the cations,

$$z_1 \frac{\mathscr{F}(\psi_0 - \psi_l)}{RT} = \frac{(+1)(96{,}500 \text{ amp-sec/mol})}{(8.31 \text{ joules/mol-}^\circ\text{K})(298^\circ\text{K})} \frac{(0.100 - 0.59 \text{ volt})}{(\text{amp-volt})/(\text{joule/sec})}$$

$$= +1.6$$

The cation flux calculated from Eq. 15.2-14 is

$$j_1 = \frac{(2.1 \cdot 10^{-5} \text{ cm}^2/\text{sec})(5.2 \cdot 10^{-4})(+1.6)}{0.0022 \text{ cm}}$$

$$\cdot \left(\frac{(10^{-4} \text{ mol/cm}^3) - (10^{-5} \text{ mol/cm}^3)e^{-1.6}}{1 - e^{-1.6}}\right)$$

$$= 9.7 \cdot 10^{-10} \text{ mol/cm}^2\text{-sec}$$

This flux is increased four times by the applied potential.

For the anions, parallel calculations show that

$$z_2 \frac{\mathscr{F}(\psi_0 - \psi_l)}{RT} = -6.2$$

and

$$j_2 = \frac{(1.6 \cdot 10^{-5} \text{ cm}^2/\text{sec})(9.4 \cdot 10^{-4})(-6.2)}{0.0022 \text{ cm}}$$

$$\cdot \left(\frac{(10^{-4} \text{ mol/cm}^3) - (10^{-5} \text{ mol/cm}^3)e^{+6.2}}{1 - e^{+6.2}}\right)$$

$$= -4.2 \cdot 10^{-10} \text{ mol/cm}^2\text{-sec}$$

This potential is large enough to reverse the direction of the anion flux. The current is found from Eq. 15.2-13:

$$i = \{[(+1)(9.7 \cdot 10^{-10}) + (-1)(-4.2 \cdot 10^{-10})] \text{ mol/cm}^2\text{-sec}\}$$
$$\cdot (96,500 \text{ amp-sec/mol})$$

$$= 0.13 \text{ ma/cm}^2$$

Remember that this result is subject to the approximations of the Goldman equation, Eq. 15.2-14.

Example 15.2-2: The flux of ion pairs

Tetrabutylammonium nitrate (Bu_4NNO_3) in aqueous solution is diffusing across a thin membrane of *n*-heptyl cyanide into pure water. At a solution concentration of 0.005 mol/liter, the flux across the membrane is $1.5 \cdot 10^{-10}$ mol/sec-cm^2 (Duffey et al., 1978). (a) Assuming that transport is ionic, estimate the flux with a solution of 0.5 mol/liter. (b) In fact, the flux in this more concentrated solution is $850 \cdot 10^{-10}$ mol/cm^2-sec because of ion-pair formation. Estimate the association constant for this information.

Solution. (a) Because this solution is dilute, we neglect water transport and use Eq. 15.2-3:

$$j_1 = \frac{DH}{l} (C_{10} - C_{1l})$$

$$1.6 \cdot 10^{-10} \text{ mol/sec-cm}^2 = \frac{DH}{l} (0.005 \text{ mol/}10^3 \text{ cm}^3)$$

Thus,

$$\frac{DH}{l} = 3.2 \cdot 10^{-5} \text{ cm/sec}$$

At the higher concentration,

$$j_1 = (3.2 \cdot 10^{-5} \text{ cm/sec})(0.5 \text{ mol/}1,000 \text{ cm}^3)$$
$$= 160 \cdot 10^{-10} \text{ mol/cm}^2\text{-sec}$$

This is five times lower than the experimental value.

(b) To estimate the association constant, we first assume that both ions and ion pairs have the same diffusion coefficient. We then have, from Eq. 15.2-19,

$$j_1 + j_{12} = \frac{DH}{l} C_{10}(1 + HKC_{10})$$

$$850 \cdot 10^{-10} \text{ mol/cm}^2\text{-sec} = (160 \cdot 10^{-10} \text{ mol/cm}^2\text{-sec})[1 + K(0.5 \text{ mol/liter})]$$

so

$$HK = 8.6 \text{ liter/mol}$$

This constant has the units of inverse concentration because ion-pair formation is a second-order process.

Section 15.3. Facilitated diffusion

In this section we discuss membrane processes involving both diffusion and reversible chemical reactions. These processes are among the most dramatic discussed in this book, for they produce highly coupled and selective transport. Such processes are believed to be common in living systems.

As an example, imagine we are studying diffusion of Na^+ and glucose across sections of rat gut. We measure the solubilities of these solutes in the lipids extracted from the gut, and we find them low, much less than in water. Accordingly, we might use our thin-film diffusion results or those in the earlier sections of this chapter to predict the following:

1   The solute fluxes should be low, because the solubilities are low.
2   Their ratio should be about the same as that of the solubilities, because the diffusion coefficients are about equal.
3   They should be linear in the concentration differences; that is, doubling the concentration differences should double the fluxes.
4   The fluxes of Na^+ and glucose should be independent, for multicomponent effects in dilute solutions are minor.

These predictions seem routine and boring.

However, experimental results on living-tissue sections show that each of these predictions is wrong, very wrong:

1   The fluxes of both Na^+ and glucose are 100,000 times greater than expected.
2   Their ratio is about 100 times different than that expected from solubilities.
3   The fluxes can reach a maximum value, independent of the concentration differences across the membrane.
4   The fluxes are strongly coupled; a flux of glucose causes a big flux of Na^+, and vice versa.

Clearly, to explain these unexpected results, we need a very different approach.

The physiologists and biochemists who first made these experiments recognized that the results could be explained by small amounts of reactive molecules, or "mobile carriers," moving within the membrane (Osterhout, 1935; Widdas, 1952; Stein, 1967). They also found that membrane transport often shows many characteristics of enzyme kinetics, and so can be interpreted with mathematical equations developed for enzyme reactions. However, these scientists were correctly cautious about their hypothesis of mobile carriers. After all, few if any of these carrier molecules had been isolated. Moreover, pretending that the mobile carrier is only a hypothesis allows it to have metaphysical properties. For example, some theories suggest that mobile carriers have different diffusion coefficients moving from left to right than they have from right to left. These metaphysical properties

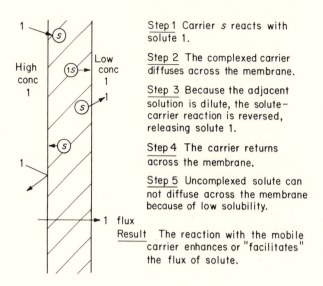

Step 1 Carrier *s* reacts with solute 1.

Step 2 The complexed carrier diffuses across the membrane.

Step 3 Because the adjacent solution is dilute, the solute–carrier reaction is reversed, releasing solute 1.

Step 4 The carrier returns across the membrane.

Step 5 Uncomplexed solute can not diffuse across the membrane because of low solubility.

Result   The reaction with the mobile carrier enhances or "facilitates" the flux of solute.

Fig. 15.3-1. The simple mobile carrier mechanism. This scheme can explain why facilitated diffusion is unexpectedly rapid, unusually selective, and nonlinear in the concentration difference across the membrane.

have disappeared as membranes containing well-defined mobile carriers have been more completely studied.

The simplest way in which one of these chemically well-defined mobile carriers operates is shown schematically in Fig. 15.3-1. The two vertical lines bounding the crosshatched area represent the membrane, which separates a concentrated solution on the left-hand side from a dilute solution on the right-hand side. Almost all of the solute transported across the membrane is transported via complex formation with the mobile carrier. This mechanism qualitatively explains the four experimental results at the beginning of this section:

1   Fluxes are much larger than expected, because the small amounts of mobile carrier solubilize the diffusing solute.

2   The transport can be selective, because the carrier–solute reaction can be selective.

3   The fluxes reach a constant value at high concentrations, when there are insufficient carrier molecules available to carry all the available solutes.

4   Fluxes can be very strongly coupled when two diffusing solutes react competitively or cooperatively with the mobile carrier.

Obviously, this explanation is qualitative, and it needs a more exact analysis before it can be applied quantitatively to experimental results.

As before, we consider the simplest case for detailed analysis, and later explore extensions to more complex situations. We assume that solute and carrier are nonionic and constantly reacting within the membrane:

$$\text{(solute 1)} + \text{(carrier } s\text{)} \rightleftarrows \text{(complex } 1s\text{)} \tag{15.3-1}$$

These three components must satisfy the continuity equations for one-dimensional steady-state transport across the membrane:

$$0 = D\frac{d^2c_1}{dz^2} - r_{1s} \tag{15.3-2}$$

$$0 = D\frac{d^2c_s}{dz^2} - r_{1s} \tag{15.3-3}$$

$$0 = D\frac{d^2c_{1s}}{dz^2} + r_{1s} \tag{15.3-4}$$

where r_{1s} is the rate of formation of the complex within the membrane. Note that we have assumed that all diffusion coefficients of all species in the membrane are equal. This assumption is not necessary, but it greatly simplifies the analysis and focuses attention on the effects of the solute–carrier reaction.

These continuity equations are subject to the restraints that

$$z = 0, \quad c_1 = HC_{10} \tag{15.3-5}$$
$$z = l, \quad c_1 = HC_{1l} \tag{15.3-6}$$

where 0 and l denote the two sides of the membrane, and

$$\frac{1}{l}\int_0^l (c_s + c_{1s})\, dz = \bar{c} \tag{15.3-7}$$

where $\bar{c}$ is the average carrier concentration in the membrane. For a complete solution, three more restraints are needed. To simplify the mathematics, we make an additional assumption about these restraints for each of three special cases. These cases then provide the more common limits of mobile carrier behavior.

Fast reaction

The first special case, that of fast reaction, has the greatest application. It satisfactorily describes carrier-assisted membrane transport in many biological systems, and it is usually successful in predicting the behavior of chemically well-defined membranes. In this case, both the forward and reverse reactions in Eq. 15.3-1 are always much faster than diffusion (Ward, 1970). As a result,

$$c_{1s} = Kc_1c_s \tag{15.3-8}$$

where K is the equilibrium constant of the reaction in Eq. 15.3-1. In addition, this means that

$$z = 0, l, \quad j_s + j_{1s} = 0$$

$$D\left(\frac{dc_s}{dz} + \frac{dc_{1s}}{dz}\right) = 0 \tag{15.3-9}$$

at the membrane boundaries. Equations 15.3-8 and 15.3-9 are the three missing restraints.

Because the case of fast reaction seems deceptively reasonable, we should briefly discuss why it cannot be exact. Equation 15.3-9 is the key. This relation implies that uncomplexed mobile carrier molecules diffuse right up to the membrane wall, instantaneously react, and move away at the same rate but in opposite direction. The molecules never have zero velocity; they instantaneously change from positive to negative velocity. This will be approximately true only when the second Damköhler number is large:

$$\frac{l^2}{Dt_{1/2}} \gg 1 \qquad (15.3\text{-}10)$$

where $t_{1/2}$ is the half-life of either the forward or reverse mobile carrier reaction (Schultz et al., 1974). For most industrial membranes, where l is at least several microns, this condition is easily satisfied. For a membrane 100 Å thick, this condition is stringent; for example, if the reaction-rate half-life is 10^{-7} sec, the derivation presented later is in error by about 10%.

With these approximations, the flux across the membrane can be found in a straightforward fashion. Equations 15.3-3 and 15.3-4 are added, integrated, combined with Eq. 15.3-9, integrated again, and combined with Eq. 15.3-7 to give

$$c_s + c_{1s} = \bar{c} \qquad (15.3\text{-}11)$$

everywhere throughout the membrane. This result and Eq. 15.3-8 are then combined with the sum of Eqs. 15.3-2 and 15.3-4 and integrated once:

$$-(j_1 + j_{1s}) = D \left[\frac{dc_1}{dz} + \frac{d}{dz} \left(\frac{K\bar{c}c_1}{1 + Kc_1} \right) \right] \qquad (15.3\text{-}12)$$

where the integration constant $j_1 + j_{1s}$ represents the total flux of solute 1. When this result is integrated from $z = 0$ to $z = l$,

$$j_1 + j_{1s} = \frac{DH}{l} (C_{10} - C_{1l})$$

$$+ \frac{DH}{l} \left[\frac{K\bar{c}}{(1 + HKC_{10})(1 + HKC_{1l})} \right] (C_{10} - C_{1l}) \qquad (15.3\text{-}13)$$

The first term on the right-hand side represents the flux due to uncomplexed solute, and the second is the flux caused by carrier-assisted diffusion.

This result explains the experimental surprises listed at the start of this section. Frequently, the diffusion of uncomplexed solute is much less than that of complexed solute, so that only the second term on the right-hand side of Eq. 15.3-13 is experimentally significant. In this case, several integrating limits occur when C_{1l} and C_{10} are small. First, the flux is

$$j_1 + j_{1s} \doteq j_{1s} = \left(\frac{DHK\bar{c}}{l} \right) (C_{10} - C_{1l}) \qquad (15.3\text{-}14)$$

This flux is larger and more selective because it is proportional to $K\bar{c}$ and thus is altered by the chemical reaction. If C_{10} becomes large, then the flux reaches a constant value:

$$j_1 + j_{1s} \doteq j_{1s} = \frac{DHK\bar{c}}{l} \qquad (15.3\text{-}15)$$

which again is consistent with experiment. However, if the solute–carrier reaction is irreversible, K becomes infinite, and

$$j_1 + j_{1s} \doteq j_1 = \frac{DH}{l}(C_{10} - C_{1l}) \qquad (15.3\text{-}16)$$

The carrier-assisted flux is effectively poisoned by this type of reaction.

Before leaving this case of fast reaction, we need to make one further comment. The flux equations developed in this section assume that the membrane chemistry is known. In fact, for most biological membranes, this chemical detail is speculative; so the flux equations with $C_{1l} = 0$ are sometimes written as

$$j_1 + j_{1s} = \frac{v_{max}C_{10}}{K_m + C_{10}} \qquad (15.3\text{-}17)$$

The quantities v_{max} and K_m, easily found by comparison with Eq. 15.3-13, can be described as the diffusion velocity and the Michaelis constant. This suggestion of enzyme kinetics is not casual, for Eq. 15.3-13 and Eq. 15.3-17 do have the mathematical form derived for the Michaelis–Menten enzyme mechanism. However, facilitated diffusion has a totally different origin and is better understood with diffusion equations.

Fast diffusion

The second special case of the mobile carrier mechanism is that of fast diffusion, the exact antithesis of the case just discussed (Ward, 1970). When the carrier–solute reaction in Eq. 15.3-1 is slow, the concentrations of carrier and complex reach constant values throughout the membrane. The concentration of solute must vary across the membrane in order to satisfy the boundary conditions Eqs. 15.3-5 and 15.3-6. As a result, the continuity equation for solute 1 becomes

$$0 = D\frac{d^2c_1}{dz^2} - \kappa_1 c_1 \bar{c}_s + \kappa_{-1}\bar{c}_{1s} \qquad (15.3\text{-}18)$$

where $\bar{c}_s$ and $\bar{c}_{1s}$ are the constant concentrations of carrier and complex in the membrane. Because c_1 is the only variable, this equation can be integrated directly. The total flux of solute 1 is then found by differentiating the concentration profile:

$$-(j_1 + j_{1s}) = H(\sqrt{\kappa_1\bar{c}_s D})\left(\frac{1 + \cosh(\kappa_1\bar{c}_s l^2/D)^{1/2}}{2\sinh(\kappa_1\bar{c}_s l^2/D)^{1/2}}\right)(C_{10} - C_{1l}) \qquad (15.3\text{-}19)$$

where

$$\bar{c}_s = \frac{\kappa_{-1}}{\kappa_1\bar{c}_1 + \kappa_{-1}}\bar{c} \qquad (15.3\text{-}20)$$

$$\bar{c}_{1s} = \frac{\kappa_1\bar{c}_1}{\kappa_1\bar{c}_1 + \kappa_{-1}}\bar{c} \qquad (15.3\text{-}21)$$

and

$$\bar{c}_1 = \frac{H}{2} (C_{10} + C_{1l}) \tag{15.3-22}$$

As before, $\bar{c}$ represents the average total carrier concentration. This situation, which also implies that the flux does not always vary linearly with the concentration difference, occurs much less frequently than the fast-reaction case discussed earlier.

Excess carrier

The third special case of the mobile carrier mechanism is that where the concentrations of both solute and complex are always much less than the concentrations of uncomplexed carrier (Donaldson & Quinn, 1975). For this case, Eq. 15.3-2 and Eq. 15.3-4 become

$$0 = D \frac{d^2 c_1}{dz^2} - \kappa_1 c_1 + \kappa_{-1} c_{1s} \tag{15.3-23}$$

$$0 = D \frac{d^2 c_{1s}}{dz^2} + \kappa_1 c_1 - \kappa_{-1} c_{1s} \tag{15.3-24}$$

In these equations, κ_1 is defined to include c_s, which is approximately a constant because it is present in excess. The boundary conditions in Eqs. 15.3-5 and 15.3-6 are now supplemented by

$$z = 0, l, \quad j_{1s} = 0 \tag{15.3-25}$$

Equations 15.3-23 and 15.3-24 can be solved by adding them, integrating, substituting the result into the first, and integrating again. The result is

$$-j_1 = \frac{DH}{l} \left(\frac{\kappa_{-1} + \kappa_1}{\kappa_{-1} + \kappa_1 (\tanh \phi)/\phi} \right) (C_{1l} - C_{10}) \tag{15.3-26}$$

where ϕ is a type of Thiele modulus:

$$\phi = \left(\frac{l^2}{4D} (\kappa_1 + \kappa_{-1}) \right)^{1/2} \tag{15.3-27}$$

Obviously, this result does not predict that the flux reaches an asymptotic value at high values of $C_{1l} - C_{10}$. However, because of its closed analytical form, this special case may be very useful in analyzing chemical systems that are not obviously in the limit of fast reaction or fast diffusion.

Example 15.3-1: The flux in facilitated diffusion

Lithium, sodium, and potassium chlorides are each diffusing in turn from a 0.1-M aqueous solution across a $32 \cdot 10^{-4}$-cm organic membrane into pure water. The membrane is largely made of a chlorinated hydrocarbon, but it also contains as a mobile carrier $6.8 \cdot 10^{-3}$-M di-benzo-18-crown-6. This synthetic carrier selectively complexes alkalai metals. For lithium chloride, the association constant is 260 liters/mol; for sodium chloride, it is $1.3 \cdot 10^4$ liters/mol; for potassium chloride, it is $4.7 \cdot 10^6$ liters/mol. The partition coefficients of the various salts are $4.5 \cdot 10^{-4}$, $3.4 \cdot 10^{-4}$, and $3.8 \cdot 10^{-4}$, respectively. Assume that all salts and complexes have diffusion coefficients of $2 \cdot 10^{-5}$ cm²/sec. Find the total flux for each of these salts.

Solution. These fluxes are easily found from Eq. 15.3-13, which for these cases reduces to

$$j_1 + j_{1s} = \frac{DHC_{10}}{l} + \frac{DHK\bar{c}C_{10}}{1 + HKC_{10}}$$

For lithium chloride, this is

$$j_1 + j_{1s} = \frac{2 \cdot 10^{-5} \text{ cm}^2/\text{sec}}{32 \cdot 10^{-4} \text{ cm}} (4.5 \cdot 10^{-4})(10^{-4} \text{ mol/cm}^3)$$

$$+ \frac{\left(\dfrac{2 \cdot 10^{-5} \text{ cm}^2}{\text{sec}}\right)(4.5 \cdot 10^{-4})\left(\dfrac{2.6 \cdot 10^5 \text{ cm}^3}{\text{mol}}\right)\left(\dfrac{6.8 \cdot 10^{-6} \text{ mol}}{\text{cm}^3}\right)\left(\dfrac{10^{-4} \text{ mol}}{\text{cm}^3}\right)}{(32 \cdot 10^{-4} \text{ cm})\left[1 + (4.5 \cdot 10^{-4})\left(\dfrac{2.6 \cdot 10^5 \text{ cm}^3}{\text{mol}}\right)\left(\dfrac{10^{-4} \text{ mol}}{\text{cm}^3}\right)\right]}$$

$$= (2.8 \cdot 10^{-10}) + (4.9 \cdot 10^{-10})$$

$$= 7.7 \cdot 10^{-10} \text{ mol/cm}^2\text{-sec}$$

Thus, ordinary diffusion is responsible for about one-third of lithium transport. For sodium chloride, the result is

$$j_1 + j_{1s} = \frac{2 \cdot 10^{-5} \text{ cm}^2/\text{sec}}{32 \cdot 10^{-4} \text{ cm}} (3.4 \cdot 10^{-4})(10^{-4} \text{ mol/cm}^3)$$

$$+ \frac{\left(\dfrac{2 \cdot 10^{-5} \text{ cm}^2}{\text{sec}}\right)(3.4 \cdot 10^{-4})\left(\dfrac{1.3 \cdot 10^7 \text{ cm}^3}{\text{mol}}\right)\left(\dfrac{6.8 \cdot 10^{-6} \text{ mol}}{\text{cm}^3}\right)\left(\dfrac{10^{-4} \text{ mol}}{\text{cm}^3}\right)}{(32 \cdot 10^{-4} \text{ cm})\left[1 + (3.4 \cdot 10^{-4})\left(\dfrac{1.3 \cdot 10^7 \text{ cm}^3}{\text{mol}}\right)\left(\dfrac{10^{-4} \text{ mol}}{\text{cm}^3}\right)\right]}$$

$$= (2.1 \cdot 10^{-10}) + (1.30 \cdot 10^{-8})$$

$$= 1.32 \cdot 10^{-8} \text{ mol/cm}^2\text{-sec}$$

The sodium chloride flux is dominated by carrier-assisted transport and is 20 times greater than that for the lithium salt. For potassium chloride,

$$j_1 + j_{1s} = \frac{2 \cdot 10^{-5} \text{ cm}^2/\text{sec}}{32 \cdot 10^{-4} \text{ cm}} (3.8 \cdot 10^{-4})(10^{-4} \text{ mol/cm}^3)$$

$$+ \frac{\left(\dfrac{2 \cdot 10^{-5} \text{ cm}^2}{\text{sec}}\right)(3.8 \cdot 10^{-4})\left(\dfrac{4.7 \cdot 10^9 \text{ cm}^3}{\text{mol}}\right)\left(\dfrac{6.8 \cdot 10^{-6} \text{ mol}}{\text{cm}^3}\right)\left(\dfrac{10^{-4} \text{ mol}}{\text{cm}^3}\right)}{(32 \cdot 10^{-4} \text{ cm})\left[1 + (3.8 \cdot 10^{-4})\left(\dfrac{4.7 \cdot 10^9 \text{ cm}^3}{\text{mol}}\right)\left(\dfrac{10^{-4} \text{ mol}}{\text{cm}^3}\right)\right]}$$

$$= (2.4 \cdot 10^{-10}) + (4.23 \cdot 10^{-8})$$

$$= 4.25 \cdot 10^{-8} \text{ mol/cm}^2\text{-sec}$$

Table 15.3-1. *Glucose uptake in human erythrocytes at 37°C*

Glucose concentration ($\times 10^{-3}$ M)	Glucose flux ($\times 10^{-3}$ mol/min)
1.0	0.09
1.5	0.12
2.0	0.14
3.0	0.20
4.3	0.25
5.0	0.28

Note that this flux is only thrice that of sodium chloride, even though complex formation is more than 300 times stronger. This is because almost all carrier molecules are involved in potassium transport, so that further facilitation is difficult.

Example 15.3-2: Analyzing data for facilitated diffusion

The data shown in Table 15.3-1 were obtained for glucose transfer across a human erythrocyte. How can these data be analyzed to determine characteristic coefficients of facilitated diffusion?

Solution. Presumably, glucose concentration on one side of the membranes is small, and facilitated diffusion is the dominant transport mechanism. In this case, we can use Eq. 15.3-13 or Eq. 15.3-17 to find

$$\frac{1}{(j_1 + j_{1s})A} = \frac{l}{DA\bar{c}} + \left(\frac{l}{DAK\bar{c}}\right)\frac{1}{C_{10}}$$

or

$$\frac{1}{(j_1 + j_{1s})A} = \frac{1}{v_{max}} + \left(\frac{K_m}{v_{max}}\right)\frac{1}{C_{10}}$$

In either case, we should plot the reciprocal of the total flux versus the reciprocal of the concentration. The plot, which is roughly linear, gives values of 0.56 M/min for $l/DA\bar{c}$ or v_{max} and 190 M^{-1} for K or for $1/K_m$. These values are consistent with a large number of experiments (Stein, 1967).

Section 15.4. Coupled facilitated diffusion

The previous section showed how a mobile carrier can greatly enhance the flux of a particular solute, facilitating its transport across a membrane. This facilitation results from a reversible carrier–solute reaction. The resulting flux is selective and nonlinear, a refreshing contrast to most diffusion processes.

This section discusses how a mobile carrier can transport a solute from a

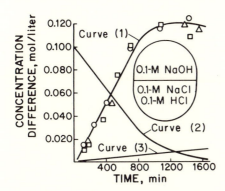

Fig. 15.4-1. Countertransport of sodium ion. Curve (1) and the circles represent sodium transported by the macrocyclic antibiotic monensin, which acts as a mobile carrier; curve (2) gives the acid flux responsible for sodium transport; curve (3) represents sodium transport without monensin. The squares illustrate sodium transport with the membrane potential set equal to zero, and the triangles represent sodium transport when the osmotic gradient across the membrane is altered with sucrose. [From Choy et al. (1974), with permission.]

dilute solution into a concentrated one. In other words, the carrier allows the solute to move against its concentration difference across the membrane. Such behavior depends on the nature of the carrier–solute reactions, which may be competitive or cooperative. When these reactions produce two fluxes in opposite directions, the phenomenon is called "countertransport." For example, for the system shown in Fig. 15.4-1, the reaction is

$$Na^+ + RCOOH \rightleftarrows RCOO^-Na^+ + H^+ \qquad (15.4-1)$$

where $RCOO^-$ represents the macrocyclic antibiotic monensin, which is the mobile carrier (Choy et al., 1974). As shown in the figure, a flux of protons from the NaCl–HCl solution into the NaOH solution engenders a flux of Na^+ in the opposite direction. For the commercially attractive flow system shown in Fig. 15.4-2 (Lee et al., 1978; Babcock et al., 1980), the key reaction is

$$Cu^{2+} + 2RH \rightleftarrows R_2Cu + 2H^+ \qquad (15.4-2)$$

where R now represents an α-hydroxy aliphatic oxime. As before, a flux of protons produces another flux in the opposite direction, this time of copper ions. Another example involves the reaction (Schultz, 1977)

$$CO + \text{hemoglobin} \underset{h\nu}{\rightleftarrows} CO\text{–hemoglobin complex} \qquad (15.4-3)$$

If one side of the membrane is illuminated, the complex will be destroyed, and CO will accumulate on that side, as shown in Fig. 15.4-3. Thus, a photon flux produces a CO flux.

In other, less common situations, the solute–mobile-carrier reactions engender two fluxes in the same direction. This phenomenon, known as "cotransport," is illustrated by the reaction (Hochhauser & Cussler, 1976)

$$2H^+ + Cr_2O_7^{2-} + 2R_3N \rightleftarrows (R_3NH)_2Cr_2O_7 \qquad (15.4-4)$$

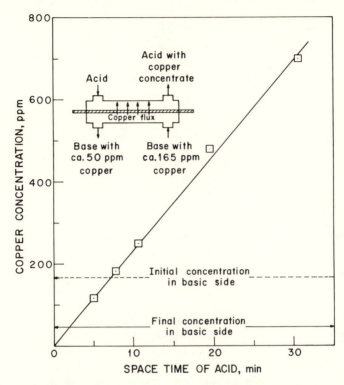

Fig. 15.4-2. Copper countertransport driven with acid. Copper is selectively concentrated by diffusion across a liquid membrane supported by a porous polymer film, shown as the crosshatched region. [From Lee et al. (1978), with permission.]

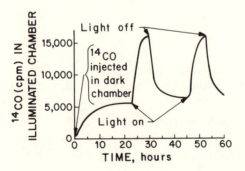

Fig. 15.4-3. Carbon monoxide transport driven with light. Radioactively tagged carbon monoxide is injected on the dark side of the membrane and allowed to equilibrate. When the other side of the membrane is illuminated, CO is transported into this compartment. [From Schultz (1977), with permission.]

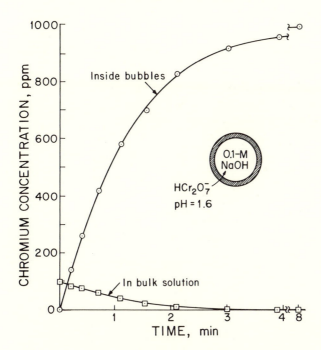

Fig. 15.4-4. Chromium cotransport. Chromium is concentrated inside droplets of a water-in-oil-in-water emulsion. This emulsion is stabilized with surfactants, and so is called a "liquid surfactant membrane." It offers an unusually large surface area of thin membranes per volume, and so has practical potential. [From Hochhauser & Cussler (1976), with permission.]

A separation based on this reaction is shown in Fig. 15.4-4, in a commercially attractive configuration.

We want to buttress these examples with a more quantitative picture of the diffusion process. This picture is similar to that for facilitated diffusion in Section 15.3, and so depends on the relative speed of diffusion and reaction. We shall consider only the case where diffusion is much slower than reaction.

For countertransport, we want to calculate the total flux of solute 1 across a membrane that contains a total of five different species: solute 1, solute 2, uncomplexed carrier s, complex $1s$, and complex $2s$. These five species are subject to two equilibria:

$$c_{is} = K_i c_i c_s \quad (i = 1, 2) \tag{15.4-5}$$

In addition, the concentrations must be consistent with five steady-state continuity equations:

$$0 = D \frac{d^2 c_i}{dz^2} - r_{is} \quad (i = 1, 2) \tag{15.4-6}$$

$$0 = D \frac{d^2 c_{is}}{dz^2} + r_{is} \quad (i = 1, 2) \tag{15.4-7}$$

$$0 = D \frac{d^2 c_s}{dz^2} - r_{1s} - r_{2s} \tag{15.4-8}$$

where r_i is the formation of the complex within the membrane. We have again assumed that all solutes have the same diffusion coefficients so that we can see more clearly the effect of the mobile carrier.

The five continuity equations are subject to a variety of restraints. At steady state, the boundary conditions on solutes 1 and 2 are

$$z = 0, \quad c_i = H_i C_{i0} \tag{15.4-9}$$

$$z = l, \quad c_i = H_i C_{il} \tag{15.4-10}$$

and the total carrier concentration must be constant:

$$\frac{1}{l} \int_0^l (c_s + c_{1s} + c_{2s}) \, dz = \bar{c} \tag{15.4-11}$$

The mobile carrier must obey the boundary conditions

$$z = 0, L, \quad j_s + j_{1s} + j_{2s} = 0$$

$$D \left(\frac{dc_s}{dz} + \frac{dc_{1s}}{dz} + \frac{dc_{2s}}{dz} \right) = 0 \tag{15.4-12}$$

These equations are analogues for countertransport to those in Section 15.3.

The solution for countertransport closely parallels the solution in the binary case (Cussler, 1971). Equations 15.4-7 and 15.4-8 are added, integrated, combined with Eq. 15.4-12, integrated again, and then rearranged with Eqs. 15.4-5 and 15.4-11. The result is

$$c_s = \frac{\bar{c}}{1 + K_1 c_1 + K_2 c_2} \tag{15.4-13}$$

This relation gives the concentration of unreacted carrier as a function of the solute concentrations in the system. The total flux of solute 1 can now be found by adding Eqs. 15.4-6 and 15.4-7, integrating twice, and using the restraints in Eqs. 15.4-9, 15.4-10, and 15.4-13:

$$j_1 + j_{1s} = - D \left(\frac{dc_1}{dz} + \frac{dc_{1s}}{dz} \right)$$

$$= \frac{DH_1}{l} (C_{10} - C_{1l}) + \frac{DH_1}{l} [R(1 + H_2 K_2 \bar{C}_2)](C_{10} - C_{1l})$$

$$- \frac{DH_1}{l} [RH_2 K_2 \bar{C}_1](C_{20} - C_{2l}) \tag{15.4-14}$$

where $\bar{C}_i$ is the average concentration of solute i between the two solutions, and R is given by

$$R = \frac{K_1 \bar{c}}{(1 + H_1 K_1 C_{10} + H_2 K_2 C_{20})(1 + H_1 K_1 C_{1l} + H_2 K_2 C_{2l})} \tag{15.4-15}$$

Again, these results depend on the assumption that diffusion across the membrane is much slower than reaction anywhere within it.

The equations for cotransport are developed in a completely parallel fashion (Schultz et al., 1974). The membrane contains four species: solute 1, solute 2, complexed carrier s, and complex 12s. These species are related by the equilibrium:

$$c_{12s} = K_{12}c_1c_2c_s \tag{15.4-16}$$

As before, the calculation of the total flux starts with the continuity equations and very similar boundary conditions and then integrates these to discover

$$j_1 + j_{12s} = - D \left(\frac{dc_1}{dz} + \frac{dc_{12s}}{dz} \right)$$

$$= \frac{DH_1}{l} (C_{10} - C_{1l})$$

$$+ \frac{DH_1H_2K_{12}\bar{c}}{l} \left(\frac{C_{10}C_{20}}{1 + H_1H_2K_{12}C_{10}C_{20}} - \frac{C_{1l}C_{2l}}{1 + H_1H_2K_{12}C_{1l}C_{2l}} \right)$$

$$\tag{15.4-17}$$

where as before, the H_i are partition coefficients, C_{i0} and C_{il} are concentrations in adjacent solutions, l is the membrane's thickness, and D is the diffusion coefficient, assumed equal for all species. Results like Eq. 15.4-14 and Eq. 15.4-17 can quantitatively explain experiments like those in Figs. 15.4-1 through 15.4-4.

Example 15.4-1: Maximum concentration differences
Estimate the maximum concentration differences that can be sustained in the steady state for countertransport and cotransport. In these estimates, assume that diffusion without mobile carriers is negligible and that solute and carrier react strongly (i.e., that the K_i are large).

Solution. The maximum concentration differences causing no net flux can be found by setting the fluxes equal to zero. For countertransport, from Eq. 15.4-14,

$$0 \doteq \frac{DH_1}{l} R(1 + H_2K_2\bar{C}_2)(C_{10} - C_{1l}) - \frac{DH_1}{l} R(H_2K_2\bar{C}_1)(C_{20} - C_{2l})$$

or

$$\frac{C_{10} - C_{1l}}{C_{20} - C_{2l}} = \frac{H_2K_2\bar{C}_1}{1 + H_2K_2\bar{C}_2} \doteq \frac{\bar{C}_1}{\bar{C}_2}$$

This can be simplified to give

$$C_{10}C_{2l} = C_{1l}C_{20}$$

Note that if one concentration is kept very small, then large concentration differences can be sustained.

For cotransport, from Eq. 15.4-17,

$$0 \doteq \frac{DH_1H_2K_{12}\bar{c}}{l} \left(\frac{C_{10}C_{20}}{1 + H_1H_2K_{12}C_{10}C_{20}} - \frac{C_{1l}C_{2l}}{1 + H_1H_2K_{12}C_{1l}C_{2l}} \right)$$

Thus,

$$C_{10}C_{20} = C_{1l}C_{2l}$$

The contrast with the previous result exemplifies the difference between cotransport and countertransport.

Section 15.5. Diffusion and phase equilibria

Thus far in this chapter we have discussed membrane diffusion. We have emphasized cases in which the diffusion is coupled with chemical reaction, for it is these cases that are dramatic and unexpected. In most instances we have emphasized fast chemical reactions, for diffusion is usually the slowest step involved.

We now turn to other cases of diffusion in two-phase systems. Like membrane diffusion, these cases show dramatic and unexpected behavior. Like membrane diffusion, this behavior is the consequence of rapid chemical reactions. Unlike membrane diffusion, the second phase is not a thin barrier, but is more typically a porous solid or an emulsion.

The effects we want to discuss are exemplified by the dissolution of slaked lime, $Ca(OH)_2$, in aqueous solutions of a strong acid like HCl:

$$Ca(OH)_2 + 2H^+ \rightleftarrows Ca^{2+} + 2H_2O \tag{15.5-1}$$

How this dissolution proceeds depends on the relative speed of diffusion and reaction. When the bulk of the solution next to the solid is rapidly stirred, the acid can diffuse to the solid's surface very quickly. It then reacts with the solid's surface. If the solid is essentially impermeable, containing a very few pores, then any ions produced by the dissolution are quickly swept back into the bulk solution. Because diffusion and chemical reaction occur sequentially, the overall dissolution rate is like that of a heterogeneous reaction depending on the sum of the resistances of diffusion and of reaction (see Chapter 13). Such a process represents an important limit of corrosion and is that usually studied.

Alternatively, the solution next to the solid may not be well stirred, and the solid may be highly porous, as shown schematically by Fig. 15.5-1. In this case, the acid concentration will drop as it approaches the solid's surface and continue to drop within the solid's pores. The ions produced as the result of the acid–solid reaction will be present in highest concentration near the solid's surface. From this maximum, they can diffuse out into the bulk solution or further into the porous solid. Within the solid, diffusion and reaction occur simultaneously, so that the overall dissolution rate is like that of a heterogeneous reaction, not a simple sum of the resistances of diffusion and reaction (see Chapter 14).

To calculate the dissolution rate of the porous solid r_1, we write continuity equations for calcium ion (species 1) and protons (species 2) (Cussler, 1982):

$$\frac{\partial c_1}{\partial t} = D \frac{\partial^2 c_1}{\partial z^2} + r_1 \tag{15.5-2}$$

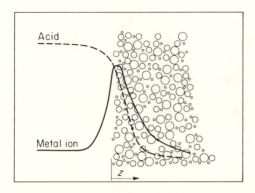

Fig. 15.5-1. Dissolution of a porous solid. In this schematic representation, acid diffuses from left to right and is consumed by chemical reaction with the solid. The metal ions produced by this reaction can, under some conditions, diffuse into the pores and precipitate as more solid. [From Cussler & Featherstone (1981), with permission.]

$$\frac{\partial c_2}{\partial t} = D \frac{\partial^2 c_2}{\partial z^2} - 2r_1 \tag{15.5-3}$$

where, as before, we assume that the diffusion coefficients are equal. We also assume that the reaction in Eq. 15.5-1 is so fast that it reaches equilibrium:

$$[Ca^{2+}] = \frac{K'[Ca(OH)_2]}{[H_2O]^2} [H^+]^2 \tag{15.5-4}$$

The $Ca(OH)_2$ is solid and of unit activity; the water is present in excess; so everything in the braces is a constant K. Thus,

$$c_1 = Kc_2^2 \tag{15.5-5}$$

$$\frac{\partial c_1}{\partial t} = 2Kc_2 \frac{\partial c_2}{\partial t} \tag{15.5-6}$$

and

$$\frac{\partial^2 c_1}{\partial z^2} = 2K \left[\left(\frac{\partial c_2}{\partial z} \right)^2 + c_2 \frac{\partial^2 c_2}{\partial z^2} \right] \tag{15.5-7}$$

Inserting these into Eq. 15.5-2 and then combining with Eq. 15.5-3, we find the dissolution rate:

$$r_1 = - \left[\frac{2Dc_1 (\partial \ln c_2/\partial z)^2}{1 + 4(c_1/c_2)} \right] \tag{15.5-8}$$

The exact values depend on the particular boundary conditions involved.

The remarkable feature about Eq. 15.5-8 is that it predicts that the dissolution rate within the pores is *negative,* because all terms in the brackets in Eq. 15.5-8 are positive. In physical terms, this means that $Ca(OH)_2$ will precipitate in front of the acid wave shown in Fig. 15.5-1. This prediction is verified by Fig. 15.5-2 (Cussler & Featherstone, 1981). Dissolution can produce precipitation.

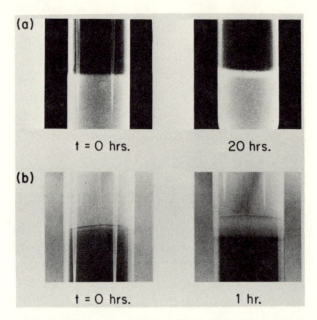

Fig. 15.5-2. Dissolution of gel-stabilized suspensions. Dilute gel-stabilized suspensions of insoluble hydroxides were dissolved by pouring acid on top of suspensions. The gel remained intact. In the $Ca(OH)_2$ suspension shown in (a), hydrochloric acid causes dissolution at the suspension's interface and a white band of precipitate below the interface. For a mixture of $Ca(OH)_2$ and Ag_2O shown in (b), nitric acid causes precipitation of $Ca(OH)_2$ near the interface, but dissolution of both species below the interface. [From Cussler & Featherstone (1981), with permission.]

To explain the origin of these effects in more physical terms, we imagine that the boundary conditions in the porous solid are those of free diffusion (see Section 2.3):

$$t < 0, \quad \text{all } z, \quad c_1 = 0 = c_2 \tag{15.5-9}$$

$$t > 0, \quad z = 0, \quad c_1 = c_{10} \tag{15.5-10}$$

$$c_2 = c_{20} \tag{15.5-11}$$

$$z = \infty, \quad c_1 = c_2 = 0 \tag{15.5-12}$$

For the moment, pretend that there is no reaction; so $r_1 = 0$. Then,

$$\frac{c_1}{c_{10}} = \frac{c_2}{c_{20}} = 1 - \text{erf} \frac{z}{\sqrt{4Dt}} \tag{15.5-13}$$

We thus know c_1 as a function of z and t. We also know c_1 as a function of c_2, and we can graph this variation as the diffusion path in Fig. 15.5-3.

However, when there is a reaction, the concentrations in the system must follow the equilibrium in Eq. 15.5-3, producing a path that is also shown in Fig. 15.5-3. This equilibrium line is essentially a phase diagram. If at fixed c_2, c_1 is below this equilibrium line, then solid will dissolve to produce more c_1. If at fixed c_2, c_1 is above this line, then solid will precipitate to reduce c_2.

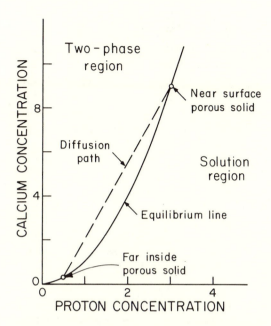

Fig. 15.5-3. How dissolution can cause precipitation. When no solid is present, the calcium and proton concentrations are approximately linearly related, as shown by the dashed diffusion path. When solid is present, they are related by the equilibrium solubility, although diffusion constantly tries to push the system into the two-phase region. The effect of these coupled effects is exemplified by the results in Fig. 15.5-2.

When we compare the equilibrium and the diffusion paths, we see that diffusion of H^+ into porous $Ca(OH)_2$ will tend to carry us into the two-phase region, and so produce precipitation in front of the acid wave. In contrast, for the reaction

$$CaCO_3 + H^+ \rightleftarrows Ca^{2+} + HCO_3^-$$ (15.5-14)

we might write

$$c_1^2 = Kc_2$$ (15.5-15)

where 1 is Ca^{2+} or HCO_3^- and 2 is H^+. Here, the equilibrium curve would show the opposite curvature, and we would expect dissolution both at the solid's surface and within the pores. In general terms, if

$$c_1 = Kc_2^\nu$$ (15.5-16)

we should expect that

$$\nu > 1 \quad \text{or} \quad \nu < 0, \quad r_1 < 0 \quad \text{(precipitation)}$$ (15.5-17)

and

$$0 < \nu < 1, \quad r_1 > 0 \quad \text{(dissolution)}$$ (15.5-18)

In other words, we should frequently get precipitation during dissolution.

This type of effect, which is intuitively surprising, is also responsible for spontaneous emulsification in liquids (Ruschak & Miller, 1972), for some phase separations in metals (Kirkaldy & Brown, 1963), and for the formation of fogs in gases (Toor, 1971). These problems are conceptually parallel, but their details are more complex. This complexity comes largely from the replacement of simple stoichiometric relations like Eqs. 15.5-3 and 15.5-14 by more complicated phase diagrams. In many interesting cases, these phase diagrams do not have a single species present in excess, and so they require representation on triangular coordinates. The phase diagrams may also imply additional interfaces. This area contains unsolved problems of practical significance.

Section 15.6. Conclusions

The material in this chapter is collected around specific applications to membranes, and less around the intellectual topics of earlier chapters. Each section could be included under some earlier heading. Section 15.1, the flux equations for membranes, could have been incorporated into the general equations developed in Chapter 3 or 8. Section 15.2 largely contains extensions of the solute–solute interactions in Chapter 6. Sections 15.3 through 15.5 are examples of coupled diffusion and homogeneous chemical reaction.

Thus, this chapter's material provides a review of much of the book. The review is interesting not only because of its major applications in the life sciences but also because the effects observed are large. Master this material, and you will understand both the basics and the nuances of diffusion.

References

Babcock, W. C., Baker, R. W., LaChapelle, E. D., & Smith, K. L. (1980). *Journal of Membrane Science, **7**,* 71, 89.
Batchelor, G. K. (1967). *An Introduction to Fluid Mechanics.* Cambridge University Press.
Choy, E. M., Evans, D. F., & Cussler, E. L. (1974). *Journal of the American Chemical Society, **96**,* 7085.
Cussler, E. L. (1971). *American Institute of Chemical Engineers Journal, **17**,* 1300.
Cussler, E. L. (1976). *Journal of Membrane Science, **1**,* 319.
Cussler, E. L. (1982). *American Institute of Chemical Engineers Journal, **28**,* 500.
Cussler, E. L., & Featherstone, J. D. B. (1981). *Science, **213**,* 1018.
deGroot, S. R., & Mazur, P. (1962). *Non-Equilibrium Thermodynamics.* Amsterdam: North Holland.
Donaldson, T. L., & Quinn, J. A. (1975). *Chemical Engineering Science, **30**,* 103.
Duffey, M. E., Evans, D. F., & Cussler, E. L. (1978). *Journal of Membrane Science, **3**,* 1.
Goldman, D. E. (1943). *Journal of General Physiology, **27**,* 37.
Hochhauser, A. M., & Cussler, E. L. (1976). *Chemical Engineering Progress Symposium Series, **71**,* 136.
Katchalsky, A., & Curran, P. F. (1967). *Non-Equilibrium Thermodynamics in Biophysics.* Cambridge, Mass.: Harvard University Press.
Kedem, O., & Katchalsky, A. (1958). *Biochimica et Biophysica Acta, **27**,* 229.
Kirkaldy, J. S., & Brown, L. C. (1963). *Canadian Metallurgical Quarterly, **2**,* 89.
Lakshminarayanaiah, N. (1969). *Transport Phenomena in Membranes.* New York: Academic Press.
Lee, K. H., Evans, D. F., & Cussler, E. L. (1978). *American Institute of Chemical Engineers Journal, **24**,* 860.

Lonsdale, H. (1982). *Journal of Membrane Science,* **10,** 81.

MacInnes, D. A. (1961). *Principles of Electrochemistry.* New York: Dover.

Mason, E. A., Wendt, R. P., & Bresler, E. H. (1972). *Journal of the Chemical Society, Faraday Transactions 2,* **68,** 1938.

Osterhout, W. J. V. (1935). *Proceedings of the National Academy of Sciences of the United State of America,* **21,** 125.

Robinson, R. A., & Stokes, R. H. (1960). *Electrolyte Solutions.* London: Butterworth.

Ruschak, K. J., & Miller, C. A. (1972). *Industrial and Engineering Chemistry Fundamentals,* **11,** 534.

Schultz, J. S. (1977). *Science,* **197,** 1177.

Schultz, J. S., Goddard, J. D., & Suchdeo, S. R. (1974). *American Institute of Chemical Engineers Journal,* **20,** 417, 625.

Schultz, S. G. (1980). *Basic Principles of Membrane Transport.* Cambridge University Press.

Sollner, K., & Shean, G. (1964). *Journal of the American Chemical Society,* **86,** 1901.

Stein, W. D. (1967). *The Movement of Molecules Across Cell Membranes.* New York: Academic Press.

Sten-Knudsen, O. (1978). In: *Membrane Transport in Biology, Vol. 1,* ed. G. Giebisch et al. Berlin: Springer-Verlag.

Toor, H. L. (1971). *American Institute of Chemical Engineers Journal,* **17,** 5; *Industrial and Engineering Chemistry Fundamentals,* **10,** 121.

van't Hoff, J. H. (1887). *Zeitschrift für Physikalische Chemie,* **1,** 481.

Ward, W. J. (1970). *American Institute of Chemical Engineers Journal,* **16,** 405.

Widdas, W. F. (1952). *The Journal of Physiology,* **118,** 23.

16 HEAT TRANSFER

In this chapter we briefly describe fundamental concepts of heat transfer. We begin in Section 16.1 with a description of heat conduction. We mount this description on three key points: Fourier's law for conduction, energy transport through a thin film, and energy transport in a semi-infinite slab. In Section 16.2 we discuss energy conservation equations that in fact are more general forms of the first law of thermodynamics. In Section 16.3 we analyze interfacial heat transfer in terms of heat transfer coefficients, and in Section 16.4 we discuss numerical values of thermal conductivities, thermal diffusivities, and heat transfer coefficients.

This material is closely parallel to the ideas about diffusion presented in the rest of this book. This parallelism is not unexpected, for heat transfer and mass transfer are described with equations that are very similar mathematically. The material in Section 16.1 is like that in Chapter 2, and the general equations in Section 16.2 are conceptually similar to those in Chapter 3. The material on heat transfer coefficients in Section 16.3 closely resembles the mass transfer material in Chapters 9 through 12, and the numerical values in Section 16.4 parallel those in Chapters 5 through 7.

Thus, we are abstracting ideas of heat transfer in a few sections, whereas we detailed similar ideas of mass transfer over many chapters. This represents a tremendous abridgment. Indeed, as those skilled in heat transfer recognize, the heat transfer literature is immense, of far greater size than the mass transfer literature. To be sure, this book is about diffusion, and so an emphasis on mass transfer is appropriate. But if the description of heat transfer is to be so terse, why include it at all?

I have included the description of heat transfer because I want to discuss simultaneous heat and mass transfer in the next chapter. This simultaneous transport process is important practically and is interesting intellectually, with implications ranging far beyond the particular problems presented. However, to discuss this simultaneous process, we need to assure a background in heat transfer. I expect that many who read this book will not have such a background, for this topic is usually buried well inside the engineering curriculum. Accordingly, this chapter is a synopsis to provide this background.

Section 16.1. Fundamentals of heat conduction

The fundamental understanding of heat conduction rests on the work of Jean Joseph Fourier, who was born March 21, 1768 (Herivel, 1975). Orphaned before 10 years of age, Fourier got an education by joining the church. He started teaching school, but then advanced rapidly through the government bureaucracy as the French Revolution eliminated those above him. It was a risky business, and Fourier spent some time in prison. Nonetheless, under Napoleon he became prefect of the department of Isère. He did his work on heat conduction while holding that position. It was as if the governor of Minnesota was doing first-rate mathematical physics in his spare time, during evenings and on weekends. In 1807, Fourier presented his work to an all-time All-Pro faculty committee, including Laplace, Lagrange, and Monge. Poisson was also involved. Lagrange was critical; so Fourier's degree was delayed.

In his 1807 paper, Fourier used the experiments of Biot to argue that the heat flux $\mathbf{q}$ should be proportional to the temperature gradient ∇T:

$$\mathbf{q} = -\ell \nabla T \qquad (16.1\text{-}1)$$

where the proportionality constant ℓ is the thermal conductivity. Note that the dimensions of ℓ are not simple, but are commonly energy per length-temperature-time. This equation is a close parallel to Fick's law; indeed, as explained in Section 2.1, Fick developed the diffusion law by analogy with Fourier's work.

To calculate heat fluxes or temperature profiles, we make energy balances and then combine these with Fourier's law. The ways in which this is done are best seen in terms of two examples: heat conduction across a thin film and into a semi-infinite slab. The choice of these two examples is not casual. As for diffusion, they bracket most of the other problems, and so provide two limits for conduction.

Steady heat conduction across a thin film

As a first example, consider a thin solid membrane separating two well-stirred fluids, as shown schematically in Fig. 16.1-1. Because one fluid is hotter than the other, energy will be conducted from left to right across the thin film. To find the amount of conduction, we make a steady-state energy balance on a thin layer located between z and $z + \Delta z$:

$$\begin{pmatrix} \text{energy} \\ \text{accumulation} \end{pmatrix} = \begin{pmatrix} \text{energy} \\ \text{conducted in} \end{pmatrix} - \begin{pmatrix} \text{energy} \\ \text{conducted out} \end{pmatrix} \qquad (16.1\text{-}2)$$

At steady state, this is

$$0 = Aq|_z - Aq|_{z+\Delta z} \qquad (16.1\text{-}3)$$

where A is the cross-sectional area and q is the heat flux in the z direction. Dividing by the layer's volume $A\Delta z$ and taking the limit as this volume goes to zero, we find

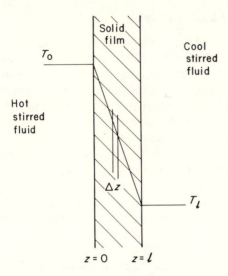

Fig. 16.1-1. Steady heat conduction across a thin film. The simplest case of heat transfer is conduction across a thin film, a process like diffusion across a membrane (see Section 2.2). The resulting temperature profile is linear, and the flux is constant and inversely proportional to the film thickness l.

$$0 = -\frac{dq}{dz} \tag{16.1-4}$$

Combining with Fourier's law,

$$0 = k\frac{d^2T}{dz^2} \tag{16.1-5}$$

This differential equation is subject to the boundary conditions

$$z = 0, \quad T = T_0 \tag{16.1-6}$$
$$z = l, \quad T = T_l \tag{16.1-7}$$

Integration to find the temperature profile is simple:

$$T = T_0 + (T_l - T_0)\frac{z}{l} \tag{16.1-8}$$

Note that this profile, shown in Fig. 16.1-1, does not depend on the thermal conductivity. Finding the heat flux is also easy:

$$q = -k\frac{dt}{dz}$$

$$= \frac{k}{l}(T_0 - T_l) \tag{16.1-9}$$

This is a complete parallel to Eq. 2.2-10.

The results in Eqs. 16.1-8 and 16.1-9 are extraordinarily useful, the basis of much thinking about heat transfer. Still, you may have trouble taking them seriously because you are not mathematically intimidated by the derivation. To test your understanding, try to answer the following questions:

1. *How are the results changed if the fluid at $z = 0$ and T_0 is replaced by a different liquid that is at the same temperature?* There is no change as long as the interfacial temperature is constant.

2. *What will the temperature profile look like across two thin slabs of different materials that are clamped together?* In steady state, the heat flux is constant. Thus, the temperature drop across the poorly conducting slab will be larger than that across the better conductor.

3. *Imagine that for the system in Fig. 16.1-1 the fluid at $z = l$ has a small volume V, but the fluid at $z = 0$ has a very large volume. How will T_l change with time?* To answer this, we write an energy balance on the fluid at $z = l$:

$$\frac{d}{dt}(\rho V \hat{C}_v T_l) = Aq|_{z=l}$$

$$= A\frac{k}{l}(T_0 - T_l) \tag{16.1-10}$$

in which ρV is the mass of fluid located at $z = l$, $\hat{C}_v$ is the specific heat capacity of this fluid, and A is the area available for heat transfer. Initially, the temperatures are known:

$$t = 0, \quad T_l = T_l(t = 0) \tag{16.1-11}$$

Integrating,

$$T_l = T_l(t = 0) + [T_0 - T_l(t = 0)](1 - e^{-(kA/l\rho V\hat{C}_v)t}) \tag{16.1-12}$$

The temperature rises to a limit of T_0. Note that in this analysis, we use the steady-state result for a thin film in conjunction with an unsteady energy balance on the fluid. The justification for this is that the film volume is much less than the fluid volume. The same justification was used for the diaphragm-cell method of measuring diffusion coefficients (see Example 2.2-4).

Unsteady heat conduction into a thick slab

Our second example involves thermal conduction into the large solid slab shown in Fig. 16.1-2. This slab is the antithesis of the thin film discussed earlier. To be sure, both the slab and the film are in contact at $z = 0$ with hot fluid at T_0; but here, the slab has no other boundary. Instead, far within the slab the temperature remains equal to the initial value T_∞.

To solve this problem, we again make an energy balance on a differential layer located between z and $z + \Delta z$:

$$\begin{pmatrix} \text{accumulation} \\ \text{of energy} \end{pmatrix} = \begin{pmatrix} \text{energy} \\ \text{conduction in} \end{pmatrix} - \begin{pmatrix} \text{energy} \\ \text{conduction out} \end{pmatrix} \tag{16.1-13}$$

When I derive these equations, I find it easiest to build up the terms I want:

$A\Delta z$ is the volume of the thin layer

$\rho A \Delta z$ is its mass

$\rho A \Delta z \hat{C}_v$ is its energy per degree temperature

$\rho A \Delta z \hat{C}_v T$ is its energy

$\frac{\partial}{\partial t}(\rho A \Delta z \hat{C}_v T)$ is the energy accumulation

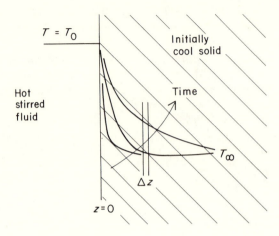

Fig. 16.1-2. Unsteady heat conduction into a semi-infinite slab. The temperature profile in this case is an error function, just like the concentration profile in Section 2.3. This profile depends on the variable $z/\sqrt{4\alpha t}$, where α (= $k/\rho \hat{C}_p$) is the thermal diffusivity.

Thus,

$$\frac{\partial}{\partial t}(\rho A \Delta z \hat{C}_v T) = Aq|_z - Aq|_{z+\Delta z} \tag{16.1-14}$$

Dividing by $\rho A \Delta z \hat{C}_v$ and taking the limit as Δz becomes small,

$$\frac{\partial T}{\partial t} = -\frac{1}{\rho \hat{C}_v}\frac{\partial q}{\partial z} \tag{16.1-15}$$

Combining with Fourier's law and making the accurate assumption that the heat capacities at constant volume and at constant pressure are the same,

$$\frac{\partial T}{\partial t} = \alpha \frac{\partial^2 T}{\partial z^2} \tag{16.1-16}$$

where α (= $k/\rho \hat{C}_p$) is the thermal diffusivity, with dimensions of length squared per time. This equation occurs so frequently that it is sometimes called "the heat conduction equation," as if there were no other forms.

For the specific case of interest here, Eq. 16.1-16 is subject to the initial condition

$$t = 0, \quad \text{all } z, \quad T = T_\infty \tag{16.1-17}$$

and to the boundary conditions

$$t > 0, \quad z = 0, \quad T = T_0 \tag{16.1-18}$$

$$z = \infty, \quad T = T_\infty \tag{16.1-19}$$

For these boundary conditions, the solution to Eq. 16.1-16 is easily obtained by combination of variables, just as was discussed in Section 2.3. The results are

$$\frac{T - T_0}{T_\infty - T_0} = \text{erf} \frac{z}{\sqrt{4\alpha t}} \tag{16.1-20}$$

$$q|_{z=0} = -\ell \frac{\partial T}{\partial z}\bigg|_{z=0}$$

$$= \sqrt{\ell \rho \hat{C}_p / \pi t} \ (T_0 - T_\infty) \tag{16.1-21}$$

In the temperature profile, the answer is the same as that for diffusion, but with the thermal diffusivity replacing the diffusion coefficient.

As in the case of the thin film, this result may be so familiar that it is difficult to think about carefully. As before, you can test your understanding by trying to answer these three questions:

1. *To what depth does the temperature change penetrate in, for example, a steel slab?* To a first approximation, the temperature changes occur to a depth where $z^2/4\alpha t$ equals unity. For steel, α equals about $0.1 \text{ cm}^2/\text{sec}$; if the steel is heated for 10 min, the temperature penetrates about 15 cm. This result is independent of heating or cooling, and it is like similar calculations for diffusion (see Section 2.6).

2. *How does the flux vary with physical properties for the thick slab as compared with the thin film?* Doubling the temperature difference doubles the heat flux in both cases. Doubling the thermal conductivity increases the flux by $\sqrt{2}$ for the thick slab and by 2 for the thin film. Doubling the heat capacity increases the flux by $\sqrt{2}$ for the thick slab, but has no effect for the steady-state conduction across a thin film.

3. *How much heat is transferred over a time t_0?* To find this, we integrate Eq. 16.1-21 over time:

$$\begin{pmatrix} \text{total heat} \\ \text{transferred} \\ \text{per area} \end{pmatrix} = \int_0^{t_0} q|_{z=0} \, dt$$

$$= \sqrt{\ell \rho \hat{C}_p t_0 / \pi} \ (T_0 - T_\infty) \tag{16.1-22}$$

To double the heat transferred in t_0, we need to wait four times as long.

The limits of heat conduction across a thin film and into a thick slab are the two most important cases of a rich variety of examples (Carslaw & Jaeger, 1959; Arpaci, 1966; Luikov, 1968). This variety largely consists of solutions of Eq. 16.1-16 for different geometries and boundary conditions. The geometries include slabs, spheres, and cylinders, as well as more exotic shapes like cones. The boundary conditions are diverse. For example, they include boundary temperatures that vary periodically, because this is important for diurnal temperature variations of the earth. They include boundary conditions in which the heat flux at the surface is related to the temperature of the surroundings, T_{surr}; for example,

$$-\ell \frac{\partial T}{\partial z}\bigg|_{z=0} = h(T|_{z=0} - T_{\text{surr}}) \tag{16.1-23}$$

This type of constraint is sometimes called a radiation condition.

At the same time, Eq. 16.1-16 is only one of a wide variety of energy

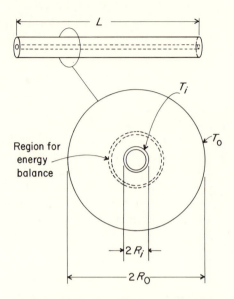

Fig. 16.1-3. Heat loss from an insulated pipe. This problem illustrates the extension of ideas of heat transfer to systems with cylindrical symmetry. When the pipe's diameter is large, the results approach the limit of the thin film shown in Fig. 16.1-1.

balances that are useful. These more general balances are the subject of the next section, which comes after some simple examples.

Example 16.1-1: Determining thermal diffusivity

A thick slab of a polymer composite at 40°C is immersed in a large stirred oil bath kept at 4°C. A thermocouple 1.3 cm below the slab's surface reads 26.2°C after 3 min. What is the thermal diffusivity of the slab?

Solution. Because the slab is thick, we can use Eq. 16.1-20:

$$\frac{T - T_0}{T_\infty - T_0} = \text{erf} \frac{z}{\sqrt{4\alpha t}}$$

$$\frac{26.2 - 4}{40 - 4} = \text{erf} \frac{1.3 \text{ cm}}{\sqrt{4\alpha(180 \text{ sec})}}$$

Thus, $\alpha = 3.1 \cdot 10^{-3}$ cm^2/sec, a typical value for this type of material.

Example 16.1-2: Heat loss from a well-insulated pipe

Imagine a well-insulated pipe used to transport saturated steam (Fig. 16.1-3). How much will the heat loss through the pipe's walls be reduced if the insulation thickness is doubled? Assume that the thermal conductivity of the pipe's walls is much higher than that of the insulation.

Solution. We begin this problem by making a steady-state energy balance on a cylindrical shell of insulation of volume $2\pi r \Delta r L$:

$$\begin{pmatrix} \text{energy} \\ \text{accumulation} \end{pmatrix} = \begin{pmatrix} \text{energy in minus energy out} \\ \text{by induction} \end{pmatrix}$$

$$0 = (2\pi r L q)_r - (2\pi r L q)_{r+\Delta r}$$

Dividing by the shell's volume and taking the limit as Δr goes to zero,

$$0 = -\frac{d}{dr}(rq)$$

Integration gives

$$rq = R_0 q_0$$

where q_0, an integration constant, is equal to the heat flux at the pipe's outer surface. Combining this with Fourier's law, we find

$$-r\ell\frac{dT}{dr} = R_0 q_0$$

or, after rearrangement,

$$q_0 = -\frac{\ell \int_{T_i}^{T_0} dt}{R_0 \int_{R_i}^{R_0} \dfrac{dr}{r}}$$

The limits of the integrals are those shown in Fig. 16.1-3. Thus,

$$q_0 = \frac{\ell(T_i - T_0)}{R_0 \ln(R_0/R_i)}$$

But we want to know the effect of doubling the insulation thickness. In other words, we want to double δ $(= R_0/R_i - 1)$. We can rearrange the foregoing equation to show

$$\begin{pmatrix} \text{flux with double insulation} \\ \text{flux with single insulation} \end{pmatrix} = \frac{(1 + \delta)\ln(1 + \delta)}{(1 + 2\delta)\ln(1 + 2\delta)}$$

which is the desired result. A good exercise is to show that this ratio approaches 0.5 as the pipe diameter R_i becomes large (i.e., as δ goes to zero).

Section 16.2. General energy balances

Energy balances can be more complicated and more difficult to understand than any mass balances. Mass balances are easy because chemical compounds are uniquely defined. For example, glucose not only has a particular ratio of atoms of carbon, hydrogen, and oxygen but also has a specific structure of these atoms. Making a mass balance on glucose is straightforward, and any appearance or disappearance of glucose is described by a chemical reaction.

Energy balances can be more difficult, because energy and work can take so many different forms. Internal, kinetic, potential, chemical, and surface energies are all important. Work can involve forces of pressure, gravity, and electrical potential. As a result, a truly general energy balance is extraordinarily complicated, so much so that it is difficult to use (Bird et al., 1960; Slattery, 1978).

As a result, people do not try to use truly general energy balances, but use simplified versions appropriate for special problems. I find this specialization reminiscent of a Tibetan painting that hangs on the wall of my study. The top of the painting shows the Buddha of the Yellow Cap, sitting like a star on a Christmas tree. Other manifestations of the god are spread out below him like the tree's ornaments. These demigods are vividly represented. Some are ferocious, some look ineffective, others seem kind and approachable.

These different forms of the Buddha are like the different forms of the energy equation. All forms of this equation are derived from the same difficult and complex spirit. These derived forms can look very different. Some are much more tractable than others and are more useful for solving specific problems.

In this section we shall focus our discussion on energy balances for a single pure component, possibly a fluid, that has internal and kinetic energy. The use of a pure fluid is equivalent to our earlier assumption of dilute solution, for the physical properties will be those of the solvent. The discussion hinges on three forms of the energy balance. The most general form is

$$\frac{\partial}{\partial t} \rho \left(\hat{U} + \frac{1}{2} v^2 \right) = -\nabla \cdot \rho \mathbf{v} \left(\hat{U} + \frac{1}{2} v^2 \right) - (\nabla \cdot \mathbf{q}) + \rho(\mathbf{v} \cdot \mathbf{g})$$

$$\begin{pmatrix} \text{energy} \\ \text{accumulation} \end{pmatrix} = \begin{pmatrix} \text{energy convection} \\ \text{in minus that out} \end{pmatrix} + \begin{pmatrix} \text{conduction} \end{pmatrix} - \begin{pmatrix} \text{work by} \\ \text{gravity} \end{pmatrix}$$

$$-\nabla \cdot \rho \mathbf{v} \left(\frac{p}{\rho} \right) - (\nabla \cdot [\boldsymbol{\tau} \cdot \mathbf{v}]) \tag{16.2-1}$$

$$-\begin{pmatrix} \text{work by} \\ \text{pressure} \\ \text{forces} \end{pmatrix} - \begin{pmatrix} \text{work by} \\ \text{viscous} \\ \text{forces} \end{pmatrix}$$

This scalar equation can be simplified by subtracting the mechanical energy balance, and thus removing the kinetic terms:

$$\frac{\partial(\rho\hat{U})}{\partial t} = -(\nabla \cdot \rho\mathbf{v}\hat{U}) - (\nabla \cdot \mathbf{q})$$

$$\begin{pmatrix} \text{energy} \\ \text{accumulation} \end{pmatrix} = \begin{pmatrix} \text{energy} \\ \text{convection} \\ \text{in minus that out} \end{pmatrix} + (\text{conduction})$$

$$- p(\nabla \cdot \mathbf{v}) - (\boldsymbol{\tau}{:}\nabla\mathbf{v})$$

$$- \begin{pmatrix} \text{reversible} \\ \text{work} \end{pmatrix} - \begin{pmatrix} \text{irreversible} \\ \text{work} \end{pmatrix} \tag{16.2-2}$$

A third useful form can be derived by combining the energy convection and the reversible work as the enthalpy:

$$\frac{\partial(\rho\hat{U})}{\partial t} = -(\nabla \cdot \rho\mathbf{v}\hat{H}) - (\nabla \cdot \mathbf{q})$$

$$\begin{pmatrix} \text{energy} \\ \text{accumulation} \end{pmatrix} = \begin{pmatrix} \text{energy} \\ \text{convection} \\ \text{in minus that out} \end{pmatrix} + (\text{conduction})$$

$$+ (\mathbf{v} \cdot \nabla p) - (\boldsymbol{\tau}{:}\nabla\mathbf{v})$$

$$- \begin{pmatrix} \text{part of the} \\ \text{enthalpy} \\ \text{definition} \end{pmatrix} - \begin{pmatrix} \text{irreversible} \\ \text{work} \end{pmatrix} \tag{16.2-3}$$

in which $\hat{H}$ $(= \hat{U} + p/\rho)$ is the specific enthalpy.

These equations are formidable, and so may best be understood by comparing them with more easily remembered results. For example, consider the form of the first law most commonly remembered by scientists (e.g., Lewis & Randall, 1923, 1961):

$$\Delta U = Q - W \tag{16.2-4}$$

This equation is for a batch system with no flow in or out, and it is restricted to changes in internal energy only. If we simplify Eq. 16.2-2 for these conditions, we have

$$\rho\,\frac{\partial\hat{U}}{\partial t} = -\nabla \cdot \mathbf{q} - [p(\nabla \cdot \mathbf{v}) + \boldsymbol{\tau}{:}\nabla\mathbf{v}] \tag{16.2-5}$$

The term ΔU corresponds to the accumulation on the left-hand side; the term Q refers to the conduction $-\nabla \cdot \mathbf{q}$; the work is represented by the quantity in brackets. Equation 16.2-4 is extensive, referring to the total system. In contrast, Eq. 16.2-5 is intensive, written on a small differential volume within the system.

Another familiar form of the first law used largely by engineers (e.g., Smith & van Ness, 1975) is that for a steady-state open system of fixed volume:

$$\Delta H = Q - W_s \tag{16.2-6}$$

where ΔH is the enthalpy change from inlet to outlet and W_s is the shaft work. For such a system, Eq. 16.2-3 becomes

$$0 = -(\nabla \cdot \rho\mathbf{v}\hat{H}) - \nabla \cdot \mathbf{q} - (\boldsymbol{\tau}{:}\nabla\mathbf{v}) \tag{16.2-7}$$

The enthalpy and conduction terms in this equation are analogous, but the work terms here reflect subtle differences.

At this point, you may justifiably wonder why the more complex equations have been introduced at all. Such wonder is legitimate, because these equations are rarely as useful as the simpler energy balances used for conduction problems in the previous section. They are much less useful than the corresponding equations used for fluid mechanics. Still, I find these equations a useful way to check my derivation of energy balances, especially in cases of simultaneous conduction and flow. Some of these are illustrated in the examples that follow.

Example 16.2-1: Conduction in a thin film and a thick slab

Derive Eq. 16.1-5 for a thin film and Eq. 16.1-16 for a thick slab from the generalized energy balance in Eq. 16.2-2.

Solution. For the thin film, the conduction is in steady state; so accumulation is zero. The film is solid; so there is no energy convection or work. Thus,

$$0 = -\nabla \cdot \mathbf{q}$$

Because transport is one-dimensional,

$$0 = -\frac{d}{dz} q$$

When we combine with Fourier's law,

$$0 = k\frac{d^2}{dz^2} T$$

This is the desired result.

For a solid slab of constant density, again there is no convection, so Eq. 16.2-2 becomes

$$\rho\frac{\partial \hat{U}}{\partial t} = -\nabla \cdot \mathbf{q}$$

Transport is still one-dimensional, and q can be restated in terms of Fourier's law. In addition, for a solid,

$$\hat{U} = \hat{C}_v[T - \text{(some reference temperature)}]$$

For solids, $\hat{C}_v$ equals $\hat{C}_p$ and is almost a constant, so

$$\rho\hat{C}_p\frac{\partial T}{\partial t} = k\frac{d^2 T}{dz^2}$$

which is the same as Eq. 16.1-16. In both cases, deriving the differential equation is easy; solving it for particular boundary conditions may be difficult.

Example 16.2-2: Heating a flowing solution

A viscous solution is flowing laminarly through a narrow pipe. At a known distance along the pipe, the pipe's wall is heated with condensing steam. Find a differential equation from which the temperature distribution in the pipe can be calculated.

Solution. Because this is a flow system, we decide to begin with the general energy balance in Eq. 16.2-3. The system is in steady state; so the accumulation is zero. We usually can anticipate that heating due to viscous dissipation is small, and so take $\tau:\nabla\mathbf{v}$ as nearly zero. Equation 16.2-3 then becomes

$$0 = -\nabla \cdot \rho\mathbf{v}\hat{H} - \nabla \cdot \mathbf{q}$$

To solve this problem, we often assume that the energy transfer along the pipe axis is largely by convection and that energy transport in the radial direction is by conduction:

$$0 = -\frac{\partial}{\partial z}\rho v\hat{H} - \frac{1}{r}\frac{\partial}{\partial r}rq$$

We expect that

$$\hat{H} = \hat{C}_p[(\text{some reference temperature})]$$

and that ρ, v, and $\hat{C}_p$ are constants. With these simplifications, we combine with Fourier's law to find

$$\frac{\partial}{\partial z}T = \left(\frac{k}{\rho\hat{C}_p v}\right)\frac{1}{r}\frac{\partial}{\partial r}r\frac{\partial T}{\partial r}$$

Solutions of the corresponding diffusion problem were discussed in Section 11.4.

Section 16.3. Heat transfer coefficients

The material presented in the first two sections focused on Fourier's law of heat conduction. This law, which is especially useful for heat conduction in solids, allows calculation of the temperature and heat flux at any position and time. It has been tremendously useful, a pillar of scientific thought. However, it can be difficult to use in fluid systems, especially when heat is transferred across phase boundaries.

Instead, we often use a different model for heat transfer, one better suited to approximate calculations of the heat transferred across interfaces. In this model, the separate phases are imagined to be well mixed, and hence isothermal. The only temperature gradients are close to the interface, in some vaguely defined interfacial region. The heat flux in this model is assumed to be

$$q = U\Delta T \tag{16.3-1}$$

where the heat flux q is taken as normal to the interface, the temperature difference ΔT is from one bulk phase across the interface into a second bulk phase, and the proportionality constant U is called the "overall heat transfer coefficient."

This new model is similar to that using mass transfer coefficients (see Chapter 9). Like the mass transfer model, it is compromised by a variety of definitions using different temperature differences. One common choice of temperature difference is based on the temperature difference at some particular position z:

$$q(z) = U\Delta T(z) \tag{16.3-2}$$

This local definition is that applied in this book except where explicitly stated otherwise. Another common choice is an arithmetic average temperature difference:

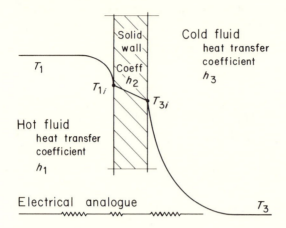

Fig. 16.3-1. Heat transfer across an interface. The overall heat transfer coefficient is a harmonic average of the individual heat transfer coefficients for the hot fluid, the wall, and the cold fluid. This averaging, which corresponds to the electrical problem of several resistances in series, is simpler than the corresponding mass transfer problem explained in Section 9.5.

$$q = U' \frac{\Delta T(\text{inlet}) + \Delta T(\text{outlet})}{2} \tag{16.3-3}$$

This definition is sometimes used for correlations of data on full-size industrial equipment (e.g., Perry & Chilton, 1973).

The overall heat transfer coefficient across an interface is often the average of several sequential steps. One classic example of this averaging is shown in Fig. 16.3-1. In this example, energy is transferred from a hot fluid to a solid wall, is conducted across the wall, and then is transferred into a cooler fluid. The heat flux in this case is

$$\begin{aligned}
q &= h_1(T_1 - T_{1i}) \\
&= h_2(T_{1i} - T_{3i}) \\
&= h_3(T_{3i} - T_3)
\end{aligned} \tag{16.3-4}$$

where the various temperatures and heat transfer coefficients are defined in the figure. The various h_i, called individual heat transfer coefficients, are characteristics of the fluids near the wall and of the wall itself.

We now want to calculate the overall heat transfer coefficient U as a function of these individual coefficients. We do this in two steps. First, we compare the second line of Eq. 16.3-4 with Eq. 16.1-9 and discover that

$$h_2 = \frac{k_2}{l_2} \tag{16.3-5}$$

where k_2 is the thermal conductivity and l_2 is the thickness of the solid wall. Second, we can combine the various equalities in Eq. 16.3-4 and eliminate the interfacial temperatures T_{1i} and T_{3i}. The final result is (after some algebra)

$$q = \frac{T_1 - T_3}{1/h_1 + l_2/k_2 + 1/h_3} \tag{16.3-6}$$

By comparison with Eq. 16.3-1, we see that

$$U = \frac{1}{1/h_1 + l_2/k_2 + 1/h_3} \tag{16.3-7}$$

The overall coefficient is a harmonic average of the individual coefficients.

Two features of this result are noteworthy. First, this heat transfer problem is a complete analogue to the electrical problem of three resistances in series. The heat flux q corresponds to the current, and the temperature difference $\Delta T \, (= T_1 - T_3)$ is like the voltage. The reciprocal of the overall heat transfer coefficient is analogous to the overall resistance. This overall resistance equals the sum of the resistances of the individual steps $1/h_1$, l_2/k_2, and $1/h_3$. This parallel is schematically suggested in Fig. 16.3-1.

At the same time, the overall heat transfer coefficient is simpler than the overall mass transfer coefficient developed in Section 9.5. Both coefficients are related to a sum of resistances, but the mass transfer case also involves weighting factors that I often find confusing. These factors relate the concentrations on different sides of the interface. In the heat transfer case, the interfacial temperature in, for example, the hot fluid at the wall, equals the temperature of the solid wall in contact with the hot fluid. This equality means no weighting factors and a simpler mathematical form.

We now illustrate the use of these coefficients by means of several simple examples.

Example 16.3-1: Finding the overall heat transfer coefficient
A total of $1.8 \cdot 10^4$ liters/hr crude oil flows in a heat exchanger with 40 tubes 5 cm in diameter and 2.8 m long (Fig. 16.3-2). The oil, which has a heat capacity of 0.43 cal/g-°C and a specific gravity of 0.9 g/cm³, is heated with 240°C steam from 20°C to 140°C. The steam is condensed at 240°C but is not cooled much below that temperature. What is the overall heat transfer coefficient based on the local temperature difference? What is it when based on the average temperature difference?

Solution. We begin with an energy balance on the volume $(\pi/4) d^2 \Delta z$, as shown in Fig. 16.3-2:

$$\begin{pmatrix} \text{energy} \\ \text{accumulation} \end{pmatrix} = \begin{pmatrix} \text{energy in minus energy out} \\ \text{by convection} \end{pmatrix} + \begin{pmatrix} \text{energy conducted} \\ \text{through walls} \end{pmatrix}$$

At steady state, this is

$$0 = \left(\frac{\pi}{4} d^2\right) (v\rho \hat{C}_p T|_z - v\rho \hat{C}_p T|_{z+\Delta z}) + (\pi d \Delta z) U (240 - T)$$

where U is the value defined by Eq. 16.3-2. The velocity in these equations is the average value in one of the tubes. Dividing by the volume $(\pi/4) d^2 \Delta z$, taking the limit as Δz goes to zero, and rearranging the result,

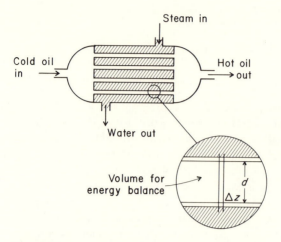

Fig. 16.3-2. A heat exchanger for crude oil. The oil flows through the tubes of the exchanger and is heated with condensing steam. The problem is to calculate the overall heat transfer coefficient from data given in the text. The exact answer depends on which definition of heat transfer coefficient is used.

$$\frac{dT}{dz} = \left(\frac{4U}{v\rho\hat{C}_p d}\right)(240 - T)$$

This is subject to the condition that

$$z = 0, \quad T = 20°C$$

Integrating,

$$\ln\left(\frac{240 - 20}{240 - T}\right) = \frac{4Uz}{v\rho\hat{C}_p d}$$

All data are given except v, which is easily found from the total flow:

$$v = \frac{(1.8 \cdot 10^7 \text{ cm}^3/\text{hr})(\text{hr}/3,600 \text{ sec})}{40(\pi/4)(5 \text{ cm})^2}$$

$$= 6.4 \text{ cm/sec}$$

Inserting this and the other values given,

$$U = \frac{v\rho\hat{C}_p d}{4z} \ln\left(\frac{240 - 20}{240 - T}\right)$$

$$= \frac{(6.4 \text{ cm/sec})(0.9 \text{ g/cm}^3)(0.43 \text{ cal/g-°K})(5 \text{ cm})}{4(280 \text{ cm})} \ln\left(\frac{240 - 20}{240 - 140}\right)$$

$$= 8.7 \cdot 10^{-3} \text{ cal/cm}^2\text{-sec-°K}$$

This value is based on the local temperature difference.

Alternatively, we might base the heat transfer coefficient on the average temperature difference (see Eq. 16.3-3):

$$\Delta T = \frac{1}{2}(\Delta T_{\text{in}} + \Delta T_{\text{out}})$$

$$= \frac{1}{2} (220°C + 100°C)$$

$$= 160°C$$

The heat flux is easily found:

$$q = \frac{(1.8 \cdot 10^7 \text{ cm}^3/\text{hr})(\text{hr}/3{,}600 \text{ sec})(0.9 \text{ g/cm}^3)(0.43 \text{ cal/g-°K})(140°C - 20°C)}{40(\pi/4)(5 \text{ cm})^2(280 \text{ cm})}$$

$$= 1.05 \text{ cal/cm}^2\text{-sec}$$

Thus, the overall heat transfer coefficient is

$$U' = 6.6 \cdot 10^{-3} \text{ cal/cm}^2\text{-sec-°K}$$

The difference between this value and that found earlier illustrates the importance of making sure which definition we are using.

Example 16.3-2: The time for tank cooling

A 100-gal tank filled with water initially at 80°F sits outside in air at 10°F. The overall heat transfer coefficient for heat lost from the water-containing tank is 3.6 Btu/hr-ft²-°F, and the tank's area is 27 ft². How long can we wait before the water in the tank starts to freeze?

Solution. As before, we begin with an energy balance on the tank:

$$\begin{pmatrix} \text{energy} \\ \text{accumulation} \end{pmatrix} = \begin{pmatrix} \text{heat loss} \\ \text{from tank} \end{pmatrix}$$

$$\frac{d}{dt} (\rho V \hat{C}_v T) = -UA(T - 10°F)$$

where ρ and C_v are the density and heat capacity of the water, V and A are the volume and area of the tank, and T is the water's temperature. This equation is subject to the initial condition

$$t = 0, \quad T = 80°F$$

We can use this condition in integrating the previous equation to find

$$\frac{T - 10°F}{80°F - 10°F} = e^{-(UA/\rho V \hat{C}_v)t}$$

We want the time at which freezing will begin:

$$\frac{32°F - 10°F}{80°F - 10°F} = \exp\left[-\left(\frac{(3.6 \text{ Btu/hr-ft}^2\text{-°F})(27 \text{ ft}^2)}{(8.31 \text{ lb/gal})(100 \text{ gal})(1 \text{ Btu/lb-°F})} \right) t \right]$$

$$t = 10 \text{ hr}$$

The tank will probably be safe overnight, but not much longer. Note that we implicitly assume that the tank's contents are isothermal and hence well mixed.

Example 16.3-3: The effect of insulation

Insulation advertisements claim that, in Minnesota, we can save 40% on our heating bills by installing 10 in. of glass wool as insulation. The glass wool has a thermal conductivity of about 0.03 Btu/hr-ft-°F; the average winter temperature in Minnesota is 15°F, and the house temperature is 68°F. If the advertisements are true, and if heat loss from doors and windows is minor, how much can we save with 2 ft of insulation?

Solution. Imagine that the heat loss in our current home is

$$q = h\Delta T$$

By adding 10 in. of glass wool, we have, from Eq. 16.3-5,

$$0.6q = \frac{1}{(1/h) + (10/12 \text{ ft})/(0.03 \text{ Btu/hr-ft-°F})} \Delta T$$

Dividing these equations, we find that

$$h = 0.014 \text{ Btu/hr-ft}^2\text{-°F}$$

With the thicker insulation, we have a heat loss q' of

$$q' = \frac{1}{(\text{hr-ft}^2\text{-°F}/0.014 \text{ Btu}) + (2 \text{ ft})/(0.03 \text{ Btu/hr-ft-°F})} \Delta T$$

$$= (0.0072 \text{ Btu/hr-ft}^2\text{-°F}) \Delta T$$

This represents a saving of about 30% over the house with 10 in. of insulation. Obviously, the additional gain should be balanced against the insulation's cost.

Example 16.3-4: Heat loss from a bar

The historian M. P. Crosland (1970) wrote that "in 1804 Biot carried out an experimental investigation of the conductivity of metal bars by maintaining one end at a high known temperature and taking readings of thermometers placed in holes along the bar. [He found] that the steady state temperature decreased exponentially along the bar." Biot could not explain this; Fourier could. Can you?

Solution. To solve this problem, we make an energy balance on a differential length Δz of the bar shown in Fig. 16.3-3:

$$\begin{pmatrix} \text{energy} \\ \text{accumulation} \end{pmatrix} = \begin{pmatrix} \text{energy in minus energy out} \\ \text{by conduction} \end{pmatrix} - \begin{pmatrix} \text{energy lost to} \\ \text{the surroundings} \end{pmatrix}$$

Because we are at steady state, the accumulation is zero, and

$$0 = Wlq|_z - Wlq|_{z+\Delta z} - 2(W + l)\Delta z h(T - T_\infty)$$

where W is the bar width, l is its vertical height, and h is the heat transfer coefficient between the bar and the surroundings. We now divide by the volume $Wl\Delta z$, take the limit as Δz goes to zero, and combine the result with Fourier's law:

$$0 = k\frac{d^2T}{dz^2} - \frac{h}{L}(T - T_\infty)$$

where L equals $Wl/(2W + 2l)$. This equation is easily integrated to give

$$T - T_\infty = ae^{\sqrt{h/kl}\,z} + be^{-\sqrt{h/kl}\,z}$$

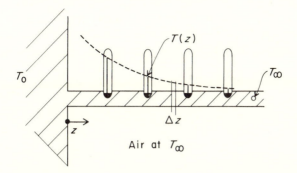

Fig. 16.3-3. Heat loss from a heated bar. The long bar shown is heated by contact with the large body at the left, which is at the high temperature T_0. The temperature in the bar drops exponentially along the bar. This situation is important historically, for it gave Fourier a major clue in developing his law for heat conduction.

where a and b are integration constants. These constants can be found from the boundary conditions

$$z = 0, \quad T = T_0$$
$$z = \infty, \quad T = T_\infty$$

We can show that a is zero by use of the second condition, equivalent to the assumption of a long bar. From the first condition, we find that

$$\frac{T - T_\infty}{T_0 - T_\infty} = e^{-\sqrt{h/kl}\,z}$$

Thus, the temperature drops off exponentially as we get farther and farther away from the bar's base. This is what Biot observed experimentally.

This problem is difficult because of the boundary condition that the bar is very long and because of the term in the energy balance for heat loss into the air. This second aspect gave Fourier himself a lot of trouble; so if it is not clear the first time, try again.

Section 16.4. Values of thermal conductivity, thermal diffusivity, and heat transfer coefficients

Up to this point, we have treated the thermal conductivity k, the thermal diffusivity α, and the heat transfer coefficient h as unknowns, adjustable parameters in any calculation. In fact, we often want to use previously measured values of these quantities to make predictions about new situations. Values for gases can be easily predicted from kinetic theory, and values for liquids and solids are best found by experiment. In this section we report a few selected values of these quantities.

Estimates of thermal conductivities of gases depend on the following result of kinetic theory (Hirschfelder et al., 1954):

Table 16.4-1. *Selected values of the collision integral Ω_ℓ for use in Eq. 16.4-1*

T/ε	Ω_ℓ
0.4	2.49
0.6	2.07
0.8	1.78
1.0	1.59
1.2	1.45
1.4	1.35
1.6	1.28
2.0	1.18
3.0	1.04
4.0	0.97
6.0	0.90
10.0	0.82

Source: Hirschfelder et al. (1954).

$$\ell = \frac{1.99 \cdot 10^{-4} \sqrt{T/\tilde{M}}}{\sigma^2 \Omega_\ell} \tag{16.4-1}$$

in which the thermal conductivity ℓ is in cal/cm-sec-°K, σ is the collusion diameter in Å, T is the temperature in °K, and $\tilde{M}$ is the molecular weight. The dimensionless quantity Ω_ℓ is of order 1, and a weak function of $\ell T/\varepsilon$, where ε is an energy of interaction. Values of σ and ε are given in Table 5.1-2; some selected values of Ω_ℓ are given in Table 16.4-1. These calculations are straightforward, completely parallel to those in Section 5.1.

Typical values of thermal conductivities and thermal diffusivities in gases, liquids, and solids are given in Table 16.4-2. Some of the values are expected from experience; for example, the thermal conductivities of metals are much higher than those of liquids or gases. Less obviously, the thermal diffusivities of nonmetallic solids and liquids are more nearly the same, indicating that unsteady heat transfer proceeds at similar rates in both phases.

The effective thermal conductivity of composite materials tends to be dominated by the continuous phase, just as in the case of diffusion (see Section 7.3). Composite materials that can partially melt show anomalous thermal diffusivities; examples include some hydrated salts and foods like ice cream. Still, thermal conductivities like those in Table 16.4-2 represent a norm from which there are few departures.

Some common correlations of heat transfer coefficients are reported in Table 16.4-3. These all refer to heat transfer across a solid–fluid interface, because other situations either are rare or are described in different terms. Like the mass transfer correlations in Section 9.3, these are best presented in

Table 16.4-2. *Thermal conductivities and thermal diffusivities of various materials*

	T (°K)	k (10^{-4} cal/cm-sec-°K)	α (cm²/sec)
Gases			
H_2	273	4.03	1.55
NH_3	273	0.53	0.14
N_2	273	0.57	0.22
O_2	273	0.58	0.22
Air	273	0.57	0.22
	573	1.08	0.69
CO_2	273	0.35	0.11
Liquids			
Water	293	14.3	0.0014
Ethanol	293	4.4	0.00093
CCl_2F_2 (Freon 12)	293	1.7	0.0056
Hexane	293	2.9	0.0011
Heptane	293	3.4	0.0011
Octane	293	3.5	0.0010
Benzene	293	3.7	0.0010
Toluene	293	3.6	0.0010
Mercury	293	210	0.46
Sodium	367	210	0.67
Bismuth	700	370	0.11
Solids			
Steel, 1% carbon	293	1,000	0.12
Steel, 1.5% carbon	293	860	0.10
Copper	293	9,500	1.17
Silver	293	10,200	1.71
AgBr	273	25	0.0055
NaCl	273	88	0.020
Brick (masonry)	293	16	0.0046
Concrete (dry)	293	3.1	0.0049
Glass wool ($\rho = 200$ kg/m³)	293	1.0	0.0028
Glass (window)	293	19	0.0034
Asbestos	293	2.7	0.0036

Source: Data from Reid et al. (1977), *Handbook of Chemistry and Physics* (1981), and *International Critical Tables* (1933).

terms of dimensionless groups. The two most important new groups are the Nusselt number and Prandtl number. The Nusselt number explicitly contains the heat transfer coefficient:

$$\begin{pmatrix}\text{Nusselt}\\\text{number}\end{pmatrix} = \frac{hl}{k} \tag{16.4-2}$$

where l is some characteristic length. The Prandtl number is more complex:

$$\begin{pmatrix}\text{Prandtl}\\\text{number}\end{pmatrix} = \frac{\mu \hat{C}_p}{k} = \frac{\nu}{\alpha} \tag{16.4-3}$$

Table 16.4-3. *A few correlations of heat transfer coefficients*

Physical situation	Correlation[a]	Specific definitions	Remarks
Turbulent flow in pipes	$\dfrac{hd}{k} = 0.027 \left(\dfrac{dv\rho}{\mu}\right)^{0.8} \left(\dfrac{\mu\hat{C}_p}{k}\right)^{0.33}$	d: tube diameter; v: velocity in tube	Widely quoted, often with slightly different constants or with small correction factors
Turbulent flow over banks of tubes	$\dfrac{hd}{k} = 0.33 \left(\dfrac{dv\rho}{\mu}\right)^{0.6} \left(\dfrac{\mu\hat{C}_p}{k}\right)^{0.33}$	d: tube diameter; v: velocity at minimum-flow area	Valid when $dv\rho/\mu > 6{,}000$, and for 10 or more rows of tubes
Flow around a solid sphere	$\dfrac{hd}{k} = 2.0 + 0.6 \left(\dfrac{dv\rho}{\mu}\right)^{0.5} \left(\dfrac{\mu\hat{C}_p}{k}\right)^{0.33}$	d: sphere diameter; v: sphere velocity	Valid only when free convection is absent
Free convection between vertical plates	$\dfrac{hl}{k} = \alpha \left(\dfrac{l^3 g\rho\Delta\rho}{\mu^2}\right)^{\beta} \left(\dfrac{l}{L}\right)^{1/9}$	l: distance between plates; L: length of plates; $\Delta\rho$: density change caused by temperature change	When $2 \cdot 10^3 < (l^3 g\rho\Delta\rho/\mu^2) < 2 \cdot 10^4$, $\alpha = 0.18$, and $\beta = 0.25$; when $2 \cdot 10^4 < (l^3 g\rho\Delta\rho/\mu^2) < 2 \cdot 10^5$, $\alpha = 0.065$, and $\beta = 0.33$

[a] hd/k is the Nusselt number; $\mu\hat{C}_p/k$ is the Prandtl number; $dv\rho/\mu$ is the Reynolds number; $l^3 g\rho\Delta\rho/\mu^2$ is the Grashöf number.
Source: Adapted from Kreith and Black (1980).

It essentially represents the relative importance of viscosity and thermal conductivity. These groups are quite closely parallel to the Sherwood and Schmidt numbers for mass transfer, a point detailed in Section 17.1. Now we turn to illustrations of heat transfer using these numerical values.

Example 16.4-1: The overall heat transfer coefficient of a heat exchanger

As part of a chemical process, we plan to use a shell-tube heat exchanger of 20 banks of 5-cm-outside-diameter steel tubes with 0.3-cm walls. Outside the tubes, we plan to use 400°C flue gas; inside, we expect to be heating aromatics like benzene and toluene at around 30°C. The gas flow will be 17 m/sec, and the liquid flow will be 2.7 m/sec. What overall heat transfer coefficient can we expect in this exchanger?

Solution. From Eq. 16.3-7, we see that

$$U = \frac{1}{1/h_1 + l_2/k_2 + 1/h_3}$$

where h_1 is the coefficient in the hot flue gas, k_2/l_2 refers to the steel wall, and h_3 is the coefficient in the liquid.

These coefficients are easily calculated. For h_1, we assume that the flue gas has the properties of nitrogen; so, from Eq. 16.4-1,

$$\begin{aligned}
k &= \frac{1.99 \cdot 10^{-4} \sqrt{T/\bar{M}}}{\sigma^2 \Omega_k} \\
&= \frac{1.99 \cdot 10^{-4} \sqrt{673°K/28}}{(3.80 \text{ Å})^2(0.87)} \\
&= 0.77 \cdot 10^{-4} \text{ cal/cm-sec-°C}
\end{aligned}$$

We then use the correlation for flow over tube banks in Table 16.4-3:

$$\begin{aligned}
h_1 &= 0.33 \left(\frac{k}{d}\right)\left(\frac{dv\rho}{\mu}\right)^{0.6}\left(\frac{\mu \hat{C}_p}{k}\right)^{0.3} \\
&= 0.33 \left(\frac{0.77 \cdot 10^{-4} \text{ cal/cm-sec-°C}}{5 \text{ cm}}\right) \\
&\quad \left(\frac{(5 \text{ cm})(1,700 \text{ cm/sec})(5.1 \cdot 10^{-4} \text{ g/cm}^3)}{3.3 \cdot 10^{-4} \text{ g/cm-sec}}\right)^{0.6} \\
&\quad \left(\frac{3.3 \cdot 10^{-4} \text{ g/cm-sec})(0.26 \text{ cal/g°C})}{0.77 \cdot 10^{-4} \text{ cal/cm-sec-°C}}\right)^{0.33} \\
&= 1.56 \cdot 10^{-3} \text{ cal/cm}^2\text{-sec-°C}
\end{aligned}$$

The value for the wall, h_2, is easily found using data from Table 16.4-2:

$$\begin{aligned}
h_2 &= \frac{k_2}{l_2} = \frac{0.10 \text{ cal/cm-sec-°C}}{0.30 \text{ cm}} \\
&= 0.33 \text{ cal/cm}^2\text{-sec-°C}
\end{aligned}$$

The value for the liquid inside the tube comes again from Table 16.4-3:

$$h_3 = 0.027 \left(\frac{k}{d}\right)\left(\frac{dv\rho}{\mu}\right)^{0.8}\left(\frac{\mu\hat{C}_p}{k}\right)^{0.33}$$

$$= 0.027 \left(\frac{3.6 \cdot 10^{-4} \text{ cal/cm-sec-}°K}{4.4 \text{ cm}}\right)$$

$$\left(\frac{(4.4 \text{ cm})(270 \text{ cm})(0.87 \text{ g/cm}^3)}{5.3 \cdot 10^{-3} \text{ g/cm-sec}}\right)^{0.8}$$

$$\left(\frac{(5.3 \cdot 10^{-3} \text{ g/cm-sec})(0.41 \text{ cal/g}°C)}{3.6 \cdot 10^{-4} \text{ cal/cm-sec-}°C}\right)^{0.33}$$

$$= 0.068 \text{ cal/cm}^2\text{-sec-}°K$$

Combining,

$$U = \cfrac{1}{\cfrac{1}{\dfrac{1.56 \cdot 10^{-3} \text{ cal}}{\text{cm-sec-}°C}} + \cfrac{1}{\dfrac{3.3 \cdot 10^{-1} \text{ cal}}{\text{cm-sec-}°C}} + \cfrac{1}{\dfrac{6.8 \cdot 10^2 \text{ cal}}{\text{cm-sec-}°C}}}$$

$$= 1.52 \cdot 10^{-3} \text{ cal/cm-sec-}°C$$

Note that the gas-side heat transfer coefficient is less than that on the liquid side, which in turn is less than that of the wall. This is the usual sequence; heat transfer in gases tends to limit the overall process. In contrast, mass transfer in liquids tends to be the rate-limiting step.

Example 16.4-2: The design of storm windows

Rising energy prices have led to a renaissance in the insulation of houses. Advertisements state that the cost of home heating can be substantially reduced by using two sets of storm windows. In other words, these advertisements urge the use of windows with three layers of glass in existing window frames, which are about 3 cm deep and 1 m long. Use your knowledge of heat transfer to decide whether or not this is a good idea. Assume that the outside temperature is $-10°C$ and the room temperature is $+20°C$.

Solution. The physical situation in this problem is illustrated schematically in Fig. 16.4-1. In the simplest case, the window consists of a single pane of glass. Heat loss through this window depends mostly on the thermal conductivity of this pane, for the adjacent air tends to be stirred by free convection.

The heat loss can be substantially reduced using storm windows of two or more panes of glass. Ideally, we would hope that this loss would now be governed by heat conduction across the gap between the panes. In fact, free convection stirs the air in this gap, so that the heat loss is much greater than that due to conduction. Still, this new resistance to heat transfer sharply reduces heat loss.

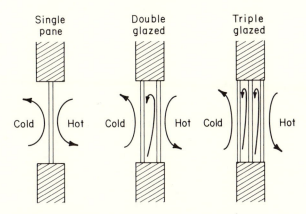

Fig. 16.4-1. Heat loss through storm windows. The heat loss through one pane of glass is much greater than the heat loss through two. Interestingly, the heat loss through two panes is greater than that through three, even though the thermal conductivity of air is less than that of the third pane of glass. This result illustrates the importance of free convection.

We want to know if the heat loss is further reduced by an additional pane of glass added between those in a conventional two-pane storm window. To answer this question, we first find the heat transfer coefficient for a two-pane storm window, using correlations for free convection given in Table 16.4-3. These correlations involve the Grashöf number, which for air is

$$\frac{b^3 g \rho \Delta \rho}{\mu^2} = \frac{b^3 g (\Delta \rho / \rho)}{\nu^2} = \frac{b^3 g}{\nu^2} \frac{\Delta(p \tilde{M}/RT)}{p \tilde{M}/RT} = \frac{b^3 g}{\nu^2} \frac{\Delta T}{T}$$

Thus, for a window with two panes 3 cm apart,

$$h \left(\genfrac{}{}{0pt}{}{\text{two}}{\text{pane}} \right) = 0.065 \left(\frac{k}{b} \right) \left(\frac{b^3 g}{\nu^2} \frac{\Delta T}{T} \right)^{1/3} \left(\frac{b}{L} \right)^{1/9}$$

$$= 0.065 \left(\frac{0.57 \cdot 10^{-4} \text{ cal/cm-sec-}^\circ K}{3 \text{ cm}} \right)$$

$$\cdot \left(\frac{(3 \text{ cm})^3 (980 \text{ cm/sec}^2)}{(0.14 \text{ cm}^2/\text{sec})^2} \right)^{1/3} \left(\frac{3 \text{ cm}}{100 \text{ cm}} \right)^{1/9}$$

$$= 0.92 \cdot 10^{-5} \text{ cal/cm}^2\text{-sec-}^\circ K$$

In contrast, the gaps in a three-pane window will be 1.5 cm. For each gap, the heat transfer coefficient will be

$$h \left(\genfrac{}{}{0pt}{}{\text{one gap of}}{\text{three pane}} \right) = 0.065 \left(\frac{0.57 \cdot 10^{-4} \text{ cal/cm-sec-}^\circ K}{1.5 \text{ cm}} \right)$$

$$\cdot \left(\frac{(1.5 \text{ cm})^3 (980 \text{ cm/sec}^2)}{(0.14 \text{ cm}^2/\text{sec})^2} \right)^{1/3} \left(\frac{1.5 \text{ cm}}{100 \text{ cm}} \right)^{1/9}$$

$$= 0.86 \cdot 10^{-5} \text{ cal/cm}^2\text{-sec-}^\circ K$$

but because there are two such gaps, the overall heat transfer coefficient is

$$U \left(\genfrac{}{}{0pt}{}{\text{three}}{\text{pane}}\right) = 0.43 \cdot 10^{-5} \text{ cal/cm}^2\text{-sec-}°K$$

Having an additional pane of glass cuts the heat loss through the windows almost in half.

Section 16.5. Conclusions

This chapter contains a synopsis of heat transfer. Although the presentation is brief, the chapter can supply a review or a summary for those already skilled in diffusion. The sections here are like those earlier in the book: a basic law, differential equations leading to fluxes, approximate models of interfacial transport, and values of the various coefficients.

The analysis of heat transfer is parallel to that for diffusion because heat transfer and diffusion are described with the same mathematical equations. Indeed, many experts argue that because of this mathematical identity, the two are identical. I do not agree with this because the two processes are, for me, so different physically. For example, heat conduction is faster in liquids than in gases, but diffusion is faster in gases than in liquids. This relation between the mathematical similarity and the physical difference affects the problems involving both heat and mass transfer, problems central to the next chapter.

References

Arpaci, V. S. (1966). *Conductive Heat Transfer*. Reading, Mass.: Addison-Wesley.

Bird, R. B., Stewart, W. E., & Lightfoot, E. N. (1960). *Transport Phenomena*. New York: Wiley.

Carslaw, H. S., & Jaeger, J. C. (1959). *Operational Methods in Applied Mathematics*. Oxford: Clarendon Press.

Crosland, M. P. (1970). "Jean-Baptiste Biot." In: *Dictionary of Scientific Biography*, ed. C. C. Gillispie. New York: Scribner.

Handbook of Chemistry and Physics (1981), ed. R. C. Weast. Boca Raton, Fla.: Chemical Rubber Publishing Co.

Herivel, J. (1975). *Joseph Fourier*. Oxford: Clarendon Press.

Hirschfelder, J. O., Curtiss, C. F., & Bird, R. B. (1954). *Molecular Theory of Gases and Liquids*. New York: Wiley.

International Critical Tables (1933). New York: McGraw-Hill.

Kreith, F., & Black, W. Z. (1980). *Basic Heat Transfer*. New York: Harper & Row.

Lewis, G. N., & Randall, M. (rev. by K. S. Pitzer & L. Brewer) (1923, 1961). *Thermodynamics*. New York: McGraw-Hill.

Luikov, A. V. (1968). *Analytical Heat Diffusion Theory*. New York: Academic Press.

Perry, R. H., & Chilton, C. H. (1973). *Chemical Engineers Handbook*. New York: McGraw-Hill.

Reid, R. C., Prausnitz, J. M., & Sherwood, T. K. (1977). *The Properties of Gases and Liquids*. New York: McGraw-Hill.

Slattery, J. C. (1978). *Momentum, Energy and Mass Transfer in Continua*. Huntington, N.Y.: Robert G. Krieger Publishing.

Smith, J. M., & van Ness, H. C. (1975). *Introduction to Chemical Engineering Thermodynamics*. New York: McGraw-Hill.

17 SIMULTANEOUS HEAT AND MASS TRANSFER

Processes involving coupled heat and mass transfer occur frequently in nature. They are central to the formation of fog, to cooling towers, and to the wet-bulb thermometer. They are important in the separation of uranium isotopes and in the respiration of water lilies.

This chapter analyzes a few of these processes. Not unexpectedly, such processes are complex, for they involve equations for both diffusion and conduction. These equations are coupled, often in a nonlinear way. As a result, our descriptions will contain approximations to reduce the complexities involved.

We begin this chapter with a comparison of the mechanisms responsible for heat and mass transfer. The mathematical similarities suggested by these mechanisms are discussed in Section 17.1, and the physical parallels are explored in Section 17.2. In Sections 17.3 and 17.4 we outline cooling-tower design and partial-condenser design as examples based on mass and heat transfer coefficients. In Section 17.5 we consider problems of fog formation as an example using diffusion coefficients and thermal diffusivities. Finally, in Section 17.6 we describe thermal diffusion and effusion.

Section 17.1. Mathematical analogies among mass, heat, and momentum transfer

Analogies among mass, heat, and momentum transfer have their origin either in the mathematical description of the effects or in the physical parameters used for quantitative description. The mathematically based analogies are useful for two reasons. First, they can save mathematical work; if the solution to a heat conduction problem is known, the solution to the corresponding diffusion problem is also known. We have already discussed this type of analogy in Section 3.5. Second, mathematical analogies often suggest dimensionless groups that are in turn helpful in correlating the results of physical experiments. It is this second use that has very broad scope and that is of interest in this section.

To explore these analogies, we remember that the diffusion of mass and the conduction of heat obey very similar equations. In particular, diffusion in one dimension is described by the following form of Fick's law:

439

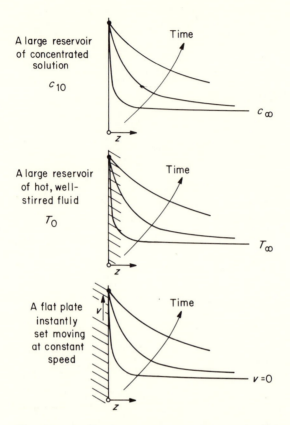

Fig. 17.1-1. Profiles for concentration, temperature, and fluid velocity. The diffusion of mass, the conduction of heat, and the laminar flow of fluids all obey laws of the same mathematical form. Accordingly, for mathematically identical boundary conditions, the profiles of concentration, temperature, and velocity are the same.

$$-j_1 = D \frac{dc_1}{dz} \tag{17.1-1}$$

where D is the diffusion coefficient. If this diffusion takes place into a semi-infinite slab, as shown in Fig. 17.1-1, the concentration profile can be shown to be (see Section 2.3)

$$\frac{c_1 - c_{10}}{c_{1\infty} - c_{10}} = \text{erf} \frac{z}{\sqrt{4Dt}} \tag{17.1-2}$$

where c_{10} and $c_{1\infty}$ are the concentrations at the slab's surface and far within the slab, respectively. Similarly, heat conduction is described by Fourier's law:

$$-q = k \frac{dT}{dz} \tag{17.1-3}$$

where k is the thermal conductivity. If heat conduction takes place into the semi-infinite slab in Fig. 17.1-1, the temperature profile is (see Section 16.1)

$$\frac{T - T_0}{T_\infty - T_0} = \text{erf} \frac{z}{\sqrt{4\alpha t}} \tag{17.1-4}$$

where α $(= k/\rho\hat{C}_p)$ is the thermal diffusivity and T_0 and T_∞ are the temperatures of the surface of the slab and far within the slab, respectively.

Although we have not discussed momentum transport in this book, we should mention that this process is also described within the same framework. The basic law is due to Newton:

$$-\tau = \mu \frac{dv}{dz} \tag{17.1-5}$$

where τ is the momentum flux or the shear stress and μ is the viscosity. If a flat plate is suddenly moved in an initially stagnant fluid, the velocity v of the fluid is

$$\frac{v - V}{0 - V} = \text{erf} \frac{z}{\sqrt{4\nu t}} \tag{17.1-6}$$

where the plate's velocity is V, the fluid's velocity far from the plate is zero, and the fluid's kinematic viscosity is ν.

At this point it has become conventional to draw an analogy among mass, heat, and momentum transfers. Each process uses a simple law combined with a mass or energy or momentum balance. If each process is described by the same mathematical equations and is subject to the mathematically equivalent boundary conditions, then each leads to results of the same mathematical form. Many believe it is more elegant to say that each process depends on combining a linear constitutive equation and a conservation relation to yield mathematically congruent results. The phenomenological coefficients of diffusion D, of thermal conductivity k, and of viscosity μ are thus analogous.

As a student, I found this conventional analogy confusing. Sure, Eqs. 17.1-1, 17.1-3, and 17.1-5 all say that a flux varies with a first derivative. Sure, Eqs. 17.1-2, 17.1-4, and 17.1-6 all have an error function in them. But D, k, and μ do not have the same physical dimensions. Moreover, D appears in both Eq. 17.1-1 and Eq. 17.1-2. In contrast, k appears in Eq. 17.1-3, but it must be replaced by the thermal diffusivity α in Eq. 17.1-4. The viscosity μ in Eq. 17.1-5 is replaced by the kinematic viscosity ν in Eq. 17.1-6. These changes confused me, and initially they undercut any value that analogies might have.

The source of my confusion stemmed from the ways in which the basic laws are written. In Fick's law (Eq. 17.1-1), the mass flux is proportional to the gradient of mass per volume, or the molar flux varies with the gradient in moles per volume. To be analogous, the energy flux q should be proportional to the gradient of the energy per volume $(\rho\hat{C}_p T)$. In other words, Eq. 17.1-3 should be rewritten as

$$-q = \frac{k}{\rho \hat{C}_p} \frac{d}{dz} (\rho \hat{C}_p T)$$

$$= \alpha \frac{d}{dz} (\rho \hat{C}_p T) \tag{17.1-7}$$

(In suggesting this alternative form, we imply that $\hat{C}_p$ equals $\hat{C}_v$.) If we use Eq. 17.1-7 instead of Eq. 17.1-3, then mass and heat transfer are truly analogous. Just as Eq. 17.1-2 follows from Eq. 17.1-1, so Eq. 17.1-4 follows from Eq. 17.1-7.

Newton's law for momentum transport can also be rewritten so that the momentum flux is proportional to the gradient of the momentum per volume (ρv), that is, by replacing Eq. 17.1-5 with

$$-\tau = \frac{\mu}{\rho} \frac{d}{dz} (\rho v)$$

$$= \nu \frac{d}{dz} (\rho v) \tag{17.1-8}$$

where ν is the kinematic viscosity. This new form, which implies a constant density, leads directly to Eq. 17.1-6.

Just as the fundamental laws for mass, heat, and momentum transfers can be made more nearly parallel, so can expressions for mass transfer coefficients and heat transfer coefficients. The interfacial mass flux already varies with the difference in mass per volume:

$$N_1 = k\Delta c_1 \tag{17.1-9}$$

The interfacial heat flux must be modified so that the energy flux varies with the energy difference per volume:

$$q \bigg|_{z=0} = h\Delta T$$

$$= \frac{h}{\rho \hat{C}_p} \Delta(\rho \hat{C}_p T) \tag{17.1-10}$$

Thus, the mass transfer coefficient k corresponds less directly to the heat transfer coefficient h than to the quantity $h/\rho \hat{C}_p$. The appropriate parallel for momentum transfer is the dimensionless friction factor f, defined as

$$\tau \bigg|_{z=0} = f\left(\frac{1}{2} \rho v^2\right)$$

$$= \left(\frac{fv}{2}\right) (\rho v - 0) \tag{17.1-11}$$

Thus, $fv/2$ is like k and $h/\rho \hat{C}_p$.

When these equations are written in these parallel forms, they automatically suggest the most common dimensionless groups. For example, the ratio of coefficients in Eqs. 17.1-8 and 17.1-1 is ν/D, the Schmidt number. The ratio of coefficients in Eqs. 17.1-9 and 17.1-11 is $(k/v)(2/f)$. Because $2/f$ is itself dimensionless, this is equivalent to k/v, the Stanton number.

Table 17.1-1. *Dimensionless analogies between heat and mass transfer*

Key properties	Heat transfer Common forms	Fundamental forms	Mass transfer Common forms	Fundamental forms
Variable	Temperature T	Energy per volume $\rho\hat{C}_pT$	Concentration c_1	Concentration c_1
Property	Thermal conductivity k	Thermal diffusivity α	Diffusion coefficient D	Diffusion coefficient D
Coefficient	Heat transfer coefficient h	$\dfrac{h}{\rho\hat{C}_p}$	Mass transfer coefficient k	Mass transfer coefficient k
Dimensionless groups often used as dependent variables[a]	Nusselt number $\dfrac{hl}{k}$	Stanton number $\dfrac{h}{\rho\hat{C}_pv}$	Sherwood number $\dfrac{kl}{D}$	Stanton number $\dfrac{k}{v}$
Dimensionless groups often used as independent variables[a]	Prandtl number $\dfrac{\mu\hat{C}_p}{k}$ Reynolds number $\dfrac{lv\rho}{\mu}$ Grashöf number[b] $\dfrac{l^3\rho g\Delta\rho}{\mu^2}$	Prandtl number $\dfrac{\nu}{\alpha}$ Reynolds number $\dfrac{lv}{\nu}$ Grashöf number[b] $\dfrac{l^3g\Delta\rho/\rho}{\nu^2}$	Schmidt number $\dfrac{\mu}{\rho D}$ Reynolds number $\dfrac{lv\rho}{\mu}$ Grashöf number[c] $\dfrac{l^3\rho g\Delta\rho}{\mu^2}$	Schmidt number $\dfrac{\nu}{D}$ Reynolds number $\dfrac{lv}{\nu}$ Grashöf number[c] $\dfrac{l^3g\Delta\rho/\rho}{\nu^2}$

[a] Remember that the characteristic length l is different in different physical situations.
[b] The density change $\Delta\rho$ is caused by a temperature difference here.
[c] The density change $\Delta\rho$ is caused by a concentration difference here.

These and other dimensionless groups formed in this way are shown in Table 17.1-1. Some of these analogies used to be surprising to me as a student; I never understood the assertion that the Prandtl number $\mu\hat{C}_p/k$ is analogous to the Schmidt number $\mu/\rho D$, although I learned to give that answer on exams. When I look at Eqs. 17.1-1, 17.1-7, and 17.1-8, I see that the Prandtl number ν/α and the Schmidt number ν/D are simply ratios of the coefficients of these equations. By similar arguments, the two Stanton numbers in the table represent the same kinds of ratios (Becker, 1976).

Thus, the parallels in the descriptions of these processes suggest not only ways to save mathematical work but also parallels between different kinds of measurements (e.g., between heat transfer coefficients and mass transfer coefficients). These similarities suggest that the numerical values of these different kinds of coefficients are also similar. This more powerful quantitative analogy is the subject of the following section.

Example 17.1-1: Cooling metal spheres

We want to quickly quench a liquid metal to make a fine powder. We plan to do this by spraying metal drops into an oil bath. How can we estimate the cooling speed of the drops?

Solution. No heat transfer correlation for this situation is given in Table 16.4-3. However, several mass transfer correlations for drops are given in Table 9.3-2. For example, for large drops without stirring,

$$\frac{kd}{D} = 0.42 \left(\frac{d^3 \Delta \rho g}{\rho \nu^2}\right)^{1/3} \left(\frac{\nu}{D}\right)^{1/2}$$

where d and ν are the dro's diameter and the fluid's kinematic viscosity. From Table 17.1-1, we see that the Sherwood number kd/D is equivalent to the Nusselt number hd/k and that the Schmidt number ν/D is analogous to the Prandtl number ν/α or $\mu \hat{C}_p/k$. Thus, we expect as a heat transfer correlation

$$\frac{hd}{k} = 0.42 \left(\frac{d^3 \Delta \rho g}{\rho \nu^2}\right)^{1/3} \left(\frac{\mu \hat{C}_p}{k}\right)^{1/2}$$

This correlation will be reliable only if the Grashöf number for the cooling falls in the same range as that used to develop the mass transfer correlation.

Example 17.1-2: Heat transfer from a spinning disc

Imagine that a spinning metal disc electrically heated to 30°C is immersed in 1,000 cm³ of an emulsion at 18°C. The disc is 3 cm in diameter and is turning at 10 rpm. The emulsion's kinematic viscosity is 0.082 cm²/sec. After an hour, the emulsion is at 21°C. What is its thermal diffusivity?

Solution. We being with an energy balance on the emulsion:

$$\begin{pmatrix} \text{energy} \\ \text{accumulation} \end{pmatrix} = \begin{pmatrix} \text{energy gained} \\ \text{from disc} \end{pmatrix}$$

$$(\rho \hat{C}_p V)\frac{dT}{dt} = (\pi R_0^2)q$$

$$= (\pi R_0^2)h(T_{\text{disc}} - T)$$

where T and T_{disc} are the emulsion and disc temperatures, respectively, R_0 is the disc radius, and h is the heat transfer coefficient. This equation is subject to the initial condition

$$t = 0, \quad T = T_0$$

Integrating,

$$\frac{T_{\text{disc}} - T}{T_{\text{disc}} - T_0} = e^{-(h/\rho \hat{C}_p)(\pi R_0^2/V)t}$$

Inserting the numbers given,

$$\frac{30 - 21}{30 - 18} = e^{-(h/\rho \hat{C}_p)[(1.5 \text{ cm})^2/(1,000 \text{ cm}^3)](3,600 \text{ sec})}$$

Thus,

$$\frac{h}{\rho \hat{C}_p} = 0.011 \text{ cm/sec}$$

We now need to relate this quantity to the thermal diffusivity. We have no direct way to do so. We do have a similar correlation for mass transfer away from a spinning disc (see Table 9.3-2):

$$\frac{kd}{D} = 0.62 \left(\frac{d^2\omega}{\nu}\right)^{1/2} \left(\frac{\nu}{D}\right)^{1/3}$$

Using Table 17.1-1, we see that the corresponding correlation must be

$$\frac{(h/\rho \hat{C}_p)d}{\alpha} = 0.62 \left(\frac{d^2\omega}{\nu}\right)^{1/2} \left(\frac{\nu}{\alpha}\right)^{1/3}$$

or

$$\alpha = \left[\frac{1}{0.62} \left(\frac{h}{\rho \hat{C}_p}\right) \nu^{1/6} \omega^{-1/2}\right]^{3/2}$$

Inserting the numerical values,

$$\alpha = \left[\frac{1}{0.62} \left(0.011 \frac{\text{cm}}{\text{sec}}\right)\left(0.083 \frac{\text{cm}^2}{\text{sec}}\right)^{1/6} \left(\frac{2\pi(10)}{60 \text{ sec}}\right)^{-1/2}\right]^{3/2}$$

$$= 1.2 \cdot 10^{-3} \text{ cm}^2/\text{sec}$$

This value is comparable to those given in Table 16.4-2.

Section 17.2. Physical equalities among mass, heat, and momentum transfer

In this section we want to discuss situations in which mass transfer, heat transfer, and fluid flow occur at the same rate. Such equivalence may be startling, for much of our earlier discussion emphasized differences between these processes. To be sure, the previous section described the parallel equations called Fick's law, Fourier's law, and Newton's law; but this parallelism was one of mathematics. The diffusion coefficient, the thermal conductivity or diffusivity, and the viscosity all had different numerical values, and so should give different rates.

The Reynolds analogy

Nonetheless, the rates of mass, heat, and momentum transfer can be essentially the same for fluids in turbulent flow. This subject was first studied by the Englishman Osborne Reynolds, who lived from 1842 to 1912. The descendant of generations of clergy who had served in the same Irish parish, Reynolds deliberately went to work in mechanical engineering before going to Cambridge University. Shortly after his graduation in 1867, Reynolds became professor of engineering at Owens College, Manchester, where he remained for his entire professional life.

Reynolds (1874) argued that mass or heat transport into a flowing fluid must involve two simultaneous processes: "1. The natural diffusion of the fluid when at rest [and] 2. the eddies caused by visible motion which mixes the fluid up and brings fresh particles into contact with the surface." He went on: "The first of these causes is independent of the velocity of the fluid [but] the second cause, the effect of eddies, arises entirely from the motion of the fluid." Note that Reynolds implies that any flowing fluid contains eddies. Nine years later, Reynolds discovered the distinction between laminar flow and turbulent flow, and that eddies occur only in the latter.

These arguments have considerable value even when restricted to turbulent flow. To see this, we write expressions for the various fluxes. For example, the mass flux should be

$$N_1 = k\Delta c_1$$
$$= [a + bv]\Delta c_1 \qquad (17.2\text{-}1)$$

where the quantity in brackets is equivalent to the mass transfer coefficient k. This coefficient has two parts: a, which is due to diffusion, the "natural internal diffusion," and bv, which represents the effect of eddies "which mix the fluid up." This seems very simple but very sensible.

In a similar fashion, we can write other flux equations. For energy, we find

$$q = h\Delta T$$
$$= [a' + b'v]\Delta(\rho\hat{C}_pT) \qquad (17.2\text{-}2)$$

where the heat transfer coefficient h reflects heat conduction a' and the effect of eddies $b'v$. For momentum, we write

$$\tau = f(\tfrac{1}{2}\rho v^2) = \left[\frac{fv}{2}\right]\rho v$$
$$= [a'' + b''v]\rho v \qquad (17.2\text{-}3)$$

where the friction factor f is made of a viscous contribution a'' and the effects of eddies $b''v$.

We now turn to the limit of very rapid turbulent flow, where the effect of eddies will dominate any diffusion, conduction, or viscosity. In other words, a, a', and a'' have very little effect. But if the eddies dominate, then all transport is due to that "mixing up" and is independent of any diffusion coefficient or thermal conductivity or viscosity. All transport is due to the same turbulent mechanism. In Reynolds's words (1874), these "various considerations lead to the supposition that"

$$b = b' = b'' \qquad (17.2\text{-}4)$$

Although this always seems to me a big intuitive leap, it does make more sense as I think about it.

This supposition provides a relation between the various transport coefficients. From Eq. 17.2-1, because a is relatively small,

$$k = bv \qquad (17.2\text{-}5)$$

We can make similar arguments for the other coefficients:

$$\frac{h}{\rho\hat{C}_p} = b'v \qquad (17.2\text{-}6)$$

$$\frac{f v}{2} = b'' v \qquad (17.2\text{-}7)$$

We then use Eq. 17.2-4 to combine the results:

$$\frac{k}{v} = \frac{h}{\rho \hat{C}_p v} = \frac{f}{2} \qquad (17.2\text{-}8)$$

This result is called the *Reynolds analogy*.

The Reynolds analogy is interesting because it suggests a very simple relation between different transport phenomena. This relation should be accurate when transport occurs by means of turbulent eddies. In this situation, we can estimate mass transfer coefficients from heat transfer coefficients or from friction factors.

The Reynolds analogy is found by experiment to be accurate for gases, but not for liquids (Sherwood et al., 1975). We can rationalize this on the basis of the transport coefficients involved. We expect turbulent mixing to take place at two levels: a macroscopic level, where eddies are dominant, and a microscopic level, where diffusion, conduction, and viscosity are important. We have buried this microscopic level in the coefficients b, b', and b''. For gases, these microscopic processes are about the same, because

$$D \doteq \alpha \doteq \nu \doteq 0.1 \text{ cm}^2/\text{sec} \qquad (17.2\text{-}9)$$

In the more dignified terms of dimensionless groups, the Schmidt and Prandtl numbers of gases are equal:

$$\frac{\nu}{D} \doteq \frac{\nu}{\alpha} \doteq 1 \qquad (17.2\text{-}10)$$

However, for liquids, these groups are significantly different; the Schmidt number is about 1,000, but the Prandtl number is around 10. Thus, the "mixing up" of turbulence may be nearly the same for gases, but it will not be for liquids.

The Chilton–Colburn analogy

Because the Reynolds analogy was practically useful, many authors tried to extend it to liquids. These extensions often included elaborate theoretical rationalizations. However, the most useful extension is the simple empiricism suggested by Chilton and Colburn (1934).

Chilton and Colburn recognized that the Reynolds number worked well for gases but not for liquids. They also believed that the changes in liquids could best be represented as Prandtl and Schmidt numbers. By an analysis of experimental data, they showed that Eq. 17.2-6 was better replaced by

$$b' = \frac{h}{\rho \hat{C}_p v} \left(\frac{\nu}{\alpha}\right)^{2/3} \qquad (17.2\text{-}11)$$

By mathematical analogy, they extended this mass transfer by replacing Eq. 17.2-5 by

$$b = \frac{k}{v} \left(\frac{\nu}{D}\right)^{2/3} \qquad (17.2\text{-}12)$$

Thus, from Eq. 17.2-4 and Eq. 17.2-7,

$$\frac{k}{v}\left(\frac{v}{D}\right)^{2/3} = \frac{h}{\rho\hat{C}_p v}\left(\frac{v}{\alpha}\right)^{2/3} = \frac{f}{2} \tag{17.2-13}$$

This *Chilton–Colburn analogy* reduces to the Reynolds analogy (Eq. 17.2-8) for gases whose Schmidt and Prandtl numbers equal unity.

Two historical asides about this result are interesting. First, the dimensionless quantities b and b' suggested by Reynolds were renamed "j factors" by Chilton and Colburn. These factors are common in the older literature, especially as j_D and j_H. Second, the exponent of $\frac{2}{3}$ on the Schmidt and Prandtl numbers is frequently subjected to theoretical rationalization, especially using boundary layer theory. Chilton is sometimes said to have cheerfully conceded that the value of $\frac{2}{3}$ was not even equal to the best fit of the data, but was chosen because the slide rules in those days had square-root and cube-root scales, but no other easy way to take exponents.

The wet-bulb thermometer

The best example of simultaneous heat and mass transfer using these analogies is the analysis of the wet-bulb thermometer. This convenient device for measuring relative humidity of air consists of two conventional thermometers, one of which is clad in a cloth wick wet with water. The unclad dry-bulb thermometer measures the air's temperature. The clad wet-bulb thermometer measures the colder temperature caused by evaporation of the water. This colder temperature is like that you feel by licking your finger and waving it about.

We want to use this measured temperature difference to calculate the relative humidity in air. This relative humidity is defined as the amount of water actually in the air divided by the amount at saturation at the dry-bulb temperature. To find this humidity, we first write equations for the mass and energy fluxes:

$$N_1 = k(c_{1i} - c_1) = kc(y_{1i} - y_1) \tag{17.2-14}$$
$$q = h(T_i - T) \tag{17.2-15}$$

where c_{1i} and c_1 are the concentrations of water vapor at the wet bulb's surface and in the bulk, y_{1i} and y_1 are the corresponding mole fractions, T_i is the wet-bulb temperature, and T is the bulk dry-bulb temperature. Note that y_{1i} is the value at saturation at T_i. The mass and energy fluxes are coupled:

$$N_1 \Delta\tilde{H}_{vap} = -q \tag{17.2-16}$$

where $\Delta\tilde{H}_{vap}$ is the heat of vaporization of the evaporating water. Thus,

$$k\Delta\tilde{H}_{vap}c(y_{1i} - y_1) = h(T - T_i) \tag{17.2-17}$$

From Eq. 17.2-13, the Chilton–Colburn analogy,

$$k = \frac{h}{\rho\hat{C}_p}\left(\frac{D}{\alpha}\right)^{2/3} = \frac{h}{c\tilde{C}_p}\left(\frac{D}{\alpha}\right)^{2/3} \tag{17.2-18}$$

For gases, the Lewis number α/D is about unity. Combining Eqs. 17.2-17 and 17.2-18 and rearranging, we find that

$$y_1 = y_{1i} - \left(\frac{\tilde{C}_p}{\Delta \tilde{H}_{\text{vap}}}\right)(T - T_i) \tag{17.2-19}$$

or

$$\left(\begin{array}{c}\text{relative}\\\text{humidity}\end{array}\right) = \frac{p_1}{p_1(\text{sat at } T)} = \frac{y_1}{y_1(\text{sat at } T)}$$

$$= \frac{1}{y_1(\text{sat at } T)}\left[y_{1i}(\text{sat at } T_i) - \left(\frac{\tilde{C}_p}{\Delta \tilde{H}_{\text{vap}}}\right)(T - T_i)\right] \tag{17.2-20}$$

Thus, the relative humidity should be independent of the flow past the thermometers and should vary linearly with the temperature difference between the wet-bulb and dry-bulb readings.

Section 17.3. Design of cooling towers

Cooling towers like those shown in Fig. 17.3-1 are the cheapest way to cool large quantities of water. As such, they are among the largest mass transfer devices in common use (Berman, 1961; American Institute of Chemical Engineers, 1972). The basic operation of one common form of cooling tower is shown schematically in Fig. 17.3-2. The tower is packed with inert material, most commonly with redwood slats. Hot water sprayed into the top of

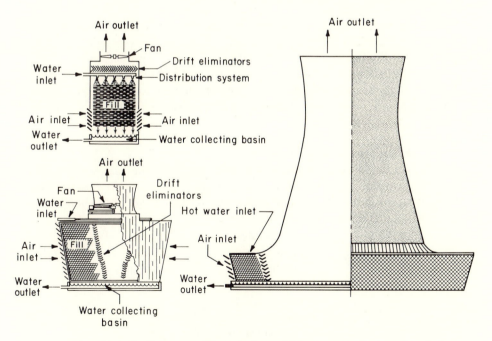

Fig. 17.3-1. Cooling towers. These devices, which are among the largest made for mass transfer, cool large quantities of water by evaporation of a small fraction of the water. They are a very common industrial sight, especially in power plants.

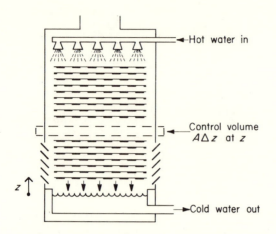

Fig. 17.3-2. Modeling a small cooling tower. We want to calculate the size of a cooling tower required to cool a given amount of water. We base this calculation on mass and energy balances on the small volume shown.

the tower trickles down through the wood, evaporating as it goes. Air enters the bottom of the tower and rises up through the packing. In smaller towers, the air can be pumped with a fan; in larger ones, it is often allowed to rise by natural convection.

We want to calculate the size of a tower required to cool a given amount of water. This calculation is roughly parallel to that for the web-bulb thermometer. There, we began with a mass balance and an energy balance; we then combined these to find how the humidity was related to these temperatures. Here, this basic strategy is the same, but with an additional step: an integration of the equations relating humidity and temperature to find the size of the cooling tower.

The mass balance

We begin by considering the differential volume $A\Delta z$ shown schematically in Fig. 17.3-2. This volume, located at z, is filled with packing having a surface area per volume equal to a. We can make a mass balance on the water vapor in this volume as follows:

$$\begin{pmatrix} \text{water} \\ \text{accumulation} \end{pmatrix} = \begin{pmatrix} \text{water} \\ \text{convection in} \\ \text{minus that out} \end{pmatrix} + \begin{pmatrix} \text{water} \\ \text{added by} \\ \text{evaporation} \end{pmatrix} \qquad (17.3\text{-}1)$$

or

$$0 = \left(Gy_{H_2O} \Big|_z - Gy_{H_2O} \Big|_{z+\Delta z} \right) + (\Delta z a)k(c_{H_2O,i} - c_{H_2O}) \qquad (17.3\text{-}2)$$

Dividing by Δz and taking the limit as this volume goes to zero,

$$0 = -\frac{d}{dz}(Gy_{H_2O}) + ka(c_{H_2O,i} - c_{H_2O}) \qquad (17.3\text{-}3)$$

In these equations, G is the molar flux of wet air, y_{H_2O} and c_{H_2O} are the mole fraction and the concentration of water in that air, respectively, $c_{H_2O,i}$ is the concentration of water in air at the air–water interface, and ka is the mass transfer coefficient times the tower's surface area per volume.

This equation is conveniently rewritten in terms of slightly different variables. First, we recognize that c_{H_2O} equals cy_{H_2O}, where c is the total molar concentration and y_{H_2O} is the water's mole fraction in the wet air. Next, we realize that G is about equal to the moles per area of dry air n_{air} flowing in the tower, which implies that the water vapor is a dilute solution in the air. Thus,

$$0 = -n_{air} \frac{d}{dz} y_{H_2O} + kac(y_{H_2O,i} - y_{H_2O}) \tag{17.3-4}$$

This form of mass balance is easier to use in the calculations that follow.

Energy balances

We now make two different energy balances on the same differential volume $A\Delta z$. First, we make a balance on the wet air alone:

$$0 = -n_{air} \tilde{C}_{p,air} \frac{dT_{air}}{dz} + ha(T_i - T_{air}) \tag{17.3-5}$$

As before, we have assumed that the flow of wet air is roughly constant and about equal to the flow of dry air. In addition, we make an energy balance on both liquid water and wet air:

$$0 = -n_{H_2O} \tilde{C}_{p,H_2O} \frac{dT_{H_2O}}{dz} - n_{air} \frac{d\tilde{H}}{dz} \tag{17.3-6}$$

where T_{H_2O} is the liquid's temperature, n_{H_2O} represents the flux of liquid water in the tower, and $\tilde{H}$ is the air's enthalpy. In words, this relation is equivalent to assuming that the tower is nearly adiabatic, for energy lost from the water is gained by the air.

In these energy balances, we usually do not know T_i, the temperature of the air–water interface. We do expect that the thermal conductivity of the water will be much higher than that of the air; so T_i will almost equal T_{H_2O}. We shall assume this equality in the rest of our development.

Combining the equations

We now want to combine the mass and energy balances in a way that allows us to size the cooling tower. To do so, we multiply Eq. 17.3-4 by the heat of vaporization of water ΔH_{vap} and add the result to Eq. 17.3-5 to obtain

$$n_{air} \frac{d}{dz} (\tilde{C}_{p,air}T_{air} + \Delta\tilde{H}_{vap}y_{H_2O})$$

$$= ha(T_{H_2O} - T_{air}) + kac \Delta\tilde{H}_{vap}(y_{H_2O,i} - y_{H_2O}) \tag{17.3-7}$$

The Chilton–Colburn analogy is

$$\frac{k}{v} \left(\frac{v}{D}\right)^{2/3} = \frac{h}{\rho\hat{C}_{p,air}v} \left(\frac{v}{\alpha}\right)^{2/3} \tag{17.3-8}$$

For gases, the Lewis number α/D is about unity, and $\rho\hat{C}_{p,\text{air}}$ equals $c\tilde{C}_{p,\text{air}}$. Thus,

$$h = (c\tilde{C}_{p,\text{air}})k \tag{17.3-9}$$

Inserting this into the previous equation and rearranging, we find

$$n_{\text{air}}\frac{d}{dz}(\tilde{H}) = kac(\tilde{H}_i - \tilde{H}) \tag{17.3-10}$$

where

$$\tilde{H} = \tilde{C}_{p,\text{air}}T_{\text{air}} + \Delta\tilde{H}_{\text{vap}}y_{H_2O} \tag{17.3-11}$$

$$\tilde{H}_i = C_{p,\text{air}}T_{H_2O} + \Delta\tilde{H}_{\text{vap}}y_{H_2O,i} \tag{17.3-12}$$

In physical terms, $\tilde{H}$ is usually described as the enthalpy of the wet air per mole of dry air. This implicitly assumes that the mole fraction y_{H_2O} approximately equals the moles of water per mole of dry air, which is true only if the water vapor is dilute. Correspondingly, the quantity $\tilde{H}_i$ is the enthalpy of wet air per mole dry air at the interface, where the air is saturated. Note that the mole fraction at this interface is the saturation value and hence is a function only of the water's temperature T_{H_2O}.

Equation 17.3-10 can be numerically integrated to find the desired height of the tower l:

$$l = \int_0^l dz = \frac{n_{\text{air}}}{kac}\int_{\tilde{H},\text{in}}^{\tilde{H},\text{out}} \frac{d\tilde{H}}{\tilde{H}_i - \tilde{H}} \tag{17.3-13}$$

Because our operating conditions are usually specified in terms of temperatures, we shall find it useful to supplement this equation with a rearranged form of Eq. 17.3-6:

$$\frac{d\tilde{H}}{dT_{H_2O}} = -\frac{\tilde{C}_{p,H_2O}n_{H_2O}}{n_{\text{air}}} \tag{17.3-14}$$

Because n_{air} is positive and n_{H_2O} is negative, this derivative is positive.

A graphical check on the numerical calculation of the required height goes as follows (Mickley, 1949). We must know the water's flow and its inlet and outlet temperatures. We also know the air flow and its inlet humidity. We plot this humidity versus the water's outlet temperature, as shown by point A in Fig. 17.3-3. Using Eq. 17.3-14, we can then calculate $\tilde{H}$ versus any T_{H_2O}, as shown by the line AB. Moreover, using vapor pressure data and values of $\Delta\tilde{H}_{\text{vap}}$, we can find $\tilde{H}_i$ versus T_{H_2O}, shown by the line CD, by using Eq. 17.3-12. We can then read off values of $\tilde{H}_i - \tilde{H}$ and numerically or graphically evaluate the integral in Eq. 17.3-13. We also need values of kc for a packed tower; some of the measured values are shown in Fig. 17.3-4 (Kelly & Swenson, 1956).

This integration is similar to those used in the gas absorption analysis in Chapter 10. Many emphasize this analogy by referring to the line CD in Fig. 17.3-3 as an equilibrium line and to the line AB as an operating line (Sherwood et al., 1975; Treybal, 1980). You may find such terminology stimulating, though you should recognize that the energy balances used here are physically different from the mass balances used in gas absorptions.

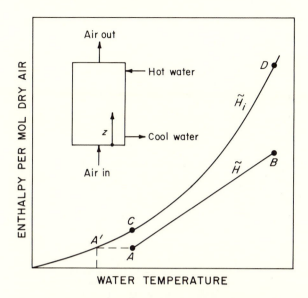

Fig. 17.3-3. Sizing a cooling tower. These calculations depend on evaluating the wet air's humidity versus the water's temperature, shown as the line *AB*. They also involve the interfacial humidity versus water temperature, shown as the line *CD*. The lines *AB* and *CD* are sometimes called the operating and equilibrium lines, by analogy with gas absorption. Note that *A'* is the dew point of the entering air.

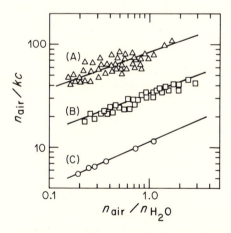

Fig. 17.3-4. Transfer coefficients for cooling towers. These coefficients are for an enthalpy driving force, and so include aspects of both heat and mass transfer. Curve (A) is for packing with a surface per volume of 40–100 m^{-1}, curve (B) is for 10 m^{-1}, and curve (C) is for 3 m^{-1}. [Data from Sherwood et al. (1975).]

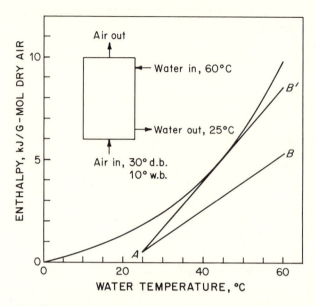

Fig. 17.3-5. Design of a specific cooling tower. This figure is a specific example of the general case in Fig. 17.3-3. The tower is to cool water from 60°C to 25°C using air supplied with the humidity at A. The line AB' represents the maximum water flow possible; anything greater would give an air enthalpy higher than the equilibrium value, which is a thermodynamic impossibility.

Example 17.3-1: Design of a cooling tower

Design a countercurrent cooling tower to cool water at 2,150 kg/min. The water enters at 60°C and is to be cooled to 25°C. The air is fed at 20°C at a rate of 60 g-mol/m²-sec, with a dry-bulb temperature of 30°C and a dew-point temperature of 10°C. The water flux should be 40% lower than the maximum allowed thermodynamically. The tower is packed with redwood slats having an area per volume of 3 m⁻¹. Find (a) the flow rate of the water per tower cross section, (b) the tower cross section, and (c) the height of tower required.

Solution. We begin by plotting the enthalpy of water-saturated air $\bar{H}_i$ versus temperature, as shown in Fig. 17.3-5. We know that $\bar{H}$ for the entering air is the saturated value at 10°C, the wet-bulb temperature; we also know that the water temperature at this point is 25°C. Thus, we can locate the point A. We can then solve for the specific values given.

(a) We then recognize from Eq. 17.3-14 that the maximum water flow must give an operating line AB'; any higher water flow would give an enthalpy of wet air higher than the equilibrium value. Thus,

$$\left(\frac{\text{slope}}{AB'}\right) = -\frac{\bar{C}_{p,\mathrm{H_2O}}n_{\mathrm{H_2O,max}}}{n_\mathrm{air}} = 230 \text{ J/g-mol-°C}$$

The flux n_{air} equals 60 g-mol/m²-sec, and $\tilde{C}_{p,H_2O}$ is 75 J/g-mol-°C; so $n_{H_2O,max}$ is 180 g-mol H_2O/m²-sec. The actual flow is to be 40% less:

$$n_{H_2O} = 110 \text{ g-mol } H_2O/\text{m}^2\text{-sec}$$

This is the desired result. Using this value, we can draw the actual operating line AB.

(b) We plan to use this tower to cool 2,150 kg H_2O/min. Thus, the tower cross section A is

$$(2{,}150 \text{ kg } H_2O/\text{min}) \left(\frac{\text{min}}{60 \text{ sec}}\right)\left(\frac{\text{g-mol}}{0.018 \text{ kg}}\right)$$

$$= (110 \text{ g-mol } H_2O/\text{m}^2\text{-sec})(A)$$

$$A = 18 \text{ m}^2$$

This corresponds to a cylindrical tower about 5 m in diameter.

(c) The tower's height can be found from Eq. 17.3-13:

$$L = \frac{n_{air}/kc}{a} \int_{\tilde{H},in}^{\tilde{H},out} \frac{d\tilde{H}}{\tilde{H}_i - \tilde{H}}$$

The value of kc can be found from Fig. 17.3-4. The value for abscissa in this figure equals 60 g-mol air/110 g-mol H_2O; thus, the ordinate n_{air}/kc equals 9. Values of $\tilde{H}_i$ and $\tilde{H}$ can easily be read off lines CD and AB, respectively; the integral found by numerical integration of these data equals 2.7. Thus,

$$L = \left(\frac{9}{3 \text{ m}^{-1}}\right)(2.7)$$

$$= 8.1 \text{ m}$$

This is the final result.

Section 17.4. Partial condensers

The design of partial condensers requires an analysis of coupled heat and mass transfer that is roughly parallel to that for cooling towers. The chief differences are that the vapor is often a concentrated solution, and that the process is not adiabatic, but is cooled with some type of coolant. These differences produce mathematical complexities that are most effectively handled by numerical calculation. In this section we derive the basic equations involved.

The physical situation involved is exemplified by that in Fig. 17.4-1. A warm mixed vapor is flowing downward through a vertical tube. The tube is surrounded by a coolant flowing countercurrently. As the vapor flows downward, it is partially condensed, so that its flux n_{1z} drops. The amount of condensation depends both on diffusion to the wall, characterized by N_{1r}, and on heat transfer to the wall, described by the flux q.

The analysis of this situation begins with a mass balance on the control volume of vapor shown in Fig. 17.4-1. We then write energy balances on the vapor and on the vapor plus coolant. Finally, we can simultaneously integrate these equations to find the size of condenser required.

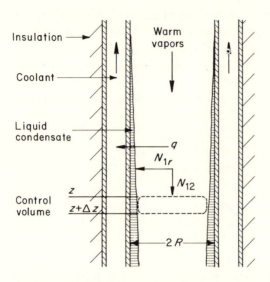

Fig. 17.4-1. Analysis of a partial condenser. A mixed vapor flowing downward is partially condensed by a coolant flowing countercurrently. The analysis of this condenser parallels that in a cooling tower, but is complicated by the concentrated vapor and by the heat conduction through the condenser wall.

The mass balance

The first step is to make a mass balance on the condensable vapor, called species 1, entering and leaving the control volume. For a cylindrical condenser operating at steady state, this is

$$\begin{pmatrix} \text{vapor} \\ \text{accumulation} \end{pmatrix} = \begin{pmatrix} \text{vapor flow in} \\ \text{minus flow out} \end{pmatrix} - \begin{pmatrix} \text{vapor} \\ \text{condensed} \end{pmatrix} \qquad (17.4\text{-}1)$$

$$0 = \pi R^2 \left(n_{1z} \Big|_z - n_{1z} \Big|_{z+\Delta z} \right) - N_{1r}(2\pi R \Delta z) \qquad (17.4\text{-}2)$$

where the variables involved are shown in Fig. 17.4-1. Dividing by the control volume $\pi R^2 \Delta z$ and taking the limit as this volume goes to zero, we find

$$0 = - \frac{dn_{1z}}{dz} - \frac{2}{R} N_{1r} \qquad (17.4\text{-}3)$$

We want to rewrite this relation in terms of the local concentration of vapor. To do this, we make a mass balance on the noncondensable material, species 2.

$$0 = - \frac{d}{dz} n_{2z} \qquad (17.4\text{-}4)$$

Thus, n_{2z} is a constant, known for the particular experimental conditions used. But, by definition,

$$\frac{n_{1z}}{n_{2z}} = \frac{c_1 v_z}{c_2 v_z} = \frac{y_1}{y_2} = \frac{y_1}{1 - y_1} \qquad (17.4\text{-}5)$$

This equation allows us to write the flux n_{1z} as a function of the concentration y_1. We can write N_{1r} as a function of concentration as well:

$$N_{1r} = kc(y_1 - y_{1i}) \tag{17.4-6}$$

Note that the mass transfer coefficient k refers to resistance in the vapor phase, implying that the condensed liquid is a pure component. In addition, the interfacial mole fraction is a function of the temperature of the condensed liquid.

We can now combine Eqs. 17.4-3, 17.4-5, and 17.4-6 to find

$$0 = -\frac{n_{2z}}{(1 - y_1)^2} \frac{dy_1}{dz} - \frac{2kc}{R} (y_1 - y_{1i}) \tag{17.4-7}$$

This equation for a partial condenser is the parallel of Eq. 17.3-4 for a cooling tower.

Energy balances

We next turn to an energy balance on the same control volume:

$$\begin{pmatrix} \text{accumulation} \\ \text{of energy} \end{pmatrix} = \begin{pmatrix} \text{energy} \\ \text{convection in} \\ \text{minus that out} \end{pmatrix} - \begin{pmatrix} \text{energy} \\ \text{lost by} \\ \text{conduction} \end{pmatrix} - \begin{pmatrix} \text{energy} \\ \text{lost by} \\ \text{mass transfer} \end{pmatrix} \tag{17.4-8}$$

$$0 = \pi R^2 \left(\tilde{C}_p(n_{1z} + n_{2z})T \Big|_z - \tilde{C}_p(n_{1z} + n_{2z})T \Big|_{z+\Delta z} \right)$$
$$- 2\pi R \Delta z q - 2\pi R \Delta z \Delta \tilde{H}_{vap} N_{1r} \tag{17.4-9}$$

where we have assumed that both vapor and inert gases have the same heat capacity. Again, we divide by the volume $\pi R^2 \Delta z$ and take the limit as this volume goes to zero:

$$0 = -\tilde{C}_p \frac{d}{dz} [(n_{1z} + n_{2z})T] - \frac{2}{R} (q + \Delta \tilde{H}_{vap} N_{1r}) \tag{17.4-10}$$

We next write this equation in terms of concentrations and temperatures. We can replace n_{1z} and N_{1r} using Eq. 17.4-5 and Eq. 17.4-6, respectively. We also write q in terms of temperatures:

$$q = h(T - T_{liq}) \tag{17.4-11}$$

where h is the heat transfer coefficient from the gas phase to the gas–liquid interface, and the condensate temperature T_{liq} is taken as equal to the interfacial temperature. With these changes, we obtain

$$0 = -\tilde{C}_p n_{2z} \frac{d}{dz} \left(\frac{T}{1 - y_1} \right) - \frac{2}{R} \{h(T - T_{liq}) + kc\Delta \tilde{H}_{vap}[y_1 - y_{1i}(T_{liq})]\} \tag{17.4-12}$$

This energy balance is for the partial condenser what Eq. 17.3-5 was for the cooling tower.

The next equation required must couple this energy balance with changes in the coolant. Now the energy balance is made on an annular ring of condensate lying outside the differential volume used in the foregoing. The result is

$$n_c \tilde{C}_{p,c} \frac{dT_c}{dz} = U(T_{\text{liq}} - T_c) \tag{17.4-13}$$

where n_c, $\tilde{C}_{p,c}$, and T_c are the molar flux, the molar heat capacity, and the temperature of the coolant, and U is the overall heat transfer coefficient from the bulk coolant into the condensed liquid, but not into the vapor. Finally, the various energy fluxes are related:

$$U(T_{\text{liq}} - T_c) = h(T - T_{\text{liq}}) + \Delta \tilde{H}_{\text{vap}} kc[y_1 - y_{1i}(T_{\text{liq}})] \tag{17.4-14}$$

This relation is like Eq. 17.2-17 for the wet-bulb thermometer.

We must now simultaneously integrate Eqs. 17.4-7, 17.4-12, 17.4-13, and 17.4-14. Although the details of such an integration are beyond the scope of this book, we can see ways in which we might try to simplify this task (McAdams, 1954; Treybal, 1980). For a dilute vapor, we can assume that $1 - y_1$ equals unity. We can then add Eq. 17.4-7 and Eq. 17.4-12, use the Reynolds analogy to remove the heat transfer coefficient h, and write the result in terms of the enthalpy. This result is similar to Eq. 17.3-10, because both are enthalpy equations that include a mass transfer coefficient. The result here can be graphically integrated simultaneously with Eq. 17.4-14 without much difficulty. However, this simplification occurs only in dilute solution. In general, we must use either numerical integration or physical experiments to design these units.

Section 17.5. Fog

Heat and mass transfer are coupled phenomena. Analysis of this coupling exploits both similarities in the mathematical description and near equalities between different transport coefficients. In previous sections we used lumped-parameter models to analyze the wet-bulb thermometer, cooling towers, and partial condensers. We used the Reynolds and Chilton–Colburn analogies to get a variety of useful results.

In this section we analyze fog formation using a distributed-parameter model. We calculate concentrations and temperatures that vary with both position and time. These variations are expressed in terms of quantities like the diffusion coefficient or the thermal conductivity, not in terms of mass or heat transfer coefficients. As such, the analysis is a more fundamental approach than that used in previous sections of this chapter.

The two cases of fog formation analyzed here are shown schematically in Fig. 17.5-1. In case (a), warm wet air initially above its dew point is suddenly contacted with a cold solid reservoir. Fog forms as the air cools. In case (b), cool wet air, also initially above its dew point, is brought into contact with a warm reservoir. Interestingly, fog forms in this case as the air is heated. In each case we want to calculate the amount of fog versus position and time.

To make this calculation, we make a variety of assumptions. Most important, we assume that the system contains abundant nuclei, so that water vapor and liquid are always in equilibrium, and any supersaturation pro-

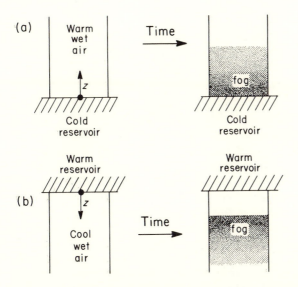

Fig. 17.5-1. Two cases of fog formation. Fog forms when the local partial pressure exceeds the equilibrium vapor pressure, which is in turn dependent on the local temperature. As such, fog formation is an example of simultaneous heat and mass transfer. In case (a), fog forms when warm wet air is cooled from below. In case (b), fog forms when cool wet air is warmed from above.

duces fog formation. Such abundant nuclei suggest that the fog formed will be present as very small droplets whose concentration can be treated as a continuous function of position and time. Phrased another way, we recognize that changes in concentration and temperature take place over a much larger scale than the size of the fog's droplets; so we neglect variations around individual droplets. Because the droplets are still much larger than air and water molecules, we assume that the fog's diffusion coefficient is zero. Finally, we assume that the system is always dilute, never has convective flow, and has a single average heat capacity.

We can use these assumptions to write mass and energy balances on the systems in Fig. 17.5-1 (Kirkaldy & Brown, 1963; Toor, 1971). For both cases, the appropriate equations in the absence of fog are

$$\frac{\partial c_1}{\partial t} = D \frac{\partial^2 c_1}{\partial z^2} \tag{17.5-1}$$

$$c_2 = 0 \tag{17.5-2}$$

$$c\tilde{C}_p \frac{\partial T}{\partial t} = k \frac{\partial^2 T}{\partial z^2} \tag{17.5-3}$$

where c_1 and c_2 are the concentrations of water vapor and fog liquid, respectively. The equations in the presence of fog are

$$\frac{\partial c_1}{\partial t} = D \frac{\partial^2 c_1}{\partial z^2} - r_2 \tag{17.5-4}$$

$$\frac{\partial c_2}{\partial t} = r_2 \tag{17.5-5}$$

$$c\tilde{C}_p \frac{\partial T}{\partial t} = k \frac{\partial^2 T}{\partial z^2} + \Delta \tilde{H}_{vap} r_2 \tag{17.5-6}$$

where r_2 is the rate of fog formation and $\Delta \tilde{H}_{vap}$ is the molar heat of vaporization. Equations 17.5-4 and 17.5-5 are similar to the case of diffusion and equilibrium chemical reaction discussed in Example 2.3-3. There, however, the reaction between c_1 and c_2 was first-order; here, the concentration c_1 is closely related to the vapor pressure and hence is a nonlinear function of the temperature T.

The rate of fog formation can be found as follows. First, we divide Eq. 17.5-6 by $c\tilde{C}_p$, and we recognize that in gases the thermal diffusivity $\alpha (= k/c\tilde{C}_p)$ is about equal to the diffusion coefficient:

$$\frac{\partial T}{\partial t} = D \frac{\partial^2 T}{\partial z^2} - \left(\frac{\Delta \tilde{H}_{vap}}{c\tilde{C}_p}\right) r_2 \tag{17.5-7}$$

The approximation that $\alpha = D$ greatly simplifies the analysis here, just as the approximation that $h/c\tilde{C}_p$ equaled k simplified calculations in earlier sections. Because the vapor concentration c_1 is a function of temperature,

$$\frac{\partial c_1}{\partial t} = \left(\frac{\partial c_1}{\partial T}\right)\left(\frac{\partial T}{\partial t}\right) \tag{17.5-8}$$

$$\frac{\partial^2 c_1}{\partial z^2} = \left(\frac{\partial c_1}{\partial T}\right)\left(\frac{\partial^2 T}{\partial z^2}\right) + \left(\frac{\partial^2 c_1}{\partial T^2}\right)\left(\frac{\partial T}{\partial z}\right)^2 \tag{17.5-9}$$

Inserting these equations into Eq. 17.5-4, and combining with Eq. 17.5-7, we find

$$r_2 = \frac{D(\partial T/\partial z)^2(\partial^2 c_1/\partial T^2)}{1 + (\Delta \tilde{H}_{vap}/c\tilde{C}_p)(\partial c_1/\partial T)} \tag{17.5-10}$$

The rate of water vapor disappearance is the negative of this.

We can calculate the rate of fog formation using this equation. We recognize that D, $\Delta \tilde{H}_{vap}$, and $c\tilde{C}_p$ are all positive and that $(\partial T/\partial z)^2$ must be positive. Moreover, $\partial c_1/\partial T$ and $\partial^2 c_1/\partial T^2$ are both positive, either by experiment or by the Clausius–Clapeyron equation. Thus, r_2 is always positive; fog always forms except at the boundaries of the fog-filled region.

To find out how much fog forms, we must solve for the concentration and temperature profiles within the region where fog is present. To do so, we multiply Eq. 17.5-4 by $\Delta \tilde{H}_{vap}/c\tilde{C}_p$, add the result to Eq. 17.5-7, and rearrange to find

$$\frac{\partial}{\partial t}(c\tilde{C}_p T + \Delta \tilde{H}_{vap} c_1) = D \frac{\partial^2}{\partial z^2}(c\tilde{C}_p T + \Delta \tilde{H}_{vap} c_1) \tag{17.5-11}$$

The quantity in parentheses is the molar enthalpy, a parallel to that used in cooling-tower design. Because fog is present, c_1 and T are not independent, but related by the vapor pressure.

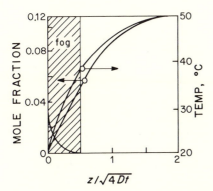

Fig. 17.5-2. Fog formation by cooling. Water vapor that is 89% saturated at 50°C is placed on top of liquid water at 20°C [see case (a) in Fig. 17.5-1]. As the vapor cools, fog forms. The water vapor and temperature profiles shown are calculated as described in the text. The concentration of fog is also shown as the solid curve totally within the crosshatched region. [Adapted from Toor (1971).]

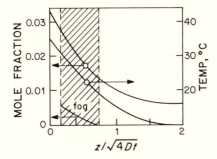

Fig. 17.5-3. Fog formation by warming. Water vapor that is 80% saturated and at 10°C is heated from above with a reservoir at 35°C [see case (b) in Fig. 17.5-1]. Interestingly, fog forms because of coupled heat and mass transfer. This situation is similar to precipitation induced by dissolution, as discussed in Section 15.5. [Adapted from Toor (1971).]

We now must solve Eq. 17.5-1 and Eq. 17.5-3 for the fog-free region and Eq. 17.5-11 for the foggy region. The boundary between these regions, called z_{sat}, is governed by the equations

$$-D \frac{\partial c_1}{\partial z} \left(\begin{array}{c}\text{no-fog}\\\text{region}\end{array}\right)\Bigg|_{z_{sat}} = -D \frac{\partial c_1}{\partial z} \left(\begin{array}{c}\text{fog}\\\text{region}\end{array}\right)\Bigg|_{z_{sat}} + (c_2|_{z_{sat}}) \frac{dz_{sat}}{dt} \qquad (17.5\text{-}12)$$

and

$$-D \frac{\partial T}{\partial z} \left(\begin{array}{c}\text{no-fog}\\\text{region}\end{array}\right)\Bigg|_{z_{sat}} = -D \frac{\partial T}{\partial z} \left(\begin{array}{c}\text{fog}\\\text{region}\end{array}\right)\Bigg|_{z_{sat}} - c_2 \left(\frac{\Delta \bar{H}_{vap}}{c \bar{C}_p}\right) \frac{dz_{sat}}{dt} \qquad (17.5\text{-}13)$$

Other boundary conditions depend on the specific problems being considered. Calculated temperature and concentration profiles for two such problems are shown in Figs. 17.5-2 and 17.5-3 (Toor, 1971).

Example 17.5-1: Progression of a fog boundary

Imagine that air at 50°C and saturated like that in Fig. 17.5-2 is placed over water at 20°C. How deep will the layer of fog be after 10 min? How deep will it be after 1 hr?

Solution. From Fig. 17.5-2, we see that the fog's boundary will be located at

$$\frac{z}{\sqrt{4Dt}} = 0.5$$

From Table 5.1-1, we see that D is about 0.23 cm²/sec. Thus,

$$z = 0.5\sqrt{4Dt}$$
$$= 0.5\sqrt{4(0.23 \text{ cm}^2/\text{sec})(600 \text{ sec})}$$
$$= 12 \text{ cm}$$

This result assumes that the fog droplets are stationary and that there is no significant free convection.

Section 17.6. Thermal diffusion and effusion

This section discusses several ways in which temperature gradients effect a solute flux. The phenomena involved occur in the absence of convection and are treated with models like those developed for diffusion in Chapters 2 and 3. Thus, the approach is again based on a distributed-parameter model like that in the previous section and is a more fundamental scheme than the mass transfer coefficients used in earlier sections of this chapter.

The first effect, thermal diffusion, is exemplified by the two experiments shown in Fig. 17.6-1. In the first, a tall column of salt solution is heated at the top and cooled at the bottom. The salt's concentration is initially uniform, but later becomes more concentrated near the bottom of the tube. This experiment was originally made in 1856 by Fick's mentor, Carl Ludwig (1859); more complete experiments were later made by Charles Soret (1879), after whom this effect is often called. A similar experiment, shown schematically in Fig. 17.6-1(b), consists of two bulbs, both of which initially contain the same gaseous mixture (Chapman & Dootson, 1917). When one bulb is heated and the other is cooled, the gas no longer has the same mole fractions throughout. This experiment was originally made to check the theoretical prediction of its existence. It is a rare occurrence when the experiment follows the theory, rather than the other way around.

Results of experiments like these are often reported in slightly different ways. For liquids, the results are usually given in terms of the flux equation:

$$-\mathbf{j}_1 = Dc(\nabla x_1 - \sigma x_1 x_2 \nabla T) \tag{17.6-1}$$

The Soret coefficient σ, which has the dimensions of reciprocal temperature, can be either positive or negative. For gases, the results are correlated using the equation

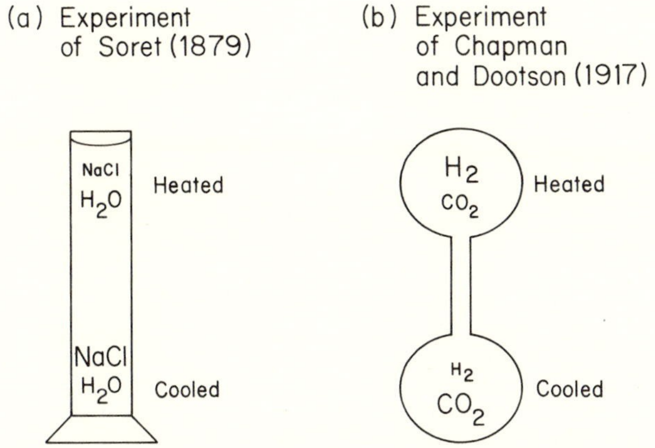

(a) Experiment
of Soret (1879)

(b) Experiment
of Chapman
and Dootson (1917)

Fig. 17.6-1. Thermal diffusion. If a tall column of initially homogeneous salt solution is heated at the top and cooled at the bottom, the mole fraction of salt will become slightly greater at the bottom. If two bulbs connected with a capillary are filled with the same gas mixture and only one bulb is heated, the gas compositions will become unequal. Both experiments are cases of a new effect: thermal diffusion.

$$-\mathbf{j}_1 = Dc \left(\nabla x_1 - \alpha x_1 x_2 \frac{\nabla T}{T} \right) \qquad (17.6\text{-}2)$$

The dimensionless thermal diffusion factor α obviously equals the Soret coefficient times the temperature. Both σ and α are positive when species 1 concentrates in the hot region.

Two aspects of these equations are interesting. First, we are now writing the diffusion flux in terms of the gradient of mole fraction, not molar concentration. This is because we know that the molar concentration varies strongly with temperature, but the mole fraction is much more nearly constant, independent of temperature. Such a flux equation implies a different reference frame than the volume average velocity emblazoned through this book (Tyrrell, 1961; deGroot & Mazur, 1963). Second, we deliberately introduce a factor $x_1 x_2$ into the expression for thermal diffusion. This anticipates observations that the effect disappears rapidly for dilute solutions and is largest when solute and solvent concentrations are similar.

A few experimental values of α are shown in Table 17.6-1. The values of α are frequently small, especially in dilute solution. They are largest for solutes of very different molecular weights or for highly nonideal solutions. They are more nearly constant for near-ideal solutions and are concentration-dependent in nonideal liquid mixtures. In short, they behave very much like the ternary diffusion coefficients discussed in Chapter 8. They are usually of minor practical importance, even though they can be used to effect surprisingly good separations (Hoglund et al., 1979).

This table of values does not explain why the thermal diffusion occurs.

Table 17.6-i. *Thermal diffusion coefficients*

Mixture[a]	Temperatures (°K)	α[b]
Gases		
(1) 50% H_2 – (2) 50% D_2	290–370	+0.17
(1) 50% H_2 – (2) 50% He	273–700	+0.15
(1) 50% H_2 – (2) 50% CH_4	300–500	+0.29
(1) 50% N_2 – (2) 50% O_2	293	+0.02
(1) 50% N_2 – (2) 50% CO_2	290–400	+0.05
Liquids		
(1) 20% cyclohexane – (2) 80% CCl_4	313	+1.3
(1) 50% cyclohexane – (2) 50% CCl_4	313	+1.3
(1) 80% cyclohexane – (2) 20% CCl_4	313	+1.3
(1) 20% cyclohexane – (2) 80% benzene	313	−0.1
(1) 50% cyclohexane – (2) 50% benzene	313	−0.4
(1) 80% cyclohexane – (2) 20% benzene	313	−0.6
(1) 25% water – (2) 75% ethanol	298	−0.9
(1) 60% water – (2) 40% ethanol	298	−1.5
(1) 90% water – (2) 10% ethanol	298	+0.3
(1) 0.01-M KCl in (2) H_2O	303	−0.6
(1) 0.01-M NaCl in (2) H_2O	303	−0.9

[a] Concentrations in mole percent, except as noted.
[b] Taken as positive if the first species given concentrates in the hot region.
Source: Tyrrell (1961).

This is not an easy explanation to give briefly, and so it is carefully avoided by authors whose knowledge of this field is much greater than mine. We can give a vague explanation by again referring to the gaseous experiment in Fig. 17.6-1(b). In steady state, the flux of molecules from left to right must equal that from right to left. These fluxes must have two parts: that due to thermal motion and that caused by the bulk flow necessary to maintain equal pressure. The thermal motion varies with the molecular weight of the particular species, but the second varies only with the average molecular weight. When the molecules interact as rigid spheres, the net flux is greater for the heavier molecules than for the lighter ones; so these heavier molecules usually will concentrate in the cooler region. Even this qualitative argument is compromised for more elaborate intermolecular potentials, for nonideal liquid solutions, and for solids. No simple, more general explanation seems possible.

Thermal diffusion, which has just been discussed, occurs in mixtures in which molecules of solute and solvent interact with each other. Thermal effusion, the effect discussed next, occurs when the molecules of a pure gas react largely with surroundings.

Thermal effusion is most clearly illustrated by the schematic drawings in Fig. 17.6-2. In these drawings, a pure gas is placed in a closed tube and separated into two volumes by a porous diaphragm. The gas on one side of the diaphragm is heated; so the gas pressure changes. After a while, the pressure reaches a constant value and can be measured.

(a) Large pores

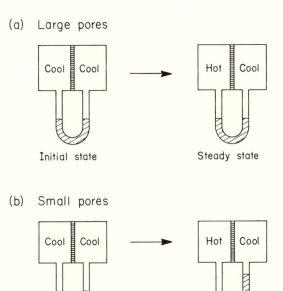

(b) Small pores

Fig. 17.6-2. Thermal effusion. Imagine a gas separated by a porous diaphragm. When the gas on one side of the diaphragm is heated, its pressure changes and eventually reaches a steady value. If the diaphragm has large pores, this pressure is the same as that on the colder side. If the diaphragm has small pores, the pressure on the hot side remains higher, an effect called thermal effusion.

Two distinctly different cases are observed (Kennard, 1938). The first occurs when the gas pressure is high and the diaphragm has large pores. When the gas is heated, its pressure initially increases, but this increase is quickly reduced by flow through the diaphragm. This flow can be described by Darcy's law, for the gas molecules collide largely with each other. At steady state, no pressure difference remains.

The second case involves a dilute gas and very small pores (Reynolds, 1879; Knudsen, 1909). Here, any gas molecules in the diaphragm collide mainly with the pore walls, not with other gas molecules. When the temperature is increased, the molecular velocity v in these collisions also increases; from kinetic theory, we can show that this velocity increases with temperature:

$$v \propto \sqrt{T/\bar{M}} \tag{17.6-3}$$

However, the concentration is decreased by temperature:

$$c = \frac{n}{V} = \frac{p}{RT} \tag{17.6-4}$$

The flux out of the hot region is the product of the hot velocity and the hot concentration; that out of the cold side is analogous. Thus, the total flux is

$$j = cv - c'v'$$

$$\propto \left(\frac{p}{RT} \sqrt{T/\tilde{M}} - \frac{p'}{RT'} \sqrt{T'/\tilde{M}} \right)$$

$$\propto \left(\frac{p}{\sqrt{T}} - \frac{p'}{\sqrt{T'}} \right) \tag{17.6-5}$$

where the primed and unprimed value refer to the cold and hot regions, respectively. At steady state, this implies

$$\frac{p}{\sqrt{T}} = \frac{p'}{\sqrt{T'}} \tag{17.6-6}$$

The pressure on the hot side will be greater than the pressure on the cold side. The key to this process is that the holes in the diaphragm be very small. Thus, thermal effusion is to Knudsen diffusion as thermal diffusion is to ordinary diffusion. Both thermal effusion and diffusion are illustrated in the following examples.

Example 17.6-1: The size of thermal diffusion

Thermal diffusion is being studied in a two-bulb apparatus like that in Fig. 17.6-1(b). Each bulb is 3 cm³ in volume; the capillary is 1 cm long and has an area of 0.01 cm². The left-hand bulb is heated to 50°C, and the right-hand bulb is kept at 0°C. The entire apparatus is initially filled with an equimolar mixture, either of hydrogen–methane or of ethanol–water. How much separation is achieved? About how long does this separation take?

Solution. At steady state, the net flux given by Eq. 17.6-1 must be zero:

$$0 = Dc \left(\nabla x_1 - \alpha x_1 x_2 \frac{\nabla T}{T} \right)$$

Thus

$$x_1|_{\text{hot}} - x_1|_{\text{cold}} = \alpha x_1 x_2 \left(\frac{T_{\text{hot}} - T_{\text{cold}}}{T_{\text{avg}}} \right)$$

For the gas mixture, this is

$$y_1|_{\text{hot}} - y_1|_{\text{cold}} = (0.29)(0.50)(0.50) \left(\frac{50°K}{298°K} \right)$$

$$= 0.012$$

For water–ethanol, we find by interpolation that

$$x_1|_{\text{hot}} - x_1|_{\text{cold}} = (-1.3)(0.50)(0.50) \left(\frac{50°K}{298°K} \right)$$

$$= -0.05$$

The separations obtained are both small; that with liquids is slightly larger, but in the opposite direction.

We now turn to the time required for this separation to occur. To find this time, we parallel the analysis given for the diaphragm cell in Example 2.2-4. In this analysis, we assume that the temperature difference is suddenly applied at time zero. We also assume that in spite of this difference, the total molar concentration c is a constant. Thus, for the left-hand bulb, we find

$$V_B \frac{dx_{1B}}{dt} = \frac{A}{c} j_1 = -\frac{AD}{l} [(x_{1B} - x_{1A}) + b]$$

where b $[= \alpha x_1 x_2 (T_{1B} - T_{1A})/T_{avg}]$ is the effect of thermal diffusion. For the right-hand bulb,

$$V_A \frac{dx_{1A}}{dt} = +\frac{AD}{l} [(x_{1B} - x_{1A}) + b]$$

Combining these equations,

$$\frac{d}{dt} (x_{1B} - x_{1A}) = -D\beta(x_{1B} - x_{1A} + b)$$

where β $[= (A/l)(1/V_B + 1/V_A)]$ was previously used as the calibration constant of the cell. Note that b depends only on average concentrations and constant temperatures; so this result can be integrated directly.

Alternatively, we note that $(D\beta)^{-1}$ is essentially the relaxation time of this cell. If the experiment takes less than this time, the steady state is still far away; if the experiment takes much longer than this time, then the steady state will be approached. For the cell used here,

$$\beta = \frac{0.01 \text{ cm}^2}{1 \text{ cm}} \left(\frac{2}{3 \text{ cm}^3}\right) = 0.007 \text{ cm}^{-2}$$

Thus, for gases, $(\beta D)^{-1}$ is about 500 sec, and the steady state is reached in a few hours. For liquids, $(\beta D)^{-1}$ is half a year, and reaching the steady state requires a very long time. These slow rates and small separations mean that thermal diffusion usually is a bad route for separations.

Example 17.6-2: Flow in water lily stems

Dacey (1980) reported that some water lilies generate flow through their hollow stems in order to facilitate oxygen transfer to their roots. This flow represents an "internal wind" that can reach 50 cm/min. It is believed to occur because of differences in pore sizes between young and old lily leaves, as shown schematically in Fig. 17.6-3. Use kinetic theory to show how the warm young leaf can generate this pressure difference.

Solution. The explanation of this effect given in the literature asserts that the flow is caused by thermal effusion. Sun strikes the lily, warming both old and young leaves. The small pores in the young leaves produce a higher pressure, as suggested by Fig. 17.6-2 and Eq. 17.6-6. The larger pores in the older leaves do not produce any pressure change. Thus, gas flows into the young leaves, down the stems, and out the older leaves. I find this delightful.

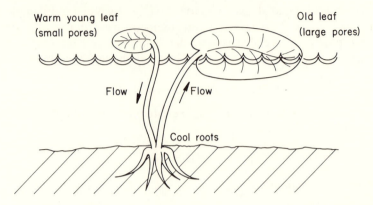

Fig. 17.6-3. Thermal effusion in water lilies. This effect apparently causes a flow of as much as 50 cm/min through the lily's hollow stems. This flow supplies the roots with oxygen.

Section 17.7. Conclusions

Simultaneous heat and mass transfer, the subject of this chapter, is a complicated process. Analyzing this process to find simple but useful results depends on making effective approximations. These approximations exploit both the similar mathematics used for the processes and the similar numerical values of the transport coefficients. This can be true for both distributed- and lumped-parameter models. More specifically, for gases, D and α are nearly equal, and k and $h/\rho\hat{C}_p$ are very similar.

This strategy for analysis immediately raises the question of the similar effects possible in liquids and solids. Here, the mathematics remain similar, but the transport coefficients are very different: D is much less than α, and k is much smaller than $h/\rho\hat{C}_p$. What should you do now?

For liquids and solids, the heat transfer is much more rapid than the mass transfer, and so proceeds as if the mass transfer did not exist. In other words, for liquids and solids, the two processes are essentially uncoupled. As an example, imagine that both mass and energy are being transferred from a well-stirred reservoir into a thick solid slab. The energy will be transferred much more rapidly than the mass. In the region where the mass flux is large enough to be interesting, the temperature will be essentially constant, equal to the reservoir temperature.

This difference between gases and other phases illustrates the different analytical strategies possible. Your success in exploring problems like these rests on your ability to make effective approximations. Good luck in your efforts.

References

American Institute of Chemical Engineers (1972). *Cooling Towers.*
Becker, H. A. (1976). *Dimensionless Parameters – Theory and Methodology.* New York: Wiley.

Berman, L. D. (1961). *Evaporation Cooling of Circulating Water*. New York: Pergamon Press.

Chapman, S., & Dootson, F. W. (1917). *Philosophical Magazine, 33,* 248.

Chilton, T. H., & Colburn, A. P. (1934). *Industrial and Engineering Chemistry, 26,* 1183.

Dacey, J. W. H. (1980). *Science, 210,* 1017.

deGroot, S. R., & Mazur, P. (1963). *Non-Equilibrium Thermodynamics*. Amsterdam: North Holland.

Hoglund, R. L., Shacter, J., & von Halle, E. (1979). Diffusion Separation Methods. In: *Encyclopedia of Chemical Technology*, ed. M. Grayson. New York: Wiley.

Kelly, N. W., & Swenson, L. K. (1956). *Chemical Engineering Progress, 52,* 263.

Kennard, E. H. (1938). *Kinetic Theory of Gases*. New York: McGraw-Hill.

Kirkaldy, J. S., & Brown, L. C. (1963). *Canadian Metallurgical Quarterly, 2,* 89.

Knudsen, M. H. C. (1909). *Annelen der Physik, 28,* 75.

Ludwig, C. (1859). *Sitzungsberichte der Akademie der Wissenschaften – Wien, 20,* 539.

McAdams, W. H. (1954). *Heat Transmission,* 3rd ed. New York: McGraw-Hill.

Mickley, H. S. (1949). *Chemical Engineering Progress, 45,* 739.

Reynolds, O. (1874). *Proceedings of the Manchester Literary and Philosophical Society, 14,* 7.

Reynolds, O. (1879). *Philosophical Transactions, 170,* 727.

Sherwood, T. K., Pigford, R. L., & Wilke, C. R. (1975). *Mass Transfer*. New York: McGraw-Hill.

Soret, C. (1879). *Archives des Sciences Physiques et Naturelles, 2* (3), 48; (1888). *Annales de Chimie et de Physique, 22* (5), 293.

Toor, H. L. (1971a). *Industrial and Engineering Chemistry Fundamentals, 10,* 121; (1971b). *American Institute of Chemical Engineers Journal, 17,* 5.

Treybal, R. E. (1980). *Mass Transfer Operations,* 3rd ed. New York: McGraw-Hill.

Tyrrell, H. J. V. (1961). *Diffusion and Heat Flow in Liquids*. London: Butterworth.

PROBLEMS

The problems that follow are intended to give insight into this book's material; they are not examples of research frontiers. These problems are organized by chapter. Within each chapter, they are roughly arranged in order of increasing difficulty. Because the problems are for learning rather than for testing, many can be solved on a single piece of paper; answers are frequently included. I have adopted suggestions of students for some problems; in such cases, the student's name is given at the end of the problem.

2. Diffusion in dilute solutions

1. Water evaporating from a pond does so as if it were diffusing across an air film 0.15 cm thick. The diffusion coefficient of water in 20°C air is about 0.25 cm²/sec. If the air out of the film is 50% saturated, how fast will the water level drop in a day? *Answer:* 1.24 cm/day.

2. In 1765, Benjamin Franklin made a variety of experiments on the spreading of oils on the pond in Clapham Common, London. Franklin estimated the thickness of the oil layers to be about 25 Å. Many more recent scientists have tried to use similar layers of fatty acids and alcohols to retard evaporation from ponds and reservoirs in arid regions. The monolayers used today usually are characterized by a resistance around 2 sec/cm. Assuming that they are the thickness of Franklin's layer and that they can dissolve up to 0.18% water, estimate the diffusion coefficient across the monolayers. *Answer:* $7 \cdot 10^{-7}$ cm²/sec.

3. The diffusion coefficient of NO_2 into stagnant water can be measured with the apparatus shown in Fig. P2.1. Although the water is initially pure, the mercury drop moves to show that 0.83 cm³ of NO_2 is absorbed in 3 min. The gas–liquid interface has an area of 36.3 cm², the pressure is 0.93 atm, the temperature is 16°C, and the Henry's law constant is 37,000 cm³-atm/mol. What is the desired diffusion coefficient? (J. Kopinsky) *Answer:* $6.5 \cdot 10^{-6}$ cm²/sec.

4. About 85.6 cm² of a flexible polymer film 0.051 cm thick is made into a bag, filled with distilled water, and hung in an oven at 35°C and 75% relative humidity. The bag is weighed, giving the following data:

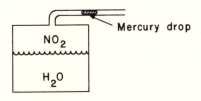

Fig. P2.1

Time (days)	Bag weight (grams)
0	14.0153
1	13.9855
4	13.9104
7	13.8156
8	13.7710
12	13.6492
14	13.5830
16	13.5256

What is the permeability (DH) of the polymer film? (R. Contravas) *Answer:* $1.8 \cdot 10^{-6}$ cm^2/sec.

5. Diaphragm cells are frequently calibrated by allowing 1-M potassium chloride to diffuse into pure water. The average diffusion coefficient in this case is $1.859 \cdot 10^{-5}$ cm^2/sec. Your cell has compartment volumes of 42.3 cm^3 and 40.8 cm^3; the diaphragm is a glass frit 2.51 cm in diameter, 0.16 cm thick, and of porosity 0.34. In one calibration experiment, the concentration difference at 36 hr 6 min is 49.2% of that originally present. (a) What is the cell's calibration constant? *Answer:* 0.294 cm^{-2}. (b) What is the effective length of the diaphragm's pores? *Answer:* 0.28 cm. (c) The current pores are about $2 \cdot 10^{-4}$ cm in diameter. What is the effect of increasing the pore diameter 10 times at constant porosity?

6. Diffusion coefficients in gases can be measured by injecting a solute gas into a solvent gas in laminar plug flow and measuring the concentration with a thermistor placed downstream. The concentration downstream is given by

$$c_1 = \frac{Q}{4\pi Dz} e^{-r^2v/2Dz}$$

where Q is the solute injection rate, z is the distance downstream, r is the distance away from the z axis, and v is the gas flow. One series of measurements involves the diffusion of helium in nitrogen at 25°C and 1 atm. In one particular measurement, the maximum concentration of helium is 0.48 wt% when z is 1.031 cm and Q is 0.045 cm^3/sec. What is the diffusion coefficient? (H. Beesley) *Answer:* 0.11 cm^2/sec.

7. Low-carbon steel can be hardened for improved wear resistance by carburizing. Steel is carburized by exposing it to a gas, liquid, or solid that

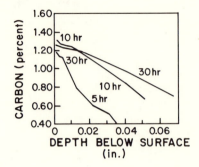

Fig. P2.2

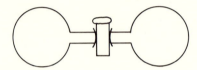

Fig. P2.3

provides a high carbon concentration at the surface. Figure P2.2 [D. S. Clark and W. R. Varney, *Physical Metallurgy for Engineers*. Princeton, N.J.: Van Nostrand (1962)] shows carbon content versus depth in steel carburized at 930°C. Estimate D from this graph, assuming diffusion without reaction between carbon and iron. (H. Beesley)   *Answer:* $2 \cdot 10^{-6}$ cm²/sec.

8. The twin-bulb method of measuring diffusion is shown in Fig. P2.3. The bulbs, which are stirred and of equal volume, initially contain binary gas mixtures of different compositions. At time zero, the valve is opened; at time t, the valve is closed, and the bulb contents are analyzed. Explain how this information can be used to calculate the diffusion coefficient in this binary gas mixture.

9. Find the steady-state flux out of a pipe with a porous wall. The pipe has an inner radius R_i and an outer radius R_0. The solute has a fixed, finite concentration c_{1i} inside of the pipe, but is essentially at zero concentration outside. As a result, solute diffuses through the wall with a diffusion coefficient D. When you have found the result, compare it with the results for steady-state diffusion across a thin slab and away from a dissolving sphere.   *Answer:* $Dc_{1i}[R_0 \ln(R_0/R_i)]^{-1}$.

10. Controlled release is important in agriculture, especially for insect control. One common example involves the pheromones, sex attractants released by insects. If you mix this attractant with an insecticide, you can wipe out all of one sex of a particular insect pest. A device for releasing one pheromone is shown schematically in Fig. P2.4. This pheromone does not sublime instantaneously, but at a rate of

$$r_0 = 6 \cdot 10^{-17}[(1 - 1.10 \cdot 10^7 \text{ cm}^3/\text{mol})c_1] \text{ mol/sec}$$

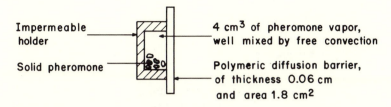

Fig. P2.4

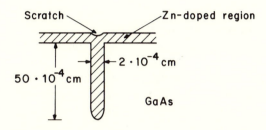

Fig. P2.5

where c_1 is the concentration in the vapor. The permeability of this material through the polymer (DH) is $1.92 \cdot 10^{-12}$ cm²/sec. Its concentration of pheromone outside of the device is essentially zero. (a) What is the concentration (mol/cm³) of pheromone in the vapor? (b) How fast is the pheromone released by this device? *Answer:* $4.8 \cdot 10^{-18}$ mol/sec.

11. Antique glass objects can be dated by measuring the amount of hydration near the object's surface. This amount can be measured using ^{15}N nuclear magnetic resonance [W. A. Lanford, *Science,* **196,** 975 (1977)]. Derive equations for the total amount of hydration, assuming that water reacts rapidly and reversibly with the glass to produce an immobile hydrate. Discuss how this amount can provide a measure of the age of the object.

12. Minnesota gophers in underground burrows must breathe; and many have wondered where the oxygen comes from. K. T. Wilson and D. L. Kilgore [*J. Theoret. Biol.,* **71,** 73 (1978)] have developed a number of models of these burrows. One model depends on the assumption that the burrow is a long cylinder of radius R_0 located a short distance l below the surface of the soil. Oxygen enters the burrow by steady diffusion through the soil; it also steadily diffuses along the burrow's axis to a gopher, who is assumed to be resting. The gopher uses oxygen at a rate M. Derive equations giving the steady-state concentration of oxygen as a function of position in the burrow.

13. Researchers in microelectronics have found that a slight scratch on the surface of gallium arsenide causes a zinc dopant to diffuse into the arsenide (Fig. P2.5). Apparently, this occurs because the scratch increases crystal defects and hence the local diffusion coefficient. When these devices are later baked at 850°C, the small pulse may spread, for its diffusion coefficient at this high temperature is about 10^{-11} cm²/sec. If it spreads enough to

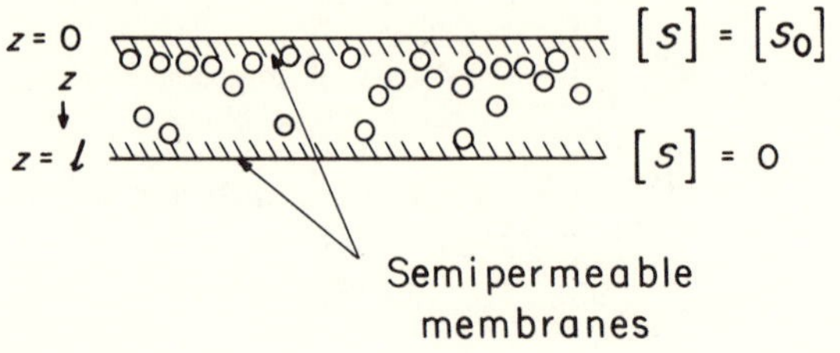

$z = 0$ $[s] = [s_0]$

z

$z = l$ $[s] = 0$

Semipermeable
membranes

Fig. P2.6

increase the zinc concentration to 10% of the maximum at $4 \cdot 10^{-4}$ cm away from the scratch, the device is ruined. How long can we bake the device? (S. Balloge) *Answer:* 30 min.

14. Adolf Fick made the experiments required to determine the diffusion coefficient using the equipment shown in Fig. 2.1-3. In these devices, he assumed that the salt concentration reached saturation in the bottom and that it was always essentially zero in the large solvent bath. As a result, the concentration profiles eventually reached steady state. Calculate these profiles.

15. Consider a layer of bacteria contained between two semipermeable membranes that allow the passage of a chemical solute S, but do not allow the passage of bacteria (Fig. P2.6). The movement of the bacteria B is described with a flux equation roughly parallel to a diffusion equation:

$$j_B = -D_0 \frac{d}{dz} [B] + \chi[B] \frac{d}{dz} [S]$$

where D_0 and χ are constant transport coefficients. In other words, the bacterial flux is affected by S, although the bacteria neither produce or consume S. If the concentrations of S are maintained at S_0 and 0 at the upper and lower surfaces of the bacterial suspension, (a) determine $S(z)$, and (b) determine $B(z)$.

16. Extraction of sucrose from food materials is often correlated in terms of diffusion coefficients. The diffusion coefficients can be calculated assuming short times and an infinite slab:

$$D = \left(\frac{\pi}{4t}\right)\left(\frac{M}{c_{10}}\right)^2$$

where M is the total extracted per area and c_{10} is the sucrose concentration at saturation. However, the diffusion coefficients found are not constant, as shown in Fig. P2.7 (H. G. Schwartzberg and R. Y. Chao, *Food Tech.*, Feb. 1982, p. 73). The reason the diffusion coefficient is not constant is not because of the failure of the approximation of an infinite slab; instead, it reflects the fact that beets and cane are not homogeneous. Instead, they have a network of cells connected by vascular channels. Diffusion across the cell wall is slow, and it dominates behavior in thin slices; diffusion through vascular channels is much faster and supplements the flux for thick slices.

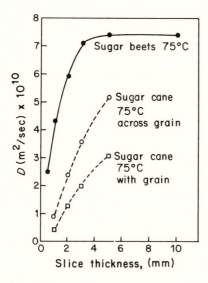

Fig. P2.7

Develop equations that justify these qualitative arguments. These equations will contain the diffusion coefficient across cell walls D_w, the diffusion coefficient in channels D_c, and the fraction of channels ε.

3. Diffusion in concentrated solutions

1. Dry ice is placed in the bottom of a capillary tube 6.2 cm long. Air is blown across the top of the tube. Calculate the ratio of the total flux to the diffusion flux halfway up the capillary for the following conditions: (a) A temperature of $-124°C$, where the vapor pressure is 5 mm Hg. *Answer:* 1.00. (b) A temperature of $-86°C$, where the vapor pressure is 400 mm Hg. *Answer:* 1.45.

2. A gas-oil feedstock is irreversibly and very rapidly cracked on a heated metal plate in an experimental reactor. The cracking reduces the molecular weight by an average factor of three. Calculate the rate of this process, assuming that the gas oil diffuses through a thin unstirred film of thickness l near the plate. Note that the reagent must be constantly diffusing against product moving away from the plate. Compare this rate with that for diffusion through a thin film and with evaporation through a stagnant solvent.

3. Imagine a long tube partially filled with liquid benzene at $60°C$. Beginning at time zero, the benzene evaporates into the initially pure air with a diffusion coefficient of about 0.104 cm^2/sec. How fast does the liquid–vapor interface move with time? *Answer:* $4 \cdot 10^{-4}$ cm/sec at 1 sec.

4. One interesting membrane reactor uses a homogeneous catalyst that cannot pass through an ultrafiltration membrane. Reagents flow continu-

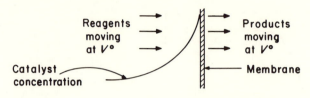

Fig. P3.1

ously toward the membrane, but the catalyst is injected only at the start of the experiment. It forms the concentration profile shown in Fig. P3.1. If the catalyst injected per membrane area is M/A, find the concentration profile of the catalyst.

5. The diffusion coefficient relative to the volume average velocity is defined by

$$-j_1 = D\nabla c_1$$

That relative to the molar average velocity is defined by

$$-j_1^* = D^* c \nabla x_1$$

Show that the diffusion coefficients in these two definitions are equal, even if the molar concentration c is not constant.

6. The most common representation of diffusion is Fick's law:

$$j_1 = c_1(v_1 - v^0) = n_1 - c_1 v^0$$
$$= -D\nabla c_1$$

where v^0 represents the volume average velocity. However, most kinetic theories lead to a very different form of flux equation:

$$\nabla \mu_1 = \frac{RTx_2}{D_0}(v_2 - v_1)$$

where

$$D = D_0 \left(1 + \frac{\partial \ln \gamma_1}{\partial \ln x_1}\right)$$

D_0/RT is called a frictional coefficient, and γ_1 is an activity coefficient. This result is sometimes referred to as the "generalized Stefan–Maxwell equation." Show that these two forms are, *without assumption*, identical.
(a) Divide Fick's law by c_1, and subtract solvent from solute to find

$$v_1 - v_2 = D\nabla \ln(c_2/c_1)$$
$$= D\nabla \ln(x_2/x_1)$$

(b) Change variables to give the gradient in ∇x_1. Then use the definition

$$\mu_1 = \mu_1^0 + RT \ln x_1 \gamma_1$$

to get the desired result.

7. You want to measure the permeability of an artificial rubber membrane to oxygen. Such membranes are often suggested as a possible means of separating air. To make this measurement, you clamp a section of the mem-

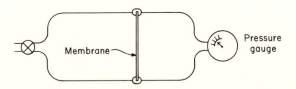

Fig. P3.2

brane in the apparatus shown in Fig. P3.2. The membrane section is 3 cm in diameter and only 35 μm thick. It is attached to a backing layer that gives it mechanical stability, but it means that only 17.3% of the membrane surface is available for diffusion. To begin an experiment, the gas volume of 68 cm^3 shown earlier is evacuated to less than 10^{-5} torr (1 torr $\equiv$ 1 mm Hg). The pressure is then measured and found to be

$$p(\mu\text{m Hg}) = 88(t - 2.3)$$

where t is the total elapsed time in seconds. Find the Henry's law coefficient and the diffusion coefficient for oxygen in this membrane. *Answer:* $D = 9 \cdot 10^{-7}$ cm^2/sec.

8. You want to measure the effectiveness of a porous solid desiccant. To do so, you attach a slab of the desiccant $0.5 \times 20 \times 20$ cm to a thin wire. You then attach the wire to an analytical balance and suspend the slab in a chamber at 45°C and 20% relative humidity. You find that the slab weight varies with time as follows:

Time (min)	Slab weight (grams)
0	166.25
10	167.03
20	167.59
30	168.07
40	168.48
50	168.88
60	169.25

Find the permeability of water vapor in this desiccant. You will find that the data fit neither a finite slab nor an infinite slab. One good alternative model is to postulate pores in the slab. *Answer:* 3.5 cm^2/sec.

9. Copper dispersed in porous low-grade ore pellets 0.2 cm in diameter is leached with 4-M H_2SO_4. The copper dissolves quickly, but diffuses slowly out of the pellets. Because the ore is low grade, the porosity can be assumed constant, and the copper concentration will be low in the acid outside of the pellets. Estimate how long it will take to remove 80% of the copper if the effective diffusion coefficient of the copper is $2.6 \cdot 10^{-6}$ cm^2/sec. *Answer:* 10 min.

10. A large polymer slab initially containing traces of solvent is exposed to excess fresh air to allow solvent to escape. Find the concentration of solvent in the slab as a function of position and time. Assume that the diffusion coefficient is a constant, but discuss how you might expect it to vary. Try to solve this problem yourself, but compare your answers with those in the literature.

11. One method of studying diffusion in liquids used by Thomas Graham is that shown in Fig. 2.1-2(b). It consists of a small bottle of solution immersed in a large bath of solvent. Calculate the solute concentration in the bath as a function of time, and show how this variation can be used to determine the diffusion coefficient.

12. Wool is dyed by dropping it into a dyebath that contains dye at a concentration C_{10} and that has a volume V. The dye diffuses into the wool, so that its concentration in the dyebath drops with time. You can measure this concentration change. You can also measure the equilibrium uptake of the dye. How can you use measurements of this change at small increments of time to find the diffusion coefficient of the dye in the wool?

13. Sows love to dig truffles, the mushroom that shows up as a condiment in French cooking. Apparently, sows do this because they smell in the truffles the sex attractant or pheromone 5a-androst-16-en-3-ol, which is secreted by boars and by human males [R. Claus, H. O. Hoppen, and H. Karg, *Experimentia,* **37,** 1178 (1981); M. Kirk-Smith, D. A. Booth, D. Carroll, and P. Davies, *Res. Comm. Psychol. Psychiat. Behav.,* **3,** 379 (1978)]. Imagine that the truffle is a point source located a distance d below the surface of the ground. Calculate the flux of pheromone leaving the ground above.

14. Imagine that two immiscible substances containing a common dilute solute are brought into contact. Solute then diffuses from one of these substances into the other. Calculate the concentration profiles of the solute in terms of each of the substances, assuming that each substance behaves as a semiinfinite slab. (S. Gehrke)

15. Find the steady-state flux away from a rapidly dissolving drop that produces a concentrated solution. Compare your result with that found for a sparingly soluble sphere and with the various results for diffusion across a stagnant film.

4. Dispersion

1. A steel mill discharges a pulse of concentrated cyanide into the center of a small stream 20 m across flowing at a velocity of 3.4 km/hr. One-half kilometer downstream, the pulse is roughly Gaussian, with a maximum concentration of 860 ppm. The concentration is 410 ppm at a distance of 5 m from this maximum, both normal to and parallel to the net flow. Estimate the shape of the pulse and the maximum concentration 10 km downstream. Assume that there is no dramatic change in the nature of the flow and no chemical interaction with the stream bed.   *Answer:* 200 ppm maximum.

2. Your plant is releasing sulfur dioxide at 680 kg-mol/hour. You expect that the plume containing this material will be neutral to slightly unstable when it is released in an 8-km/hr wind. Estimate the range of maximum concentrations under these conditions in a town 17.5 km downwind.   *Answer:* $y_1 = 10^{-15}$.

3. You are pumping 1.7 kg/sec of a cold stream of monomer through 72 m of 2.5-cm-diameter pipe to a reactor. At the entrance of the pipe, you inject 30 pulses/sec of catalyst with a small piston pump. When this stream reaches the reactor, the total stream is quickly heated, and polymerization begins. The cold stream has a specific gravity of 0.83 and a viscosity of 3.7 centipoises. How well will the catalyst be mixed by flow through the pipe?

4. You are studying dispersion in a small air-lift fermentor. This fermentor is 1.6 m tall, with a 10-cm diameter. Air and pure water are fed into the bottom at superficial velocities of 11 and 0.78 cm/sec; under these conditions the gas bubbles occupy 45% of the column volume. You continuously add 15 cm³/min of 1-M NaCl solution near the top of the column. You find by conductance that the salt concentration halfway down the column is $2.32 \cdot 10^{-3}$ M. What is the dispersion coefficient?   *Answer:* 54 cm²/sec.

5. The best marathon in Minnesota is run by Grandma's, a reformed brothel in Duluth. In the 1981 race, 3,202 persons finished. One-quarter of the runners finished within 3 hr 6 min and half within 3 hr 26 min. If I ran the race in 2 hr 54 min 42 sec, what place did I come in?   *Answer:* 460 by experiment.

6. A handful of pheromone-impregnated pellets are being used to give an overall release of 1.3 mol/hr into a 15-km/hr wind blowing in the z direction. In this case, the pheromone concentration is given by

$$c_1 = \frac{Q}{2\pi Ez}\, e^{-vr^2/4Ez}$$

where r is the width of release. Gypsy moths respond to this release over an area 25 km long, with a maximum width of 8 km. What is the dispersion coefficient E of the pheromone?

7. Harvest ants inform each other of danger by releasing a pulse of pheromone. The dispersion of this pheromone can be modeled using results like those in this chapter. For harvest ants, the maximum distance over which this chemical alarm is effective is 6 cm; this occurs at a time of 32 sec. The alarm is no longer effective after 35 sec [E. O. Wilson, *Psyche,* **65,** 41 (1958)]. Assume that the pheromone is dispersed in a hemispherical volume; so its concentration is

$$c_1 = \frac{2M}{(4\pi Dt)^{3/2}}\, e^{-r^2/4Et}$$

Also assume that neighboring ants respond only when c_1 exceeds c_{10}. Then show that

$$R = \left[6Dt \ln\!\left(\frac{t_{\text{final}}}{t}\right)\right]^{1/2}$$

where R is the radius of communication at time t, and t_{final} is the time when the signal is ignored. Discuss how R varies with t. Estimate the dispersion coefficient E from the values given, and compare it with your guess of a diffusion coefficient. *Answer:* $E = 2$ cm^2/sec.

8. In 1905, five muskrats escaped in Bohemia. These animals quickly spread over Europe as shown in Fig. P4.1 [J. G. Skellam, *Biometrika,* **38,** 196 (1951)]:

Year	Area inhabited
1905	0
1909	50
1911	120
1915	300
1920	670
1927	1,720

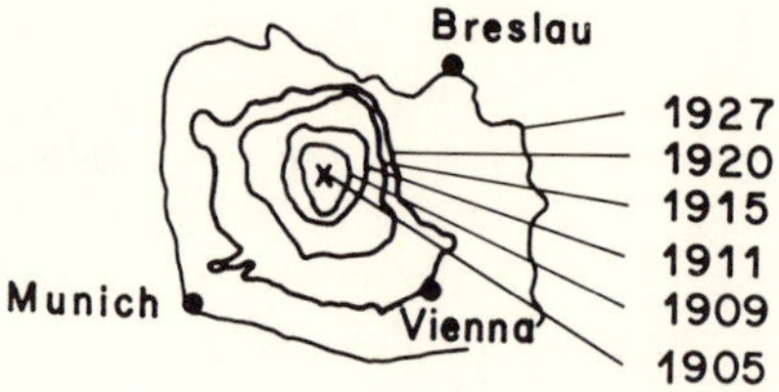

Fig. P4.1

Show that these results are consistent with a two-dimensional dispersion model

$$c_1 = \frac{M_0}{4\pi Et} e^{\alpha t - r^2/4Et}$$

where

$$\left(\begin{array}{c}\text{growth rate}\\ \text{of } [M]\end{array}\right) = \alpha[M]$$

E is the dispersion coefficient, and M_0 is the original number of animals.

9. Work out the details of the derivation for Taylor dispersion given in Section 4.2. More specifically, do the following: (a) Show how Eq. 4.2-7 is converted into Eq. 4.2-13. Then show that Eq. 4.2-14 is a potential solution. (b) Use Eqs. 4.2-14, 4.2-15, and 4.2-19 to derive Eq. 4.2-20.

5. Values of diffusion coefficients

1. Estimate the diffusion coefficient of carbon dioxide in air at 740 mm Hg and 37°C. How does this compare with the experimental value of 0.177 cm^2/ sec? *Answer:* about 4% low.

2. As part of a course on diffusion, you are to measure the diffusion coefficient of ammonia in 25°C air, using the two-bulb capillary apparatus shown in Fig. 3.1-2. In your apparatus, the bulbs have volumes of about 17 cm^3, and the capillary is 2.6 cm long and 0.083 cm in diameter. You are told that you should make your measurements when the concentration difference is about half the initial value. (a) Use the Chapman–Enskog theory to estimate how long you should run your experiment. *Answer:* 3.6 hr. (b) Why are you told to make your measurement near this particular concentration difference?

3. Estimate the diffusion coefficient at 25°C of traces of ethanol in water

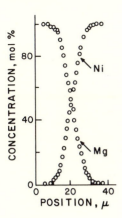

Fig. P5.1

and of traces of water in ethanol. Compare your estimates with the experimental values of $1.24 \cdot 10^{-5}$ cm²/sec and $1.31 \cdot 10^{-5}$ cm²/sec, respectively.

4. Tobacco mosaic virus has been shown by electron microscopy to be shaped like a cylinder 150 Å in diameter and 3,000 Å long. Its molecular weight is about 40,000,000, and its partial specific volume is 0.73 cm³/g. Estimate the diffusion coefficient of this material, and compare with the experimental value at 25°C of $3 \cdot 10^{-8}$ cm²/sec. *Answer:* $2.7 \cdot 10^{-8}$ cm²/sec.

5. Estimate the diffusion coefficient of lactic acid under each of the following conditions: (a) in air at room temperature and pressure; (b) in milk in the refrigerator; (c) through the wall of a plastic milk bottle.

6. In an experiment to determine the diffusion coefficient of urea in water at 25°C with the diaphragm cell, you find that a density difference of 0.01503 g/cm³ decays to 0.01090 g/cm³ after a time of 16 hr and 23 min. The cell's calibration constant is 0.397 cm⁻². If the density of these solutions varies linearly with concentration, what is the diffusion coefficient? Compare your answer with the value of $1.373 \cdot 10^{-5}$ cm²/sec obtained with the Gouy interferometer. *Answer:* $1.37 \cdot 10^{-5}$ cm²/sec.

7. The concentration profiles of Ni_2SiO_4 diffusing into Mg_2SiO_4 are given in Fig. P5.1 [M. Morioka, *Geochim Cosmochim Acta,* **45,** 1573 (1981)]. These data were found after 20 hr using an infinite couple at 1,350°C. Calculate the diffusion coefficient in this system. *Answer:* $1.2 \cdot 10^{-11}$ cm²/sec.

8. You are studying the diffusion of a 5% solution of a highly fractionated polystyrene in toluene using the Gouy interferometer. This apparatus allows measurement of the intensity minima of the interference fringes Y_j versus fringe number j. The diffusion coefficient can be estimated from these data using the equation

$$\left(\frac{1}{c_{1\infty} - c_{10}}\right) \frac{\partial c_1}{\partial z} = \frac{1}{\sqrt{\pi Dt}} e^{-z^2/4Dt}$$

which in terms of the experimentally determined quantities may be written as

$$\left(\frac{1}{n_\infty - n_0}\right) \frac{\partial n}{\partial z} = \left(\frac{2Y_j}{bj\lambda}\right) = \frac{1}{\sqrt{\pi Dt}} e^{-\zeta^2}$$

where n is the refractive index, b is the magnification of the apparatus, λ is the wavelength of light used, and $e^{-\zeta^2}$ is a function found from the following table:

$(j + \frac{3}{4})/J$	$e^{-\zeta^2}$	$(j + \frac{3}{4})/J$	$e^{-\zeta^2}$
0.0000	1.0000	0.1939	0.6126
0.0030	0.9747	0.3050	0.4855
0.0158	0.9246	0.4027	0.3903
0.0351	0.8721	0.5019	0.3048
0.0646	0.8093	0.6198	0.2149
0.1047	0.7390	0.7042	0.1573
0.1551	0.6639	1.0000	0.0000

More accurate values of this function are available elsewhere [G. Kegeles and L. J. Gosting, *J. Amer. Chem. Soc.* **69**, 2516 (1947); L. J. Gosting and M. S. Morris, *J. Amer. Chem. Soc.*, **71**, 1998 (1949)].

In your particular experiment (#W24E), you are allowing a solution of polystyrene to diffuse into the solvent, pure toluene. The polymer is a sharp fraction, with a ratio of weight average to number average molecular weights of 1.1. You obtain the following data:

Fringe minimum number j	Fringe position y (cm)
0	1.589
1	1.482
2	1.396
3	1.320
4	1.253
6	1.131
10	0.926
14	0.751
18	0.599
26	0.341
34	0.133

$\lambda = 5462$ Å, $J = 41.3$, $b = 305.0$ cm, $t = 14{,}300$ sec. Calculate the diffusion coefficient from these data.   *Answer:* $8.89 \cdot 10^{-7}$ cm^2/sec.

9. The ionic diffusion coefficient D or, more exactly, the ionic conductivity λ can frequently be described by the equation

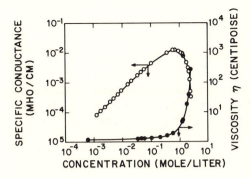

Fig. P5.2

$$\lambda = \frac{a_0}{T} e^{-E/RT}$$

For β-alumina, the following values are obtained:

	R_0 (Å)	a_0 (°K/ohm-cm)	E (kcal/mol)
Li^+	0.68	54	2.9
Na^+	0.98	2,500	2.4
H_3O^+	1.32	81,000	11.9
K^+	1.33	1,500	4.6

[G. C. Farrington and J. L. Briant, *Science,* **204,** 1371 (1979)]. (a) Calculate the ionic conductivity at 25°C for each of these ions. (b) Show that these conductivities can be as large as that in 1-M KCl, in which the diffusion coefficient is $2.0 \cdot 10^{-5}$ cm²/sec. (c) Because we usually expect transport in solids to be much slower than transport in liquids, we recognize that β-alumina is an exceptional material. Discuss the factors that might cause this effect.

10. Jeng-Ping Yao and D. N. Bennion [*J. Phys. Chem.,* **75,** 3586 (1971)] measured the electrolytic conductance of aqueous solutions of tetra-*n*-amylammonium thiocyanate at 55°C. The data are most easily presented graphically (Fig. P5.2). Note that this salt is a liquid at this temperature and is completely miscible with water; so the measurements go all the way from mass transfer at infinite dilution through to mass transfer in the molten salt. As detailed in Section 6.1, specific conductance is approximately equivalent to the diffusion coefficient times the ionic concentration. Use your knowledge of diffusion to suggest how the data at high salt concentration might be conveniently correlated.

11. Diffusion in molten silicate deep within the earth is central to many of the chemical processes that take place there. However, the diffusion coefficients in such magma seem to vary widely. For example, for cesium ion dissolved in obsidian at 2 kilobars pressure,

$$D = 8 \cdot 10^{-2} e^{-49.9 \text{ kcal}/RT} [=] \text{ cm}^2/\text{sec}$$

For cesium ion dissolved in obsidian containing 6 wt% water,

$$D = 7 \cdot 10^{-5} e^{-19.52 \text{ kcal}/RT} [=] \text{ cm}^2/\text{sec}$$

[E. B. Watson, *Science, 205*, 1259 (1979)]. (a) How much does the diffusion coefficient at 800°C differ in the dry and the water-saturated samples? (b) The reason for this difference is not known. Assume that the water causes thin pores to form, and diffusion in the pores is that in bulk water. What is the pore area per obsidian area?

6. Solute–solute interactions

1. You are studying a thin film of 310 stainless steel that apparently is without pores. You clamp this film in a diaphragm cell, put a hydrogen pressure of 0.43 atm on one side, and measure the much smaller hydrogen pressure on the other side. You find the data shown in Fig. P6.1 [N. R. Quick and H. H. Johnson, *Metal Trans. A.*, **10A,** 67 (1979)]. These data show that the flux depends on the square root of hydrogen pressure, a dependence known as Sievert's law. This is believed to occur because molecular hydrogen dissociates into atomic hydrogen within the film. Use your knowledge of diffusion to justify this conclusion.

2. The ion diffusion coefficients at 25°C of Na^+, K^+, Ca^{2+}, and Cl^- are 1.33, 1.9, 0.79, and 2.0 (all times 10^{-5} cm^2/sec). Find the diffusion coefficients and transport numbers for NaCl, KCl, and $CaCl_2$ in water and in excess KCl.   *Answer:* In water, the diffusion coefficients are, respectively, 1.60, 1.95, and 1.32 all in 10^{-5} cm^2/sec.

3. Calculate the diffusion coefficient at 25°C of NH_4OH versus concentration. The relevant ionic diffusion coefficients are $D_{NH_4} = 1.96$ and $D_{OH} = 5.28$ cm^2/sec. The pK_a of the NH_4^+ is 9.245. Estimate the diffusion coefficient of the NH_4OH molecule from the Wilke–Chang correlation.

4. The uptake of drugs from the intestinal lumen is often strongly influenced by diffusion. For example, consider a water-insoluble steroid for birth control that is solubilized in detergent micelles. These micelles, aggregates of steroid and soap, have a molecular weight of 24,000, an aggregation number of 80, a diameter of 26 Å, and a charge of -27. The counterion is Na^+. (a) What is the diffusion coefficient of the micelle in water at 37°C? *Answer:* $1.0 \cdot 10^{-5}$ cm^2/sec. (b) What is it in 0.1-M NaCl? Assume the micelle concentration is relatively low.   *Answer:* $2.5 \cdot 10^{-6}$ cm^2/sec.

5. Electrolyte solutions can be highly nonideal. In these solutions the flux equation for a 1-1 univalent electrolyte is often written as

$$-j_T = \frac{D_0 c_T}{RT} \nabla \mu_T$$

where the chemical potential μ_T is given by

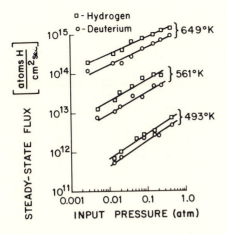

Fig. P6.1

$$\mu_T = \mu_T^0 + RT \ln c_T \gamma_T$$

and the activity coefficient γ_T in water at 25°C is estimated from the Debye–Hückel theory:

$$\gamma_T = -1.02 \, c_T^{1/2}$$

where c_T is in moles per liter. Using the values in Table 6.1-1, estimate the variation with concentration of the diffusion coefficient of potassium chloride, and compare it with the experimental values. (J. Zasadzinski)

6. The following data have been reported for ε-caprolactam diffusing in water at 25°C [E. L. Cussler and P. J. Dunlop, *Austral. J. Chem.* **19**, 1661 (1966)]:

C (mol/dm^3)	D (10^{-5} cm^2/sec)
0.0514	0.8671
0.0515	0.8669
0.500	0.6978
0.991	0.5254
1.998	0.4160
3.003	0.3311

This solute is believed to dimerize by forming hydrogen bonds:

Estimate the equilibrium constant K for this reaction. (G. Jerauld)   *Answer:* about 0.5 M^{-1}.

7. Solute–solvent and solute–boundary interactions

1. Each molecule of sucrose in dilute aqueous solution is believed to combine with about four molecules of water. Such hydration has two effects. First, it increases the size of the sucrose solute and thus retards diffusion. Second, it increases the mole fraction of sucrose and hence may accelerate diffusion. (a) Estimate how the measured diffusion coefficient of sucrose differs from that of the unhydrated sucrose. In this estimate, take the hydrated sucrose diffusion coefficient at infinite dilution as $5.21 \cdot 10^{-6}$ cm^2/sec, its molecular weight as 342, and its solid density as 1.59 g/cm^3. *Answer:* about $5.7 \cdot 10^{-6}$ cm^2/sec. (b) Assume that the diffusion coefficient is given by Eq. 7.1-6 times a viscosity correction. Find how the coefficient varies with concentration.

2. Porous catalyst particles are often made by compressing the powdered catalyst into a particle, the pore structure of which can be controlled by the compression process. You are the engineer in charge of quality control at a catalyst manufacturing facility. You are making catalyst with pore sizes around 3 μm in diameter. Unfortunately, electron micrographs show one batch of product with pore sizes that are much smaller – about 550 Å in diameter. Paradoxically, the catalyst still has the same surface area per volume. The bad batch of particles was to be used in a diffusion-controlled oxidation at 400°C and 1 atm total pressure. As an estimate of the extent to which these particles will perform off-standard, calculate the diffusion coefficient of O_2 in the two different cylindrical pores. (S. Balloge)   *Answer:* 0.13 cm^2/sec in small pores.

3. The diffusion coefficients in water at 20°C of hemoglobin and of catalase are $6.9 \cdot 10^{-7}$ cm^2/sec and $4.1 \cdot 10^{-7}$ cm^2/sec, respectively. They are $4.3 \cdot 10^{-10}$ cm^2/sec and $1.8 \cdot 10^{-10}$ cm^2/sec across a porous membrane. Estimate the pore size in this membrane.   *Answer:* 300 Å.

4. An emulsion of an aqueous electrolyte solution contains an averge salt concentration equivalent of 0.1-M KCl. The aqueous phase is 40% of the total volume. The oil phase, which has a viscosity comparable to that of water, dissolves KCl until the ionic concentration is 0.0034 of that in the aqueous phase; the association constant for ion-pair formation in this oil is $3.6 \cdot 10^5$ liters/mol. The emulsion is stabilized with a nonionic detergent. This emulsion inverts from a water-continuous phase to an oil-continuous phase at 41°C. Estimate the change in electrical conductivity accompanying this inversion by assuming that both emulsions consist of a periodic array of spherical particles.   *Answer:* about 80 times.

5. A. Vignes [*Ind. Engr. Chem. Fund.*, **5**, 189 (1966)] suggested that the concentration dependence of many liquid diffusion coefficients can be predicted with the equation

$$D = D_0 \left(1 + \frac{\partial \ln \gamma_1}{\partial \ln x_1} \right)$$

$$D_0 = D_1^{x_2} D_2^{x_1}$$

where D_1 is the diffusion coefficient of a trace of species 1 in excess species 2 and D_2 is that of a trace of species 2 in excess species 1. (a) Test the Vignes equation using the following data for ethanol 1 and water 2 at 25°C [B. R. Hammond and R. H. Stokes, *Trans. Faraday Soc.*, **49**, 890 (1953)]:

x_1	D $(10^{-5}$ cm^2/sec)	$1 + \dfrac{\partial \ln \gamma_1}{\partial \ln x_1}$
0.0	1.24	1.00
0.1	0.66	0.76
0.2	0.41	0.41
0.4	0.42	0.355
0.6	0.64	0.53
0.8	0.94	0.77
1.0	1.31	1.00

(b) Using these same data, calculate D_0 from Eq. 7.3-11. Compare how these quantities vary with concentration.

6. At the consolute temperature of a regular solution, the diffusion coefficient is approximately given by

$$D = D_0(1 - 4x_1x_2)^\gamma$$

where D_0 is a constant and x_i is the mole fraction of species i. Assume that the exponent γ equals 0.5. Note that D is zero when $x_1 = x_2 = 0.5$. The volume average velocity is zero, and the total concentration c is constant. Imagine that you are letting pure solute $(x_1 = 1)$ diffuse through a long thin capillary into an equally large volume of pure solvent $(x_1 = 0)$. You analyze your data as

$$j_1 = \frac{\bar{D}}{l}(c_{10} - c_{1l})$$

Show that $\bar{D}$ equals $D_0/2$.

8. Multicomponent diffusion

1. Calculate the diffusion coefficient at 1 atm pressure for each of the following binary gas mixtures at 25°C: (a) C_6H_6 and carbon-14-labeled C_2H_6; (b) C_6H_6 and unlabeled C_2H_6; (c) carbon-14-labeled C_2H_6 and unlabeled C_2H_6. Recognizing that the tracer concentration is very small, find the ternary diffusion coefficients in an equimolar mixture of benzene and ethane. Define carbon-14-labeled C_2H_6 as species 1, unlabeled C_2H_6 as species 2, and C_6H_6 as species 3. *Answer:* $D_{11} = 0.074$; $D_{12} = 0$; $D_{21} = 0.007$; $D_{22} = 0.059$ (all in cm^2/sec).

2. Imagine a thin membrane separating two large volumes of aqueous solution. The membrane is 0.014 cm thick and has a void fraction of 0.32. One solution contains 2-M H_2SO_4 and the other 2-M $NaSO_4$. As a result, there is no gradient of sulfate across the membrane. Ternary diffusion coefficients for this system are given in Table 8.4-2. What is the sulfate flux? *Answer:* $5.6 \cdot 10^{-8}$ mol/cm^2-sec.

3. A solution of 12 mol% hexadecane (1), 55 mol% dodecane (2), and 33% hexane (3) is diffusing at 25°C in a diaphragm cell into a solution of 52 mol% hexadecane (1), 15 mol% dodecane (2), and 33 mol% hexane (3). The cell constant of the cell is 3.62 cm^{-2}, and the ternary diffusion coefficients are

$$D_{11} = 1.03, \quad D_{12} = 0.23,$$
$$D_{21} = 0.27, \quad D_{22} = 0.97$$

all times 10^{-5} cm^2/sec. Plot the concentration differences Δc_1 and Δc_2 versus time.

4. In a two-bulb capillary diffusion apparatus like that in Fig. 3.1-2, one bulb contains 75% H_2 and 25% C_6H_6, and the other contains 65% H_2, 34.9% C_6H_6, and 0.1% radioactively tagged C_6H_6. The system is at 0°C. We can measure diffusion in one of two ways. First, we can measure the concentration change of all the benzene using a gas chromatograph. Second, we can measure the concentration difference of the radioactive isotopes. How different are these results? To answer this problem, let C_1 = tracer, C_2 = untagged benzene, and C_3 = hydrogen solvent. (a) Find the ternary diffusion coefficients assuming that the radioactive concentration is much less than the nonradioactive. (b) Using the binary solution, write out the ternary one. (c) Combine parts (a) and (b) to find Δc_1 and Δc_2 versus βt, where β is the cell constant of this apparatus.

5. An iron bar containing 0.86 mol% C is joined with a bar containing 3.94 mol% Si. The two bars are then heated to 1,050°C for 13 days; under these conditions, there is only one equilibrium phase, fcc austenite. Calculate the carbon concentration profile under these conditions using the values in Table 8.4-3. Remember that these coefficients are relative to the *solvent average* velocity.

6. In practical work, air is often treated as if it is a pure species. This problem tests the accuracy of this assumption for diffusion. Imagine a large slab of an isotropic porous solid centered at $z = 0$. To the left, at $z < 0$, the solid's pores initially contain pure hydrogen; to the right, at $z > 0$, they initially contain pure air. If air were really a single component, then the mole fraction of hydrogen y_1 would vary as follows (see Section 2.3):

$$y_1 = \tfrac{1}{2} \left(1 - \text{erf} \, \frac{z}{\sqrt{4Dt}} \right)$$

Because air is really a mixture, the exact solution involves ternary diffusion coefficients that can be calculated from Table 8.1-1. Calculate the ternary concentration profile and compare it with the binary one. (S. Gehrke)

7. You are using the diaphragm cell to study diffusion in the ternary system sucrose(1)–KCl(2)–water(3). Instead of measuring the concentration differences of each species in these experiments, you find it convenient to measure the overall density and refractive-index differences, defined as

$$\Delta \rho = H_1 \Delta \rho_1 + H_2 \Delta \rho_2$$
$$\Delta n = R_1 \Delta \rho_1 + H_2 \Delta \rho_2$$

In separate experiments, you find $H_1 = 0.379$, $H_2 = 0.602$, $R_1 = 0.1414$, and $R_2 = 0.1255$. You find the calibration constant of the cell to be 0.462 cm^{-2}. Other relevant data are as follows [E. L. Cussler and P. J. Dunlop, *J. Phys. Chem.*, **70**, 1880 (1966)]:

	Exp. 20	Exp. 26	Exp. 24	Exp. 22
$\Delta\rho_{10}$	0.0000	0.00277	0.01111	0.01500
$\Delta\rho_{20}$	0.0150	0.01250	0.00313	0.00000
Δn_0	86.33	89.88	89.96	97.21
$\Delta\rho_0$	0.00904	0.00856	0.00609	0.00569
Δn	28.24	33.56	46.34	55.38
$\Delta\rho$	0.00293	0.00299	0.00279	0.00315
$10^{-5}\,\beta t$	0.627	0.620	0.9526	1.0598

Use these data to calculate the four ternary diffusion coefficients, and compare them with the following values found with the Gouy interferometer: $D_{11} = 0.497$, $D_{12} = 0.021$, $D_{21} = 0.069$, $D_{22} = 1.775$ (all times 10^{-5} cm^2/sec). *Answer:* $D_{11} = 0.498$, $D_{12} = 0.022$, $D_{21} = 0.071$, $D_{22} = 1.776$ (all times 10^{-5} cm^2/sec).

9. Fundamentals of mass transfer

1. Water flows through a thin tube, the walls of which are lightly coated with benzoic acid. The benzoic acid is dissolved very rapidly, and so is saturated at the pipe's wall. The water flows slowly, at room temperature and 0.1 cm/sec. The pipe is 1 cm in diameter. Under these conditions, the mass transfer coefficient varies along the pipe:

$$\frac{kx}{D} = 0.3 \left(\frac{xv\rho}{\mu}\right)^{1/2}\left(\frac{\mu}{\rho D}\right)^{1/3}$$

where x is the distance along the pipe and v is the average velocity in the pipe. What is the average concentration of benzoic acid in the water after 2 m of pipe? *Answer:* 10% saturated.

2. Water containing 0.1-M benzoic acid flows at 0.1 cm/sec through a 1-cm-diameter rigid tube of cellulose acetate, the walls of which are permeable to small electrolytes. These walls are 0.01 cm thick; solutes within the walls diffuse as through water. The tube is immersed in a large well-stirred water bath. Under these conditions, the flux of benzoic acid from the bulk to the walls can be described by

$$\frac{kx}{D} = 0.3 \left(\frac{xv\rho}{\mu}\right)^{1/2}\left(\frac{\mu}{\rho D}\right)^{1/3}$$

After 50 cm of tube, what fraction of a 0.1-M benzoic acid solution has been removed? Remember that there is more than one resistance to mass transfer in this system. *Answer:* 5% removed.

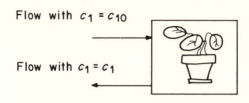

Flow with $c_1 = c_{10}$

Flow with $c_1 = c_1$

Fig. P9.1

3. How much is the previous answer changed if the benzoic acid solution in the tube is in benzene, not water? *Answer:* 4% removed.

4. A disc of radioactively tagged benzoic acid 1 cm in diameter is spinning at 20 rpm in 94 cm³ of initially pure water. We find that the solution contains benzoic acid at $7.3 \cdot 10^{-4}$ g/cm³ after 10 hr 4 min and $3.43 \cdot 10^{-3}$ g/cm³ after a long time (i.e., at saturation). (a) What is the mass transfer coefficient? *Answer:* $8 \cdot 10^{-4}$ cm/sec. (b) How long will it take to reach 14% saturation? (c) How closely does this mass transfer coefficient agree with that expected from the theory in Example 3.4-3?

5. As part of the manufacture of microelectronic circuits, silicon wafers are partially coated with a 5,400-Å film of a polymerized organic film called a "photoresist." The density of this polymer is 0.96 g/cm³. After the wafers are etched, this photoresist must be removed. To do so, the wafers are placed in groups of 20 in an inert "boat," which in turn is immersed in strong organic solvent. The solubility of the photoresist in the solvent is $2.23 \cdot 10^{-3}$ g/cm³. If the photoresist dissolves in 10 min, what is its mass transfer coefficient? (S. Balloge) *Answer:* $4 \cdot 10^{-5}$ cm/sec.

6. Find the time that it takes for a 1-mm water drop to evaporate to a 0.1-mm diameter. The drop is at 18°C, and the surrounding air is dry and at 30°C. The drop is falling under the influence of gravity.

7. To study mass transfer coefficients of oxygen in green plants, you use the apparatus shown in Fig. P9.1. At an air flow of 1 mol/hr at 27°C and 1 atm, the inlet oxygen concentration is 21%, and the outlet concentration is 16%. At an air flow of 3 mol/hr, the inlet oxygen concentration is the same, and the outlet is 19%. The total volume of the apparatus corresponds to 1 mol, and the leaves have a total area of 400 cm². You expect that the overall mass transfer coefficient is the average of mass transfer to the leaves' surface and mass transfer across the leaves' surface. You also expect that, within the leaves, oxygen is quickly metabolized. Find the mass transfer coefficient across the leaves' surface. (a) Write a mass balance on the oxygen in the gas within the tank. (b) Find the overall mass transfer coefficients (in cm/sec) involved under various flows. (c) Find the mass transfer coefficient across the surface of the leaves (including any partition coefficient). Assume that any external mass transfer is proportional to the square root of flow. *Answer:* $5.5 \cdot 10^{-3}$ cm/sec.

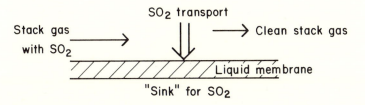

Fig. P9.2

8. Calculate the fraction of the resistance to SO_2 transport in the gas and liquid membrane phases for an SO_2 scrubber operating at 100°C (Fig. P9.2). The membrane liquid, largely ethylene glycol, is $5 \cdot 10^{-3}$ cm thick. In it, the SO_2 has a diffusion coefficient of about $0.85 \cdot 10^{-5}$ cm^2/sec and a solubility of

$$\frac{0.026 \text{ mol } SO_2/\text{liter}}{\text{mm Hg of } SO_2}$$

In the stack gas, there is an unstirred film adjacent to the membrane 0.01 cm thick, and the SO_2 has a diffusion coefficient of 0.13 cm^2/sec. (W. J. Ward)

9. Estimate the average mass transfer coefficient for water evaporating from a film falling at 0.82 cm/sec into air. The air is at 25°C and 2 atm, and the film is 186 cm long. Express your result in cm^3 H$_2$O vapor at STP/(hr-cm^2-atm). *Answer:* 120 in units given.

10. Find the dissolution rate of a cholesterol gallstone 1 cm in diameter immersed in a solution of bile salts. The solubility of cholesterol in this solution is about $3.5 \cdot 10^{-3}$ g/cm^3. The density difference between the bile saturated with cholesterol and that containing no cholesterol is about $3 \cdot 10^{-3}$ g/cm^3; the kinematic viscosity of this solution is about 0.06 cm^2/sec; the diffusion coefficient of cholesterol is $1.8 \cdot 10^{-6}$ cm^2/sec. *Answer:* 0.2 g/month.

11. Air at 100°C and 2 atm is passed through a bed 1 cm in diameter composed of iodine spheres 0.07 cm in diameter. The air flows at a rate of 2 cm/sec, based on the empty cross section of bed. The area per volume of the spheres is 80 cm^2/cm^3, and the vapor pressure of the iodine is 45 mm Hg. How much iodine will evaporate from a bed 13 cm long, assuming a bed porosity of 40%? *Answer:* 5.6 g/hr.

12. The largest liquid–liquid extraction process is probably the dewaxing of lubricants. After they are separated by distillation, crude lubricant stocks still contain significant quantities of wax. In the past, these waxes were precipitated by cooling and separated by filtration; now, they are extracted with mixed organic solvents. For example, one such process uses a mixture of propane and cresylic acid. You are evaluating a new mixed solvent for dewaxing that has physical processes like those of catechol. You are using a model lubricant with properties characteristic of hydrocarbons. Waxes are 26.3 times more soluble in the extracting solvent than they are in the lubricant. You know from pilot-plant studies that the mass transfer coefficient based on a lubricant-side driving force is

$$K_L a = 16,200 \text{ lb/ft}^3\text{-hr}$$

What will it be (in sec^{-1}) if the driving force is changed to that on the solvent side? *Answer:* $7.6 \cdot 10^{-6}$ sec^{-1}.

13. Digestion from the lumen (the inside) of the small intestine involves two sequential steps: (1) mass transfer from the lumen bulk to the intestinal wall and (2) diffusion across the wall itself. The first step apparently controls the overall rate [F. A. Wilson and V. L. Sallee, *Science,* **174,** 1031 (1971); A. B. R. Thomson, *J. Theoret. Biol.,* **64,** 277 (1977)]. Assume that step 1 is characterized by a mass transfer coefficient D/l and that the second step is characterized by a permeability P. (a) Show that the overall rate j_1 is given by

$$j_1 = \frac{c_1}{l/D + 1/P}$$

where c_1 is the concentration in the lumen. (b) Show that this equation is close to

$$j_1 = \frac{c_1}{(\text{constant})(\tilde{M})^{1/2} + 1/P}$$

where $\tilde{M}^{1/2}$ is the molecular weight of the solute being digested. This result is observed experimentally.

14. A horizontal pipe 10 in. in diameter is covered with an inch of insulation that is 36% voids. The insulation has been soaked with water. The pipe is now drying slowly and hence almost isothermally in 80°F air that has a relative humidity of about 55%. Estimate how long it will take the pipe to dry, assuming that capillarity always brings any liquid water to the pipe's surface. (H. A. Beesley)

15. A 500-gal tank 8 ft deep is to be saturated with air using a small sparger 2 in. in diameter with a flow of 1 ft^3/min. The sparger produces 0.3-cm bubbles that rise through the tank at 10 cm/sec; other properties of the fluid in the tank are essentially those of water. How long does it take to reach 50% saturation? Useful correlations are as follows. Within the bubbles, the mass transfer is faster than the Nusselt limit:

$$k = \frac{2D}{d}$$

where d is the bubble diameter; outside the bubbles, one good correlation is given by P. H. Calderbank [In: *Mixing,* ed. V. Uhl. New York: Academic Press (1967)]:

$$k \left(\frac{\mu}{\rho D}\right)^{2/3} = 0.31 \left(\frac{\Delta \rho \mu g}{\rho^2}\right)^{1/3}$$

16. One type of heart-lung machine is shown schematically in Fig. P9.3. The discs slowly turn, exposing a fresh layer of blood to the air. This blood concentration c_b changes between inlet and outlet. However, the air flow is

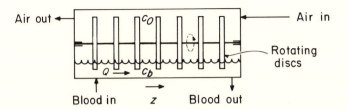

Fig. P9.3

high; so its concentration c_0 does not change. The mass transfer coefficient in the blood is $\sqrt{\kappa D}$, where κ is a pseudo-first-order reaction-rate constant for the oxygen–hemoglobin reaction. The mass transfer coefficient in the air is essentially infinite. (a) Find the oxygen transferred in one rotation of a disc. Assume that only the outer part of the disc is wet and that blood on the disc does not flow much. Write your answer in the notation

c = concentration on the disc
ω = disc rotation
A = wet area of disc
a = area per volume blood on the disc

(b) Find the oxygen concentration c_b in the bulk as a function of z. In this, assume that the number of discs is large and that any oxygenated blood is completely removed from the disc. Write your answer in the following notation:

c_b = blood concentration
n = number of discs
L = oxygenator length
q = volume per time coming off one disc
Q = overall blood flow

(c) Compare your results with those found by dimensional analysis. Comment in detail about any new groups that appear.

10. Absorption and distillation

1. A packed tower is being used to scrub ammonia out of a stream containing only 3% of that gas. The tower contains 10-cm Raschig rings; it is 50 cm in diameter and 4.3 m high. The gas flow of 0.93 kg/cm is at 30°C and is largely air at 100% relative humidity and 1,100 mm Hg; it leaves the tower with only $2.2 \cdot 10^{-6}\%$ NH_3. The liquid flow of 6.7 kg/sec is also at 30°C. The Henry's law constant under these conditions is $y_{NH_3} = 0.85 x_{NH_3}$. What is the mass transfer coefficient K_G in this tower?

2. A process gas containing 4% chlorine (average molecular weight 30) is being scrubbed at a rate of 14 kg/min in a 13.2-m packed tower 60 cm in diameter with aqueous sodium carbonate at 850 kg/min. Ninety-four percent of the chlorine is removed. The Henry's law constant (y_{Cl_2}/x_{Cl_2}) for this case is 94; the temperature is a constant 10°C, and the packing has a surface area of 82 m²/m³. (a) Find the overall mass transfer coefficient K_G. (b) Assume that this coefficient results from two thin films of equal thickness, one on the gas side and one on the liquid. Assuming that the diffusion coefficients in the gas and in the liquid are 0.1 cm²/sec and 10^{-5} cm²/sec, respectively, find this thickness. (c) Which phase controls mass transfer?

3. Find the height of a packed tower that uses air to strip hydrogen sulfide out of a water stream containing only 0.2% H₂S. In this design, assume that the temperature is 25°C, the liquid flow is 58 kg/sec, the liquid out contains only 0.017 mol% H₂S, the air enters with 9.3% H₂S, and the entire tower operates at 90°C. The tower diameter and the packing are 50 cm and 1.0-cm Raschig rings, respectively, and the air flow should be 50% of the value of flooding. The value of $K_L a$ is 0.23 sec⁻¹, and the Henry's law constant (y_{H_2S}/x_{H_2S}) is 1,440.

4. Design a countercurrent absorption tower to scrub 800 ft³/min (corrected to STP) of a process stream containing 27% acetone and 63% air. The gas stream at 1 atm may be assumed to be isothermal at 100°F, but should exit with only 0.7% acetone. The absorbing liquid enters as water containing 0.006 mol% acetone and leaves containing 3.1 mol%. The mass transfer coefficient is about 1.6 cm/sec; the total pressure is 1 atm; and y^* can be taken to be $(xe^{1.95(1-x)^2})$. Design the tower to operate at 50% flooding velocity with a packing of 10-mm Berl saddles.

5. A saturated liquid solution of 40 mol% acetone and 60 mol% acetic acid is fed at a rate of 100 lb-mol/hr to a distillation column. The desired separation of acetone is a 96-mol% distillate concentration and a 5-mol% bottoms concentration. The column is operated at 1.6 times the minimum reflux ratio, so $R_D = 0.44$. The average column temperature is 95°C and the average liquid density is 0.95 g/cm³. The liquid depth on a tray is 1.5 cm, and the tower diameter is 30 cm. The mass transfer coefficients are $k_G a = 505$ sec⁻¹ and $k_L a = 1.8$ sec⁻¹. Equilibrium data for acetone and acetic acid are:

x acetone	y acetone
0.05	0.162
0.10	0.306
0.20	0.557
0.30	0.725
0.40	0.840
0.50	0.912
0.60	0.947
0.70	0.969
0.80	0.984
0.90	0.993

Determine the Murphee efficiency from Eq. 10.4-10, and then find the number of stages and the feed tray location for this column. (D. McCullum).

11. Forced convection

1. As part of a study of O_2 absorption in water in a small packed tower, you find that the outlet concentration of O_2 is $1.1 \cdot 10^{-6}$ molar. The partial pressure of O_2 in the tower is about 0.21 atm; the total area in the tower is 1.37 m^2; the liquid flow rate is 1.62 liters/min. (The diffusion coefficient of oxygen in water is $1.8 \cdot 10^{-5}$ cm^2/sec.) Find (a) the film or unstirred-layer "thickness," (b) the "contact time," and (c) "average residence time on the surface." Assume no gas-phase resistance. *Answer:* 0.005 cm, 1.6 sec, 1.2 sec.

2. The mass transfer coefficient in gas–solid fluidized beds has been correlated with the dimensionless expression [J. C. Chu, J. Kalil, and W. A. Wetteroth, *Chem. Engr. Prog.*, **49**, 141 (1953)]

$$\frac{kp}{G} \left(\frac{\mu}{\rho D} \right)^{2/3} = 1.77 \left(\frac{dG\bar{M}}{\mu(1 - \varepsilon)} \right)^{-0.44}$$

where k is the mass transfer coefficient (mol/L^2-t-atm), p is mean pressure of the fluidized gas (atm), G is total convective flux (mol/L^2-t), d is particle diameter, and ε is bed porosity. You are studying 0.1-cm particles of coal burning in a bed with $\varepsilon = 0.42$ fluidized by air at 1,250°C. The air flux at 1.7 atm is 380 lb/ft^2-hr. What is the film or unstirred-layer thickness around these particles? (H. Beesley) *Answer:* 0.03 cm.

3. Copper is absorbed from a stream flowing at 5.1 liters/hr using a countercurrent flow of 23 lb/hr ion-exchange resin. The bed has a volume of 160 liters. The resin area per volume is 40 cm^2/cm^3, and its density is 1.1 g/cm^3. The equilibrium concentration of copper in solution varies with that in the resin as follows:

c_1(solution) (mol/liter)	c_2(beads) (g Cu/g beads)
0.02	0.011
0.10	0.043
0.25	0.116
2.40	0.200

The resin enters without copper. The copper solution flows in at 0.40 M, but leaves at 0.056 M. What is the mass transfer coefficient and the penetration time into the beads? (R. Contraras) *Answer:* $7 \cdot 10^{-3}$ cm/sec and 0.2 sec.

4. Ether and water are contacted in the Lewis cell shown in Fig. P11.1. An iodinelike solute is originally present in both phases at $3 \cdot 10^{-3}$ M. However, it is 700 times more soluble in ether. Diffusion coefficients in both phases are around 10^{-5} cm^2/sec. Resistance to mass transfer in the ether is across a 10^{-2}-cm film; resistance to mass transfer in the water involves a surface renewal time of 10 sec. What is the solute concentration in the ether after 20 min? *Answer:* $5 \cdot 10^{-3}$ mol/liter.

5. To grow, embryos in eggs must breathe, a process that is believed to be influenced strongly by the diffusion of oxygen through pores in the shell.

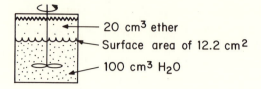

Fig. P11.1

Data from a large number of birds, from warblers to ostriches, suggest that

$$\left(\begin{array}{c}\text{oxygen uptake,}\\ \text{cm}^3/\text{day-mm Hg}\end{array}\right) = 0.2\left(\begin{array}{c}\text{egg mass,}\\ \text{g}\end{array}\right)$$

and

$$\left(\begin{array}{c}\text{pore area,}\\ \text{mm}^2\end{array}\right) = 0.04\left(\begin{array}{c}\text{egg mass,}\\ \text{g}\end{array}\right)$$

Pore length varies from 0.1 to 1 mm and is correlated by

$$\left(\begin{array}{c}\text{pore length,}\\ \text{mm}\end{array}\right) = 0.03\left(\begin{array}{c}\text{egg mass,}\\ \text{g}\end{array}\right)^{1/2}$$

[H. Rahn, A. Ar, and C. V. Paganelli, *Sci. Amer.*, **240** (2), 46 (1979)]. Do these results make sense in terms of the film theory of mass transfer?

6. Develop the intermediate steps from which you can use the film theory for fast mass transfer in concentrated solution. More specifically, (a) integrate Eq. 11.5-4 to find Eq. 11.5-7, (b) rearrange this result to give Eq. 11.5-8, and (c) show that Eq. 11.5-10 can be expanded to give Eq. 11.5-11.

7. A large pancake-shaped drop of water loses 18% of its area in 2 hr of sitting at 24°C in air at 10% relative humidity. The drop's thickness is about constant. If the same drop were placed in an 80°C oven containing air at the same absolute humidity, how long would it take to lose the same fraction of its area? (G. Jerauld)

8. In this book we routinely assume that there is no slip at a solid–fluid interface (i.e., that at a stationary solid interface, the fluid velocity is zero). This leads to expressions for mass transfer in laminar flow in short tubes like

$$k = \frac{D}{\Gamma(\tfrac{4}{3})}\left(\frac{4v°}{9DRz}\right)^{1/3}$$

(see Eq. 11.4-17). However, for some polymer solutions, this assumption is not valid, and the solution will slip along a smooth wall [A. M. Kraynik and W. R. Schowalter, *J. Rheology,* **25,** 95 (1981)]. Find the mass transfer coefficient in the case of total slip, and calculate the ratio of this slippery case and the more common one.

9. You are studying the thermal cracking of a gas oil flowing inside a reactor of cylindrical tubes. The cracking, a heterogeneous reaction at the tube's wall, reduces the average molecular weight of the hydrocarbons by a factor of 2.9. You expect that the cracking rates are controlled by mass transfer, because the rates increase only slightly with temperature. When you dilute the gas oil with 85% nitrogen, you do get results in rough agreement with predictions from mass transfer coefficients like those in Table 9.3-2. You now want to decrease the amount of nitrogen to just 20%. You plan to keep conversion constant at 73%. You recognize this will be compromised by diffusion-induced convection. How much should you change the flow?

12. Free convection

1. Water at 60°C is steadily evaporating upward into dry air through a porous solid 0.1 cm thick. How small must the pores in this solid be to avoid free convection? *Answer:* about 2 mm.

2. Water is diffusing upward into sodium chloride brine through a porous solid 0.1 cm thick. The bottom of the porous solid is covered with a thin ion-exchange membrane that stops any downward flux of the salt. How small must the pores in the solid be to avoid free convection now? *Answer:* about 0.1 mm.

3. When there is wind, the equations for plumes that are developed in this chapter must be modified to include the additional dispersion. Specifically, Eq. 12.3-14 must be replaced by

$$v = \left(\frac{4}{9\alpha^2\gamma} \frac{Q}{v_0 t} \right)^{1/3}$$

where v_0 is the wind velocity and γ is a dimensionless constant characterizing dispersion caused by the wind. For this type of plume, G. Tsang [*Atmos. Environ.*, **5**, 445 (1971)] obtained the following results:

Initial buoyancy $Ag\Delta\rho/\rho_0$ (cm³/sec)	Time out of stack (sec)	Leading plume edge (cm)
480	9	33
	21	60
	35	80
880	26	82
1,400	6	37
	11	55

Show that these results are consistent with the modified theory.

4. In reverse osmosis, water is forced under pressure through a thin membrane. Any salt in the water is retarded, so that the extractant is purified water and the raffinate is a concentrated brine. The salt concentration directly adjacent to the membrane can be especially high because of concentration polarization (see Example 15.1-3). Imagine an experiment using a horizontal membrane. When water is forced upward through the membrane, the permeability is greater than when the water is forced downward. The increase in the former case is attributed to free convection caused by the dense salt solution close to the membrane. Studies of this effect have used a crystal of KCl dissolving in ethanol–water mixtures [D. G. Thomas and

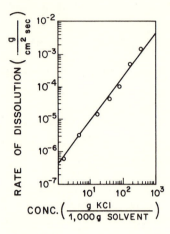

Fig. P12.1

R. A. Armistead, *Science,* **160,** 995 (1968)]. When the liquid mixture is 14 cm deep and below the crystal, the data in Fig. P12.1 are obtained. (a) Use these data to develop an expression giving the Sherwood number as a function of Rayleigh number. (b) Show that the Rayleigh number in this case is well above the critical value.

5. You are interested in studying the dissolution in acid of hydroxy-apatite, the basic mineral of bones and teeth. You have already studied these effects using a spinning disc of radioactively tagged mineral, but now want to repeat these experiments at much lower flows. To do so, you slide a tube containing acid onto a flat surface of the hydroxyapatite and measure the amount of mineral that dissolves after 30 min. Under your experimental conditions, the mineral has a solubility of 0.065 g/cm^3, a density of 2.3 g/cm^3, and a diffusion coefficient of $0.55 \cdot 10^{-5}$ cm^2/sec. The cylindrical tube is 0.26 cm in diameter and contains solution to a depth of 4.20 cm. You expect that the dissolution will be diffusion-controlled. (a) If the mineral surface is horizontal and below the solution, what will be the hydroxyapatite concentration in solution after the experiment? (b) If the mineral surface is horizontal and above the solution, what will be the concentration after this experiment?

13. General questions and heterogeneous chemical reactions

1. The solubilization rates of ^{14}C-tagged linoleic acid in solutions containing 1 wt% of the detergent sodium taurodeoxycholate are shown in Fig. P13.1. [C. Huang, D. F. Evans, and E. L. Cussler, *J. Colloid Interface Sci.,* **82,** 499 (1981)]. These data were found using a spinning liquid disc, for which

$$\frac{kd}{D} = 0.62 \left(\frac{d^2\omega}{\nu}\right)^{1/2} \left(\frac{\nu}{D}\right)^{1/3}$$

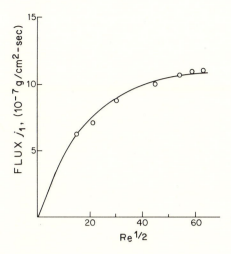

Fig. P13.1

where k is the mass transfer coefficient, D is the diffusion coefficient, d and ω are the disc diameter and rotation speed, and ν is the kinematic viscosity. These data can be explained as a heterogeneous reaction followed by mass transfer. The solubility of linoleic acid in this solution is $2.23 \cdot 10^{-3}$ g/cm³. Find the rate constant of the heterogeneous chemical reaction.   *Answer:* $7 \cdot 10^{-4}$ cm/sec.

2. The oxidation

$$Ce^{3+} \rightarrow Ce^{4+} + e^-$$

has a rate constant of $4 \cdot 10^{-4}$ cm/sec when effected on platinum in 1-M H_2SO_4. You carry out this reaction by suddenly applying a potential across a large stagnant volume of this solution. Estimate how long you can reliably measure the kinetics before diffusion becomes important.   *Answer:* 20 sec.

3. As part of a study of electrochemical kinetics, you insert a gold electrode into a solution at 25°C containing 0.5-M H_2SO_4 and small amounts of ferrous ion. You then apply a potential between this electrode and a second, reversible electrode, so that at the gold the ion is oxidized:

$$Fe^{2+} \rightarrow Fe^{3+} + e^-$$

You measure the current density i (amp/cm²) under a fixed potential and find that, to a first approximation,

$$\frac{c_1}{i} = 2 + 50t^{1/2}$$

where c_1 is the ferrous concentration in moles/liter and t is the time in minutes. Find the rate constant of the surface reaction and the diffusion coefficient of the ferrous ion.   *Answer:* $D = 0.81 \cdot 10^{-5}$ cm²/sec.

4. A single potassium chloride crystal about 0.063 cm in diameter, which is immersed in a 5.2% supersaturated solution containing about 25 wt%

potassium chloride, is growing at a rate of 0.0013 cm/min. If the system is well mixed, this growth is second-order, presumably because both potassium and chloride ions are involved. In our case, the solution may not be well mixed; it flows past the crystal at 6 cm/sec. The solution's viscosity is about 1.05 centipoise (cP); its density is 1.2 g/cm^3, and the crystal's density is 1.984 g/cm^3. Does diffusion influence the rate of crystal growth? *Answer:* Diffusion is about 25% of the total resistance.

5. When copper and silicon are placed together, a layer of Cu_3Si grows at the interface. W. J. Ward and K. M. Carroll [*J. Electrochem. Soc.,* **129,** 227 (1982)] reported that the thickness of this layer *l* at 350°C is

$$l = (1.4 \cdot 10^{-4} \text{ cm/sec}^{1/2})t^{1/2}$$

They argued that the layer forms by reaction of Cu at the Cu_3Si–Si interface. They also maintained that this reaction is controlled by diffusion of Cu through Cu_3Si. (a) Show that the variation of thickness versus time is consistent with a diffusion mechanism. (b) Discuss what reaction stoichiometry would be required to produce this same variation. (c) Calculate the diffusion coefficient and compare it with other values for diffusion in solids. In this experiment, the driving force of Cu_3Si is believed to be 1.10 mol% Cu. *Answer:* $2 \cdot 10^{-6}$ cm^2/sec.

6. One frequently derived result, published in more than a dozen independent papers, concerns the reaction rate of supported enzymes. In these systems, enzyme is adsorbed or covalently bound to small solid spheres. These spheres are suspended in a solution containing a reagent or "substrate." Substrate diffusion is described with a mass transfer coefficient; surface reaction is described by

$$r_1 = \frac{v_{max}c_{1i}}{K_m + c_{1i}} \, [=] \text{ mol}/L^3\text{-}t$$

where v_{max} and K_m are constants, and c_{1i} is the interfacial concentration of substrate. Note that r_1 is defined per volume rather than per area; this is done to keep the analogy with homogeneous enzyme kinetics. (a) Find the overall rate of this reaction. (b) Because the general result is complex, express your answer as a power series in the bulk concentration c_1.

7. Diffusion out of the intestinal lumen may be governed by two resistances in series [K. W. Smithson, D. B. Millar, L. R. Jacobs, and G. M. Gray, *Science,* **214,** 1241 (1981)]. The first is mass transfer in the lumen itself, which is described by $j_1 = k(c_1 - c_{1i})$, where k is the mass transfer coefficient. The second is mass transfer across the intestinal wall, which in this case is governed by a rate equation:

$$j_1 = \frac{v_{max}c_{1i}}{K_m + c_{1i}}$$

where v_{max} and K_m are parameters measured in well-mixed experiments unaffected by mass transfer. (a) To avoid the complexities of the previous problem, these authors assumed that

$$j_1 = \frac{v_{max}c_1}{K_a + c_i}$$

where K_a is found from the slope of a plot of $1/j_1$ versus $1/c_1$. Discuss the approximations in this plot. (b) These authors then found the mass transfer coefficient k from "Winne's equation":

$$k = \frac{0.5v_{max}}{K_a - K_m}$$

Justify this equation and comment on its accuracy.

8. A gaseous stream of excess pure reagent c_1 is in contact with a flat catalytic surface. It diffuses to this surface with a diffusion coefficient D. At the surface, the reagent is split in two:

$$c_{1i} \rightleftarrows 2c_{2i}$$

This reaction is reversible and fast, characterized by the equilibrium constant K. The product then diffuses away, a step characterized by the same diffusion coefficient D. You are interested in the overall rate of reaction. In this, remember that the mixture is *not* dilute. The calculation may be easier if you use the following steps: (a) Assume a thin film near the surface. Write boundary conditions across this film. (b) Write mass balances valid within the film. Combine these with the appropriate form of Fick's law. (c) Integrate these mass balances to find the flux of species 1 in terms of any necessary interfacial concentrations. (d) Simplify your result for the case in which an excess of an inert diluent is added to the reagent stream. Remember that

$$\ln(1 + x) = x - \frac{x^2}{2} + \cdots$$

9. Phosphate nodules off the shore of Peru are believed to grow downward into the sediment. The phosphate is first generated by chemical reactions in the sediments and then diffuses to the growing nodule, where it precipitates. The precipitated phosphate also contains uranium and thorium; measuring the amounts of these elements provides a means of determining the age of that part of the nodule. For one nodule, the following data were obtained [W. C. Burnett, M. J. Beers, and K. K. Rose, *Science,* **215**, 1616 (1982)]:

Position from the bottom of one nodule	Thorium concentration
0–4 mm	4.40 ppm
4–10 mm	3.53 ppm
10–16 mm	2.77 ppm
16–20 mm	2.01 ppm

The thorium concentrations shown change because of radioactive decay:

$$[Th] = [Th]_0 e^{-\gamma t}$$

where γ equals $9 \cdot 10^{-6}$ yr^{-1}. Use these data to estimate the diffusion flux reaching the phosphate crystals, assuming that these crystals are largely $Ca_3(PO_4)_2$.

10. Obsidian is a volcanic glass used by primitive peoples for arrowheads, knife blades, and the like. The depth of water penetration into artifacts made of obsidian can be measured by cleaving the object and examining its surface under a microscope. Because the hydrate has a greater specific volume, there is a stress crack at the hydrated–unhydrated interface. This depth of the hydrate is a measure of the age of the artifact; so obsidian is sometimes called "the dating stone." Most investigators [I. Friedman and F. W. Trembour, *Amer. Sci.*, **66**, 44 (1978)] have reported that the depth of hydration is proportional to the square root of time. Show that this is consistent with diffusion of water through the hydrate followed by very fast heterogeneous chemical reaction at the hydrated–unhydrated interface. In this, assume that the humidity outside the obsidian is constant, but that at the interface it is essentially zero. Compare your results with those of Problem 11 for Chapter 2.

14. Homogeneous chemical reactions

1. You have been recovering an antibiotic from a fermentation broth by adsorption on activated carbon and have found that the mass transfer coefficient for this adsorption is $6.1 \cdot 10^{-4}$ cm/sec. In an effort to accelerate this adsorption, you switch to a cation-exchange resin of the same size beads and keep all details of your experiment the same. You find that the coefficient is now $1.03 \cdot 10^{-2}$ cm/sec. Because this coefficient is highly temperature-dependent, you suspect that it is influenced by chemical reaction. The antibiotic has a diffusion coefficient of about $9.4 \cdot 10^{-7}$ cm^2/sec. What is the rate constant for the reaction? (E. Frieden)

2. (a) Estimate the change in mass transfer coefficient $k/k°$ for 1.2-atm partial pressure H_2S being quickly absorbed by large quantities 0.1-M monoethanol amine in water instead of by pure water. The chief reaction

$$H_2S + RNH_2 \rightleftarrows HS^- + RNH_3^+$$

has an equilibrium constant of 275; the Henry's law constant for H_2S in water is 545. (H. Beesley) (b) How large an error is made if the reaction is assumed to be irreversible? (S. Gehrke)

3. The hydrocracking of a heavy oil at 1,080°C and 1 atm uses a porous catalyst and excess hydrogen. Batch experiments with finely divided catalyst show a half-life for the oil of 0.082 sec. Moreover, the oil has a diffusion coefficient of 0.014 cm^2/sec in the catalyst. Estimate the apparent rate constant in 0.1-cm spherical catalyst pellets.

4. The reaction

$$CO + 3H_2 \rightarrow CH_4 + H_2O$$

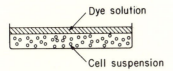

Fig. P14.1

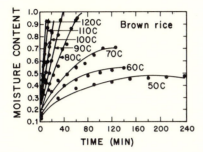

Fig. P14.2

has a rate on a single nickel crystal given by [D. W. Goodman, R. D. Kelley, T. E. Madey, and J. T. Yates, Jr., *J. Catal.*, **63**, 226 (1980)]

$$\text{rate} \left([=] \frac{\text{molecules}}{\text{sec-site}} \right) = 2.2 \cdot 10^8 \; e^{-(24.7 \text{ kcal/mol})/RT} p_{H_2}{}^{0.77} p_{CO}{}^{-0.31}$$

where the pressures are in atmospheres. This rate is known to be the same as that for a supported nickel catalyst [M. A. Vannice, *J. Catal.*, **37**, 449 (1975)]. With this in mind, estimate the reaction rate at 430°C for a feed containing equal concentrations of CO and H_2 and a catalyst containing $1.1 \cdot 10^{14}$ sites/cm^3 in 1.6-cm spherical pellets.

5. The plasma membranes of cells in a suspension 0.7 cm deep are stained with an organic dye by layering the dye in a tetradecane solution on top of an agar-stabilized suspension in a 10-cm Petri dish (Fig. P14.1). The dye has a dissusion coefficient of $2.8 \cdot 10^{-6}$ cm^2/sec. Its concentration in the tetradecane solution is 0.1 M, and the partition coefficient between the tetradecane and the suspending medium is 13. The cells are roughly spherical, approximately $2.1 \cdot 10^{-6}$ m in diameter, and present at a volume fraction of 0.05. In the range of dye concentrations being considered, the concentration of dye on the cells c_2 is proportional to the dye concentration in the suspending medium c_1:

$$c_2 = 3.4 \cdot 10^3 \; c_1$$

The dye cannot penetrate the cell membrane. How much dye will have entered the solution after 30 min? (K. H. Keller)

6. The data in Fig. P14.2 give the moisture content of rough brown rice as a function of time [Bakshi and Singh, *J. Food Sci.*, **45**, 1387 (1980)]. Estimate

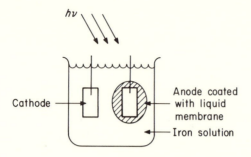

Fig. P14.3

the mass transfer coefficient versus temperature in the rice. Compare this temperature dependence with that expected if the water did not react with starch in the rice.

7. Imagine mass transfer across a thin film like that shown in Fig. 14.1-1. A zero-order reaction is taking place with the film. (a) What is the mass transfer coefficient in the absence of chemical reaction? (b) What is the differential equation and boundary conditions for the concentration in the film? The reaction is

$$r_1 = -\kappa \quad \text{when} \quad C_1 > 0$$
$$r_1 = 0 \quad \text{when} \quad C_1 = 0$$

(c) What is the mass transfer coefficient in this second situation?

8. One type of solar cell based on the dye thionine is shown in Fig. P14.3 [Gokalp and Cussler, *J. Memb. Sci.*, **11**, 53 (1982)]: The thionine is contained in a liquid membrane that coats the anode. The cathode is uncoated. Both electrodes are immersed in an acid solution of ferrous and ferric ion. There are four important processes in this cell. At the cathode, the key reaction is

$$Fe^{3+} + e^- \xrightarrow{\text{fast}} Fe^{2+}$$

At the water–membrane interface, the chief reaction is

$$Fe^{2+}(\text{aqueous}) + TH^+(\text{membrane}) + H^+ + h\nu$$
$$\xrightarrow{\kappa} Fe^{3+}(\text{aqueous}) + TH_2^+(\text{membrane})$$

In this, TH^+ is the low-energy form of thionine and TH_2^+ is the high-energy form; $[TH^+]$ and $[H^+]$ are present at much higher concentrations than anything else. Once this reaction occurs, TH_2^+ diffuses to the anode

$$TH_2^+(\text{membrane}) \xrightarrow{D/l} TH_2^+(\text{at anode})$$

and reacts there

$$TH_2^+ \xrightarrow{\text{fast}} TH^+ + H^+ + e^-$$

Show that these equations predict that if κ is large, the current varies with the square root of the concentration of $[TH^+]$.

9. Studies of air pollution show that sulfur oxides are much more harmful when they are present with particulates. The reason that this is true may be that the particulates first adsorb sulfur oxides and then are deposited in the lungs. As a result, they cause very high local concentrations of sulfite and sulfate. To investigate this idea, you are to develop a model of sulfur concentration in the lung. Assume that the particle is solid and small, that the total sulfur on the particle is initially M, and that sulfur compounds on the particle are desorbed quickly. The sulfur compounds then diffuse over the surface in a thin, liquidlike layer on the lung's surface. While this diffusion occurs, the sulfur compounds react with the lung by means of mass transfer followed by a first-order reaction, characterized by an overall coefficient k. It is this reaction that damages the lung. Find the local sulfur concentration on the lung's surface versus position and time.

10. Under one set of experimental conditions, oxygen diffuses to the surface of carbon particles and reacts to form carbon monoxide:

$$O_2 + 2C(solid) \xrightarrow{\kappa_2} 2CO$$

As the carbon monoxide diffuses away, it rapidly reacts to form carbon dioxide:

$$O_2 + 2CO \xrightarrow{\kappa_3} CO_2$$

In the region where this second reaction occurs, carbon monoxide has a much higher concentration than oxygen. Use the film theory to estimate the flux of oxygen as a function of both rate constants.

11. In the text, we used the film theory to show that for a fast, irreversible second-order reaction,

$$\frac{k}{k^0} = 1 + \frac{D_2 C_{20}}{\nu D_1 C_{10}}$$

This derivation assumed implicitly that the system was always dilute, so that there was no diffusion-engendered convection (i.e., $n_1 = j_1$). Imagine, instead, that the solution is concentrated, diluted only with product c_3. The stoichiometry is simple:

(species 1) + (species 2) → (species 3)

Find the mass transfer coefficient in this case, assuming that diffusion obeys binary equations, with diffusion coefficients that are the same for all species.

12. In 1874, Louis Pasteur was commissioned by Napoleon III to determine why opening a bottle of full-bodied red wine hours before it is to be consumed gives the wine a more "mature" flavor than the same wine when drunk immediately after opening. Pasteur reported that the "aging" process is associated with a reaction consuming oxygen. More recently, some investigators have argued that this process involved oxygen diffusion into the wine. They used the oxygen concentration profile for unsteady diffusion without reaction in the wine and claimed justification of Pasteur's conclusions. (a) Write the differential equation and boundary conditions for this problem without chemical reaction. (b) Solve this equation, or determine the key dimensionless variables. (c) Use typical values for diffusion coefficients and wine bottles to discuss this result. (d) Qualitatively describe the effect of chemical reaction.

15. Membranes and other diffusion barriers

1. Imagine a thin membrane, of thickness l and largely of water, separating two well-stirred organic solutions. One of the solutions contains acetic acid, which diffuses across the membrane. (a) If the acetic acid concentration is very high, it will largely be un-ionized everywhere. What is the flux across the membrane in terms of the concentrations in the organic phases? (b) If the acetic acid concentration is moderate, it will be un-ionized outside the membrane, but ionized in it. Again, what is the flux? (c) If the acetic acid concentration is very low, it will be ionized everywhere. What is the flux now? (d) Use the results of (a)–(c) to plot log(flux) versus log(concentration difference). What is the slope on this graph?

2. The flux at 37°C of a 1-1 electrolyte across an uncharged membrane, but under an electric field, is approximately (see Eq. 15.2-14)

$$j_1 = \frac{D_1 H_1 z_1 \mathfrak{F}(\psi_0 - \psi_l)}{lRT} \left(\frac{c_{10} - c_1 e^{z_1 \mathfrak{F}(\psi_l - \psi_0)/RT}}{1 - e^{z_1 \mathfrak{F}(\psi_l - \psi_0)/RT}} \right)$$

Imagine that this membrane separates two large, well-stirred volumes of solution, one of which has $c_{10} = 0.1$ M. If $\psi_l - \psi_0$ is 62 mV and $z_1 = +1$, what is c_{1l} at steady state?   *Answer:* 0.01 mol/liter.

3. One commercially available ultrafiltration membrane is claimed to have a permeability of 0.62 m^3/m^2-day under a pressure difference of 3.4 atm. This membrane initially rejects 96% of a 3 wt% suspension of partially hydrolyzed starch (mol. wt. 17,000). However, if 4.2 cm^2 of membrane separates 65 cm^3 of a starch solution from the same volume of pure water, the volumetric flow is zero, and the osmotic difference is 85% of the original value in 1 week. Assuming the temperature is 25°C, find the Darcy's law permeability L_p, the solute permeability ω, and the reflection coefficient σ.   *Answer:* $L_p = 0.2$ m/day-atm, $\omega = 0.4$ mol/m^2-day-atm, $\sigma = 0.985$.

4. The flux of a solute across an inhomogeneous membrane of thickness l can be described by the flux equation

$$-j_1 = \frac{Dc_1}{RT} \frac{d\mu_1}{dz}$$

This equation can be derived from thermodynamic arguments like those in Section 8.2. If the solute is dilute, the chemical potential may be shown to be

$$\mu_1 = \mu_1^0 + RT \ln c_1$$

where μ_1^0 is the chemical potential in the standard state of infinite dilution and c_1 is the concentration in the membrane. (a) If μ_1^0 is a constant, show that the foregoing flux equation leads to Eq. 15.1-5. (b) If $\mu_1^0 = \mu_{10}^0 + (\mu_{1l}^0 - \mu_{10}^0)(z/l)$, show that the flux has the same form as Eq. 15.2-14. (c) Imagine that the concentration difference across this membrane is 0.1 M. What is the effect of reversing the gradient in (a) and (b)? ·

5. R. L. Sparks has made membranes with unusually sharp cutoffs by

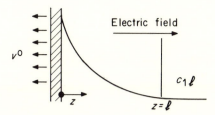

Fig. P15.1

deliberately plugging the pores in an ultrafiltration membrane with a solute of known molecular weight. Such membranes promise much more selective separations, especially for proteins. Unfortunately, these membranes show tremendous concentration polarization and very easily plug. Sparks used an electrophoretic field to reduce this effect (Fig. P15.1). Such a field alters Fick's law for the protein:

$$-j_1 = D \left(\frac{dc_1}{dz} + \frac{c_1 z_1}{RT/\mathcal{F}} \frac{d}{dz} \psi \right)$$

Find the steady-state concentration profile for protein assuming that the field $d\psi/dz$ is a constant $\Delta\psi/l$.

6. In the text, we calculate the flux in facilitated diffusion by homogeneous chemical reaction with a mobile carrier. Rework this analysis, assuming that the carrier reaction is very fast but heterogeneous, occurring only at the membrane boundaries. (J. Zasadzinski)

7. The copper separation shown in Fig. 15.4-2 uses a membrane containing a liquid ion exchanger that reacts reversibly with copper according to the equation

$$Cu^{2+} + 2RH \rightleftarrows R_2Cu + 2H^+$$

Most of this reaction is fast and takes place at the membrane's interfaces; almost none takes place within the membrane, where the copper concentration is extremely small. In addition, the acid concentration on one side of the membrane is close to zero, the copper concentration on the other side is zero, and all diffusion coefficients in the membrane are about equal. (a) Find the copper flux across the membrane in terms of the average carrier concentration and the copper concentrations in the solutions adjacent to the membrane. (b) Explain how this result could be checked experimentally. (c) In fact, the flux in (a) varies as predicted only at low carrier concentration; in concentrated carrier, the flux reaches a constant value. Speculate on why this may be true.

8. Imagine a thin liquid film bounded by porous electrodes. The film contains Fe^{2+} and Fe^{3+}. At one electrode, there is the reaction

$$Fe^{3+} + e^- \xrightarrow{\text{fast}} Fe^{2+} \qquad \text{(a)}$$

At the other electrode this reaction is quickly reversed. Thus, the electrodes engender fluxes of Fe^{2+} and Fe^{3+} within the film. Both sides of the film are exposed to equal pressures of nitric oxide (NO) gas. The following reversible reaction occurs in the liquid:

$$NO + Fe^{2+} \rightleftharpoons FeNO^{2+} \tag{b}$$

The following reaction *does not* occur:

$$NO + Fe^{3+} \not\rightleftharpoons FeNO^{3+} \tag{c}$$

The net result will be a flux of NO, even though there is no NO concentration difference between the gases. Find the size of this flux and how it is related to the electrode current.

16. Heat transfer

1. Find the heat lost per external area from a house at 18°C on a winter day at −14°C. The house is insulated with the equivalent of 8 cm of asbestos. The resistance to heat transfer at the inner walls is negligible, but that at the outer walls is controlled by an air layer equivalent to 0.2 cm.   *Answer:* 10^{-3} cal/cm²-sec.

2. Find the heat loss per external area from a water pipe containing water at 18°C on a winter day at −14°C. The pipe has a diameter of 5 cm and is insulated with an 8-cm layer of asbestos. The resistances to heat transfer inside the pipe and of the pipe wall are negligible. That outside the pipe is controlled by an air layer equivalent to 0.2 cm.   *Answer:* $5 \cdot 10^{-4}$ cal/cm²-sec.

3. The values given below are abstracted from a table [*Wind Chill: Equivalent Temperatures.* NOAA (1974)] of windchill versus temperature.

Dry-bulb temperature (°F)

Wind velocity (mph)	20	0	−20	−40
4	20	0	−20	−40
10	3	−22	−46	−71
20	−10	−39	−67	−95
30	−18	−49	−79	−109
40	−21	−53	−84	−115

Windchill is popularly interpreted as how cold the weather "feels" at the true temperature and the given wind. In fact, these values are based on the time to freeze water in a sausage casing hung over a Quonset hut in Antarctica. Since these results were published, several have asserted that they are equivalent to the heat loss from a cylinder, and so can be predicted from standard engineering correlations. Test this assertion using the values in the table.

4. The energy balance on a differential volume can be written in a variety of ways, including

$$\frac{\partial}{\partial t} \rho(\hat{U} + \frac{1}{2}v^2) = -\nabla \cdot \rho\mathbf{v}(\hat{U} + \frac{1}{2}v^2) - \nabla \cdot \mathbf{q} + \rho(\mathbf{v} \cdot \mathbf{g})$$
$$- \nabla \cdot p\mathbf{v} - \nabla \cdot [\boldsymbol{\tau} \cdot \mathbf{v}]$$

$$\rho\frac{D\hat{U}}{Dt} = \rho\left(\frac{\partial\hat{U}}{\partial t} + \mathbf{v} \cdot \nabla\hat{U}\right) = -(\nabla \cdot \mathbf{q}) - p(\nabla \cdot \mathbf{v}) - (\boldsymbol{\tau}:\nabla\mathbf{v})$$

$$\frac{\partial}{\partial t} \rho\hat{U} = -(\nabla:\rho\mathbf{v}\hat{H}) - (\nabla \cdot \mathbf{q}) - (\boldsymbol{\tau}:\nabla\mathbf{v}) + \mathbf{v} \cdot \nabla p$$

Prove that these equations are equivalent.

5. Assume that a straight wire in a large volume of fluid is suddenly connected to an electrical power supply that puts a constant wattage through the wire. The resistance of the wire is then measured as a function of time. Because this resistance is a function of temperature, and the wire's temperature depends on the thermal conductivity, this measurement of resistance provides a way of determining the thermal conductivity. Derive an equation that allows calculation of thermal conductivity from this resistance [as references, see H. Ziebland, in: *Thermal Conductivity,* ed. R. P. Tye. London: Academic Press (1969); J. K. Horrocks and E. McLaughlin, *Proc. Roy. Soc.,* **A273,** 259 (1963)].

6. Polymer fibers are often melt-spun by forcing a polymer melt through small holes into cold air. The specific polymer is first a rubber and then becomes a glass at T_c; this transition involves a negligible enthalpy of fusion, but does result in altering thermal diffusivity from α_1 to α_2. At the same time, the polymer surface quickly reaches the temperature of the surrounding air. Find the radius where the transition occurs as a function of time.

17. Simultaneous heat and mass transfer

1. You want to use a wet-bulb thermometer wet with carbon tetrachloride to determine the carbon tetrachloride concentration in air at 2 atm flowing at 62°C. The wet bulb reads 23°C. What is the carbon tetrachloride concentration? *Answer:* 0.2 atm.

2. Predict the mass transfer coefficient in cm/sec for liquid *n*-butyl alcohol vaporizing into air at 80°F and 1 atm. You know that the heat transfer coefficient in the same system is 56 Btu/hr-ft²-°F. *Answer:* 25 cm/sec.

3. A meteor is falling through the earth's atmosphere and burning as it falls. The burning can be approximated as a diffusion-controlled first-order chemical reaction oxidizing iron at the meteor's surface. Find the meteor's temperature in terms of only the heat of this reaction, the concentration of iron oxide vapor near the surface, and the properties of the air.

4. A layer of wet air at 4°C and 80% relative humidity lies on top of a layer of wet air at 0°C and 90% relative humidity. The two layers diffuse together. Will fog form? *Answer:* yes.

5. (a) Heat is being lost from the human body at 37°C to the surrounding air at 25°C. The heat transfer coefficient from air to skin is 12 kcal/m²-hr-°C;

that across skin is 380 kcal/m²-hr-°C, and that within the flesh is about 32 kcal/m²-hr-°C. Find the overall heat transfer coefficient. *Answer:* 8.5 kcal/m²-hr-°C. (b) A gaseous nerve agent is diffusing from the surrounding air at 18°C into the skin. The conditions are such that the heat and mass transfer coefficients in the air have approximately equal Stanton numbers:

$$\frac{h}{\rho \hat{C}_p v} = \frac{k}{v}$$

where v is the wind's velocity. Across the skin and in the flesh, the conditions are such that the effective film thicknesses are unchanged. The nerve agent dissolves in flesh according to the equation

$$p(\text{agent in air, atm}) = (3 \cdot 10^6) \cdot (\text{concentration of flesh, mol/cm}^3)$$

Moreover, the agent is about 50 times more soluble in skin than in flesh. It has the same diffusion coefficient of $3.1 \cdot 10^{-6}$ cm²/sec in both skin and flesh. Both skin and flesh have thermal conductivities that are close to those of water. Find the overall mass transfer coefficient. *Answer:* $2.3 \cdot 10^{-5}$ cm/sec.

6. Imagine that you fill the two-bulb capillary apparatus (see Fig. 3.1-2) with an equimolar mixture of hydrogen in methane. Each bulb has a volume of 270 cm³; the vertical capillary is 6 cm long. You place the lower bulb in ice water and heat the upper one with steam. (a) What is the maximum concentration difference due to thermal diffusion? *Answer:* $\Delta y_1 = 0.02$. (b) How many moles of hydrogen are there in the hot bulb? How many in the cold?

7. Porous catalyst spheres that are 0.3 cm in diameter have a void fraction of 0.32 and are saturated with water. The spheres are dried adiabatically by placing them in a dumped bed 54 cm in diameter and 280 cm deep. Air at 50°C and 10% relative humidity is blown into the bed at a rate of 165 cm/sec. (a) Soon after the drying is begun, the temperature 5 cm into the bed is 33°C. What is the concentration at that point? *Answer:* $y_1 = 0.024$. (b) Adiabatic driers are often assumed to operate at constant wet-bulb temperature. Use a humidity chart to estimate the concentration, halfway through, and compare your answer with part (a). (c) What are the heat and mass transfer coefficients in the bed? *Answer:* $k = 2$ cm/sec, $h = 2.6 \cdot 10^{-3}$ J/sec-cm²-°K.

8. Track-etched membranes are made by exposing mica or polycarbonate sheets $15 \cdot 10^{-4}$ cm thick to an α-radiation source and then etching the sheet in hydrofluoric acid. The resulting membrane can have about 0.4% of its area pierced by 120-Å cylindrical pores. Imagine that you place a track-etched membrane across one end of a 2-cm-diameter glass pipe 36 cm long. You cover the other end with a filter that has a high Darcy's law permeability. If you set the pipe in the sun, air will flow into the pipe by thermal effusion through the track-etched membrane and out of the pipe by Darcy's law flow through the filter. How fast will the air flow if the air in the pipe is 47°C and the surrounding air is 23°C?

9. Your plant has available a countercurrent cooling tower 10 m high and 6 m in diameter. The tower packing has a surface area per volume of about

63 m^{-1}. At present, it is effectively cooling water at 3,200 kg/min from 66°C to 20°C, using air at 80 g-mol/m-sec at 18°C and 20% relative humidity. (a) What is the mass transfer coefficient in this tower? (b) You need 1,000 kg/min water at 15°C or less to cool a new chemical reactor. By how much should you reduce the water flow to get this output? Assume that the mass transfer coefficient varies as do the values in Fig. 17.3-4.

10. Extend the analysis for fast mass transfer given in Section 11.5 to include the effect of diffusion-engendered convection on heat transfer. Use the film theory in this extension. (a) Show that the energy equation in this situation is

$$0 = -\frac{d}{dz}(q + \rho \hat{C}_p T v)$$

subject to

$$z = 0, \quad T = T_0$$
$$z = l, \quad T = T_l$$

(b) Integrate this to find

$$q|_{z=0} = \rho \hat{C}_p v \left(\frac{T_l - T_0 e^{vl/\alpha}}{1 - e^{vl/\alpha}} \right)$$

(c) Defining

$$q|_{z=0} = h(T_o - T_l) + \rho \hat{C}_p T_0 v$$

show that

$$h = \frac{\rho \hat{C}_p v}{e^{vl/\alpha} - 1}$$

11. The thermal conductivity of reacting gas mixtures is sometimes found to be larger than would be expected from molecular considerations. If a temperature gradient exists in a gas, then in different temperature regions, the concentrations of reactive species may be different; this concentration gradient can augment conduction because there is a transport of energy by molecular diffusion. A convenient system to study this phenomenon utilizes nitrogen dioxide. The reaction

$$2NO_2 \rightleftharpoons N_2O_4$$

is very rapid in both directions, and for most studies, the mixture may always be assumed to be in chemical equilibrium. (a) Assume two horizontal parallel plates separated by a distance of 0.16 cm. The gap is filled with a NO_2–N_2O_4 mixture. The lower plate is at 40°C; so the mole fraction of [NO_2] next to this plate is 0.48. The upper plate is at 80°C; so the mole fraction of [NO_2] adjacent to this plate is 0.85. If the diffusion coefficient of both species is 0.07 cm^2/sec, find the flux of [NO] across the gap. (b) Find the molar average velocity across this gap. (c) Calculate the temperature profile, including that due to diffusion. (d) The thermal conductivity of these mixtures is about $4 \cdot 10^{-5}$ cal/cm-sec-°K, and their heat capacity is about 7 cal/mol-°K. How much will the reaction in this system increase the heat flux?

12. Imagine a thin layer of gas of thickness l across which heat transfer occurs. The heat transfer is facilitated by a reactive gas 1 within the film. At the hot *surface* at T_0 and $z = 0$, the gas reacts catalytically, rapidly, and endothermally:

(2 moles of gas 1) + (heat) → (1 mol of gas 2)

At the cold *surface* at T_l and $z = l$, the reaction is rapidly reversed:

(1 mol of gas 2) → (2 moles of gas 1) + (heat)

Heat conduction also occurs, but free convection does not. Only gas 1 and gas 2 are in the film, and the thermal conductivity is constant. Also assume that the thermal conductivity at the boundaries is much greater than in the bulk. Find the heat transfer coefficient across this thin film in three steps: (a) Find the concentration profiles in the film. (b) Find the temperature profile corrected for mass transfer. (c) Find the heat flux at the boundary $z = 0$.

LIST OF SYMBOLS

a	surface area per volume
a	major axis of ellipsoid (Section 5.2)
a, a_i	constant
a_0	atomic spacing (Section 5.3)
A	area
b	constant
b	minor axis of ellipsoid (Section 5.2)
B	bottoms (Chapter 10)
B, B'	boundary positions (Section 8.3)
c	total molar concentration
c_i	concentration of species i, in either moles per volume or mass per volume
c_{CMC}	critical micelle concentration (Section 6.2)
c_T	total ion concentration (Eq. 6.1-22)
$\bar{c}_1$	concentration of species 1 averaged over time (Chapter 4)
c_1'	concentration fluctuation of species 1 (Sections 4.3, 12.1, and 14.5)
C_i	concentration of species i at a boundary
$\tilde{C}_p, \hat{C}_p$	molar and specific heat capacities at constant pressure, respectively
$\tilde{C}_v, \hat{C}_v$	molar and specific heat capacities at constant volume, respectively
d	diameter or other characteristic length
D	binary diffusion coefficient
D	distillate (Chapter 10)
D_{eff}	effective diffusion coefficient, for example, in a porous solid
D_i	binary diffusion coefficient of ion i (Chapter 6, Section 14.2)
D_0	binary diffusion coefficient corrected for activity effects (Chapter 7)
D_{ij}	multicomponent diffusion coefficient (Chapter 8)
$\mathcal{D}_{ij}$	binary diffusion coefficient in a dilute gas mixture (Chapter 8)
D_{Kn}	Knudsen diffusion coefficient of a gas in a small pore
D_m	micelle diffusion coefficient (Section 6.2)
D^*	intradiffusion coefficient (Section 7.2)
E	dispersion coefficient (Chapter 4)

E	constant in conductance theory (Eq. 6.1-32)
$E(t)$	residence-time distribution (Section 11.2)
E_x, E_y, E_z	dispersion in directions x, y, and z, respectively (Chapter 4)
f	friction coefficient (Chapter 5)
ℓ	friction factor
F	feed (Chapter 10)
$\mathcal{F}$	Faraday's constant
$F(D)$	solution to a binary diffusion problem (Eq. 8.3-1)
$\mathbf{g}$	acceleration due to gravity
G	molar flux of gas (Chapter 10)
$\hat{G}$	specific Gibbs free energy
h, h_i	heat transfer coefficients (Chapters 16–17)
H	partition coefficient or Henry's law coefficient
$\bar{H}, \hat{H}$	molar and specific enthalpies (Chapters 16–17 and Chapter 8, respectively)
$\bar{H}_i$	partial specific enthalpy (Eq. 8.2-2)
HTU	height of transfer unit (Chapter 10)
$\mathbf{i}$	current density
j	fringe number (Example 2.3-4)
j_D	solute flux across a membrane (Eq. 15.1-3)
j_v	volume flux across a membrane (Eq. 15.1-12)
$\mathbf{j}_T$	total electrolyte flux (Eq. 6.1-21)
$\mathbf{j}_i$	diffusion flux of species i relative to the volume average velocity
$\mathbf{j}_i^m$	diffusion flux of i relative to the mass average velocity
$\mathbf{j}_1^{(2)}$	diffusion flux of solute (1) relative to velocity of solvent (2)
$\mathbf{j}_i^a$	diffusion flux of solute i relative to reference velocity a
J	total number of interference fringes (Example 2.3-4)
J, J'	constants in conductance theory (Eq. 6.1-32)
$\mathbf{J}_s$	entropy flux (Eq. 8.2-3)
J_T	total solute flux in different chemical forms (Section 6.2)
J_1	dispersion flux (Section 4.2)
k, k_i	mass transfer coefficient
k_B	Boltzmann's constant
ℓ	thermal conductivity (Chapters 16–17)
k^0	mass transfer coefficient at low transfer rate (Section 11.5)
k^0	mass transfer coefficient without chemical reaction (Chapter 14)
k', k'', k'''	alternative forms of mass transfer coefficient (Table 9.2-2)
$\bar{k}$	average mass transfer coefficient
k_G	mass transfer coefficient in gas
k_L	mass transfer coefficient in liquid
$k_{\log}$	log mean mass transfer coefficient (Example 9.2-3)
K	equilibrium constant for chemical reaction; overall rate constant
K_{cell}	conductance cell constant (Eq. 6.1-26)
K_G	overall mass transfer coefficient based on driving force in gas
K_L	overall mass transfer coefficient based on driving force in liquid

K_m	Michaelis constant for membrane transport (Eq. 15.3-17)
Kn	Knudsen number (Eq. 7.4-6)
l	characteristic length, e.g., membrane thickness
L	length, e.g., of a pipe
L	molar flux of liquid (Chapter 10)
L_{ij}	Onsager phenomenological coefficient (Eq. 8.2-15)
L_p	Darcy's law permeability (Section 15.1)
m	stoichiometric coefficient (Section 13.5)
m	magnification
m_i	mass of species i
M	mass
M	total solute in dispersion (Sections 4.2 and 5.5)
$\tilde{M}_i$	molecular weight of species i
n	refractive index
n	micelle aggregation number (Section 6.2)
$\mathbf{n}_i$	flux of species i relative to fixed coordinates
N	fraction of vacant sites (Section 5.3)
N	number of transfer units (Chapter 10)
$\tilde{N}$	Avogadro's number
N_i	flux of species i at an interface
$\mathfrak{N}_i$	number of moles of species i
p	pressure
P	power
P_{ij}	weighting factor (Table 8.3-1)
$\mathbf{q}$	energy flux by conduction
Q	buoyancy factor (Eq. 12.3-1)
r	radius
r_i	rate of chemical reaction
R	gas constant
R_D	reflux ratio (Chapter 10)
R_0	characteristic radius
s	distance from pipe wall (Section 11.4)
S	source in dispersion (Chapter 4)
S	constant in conductance theory (Eq. 6.1-32)
$\hat{S}$	specific entropy (Chapter 8)
t	time
$\mathbf{t}$	modal matrix (Chapter 8)
t_i	transference number of ion i
$t_{1/2}$	reaction half-life
T	temperature
u_i	ionic mobility (Chapter 6)
U	overall heat transfer coefficient
$\hat{U}$	specific internal energy
$v^{(i)}$	velocity of interface (Section 11.5)
v_{max}	maximum speed of membrane transport (Eq. 15.3-17)

v_r, v_θ	velocities in the r and θ directions
v_x, v_y	velocities in the x and y directions
$\hat{v}$	average molecular velocity (Section 5.1)
v	mass average velocity
v^a	velocity relative to reference frame a
v^0	volume average velocity
v$'$	velocity fluctuation (Sections 4.3, 12.1, and 14.5)
v*	molar average velocity
v$_i$	velocity of species i
V	volume
$\hat{V}_i$	partial molar or specific volume of species i
V_{ij}	fraction of molecular volume (Eq. 5.1-9)
W	width
W	work (Eq. 16.2-4)
W_s	shaft work (Eq. 16.2-6)
x	mole fraction in liquid of species being transferred (Chapter 10)
x_i	mole fraction of species i, especially in a liquid or solid phase
X$_i$	generalized force causing diffusion (Eq. 8.2-14)
y	mole fraction in vapor of species being transferred (Chapter 10)
y_i	mole fraction of species i in a gas
z	position
$|z|$	magnitude of charge (Sections 6.1 and 15.2)
z_i	charge on species i
Z	position in diffusion cell (Example 2.3-4)
α	thermal diffusivity (Chapters 16–17)
α	thermal diffusion factor (Section 17.5)
α	entrainment coefficient (Eq. 12.3-12)
α_{ij}	conversion factor (Eq. 8.1-4)
β	diaphragm cell calibration constant (Example 2.2-4)
β	density change with concentration (Eq. 12.1-14)
δ	boundary layer thickness (Section 11.3)
$\delta(z)$	Dirac function of z
δ_{ij}	Kronecker delta
ε	void fraction
ε_{ij}	interaction energy between colliding molecules
η	effectiveness factor (Section 14.3)
η	dimensionless distance (Eq. 4.2-11)
η	efficiency of distillation (Section 10.4)
ζ	combined variable (Section 2.3)
θ	fraction of surface elements remaining at time t (Section 11.2)
κ_i, κ_{-i}	forward and reverse reaction rate constants of reaction i, respectively
λ	length ratio (Eq. 7.4-10)
λ_i	equivalent ionic conductance of species i
Λ	equivalent conductance

μ	viscosity
μ_i	chemical potential of species i
μ_i	partial specific Gibbs free energy of species i, i.e., the chemical potential divided by the molecular weight (Section 8.2)
ν	kinematic viscosity
ν	stoichiometric coefficient (Section 14.2)
ξ	combined variable (Section 2.4) or $(z - v^0 t)/R_0$ (Section 4.2)
Π	osmotic pressure (Section 15.1)
ρ	total density
ρ_i	density or mass concentration of species i
σ	rate of entropy production (Section 8.2)
σ	standard deviation of dispersion profile (Chapter 4)
σ	Soret coefficient (Chapter 17)
σ, σ'	reflection coefficients (Section 15.1)
σ	diagonal matrix of eigenvalues (Chapter 8)
σ_i	eigenvalue (Section 8.3)
σ_{ij}	collision diameter (Sections 5.1 and 14.4)
τ	relaxation time (Section 14.4)
τ	dimensionless time (Section 2.4)
τ	residence time for surface element (Section 11.2)
τ	tortuosity (Section 7.4)
$\boldsymbol{\tau}$	shear stress
τ_0	shear stress at wall (Section 11.3)
ϕ	Thiele modulus (Section 14.3)
ϕ_i	volume fraction of species i
Φ	dimensionless interfacial velocity (Eq. 3.3-35)
ψ	electrostatic potential
Ψ	dispersion function (Eq. 14.5-6)
$\boldsymbol{\Psi}$	combined concentration (Eq. 8.3-19)
ω	jump frequency (Section 5.3)
ω	coefficient of solute permeability (Section 15.1)
ω_i	mass fraction of species i
Ω	collision integral in Chapman–Enskog theory

SUBJECT INDEX

All italic references refer to problems

MATERIALS INDEX

References to air and water are not given.
All italic references refer to problems

524